Biotechnological Advances in Bamboo

Zishan Ahmad • Yulong Ding • Anwar Shahzad
Editors

Biotechnological Advances in Bamboo

The "Green Gold" on the Earth

Springer

Editors
Zishan Ahmad
Bamboo development laboratory
Bamboo Research Institute, Nanjing Forestry University
Nanjing, China

Yulong Ding
Bamboo development laboratory
Bamboo Research Institute Nanjing Forestry University
Nanjing, China

Anwar Shahzad
Department of Botany
Aligarh Muslim University
Aligarh, India

ISBN 978-981-16-1309-8 ISBN 978-981-16-1310-4 (eBook)
https://doi.org/10.1007/978-981-16-1310-4

This Springer imprint is published by the registered company Springer Nature Singapore Pte Ltd.
The registered company address is: 152 Beach Road, #21-01/04 Gateway East, Singapore 189721, Singapore

Professor Xiong Wenyu
(July 1915–September 2014)

Founder of Scientific Cultivation of Bamboo Plantation
Known as "Mr. Bamboo" *and* "Bamboo Pioneer"

Foreword

Bamboos are considered as the "green gold of the forest" as these have improved many facets of rural livelihood while strengthening urban sector. Bamboos have been traditionally used as construction material for rural housing, food, and, last but not least, handicraft products. In today's times, however, its uses have become more diverse. Presently, bamboos have emerged as one of the most important renewable resources for food, fuel, paper, pulp, textile, plywood, etc. Therefore, the annual demands for bamboos have already outcrossed the annual yields across the world. And the current scenario has forced scientists to pay more attention to the utilization of biotechnological tools for better understanding and improving bamboos. Modern biotechnologies are more and more applied in bamboo research and practice and will bring deep influence on the development of bamboo. The concerns in this book initially begin with a chapter introducing the global distribution of the bamboo. The book provides an overview of the adoption of biotechnological approaches to advance bamboo research for better utilization of bamboo resources for human beings. Various applications of biological techniques related to bamboo have been discussed in detail, for example, plant tissue culture techniques, somatic embryogenesis, germplasm conservation techniques, utilization of molecular markers, transcriptomics, polymorphism, and phylogenetic analysis. The book also addresses novel industrial applications of bamboo in structural, food and pharmaceutical applications along with traditional uses in detail. The book is a reference text with essential information for various scientists, including advanced students, teachers as well as research scientists working in different areas of bamboo research.

University of Hamburg, Hamburg
Germany

Walter Liese

Preface

Bamboo is considered as a multipurpose plant and has a prolonged history as an adaptable and extensively used renewable resource in conventional and commercial applications. Therefore, the annual demands for bamboos have already outcrossed the annual yields across the world. Increasing population pressure; indiscriminate exploitation by paper, pulp, and fuel industry; and insufficient attempts to replenish and cultivate bamboos are further widening the gap between demand and supply. And the current scenario has forced scientists to pay more attention to the utilization of biotechnological tools for better understanding and improving bamboos. As a result, new insights into bamboos were gained through various biotechnological approaches. The book provides an overview of the adoption of biotechnological approaches to advance bamboo research for better utilization of bamboo resources for human beings. Various applications of biological techniques related to bamboo have been discussed in detail, for example, plant tissue culture techniques, somatic embryogenesis, germplasm conservation techniques, utilization of molecular markers, transcriptomics, polymorphism, and phylogenetic analysis. Being involved in this area, we comprehend that information on the application of the biotechnological approaches in bamboo is still obscure, and there is no single book available on this aspect. This volume comprises several chapters on relevant topics contributed by experts working in the field of plant biotechnology so as to make available a comprehensive treatise designed to provide an in-depth analysis of the subject in question. The book is a compilation of 20 chapters with relevant text, tables, and illustrations describing the experimental work on bamboo biotechnology, which will be useful in the planning and execution of various experiments smoothly and effectively. The present book aims to induce new outlooks to scientists/researchers who are unfamiliar with bamboo biotechnology and will be very helpful in various present and future researches in different areas of plant biotechnology, molecular biology, and plant physiology. We are extremely thankful to all the contributors who wholeheartedly welcomed our invitation and agreed to contribute chapters to embellish information on bamboo biotechnology, thus helping in this endeavor.

Nanjing, China Zishan Ahmad
Nanjing, China Yulong Ding
Aligarh, India Anwar Shahzad
17 July 2021

Acknowledgments

We, the editors, are thankful to our publisher, Springer Singapore, for continuous support throughout this project. We acknowledge the generous support of all the contributors who spared their precious time in scripting the chapters for this edited volume. We are grateful to the reviewers for their thoughtful comments and suggestions, which have been very helpful for improving the manuscripts. We express our special thanks to the Bamboo Research Institute, Nanjing Forestry University, and Aligarh Muslim University for all their support.

Zishan Ahmad
Yulong Ding
Anwar Shahzad

Contents

Chapter 1
Bamboo: Origin, Habitat, Distributions and Global Prospective

Zishan Ahmad, Anamica Upadhyay, Yulong Ding, Abolghassem Emamverdian, and Anwar Shahzad

Abstract Bamboo is the multifunctional and fastest-growing plant on Earth. Bamboo has played a crucial role in the daily life of millions of people in tropical countries, where it provides environmental, social and economic benefits. Bamboo belongs to the subfamily Bambusoideae of the grass family Poaceae. Today, the native bamboo distributes mainly in Asia, America and Africa, but Europe is unable to claim any native bamboo species. Approximately 123 genera and more than 1500 species of bamboo plant have been identified across the world. The natural distribution of bamboo across the tropical and subtropical regions is dependent on the type of soil, rainfall, temperature and altitude. The latitudinal distribution of bamboo ranges from 47° S to 50° 30′ N, while the altitudinal distribution from sea level to 4300 m. The occurrence of bamboo is therefore associated with the area mostly mesic to wet forest types in both temperate and tropical regions. However, there are some other bamboos reported to be adapted for open grasslands or narrow habitat. The present chapter is aiming to summarize the details of bamboo origin, habitat and global distribution. The classification of Bambusoideae has been also given. Moreover, an overview of important bamboo genera and species has been discussed.

Keywords Bamboo diversity · Bamboo habitats · Bamboo classification · Woody bamboo

Z. Ahmad · Y. Ding (✉) · A. Emamverdian
Bamboo Research Institute, Nanjing Forestry University, Nanjing, Jiangsu, People's Republic of China

Co-innovation Center for Sustainable Forestry in Southern China, Nanjing Forestry University, Nanjing, Jiangsu, People's Republic of China
e-mail: ylding@vip.163.com

A. Upadhyay · A. Shahzad
Plant Biotechnology Section, Department of Botany, Aligarh Muslim University, Aligarh, India

Z. Ahmad et al. (eds.), *Biotechnological Advances in Bamboo*,
https://doi.org/10.1007/978-981-16-1310-4_1

1.1 Introduction

Bamboo is a versatile group of plants that comes under the subfamily Bambusoideae of the family Poaceae. From the beginning of civilization up to nowadays, bamboo has played an immensely important part in the daily lives of human beings, and therefore, this plant is known as "the plant of multifunctional uses" (Amada and Untao 2001; Wooldridge 2012; Sharma et al. 2014). Bamboo grows fast and matures early, due to which, its plantation is good over other forestry species. One more significant advantage in comparison to other forest plant species is that bamboo can harvest multiple times within a few years because of its fast growth rates. Nowadays, the native bamboo distributes mainly in Asia, America and Africa. There is no native bamboo species in Europe; nevertheless, a lot of species were introduced and cultivated in Europe. The use of bamboo stem is wide, and its major quantity is utilized as raw material for housing, utensils, agricultural applications, handicraft items, packing materials, paper and pulp industry, etc. (Singh 2008). Bamboo is inextricably linked to human beings as it fulfils the need for shelter, clothing and many other items (Wooldridge 2012; Chaowana 2013). Therefore, commonly, bamboo described as "friend of people", "poor man's timber", "green gasoline" and "the cradle to coffin timber" (Singh 2008). Bamboo serves as an important resource for living organisms like pandas and many other organisms.

The distribution of bamboo plants over the subtropical and tropical regions naturally depends on several conditions like rainfall, soil type, temperature and altitude. Bamboos naturally grow in those areas that receive annual rainfall and vary from 1200 to 4000 mm approximately, and the annual average temperature varies from 8 to 36 °C approximately. They grow well in different types of soils like sandy soils, loamy soils, hard lateritic soils and rich alluvium soils. Except Antarctica and Europe, bamboos are native to all continents and distributed 47° S to 50° 30′ N latitudinal and 4, 300 m from the sea level in altitude (Judziewicz et al. 1999; Ohrnberger 1999). The multipurpose application of bamboo hardly rivalled with other plants of plant kingdom. However, bamboos are complex plants and difficult to identify and classify, but considering the ecological and economic importance, the correct identification and classification are a critical step. This chapter deals with the up-to-date information of bamboo origin, habitat and its global prospective. Bamboo taxonomical identification and classification system on the basis of morphological and molecular characteristics has been also discussed. Moreover, the end section of this chapter provides an overview of important bamboo genera and species.

1.2 Bamboo Origin, Habitat and Diversity

The origin of bamboo will be discussed based on the evolution of the grass, distribution centre, fossil evidence and the basal lineage of grass. It deals with where and when its origin and the possible ancestor. The utilization of bamboo can be dated back to 3500 years ago. In order to better use bamboo, the ancient people started to give names of different bamboos. The first monograph of bamboo named "Zhu Pu" in the world was written by Dai Kai-zhi (317–420 A.C.) from Jin Dynasty. In his book, the morphological and biological characters, distributions and usages of more than 60 bamboos were described. He also modified the definition of bamboo from grass to "Bamboo is a non-wood, non-grass strange plant".

In fact, the English word bamboo" in west literature was originated from the Indian name "Mambu". In India, the local people collected one kind of substance from a plant named "Mambu" or "Saccar Mambu" as medication which was called "Tabaxir" later. In fact, the first bamboo classification in Europe was done by Georg Eberhard Rumpf (1750) who divided the bamboos into eight classes, all with the name Arundo. On this basis, Linnaeus in 1753 used the name *Arundo bambos* which included all bamboos up to his time, from which the genus name *Bambusa* was later adapted. This classification system was used with little change by Linnaeus (1753). The number of bamboo species known to science has risen sharply since Linnaeus published only one species, *Arundo bambos* (*Bambusa bambos*), in his *Species Plantarum*. One hundred fifteen years later (1868), a new list of 170 species and 21 genera of bamboo was issued by Munro who published a monograph of the Bambuseae and classified the bamboo in three divisions.

In 1881, Gamble proposed a classification system in which he recognized 4 subtribes, 14 genera and 151 species. In 1883, Bentham and Hooker, *Genera Plantarum* 3 (2), Bambuseae, modified the system of Munro, so instead of three divisions, four subtribes were introduced. About 30 years later (1913), French botanist E.G. Camus adopted Munro's system of classification. In his publication, he listed 33 genera and 490 species of bamboo, including many of the herbaceous bamboos. During the 1980s–1990s, a few bamboo classification systems were published by Clayton and Renvoize (1986), Soderstrom and Ellis (1987), Dransfield and Widjaja (1995) and Stapleton (1998). More recently, a new system based on the molecular data was built by Kelchner and Bamboo Phylogeny Group (2013). According to Ohrnberger (1999), a number of species of woody bamboos purely occurring in the countries of the Asia-Pacific region, Africa and Americas have been listed in Table 1.1.

Bamboo plant species is naturally distributed globally in all continents except Antarctica and Europe; however, a few years ago, some species of bamboo plants had been introduced in Europe (Akinlabi et al. 2017; Wu et al. 2020). In terms of altitudinal and latitudinal, distribution of bamboo plants ranges from sea level to 4300 m and 47°S to 50°30′N, respectively (Soderstrom and Calderon 1979, Ohrnberger 1999, Judziewicz et al. 1999, Liese and Tang 2015). In an ecological area, bamboo plants are dominant in subtropical and tropical regions (Ohrnberger 1999, Jiang 2007, Akinlabi et al. 2017, Wu et al. 2020). However, its occupancy also includes coniferous forests, temperate deciduous forests, mountainous forests, lowland tropical forests, wetter forests, grasslands, etc., and therefore bamboo plants occupied a broad range of habitat types. A description of bamboo species occurrence and habitats has been summarized in Table 1.2.

Bamboo has the richest diversity in Asia. There are about 50 genera and 900 species which is about 75% of the total species in the world. The largest national complement of species is for China, which had about 600 described species,

Table 1.1 Naturally occurring bamboo species

	Country	Number of naturally occurring bamboo species
Asia-Pacific	Australia	2
	Bhutan	1
	Brunel	1
	China	583
	Hong Kong	3
	India	40
	Indonesia	29
	Japan	75
	Republic of Korea	2
	Laos	4
	Malaysia	26
	Myanmar	30
	Nepal	6
	Papua New Guinea	15
	Philippines	14
	Sri Lanka	6
	Thailand	4
	Vietnam	38
Africa	Madagascar	32
	Tanzania	1
North, Central and South America	Argentina	2
	Belize	1
	Bolivia	2
	Brazil	110
	Chile	10
	Colombia	15
	Costa Rica	13
	Cuba	6
	Ecuador	14
	El Salvador	1
	Guyana	2
	Martinique	1
	Mexico	17
	Panama	1
	Paraguay	1
	Peru	15
	Trinidad and Tobago	2
	United States	1
	Uruguay	1
	Venezuela	32

Table 1.2 Bamboo species occurrence and habitats

Bamboo species	Occurrence	Habitats	References
Acidosasa	Asia	Dry or evergreen subtropical forests	Stapleton (1994a, b, c), Li and Xue (1997), BPG (2012)
Actinocladum	Brazil	Lowland bamboos	Soderstrom and Calderon (1979), Judziewicz et al. (1999)
Alvimia	America, Asia	Lowland tropical bamboos	Soderstrom and Londoño (1988), Dransfield (1992, 1994)
Arundinaria	Eastern United States, China, Japan, North America	Woodlands and forests, mostly along rivers	Yang and Xue (1990), Triplett et al. (2006), Dai et al. (2011)
Aulonemia	Brazil	Andean montane forests, mono-dominant stands at high elevations	Judziewicz et al. (1999), Viana et al. (2013)
Bambusa	China, Mexico, India, Vietnam	Lowland moist tropical forests, lower montane forests	Soderstrom and Calderon (1979), Seethalakshmi and Kumar (1998), Judziewicz et al. (1999), BPG (2012)
Bashania	China	Montane forests	Taylor and Qin (1997), Li and Xue (1997)
Bergbambos	India, South Africa, Sri Lanka	Tropical mountain grasslands and shrublands	Soderstrom and Ellis (1982); Soderstrom et al. (1988)
Cambajuva	Brazil	Mono-dominant stands at high elevations	Judziewicz et al. (1999), Viana et al. (2013)
Chimonobambusa	China, Myanmar	Montane forests, wetter side of mountain ranges	Stapleton (1994a, b, c), Taylor and Qin (1997), Li and Xue (1997)
Chusquea acuminatissima, *C. aristata*, *C. tessellate*, *C. guirigayensis*, *C. villosa*	America, Argentina, Brazil, Chile, Mexico,	Andean montane forests, *Araucaria* forests, Atlantic forests, cloud forest, highest elevation (4000–4400 m), montane forests, *Nothofagus*, pine oak fir forests	Soderstrom and Londoño (1988), Dransfield (1992, 1994), Judziewicz et al. (1999), Fisher et al. (2009), BPG (2012)
Cryptochloa	Argentina, Brazil, America, Mexico	Lower montane forests	Judziewicz et al. (1999), Judziewicz and Clark (2007)

(continued)

Table 1.2 (continued)

Bamboo species	Occurrence	Habitats	References
Dendrocalamus strictus	America, China, Colombia, India, Madagascar, Mexico	Lowland moist tropical forests or lower montane forests	Soderstrom and Calderon (1979), Gadgil and Prasad (1984), Rao and Ramakrishnan (1988), Seethalakshmi and Kumar (1998), Judziewicz et al. (1999), Ruiz-Sanchez (2011), BPG (2012)
Dinochloa	America, Asia	Lowland tropical bamboos	Soderstrom and Londoño (1988), Dransfield (1992, 1994)
Drepanostachyum	Asia, Central Himalayas	Evergreen subtropical forests, seasonally dry forests	Stapleton (1994a, b, c), Li and Xue (1997), BPG (2012)
Ekmanochloa	America, Argentina, Brazil, Mexico, West Indies	Savannas (semi-deciduous seasonal forests)	BPG (2012), Ferreira et al. (2013)
Eremocaulon	Mexico	Lowland moist tropical forests or lower montane forests	Soderstrom and Calderon (1979), Seethalakshmi and Kumar (1998), Judziewicz et al. (1999), BPG (2012)
Fargesia	Asia, China, Japan, Madagascar, Sri Lanka	Montane forests, temperate mountains	Li and Xue (1997), Taylor and Qin (1997), Li et al. (2006), BPG (2012)
Fargesia yulongshanensis	China, Himalaya	Might elevations	Li et al. (2006)
Filgueirasia	Brazil	Lowland bamboos	Soderstrom and Calderon (1979), Judziewicz et al. (1999)
Gigantochloa	China, Mexico	Lowland moist tropical forests or lower montane forests	Soderstrom and Calderon (1979), Seethalakshmi and Kumar (1998), Judziewicz et al. (1999), BPG (2012)
Guadua paniculata	Amazon Basin, America, China, Colombia, India, Mexico, Madagascar	Lowland moist tropical forests or lower montane forests	Soderstrom and Calderon (1979), Gadgil and Prasad (1984), Rao and Ramakrishnan (1988), Seethalakshmi and Kumar (1998), Judziewicz et al. (1999), Ruiz-Sanchez (2011), BPG (2012)

(continued)

Table 1.2 (continued)

Bamboo species	Occurrence	Habitats	References
Hickelia	America, Asia	Lowland tropical bamboos	Soderstrom and Londoño (1988), Dransfield (1992, 1994)
Holttumochloa	Southeast Asia	Montane forests	Dransfield (1992), Wong (1993)
Indosasa	Asia, China	Montane forests, dry or evergreen subtropical forests	Stapleton (1994a, b, c), Li and Xue (1997), Taylor and Qin (1997), BPG (2012)
Kuruna	India, South Africa, Sri Lanka	Tropical mountain grasslands and shrublands	Soderstrom and Ellis (1982); Soderstrom et al. 1988
Lithachne	Brazil, America, Colombia, Cuba, Mexico, Panama	Near the equator, lower montane forests	Judziewicz et al. (1999), Judziewicz and Clark (2007), BPG (2012)
Merostachys	Brazil	Atlantic forests	Judziewicz et al. (1999)
Nastus borbonicus	Argentina, Chile	Montane forests	Judziewicz et al. (1999), BPG (2012)
Neomicrocalamus	America, Asia	Lowland tropical bamboos	Soderstrom and Londoño (1988), Dransfield (1992, 1994)
Ochlandra	Asia, Africa, Australia, India, Sri Lanka	Reed-like thickets along stream banks	Seethalakshmi and Kumar (1998), BPG (2012), Gopakumar and Motwani (2013)
Olyra (*O. latifolia*)	Argentina, America, Brazil, Colombia, Cuba, Panama Mexico	Near the equator, lower montane forests	Judziewicz et al. (1999), Judziewicz and Clark (2007), BPG (2012)
Otatea	America, Colombia, India, Mexico, Madagascar	Drier forests	Soderstrom and Calderon (1979), Gadgil and Prasad (1984), Rao and Ramakrishnan (1988), Seethalakshmi and Kumar (1998), Ruiz-Sanchez (2011)
Pariana	Amazonian, America, Argentina, Brazil, Bolivia, Costa Rica, Mexico	Lowland tropical montane forests	Judziewicz et al. (1999), Judziewicz and Clark (2007)
Perrierbambus	America, Colombia, India, Mexico, Madagascar	Drier forests	Soderstrom and Calderon (1979), Gadgil and Prasad (1984), Rao and Ramakrishnan (1988), Seethalakshmi and Kumar (1998), Ruiz-Sanchez (2011)

(continued)

Table 1.2 (continued)

Bamboo species	Occurrence	Habitats	References
Raddiella (*R. esenbeckii*)	America, Argentina, Brazil, Mexico	Lower montane forests	Judziewicz et al. (1999), Judziewicz and Clark (2007)
Racemobambos	Southeast Asia, America	Montane forests, lowland tropical bamboos	Soderstrom and Londoño (1988), Dransfield (1992, 1994), Wong (1993)
Reitzia	Brazil	Atlantic forests	Judziewicz et al. (1999)
Rhipidocladum	Brazil	Andean montane forests	Judziewicz et al. (1999)
Sasa	China, Korea and Japan	Wetter forests	Noguchi and Yoshida (2005), Tsuyama et al. (2011)
Sasamorpha	China, Korea and Japan	Wetter forests	Noguchi and Yoshida (2005), Tsuyama et al. (2011)
Sarocalamus	China	Wetter side of mountain ranges	Stapleton (1994a, b, c)
Schizostachyum	China, Mexico	Lowland moist tropical forests or lower montane forests	Soderstrom and Calderon (1979), Seethalakshmi and Kumar (1998), Judziewicz et al. (1999), BPG (2012)
Sinobambusa	Asia	Dry or evergreen subtropical forests	Stapleton (1994a, b, c), Li and Xue (1997), BPG (2012)
Thamnocalamus	Central Himalayas	Seasonally dry	Stapleton (1994a, b, c)
Vietnamosasa	Indochina	Grassland	Stapleton (1998)
Yushania	Asia, China, Japan, Madagascar, Sri Lanka	Montane forests, wetter side of mountain ranges, temperate mountains	Stapleton (1994a, b, c), Li and Xue (1997), Taylor and Qin (1997), Li et al. (2006), BPG (2012)

followed by India (102 species) and Japan (84 species). The bamboo area in Asia is about 25 million hm^2; among them, India ranks first with about nine million hm^2, while China has about 6.4 million hm^2. Myanmar, Indonesia, Malaysia, Vietnam, Laos, Cambodia, Philippines, Thailand, Japan, Bangladesh, South Korea, Sri Lanka, Nepal and other countries have about ten million hm^2. In America, there are 21 genera and 345 species, but it mainly distributes in Latin America, with only 3 species in southeast part of the United States. However, the biodiversity of bamboo is the richest in this region. In Africa, about 13 genera and 40 species were recognized. But the bamboo forests distribute mainly in East Africa, like Tanzania, Kenya, Zambia, Ghana, Ethiopia, Uganda, Mozambique and Madagascar.

1.3 Taxonomical Identification of Bamboo

As bamboo is a highly adaptable plant in different environmental conditions, it is difficult to identify or classify. Moreover, on the basis of ecological characteristics and the economic importance, identification of bamboo species was done. In the whole world, approximately 123 genera and more than 1500 species of bamboo plant have been identified (Steinfeld 2001; Nguyen 2006; Chaowana 2013). Bamboo belongs to the monocotyledonous flowering plants in the group angiosperms. Bamboo plant is basically divided into two major parts, the underground part of stem known as rhizomes and the upper portion, i.e. stem known as culms (Steinfeld 2001). Rhizomes of the bamboo are mostly sympodial, and they store the nutrients for their growth and development along with the sustenance of the plant in the ground. Rhizomes also contain the meristematic buds which further grow into shoots and emerge from the ground to form the group of culms. The culm of the bamboo plant is cylindrical in shape and contains most of the woody material of the plant (Ahmed and Kamke 2005). The culm is subdivided by multiple sections which are called nodes or diaphragms, and the part of the culm is between the two adjacent nodes called internodes (Amada and Untao 2001). Bamboo culm is hollow and has a thick wall, due to which, it ideals with the formation of household products, etc. The diameter of bamboo culm is ranging from 0.64 to 30.48 cm, and its height ranges from 1 foot to 120 feet (Amada and Untao 2001). Bamboo culm does not have bark, but it has smooth outer skin which is hard in nature because of the presence of silica. Bamboo culms bear branches and foliage leaves. Species of bamboo differ from each other in the sense of growth style of culm, i.e. it may be erect with droopy tips or simple erect, and also, it may be arched or clambering type (Akinlabi et al. 2017). Bamboo has the main characteristic of having high tensile strength compared to mild steel for the load-bearing capacity, and this is because bamboo has a natural composite material made up of cellulose fibres which immersed in a matrix of lignin that gives of an average 700 MPa tensile resistance (Janssen 2000; Li and Shen 2011).

1.4 Classification of Bamboo

On the basis of morphological as well as molecular characteristic features, bamboo is classified into three tribes, namely, Arundinarieae, Bambuseae and Olyreae. Arundinarieae includes temperate woody bamboos, Bambuseae includes tropical bamboos, and Olyreae includes herbaceous bamboos (Sungkaew et al. 2009; BPG 2012; Kelchner and Bamboo Phylogeny Group 2013). Recently, bamboos (i.e. Bambusoideae) is considered as part of the subfamily of Poaceae, i.e. Grasses family (GPWG 2001; GPWG II 2012), and these are found to be the most diversified grasses in the global forests (Zhang and Clark 2000; Judziewicz and Clark 2007; Sungkaew et al. 2009). According to GPWG (2001) and BPG (2012), there are approximately 115 to 119 genera and more than 1500 species of bamboos where

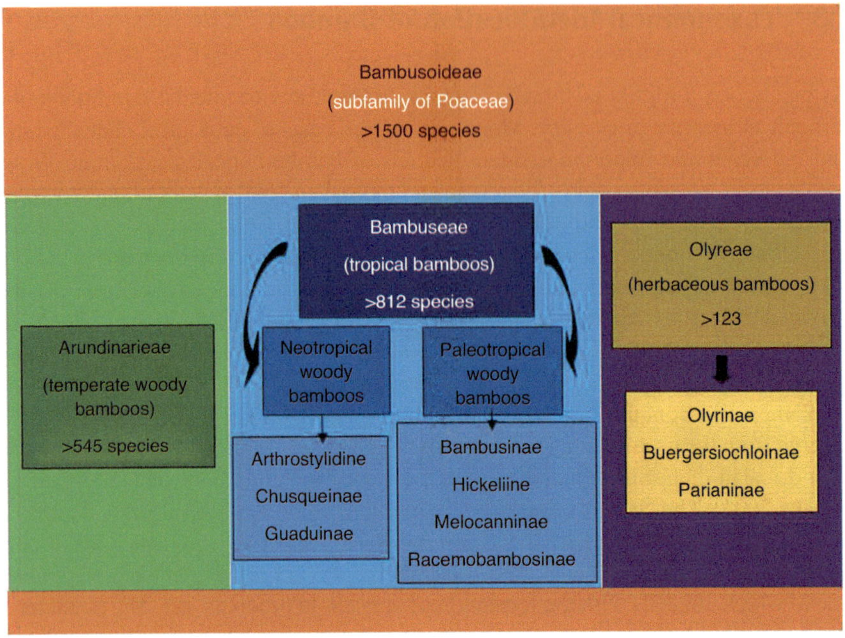

Fig. 1.1 Classification of Bambusoideae

Arundinarieae consists of about 533 to 546 bamboos, Bambuseae contains about 784–813 bamboos and Olyreae has only about 122–124 bamboos (Fig. 1.1). Tribe Bambuseae is further subdivided into two clades, namely, neotropical woody bamboos and paleotropical woody bamboos. Neotropical woody bamboos are again categorized into three subtribes, namely, Arthrostylidium, Chusqueinae and Guaduinae. Paleotropical woody bamboos are categorized into four subtribes, namely, Bambusinae, Hickeliinae, Melocanninae and Racemobambosinae. Further, tribe Olyreae are subdivided into three subtribes, namely, Olyrinae, Buergersiochloinae and Parianinae. Morphologically, woody bamboos can be distinguished from the herbaceous bamboos as they have lignified culms, culm leaves and outer ligules, while herbaceous bamboos are made up of weekly lignified culms only (BPG 2012). Moreover, woody bamboos have the characteristic feature of monocarpic flowering with bisexual spikelets, while herbaceous bamboos have unisexual spikelets with seasonal flowering. Furthermore, olyroid silica found in the herbaceous bamboos is absent in the case of woody bamboos.

1.5 Bamboo Culm and Its Development

The culm is the main material for utilization. It can reach the maximum height in about 60 days. So, bamboo is considered the most fast-growing plant. No matter it is a sympodial or monopodial bamboo, a culm is only one piece of the branch of one ramet. It developed from the buds on the rhizome or on the culm prope. Some bamboo species are shooting in spring, some in summer and some in autumn. The differentiation of the buds may be affected by temperature and the endogenous hormone.

The most important process of culm development includes bud differentiation, primary thickening growth, elongation growth and maturation. There are already quite a lot of publications that deal with the development of the culm, but there still remain a lot of questions which need to continue the investigation. For example, the culm sheath plays an important role by shoot elongation growth. But there are very few little publications that deal with this aspect. During the primary thickening, how the vascular bundles differentiate remains unknown. The mechanism of the fibre thickening during the maturation of culm is still a mysterious matter. Culm development occurs in two phases: (1) Newly unbranched shoots having culm leaves that elongate to their full length takes place. (2) Elongated culm lignification and development of branches along with the production of foliage leaves take place. The longevity of culms is varied among species to species, but generally they persist for 5–10 years (McClure 1966), and usually 3–5-year-old culms are treated as mature culms. The age of culms is an important parameter in the management of bamboo forest. Moreover, the age of bamboo can also be determined by counting the base node (leaf scar) on the leaf sheath present in the twig of a culm of bamboo (Banik 2000). Leaves of a bamboo fall in 1 year or 1 ½ year, and new leaves are developed from the near one node of leaf fall region keeping leaf scar marks on twig. Therefore, within 12–15 months of culm age, mark of one leaf scar is formed, and after in the next 24–30 months of culm age, another mark of leaf scar is formed. Therefore, in the third year of culm age, total three marks of leaf scar are formed in the twig.

1.6 Difference in Characteristics of Bamboo in Reference to Trees

"In the plant kingdom, there is one kind of plant called bamboo, which is different from both trees and grasses. It is a strange plant". This definition was described by Dai Kaizhi (AC 420–479), who write the first monograph of bamboo. Bamboo is evergreen with a special leaf blade. The anatomical structure is also different from those of grass. Bamboo has no secondary growth. Bamboo has a special branching pattern and has no main trunk. Bamboo has a very long vegetative growth cycle and propagates mainly by clones. A differential description of bamboo and trees is summarized in Table 1.3.

Table 1.3 Differences in the characteristic features of bamboos and trees

S. No.	Bamboos characteristics	Tree characteristics
1	Stem, i.e. culm is hollow and segmented	Stem is solid and unsegmented
2	Peripheral region of the culm is the hardest part of the plant	Central region of the stem is the hardest part of the plant
3	Culm has no bark	Stem has bark (secondary phloem + cork)
4	Culm grows very fast and reaches its full height in a single season (as its height reaches up to 36 m tall within 4–6 months)	Stem grows slowly in height as well as in diameter for many seasons
5	Conducting tissues, i.e. xylem and phloem lie together inside the vascular bundle	Conducting tissues, i.e. xylem and phloem are separated by vascular cambium inside the vascular bundle
6	Due to the absence of vascular cambium, culm does not grow in diameter with the age	Due to the presence of vascular cambium, stem grows in diameter with the age
7	Harvesting of culms directly effects on the clump community	Harvesting of stems does not directly affect on the remaining tree
8	Culm depends on the other in a clump because it grows in association with the network of rhizome	Stem grows independently
9	Culm has no radial, i.e. lateral, communication except at the internodes	Stem has radial communication throughout its length
10	Underground part consists of both rhizome and roots	Underground part does not have rhizome and consist of only roots

1.7 An Overview of Important Genera and Species of Bamboos

There are 123 genera and more than 1500 species of bamboos in the world (GPWG 2001; BPG 2012; Steinfeld 2001; Nguyen 2006; Chaowana 2013). They are native to Asia, America, Africa and Oceania. The most important feature of bamboos is being sympodial and its discrete clump form. A brief description of some important bamboo genera and species has been given in Table 1.4. Similarly, the important species of high economic use in China, South America and Africa has been listed below.

Phyllostachys edulis, the most important bamboo in China (Fig. 1.2). *Phyllostachys violascens*, an important shoot production bamboo in Yangtze Delta region (Fig. 1.3). *Dendrocalamus latiflorus* and *Bambusa oldhamii*, two important shoot production bamboos in South China (Figs. 1.4 and 1.5). *Dendrocalamus brandisii* an important shoot production bamboo in south Yunnan Province (Fig. 1.6). *Chimonobambusa utilis*, an important shoot production in southwest China (Fig. 1.7). *Dendrocalamus farinosus*, *Bambusa rigida* and *Bambusa emeiensis*, important species for pulp industry in China (Figs. 1.8, 1.9 and 1.10). *Guadua angustifolia*, the most important species for construction in South America

Table 1.4 Overview of some important genera and species of bamboos

Genus	Species	Characteristic features		
		Geographical distribution	Brief description	Uses
Bambusa	*B. arundinacea*	Bangladesh, China, India, Indonesia, Myanmar, Nepal, Philippines, Thailand, Vietnam	It is a versatile bamboo, having thick culm internodes. It is thorny in nature. Its culm height is up to 30 m, thickness about 18 cm and wall thickness approx. 2.5 to 5 cm. It can withstand 2 °C temperature	Culm used in artefacts, pulp and paper industry, handicrafts, river bank stabilization, land rehabilitation
	B. balcooa	Australia, Bangladesh, India, Indonesia	Its culm height is up to 24 m, diameter up to 15 cm and wall thickness approx. 2.5 cm	Culm used in construction of artefacts, pulp and paper, furniture, handicrafts, river bank stabilization, land rehabilitation
	B. blumeana	China, Borneo, Java, Malaysia, Philippines, Papua Guinea, Sumatra, Thailand	It is a giant thorny bamboo and green in colour with prominent nodes. Its culm height is up to 15–25 m and diameter up to 20 cm. They can be propagated by seeds, layering, culm and rhizome cuttings	Culms used in construction of artefacts, handicrafts, furniture, pulp and paper, chopsticks, riverbank stabilization, rehabilitation of degraded land
	B. polymorpha	Bangladesh, India, Myanmar, Thailand	Its culm height up to 16 and 25 m, diameter up to 15 cm and its wall thickness up to 1 cm. It can be propagated by seeds, layering, culms, offset and rhizome cuttings	Culms used in construction of artefacts, handicrafts, furniture and buildings. Its shoots are edible also
	B. textilis	China	It is a medium-sized bamboo. Its cum height is up to 15 m and diameter up to 3–5 cm. Its wall is thin and delicate in nature. It can be propagated by seeds, culms and offset cuttings	It is a high-quality bamboo. Culm is used in artefacts, handicrafts and weaving purpose because it splits easily. Its shoots are edible also

(continued)

Table 1.4 (continued)

Genus	Species	Characteristic features		
		Geographical distribution	Brief description	Uses
	B. tulda	Bangladesh, India, Indonesia, Myanmar, Nepal, Thailand, Vietnam	Its culm height is up to 30 m, diameter up to 5–10 cm and wall thickness approx. 1 cm. It can be propagated by seeds, marcotting, culms, rhizomes and offset cuttings	Culms used in construction artefacts, paper and pulp, handicrafts, furniture and architectural works
	B. vulgaris	Found in all regions where bamboo can grow easily, i.e. temperate, tropical and subtropical regions	This bamboo is strong and medium to large with having open clumps. Its culm height is up to 20 m, diameter up to 5–10 cm and wall thickness approx. 1.5 cm. This bamboo is easy to propagate because it is tremendously vegetative in nature, and therefore it mostly propagated by branch cuttings, layering, marcotting and culm, offset and rhizome cuttings. Its culms and internodes are curved. This species is of two varieties. One is *B. vulgaris vittata* and characterized by having yellow culm, and the other is *B. vulgaris wamin* and characterized by having green culm. These are grown as ornamental plants	Culms used in construction of artefacts, furniture, handicrafts, architectural works and manufacturing of pulp and paper
Cephalostachyum	*C. pergracile*	China, India, Java, Myanmar, Thailand	It is a medium-sized bamboo and is having straight culms. Its culm height is up to 30 m and its wall is thin. They can be propagated by using seeds but mostly propagated	Culms used in construction of artefacts, handicrafts and weaving of baskets. It used as an ornamental plant also

			by culm and offset cuttings because it is tremendously vegetative species	
Dendrocalamus	*D. asper*	Bangladesh, China, India, Indonesia, Laos, Malaysia Myanmar, Nepal, Philippines, Thailand, Vietnam	These bamboos possess large culms whose height is up to 20–30 m, diameters up to 8–20 cm and wall thickness approx. 2 cm. It can be propagated by culm, offset and branch cuttings	Culms used in construction of artefacts, handicrafts, furniture, musical instruments and architectural works Its shoots are edible also
	D. giganteus	Bangladesh, China, Ghana, India, Indonesia, Kenyan, Myanmar, Philippines, Thailand, Vietnam	This bamboo is made up of huge and large culm and green to dark bluish green in colour. Its culm height is up to 25–60 m, diameter up to 10–20 cm and wall thickness approx. 2.5 cm. It can be propagated by layering, marcotting, macroproliferation, culms, branch and rhizome cuttings	Culms used in construction of artefacts, handicrafts, structural works, building,bamboo boards and pulp. Its shoots are edible also
	D. lactiferous	China, India, Japan, Myanmar, Philippines, Taiwan, Thailand, Vietnam	This bamboo is having medium-sized culms. It grows well in areas with high rainfall. Its culm height is up to 14–25 m, diameter up to 8–20 cm and wall thickness approx. 0.5–3 cm. It can be propagated by marcotting, layering and culm cuttings	Culm used in construction of artefacts, handicrafts, furniture, pulp and architectural works
	D. strictus	India, Myanmar, Nepal, Thailand	This bamboo has medium-sized culm. Its culm height up to 8–20 m and diameter up to 2.5–8 cm, and its wall is thick but not straight. It can be propagated by seeds,	Culms used in construction of artefacts, handicrafts, structural works, bamboo boards, building, pulp and household utensils. Its shoots are also edible

(continued)

Table 1.4 (continued)

Genus	Species	Characteristic features		
		Geographical distribution	Brief description	Uses
			layering, marcotting, macroproliferation and culm and rhizome cuttings	
Gigantochloa	*G. apus*	India, Indonesia, Malaysia, Myanmar, Thailand	This is a large bamboo. Its culm height is up to 8–30 m, diameter up to 4–13 cm and wall thickness approx. 1.5 cm which is strongly tufted in nature. It is propagated by culm and offset cuttings	Culm used in construction of artefacts, handicrafts, musical instruments, structural works, building, pulp and household utensils. Its shoots are edible, but it has bitter in taste
	G. levis	China, Indonesia, Kalimantan, Malaysia, Philippines, Vietnam	This bamboo has large culm. Its culm height is up to 30 m, diameter up to 5–16 cm and wall thickness approx. 1–1.2 cm. It is propagated by culm and offset cuttings	Culms used in construction of artefacts, handicrafts, walls, furniture, pulp and paper, architectural works and household utensils. Its shoots are having high edible quality
	G. pseudoarundinacea	China, India, Java, Malaysia, Sumatra	Its culm height is up to 7–30 m, diameter up to 5–13 cm and wall thickness approx. 2 cm which is strong in nature. It is propagated by culms and branch cuttings	Culms used in construction of artefacts, handicrafts, toothpicks, structural works and household utensils. Its shoots are edible also
Guadua	*G. angustifolia*	Argentina, China, India, Mexico	This is a large bamboo, and it can be easily identified by their aesthetic feature like its dark green culm having white node bands. Its culm height is up to 30 m, diameter up to 20 cm and wall thickness approx. 1.5 cm with strongly tufted	Culms used in construction of artefacts, handicrafts, furniture, pulp, structural works, building and bamboo boards

			in nature. It is propagated by culms and offset cuttings	
Melocanna	*M. baccifera*	Bangladesh, China, India, Indonesia, Myanmar, Vietnam	This is a large bamboo species. Its culm height is up to 10–25 m and diameter up to 5–9 cm, but its wall is thin approx. 0.5–1.2 cm. The important feature of this species is its culm tips are pendulous in nature and this produces the largest fruit in the grass family. It is propagated by seeds and culm cuttings	Culms used in construction of handicrafts, structural works, roofing, walls, weaving of mats, pulp and paper. Its shoots are used in the preparation of liquor, and also it is edible
Phyllostachys	*P. edulis*	China, Europe, Japan, Korea, USA, Vietnam	This species of bamboo is medium to large in size. Its culm height is up to 10–20 m and diameter up to 18–20 cm, having a white waxy covering. It is propagated by seeds and offset cuttings	Culm used in construction of handicrafts and in structural works
Thyrsostachys	*T. siamensis*	Indochina and Myanmar	This bamboo has densely clumped, and its sheath on the culm is obstinate with single white ring below nodes. Its culm height is up to 8–16 m and diameter up to 2–6 cm, and its wall is thin in nature. It is propagated by seeds, macroproliferation and offset cuttings	Culms used in construction of handicrafts, walls, furniture, structural works, pulp and paper. It is also used as an ornamental plant. Its shoots are edible also

Fig. 1.2 *Phyllostachys edulis*. (**a**, **b**) plantation. (**c**, **d**) Shoot culms. PC: Prof. Yulong Ding

(Fig. 1.11). *Oxytenanthera abyssinica*, an important bamboo in lowland found in Africa (Fig. 1.12).

Fig. 1.3 *Phyllostachys violascens.* (**a**, **b**) plantation. (**c**) New shoot. PC: Prof. Yulong Ding

Fig. 1.4 *Dendrocalamus latiflorus*. (**a**) Plantation. (**b**) Shoot. (**c**) Culm with new culms. PC: Prof. Yulong Ding

Fig. 1.5 *Bambusa oldhamii*. (**a**) Plantation. (**b**) New shoot. (**c**) Harvesting shoot. PC: Prof. Yulong Ding

Fig. 1.6 *Dendrocalamus brandisii*. (**a**) Plantation. (**b**) New shoot. (**c**) Culm. PC: Prof. Yulong Ding

Fig. 1.7 *Chimonobambusa utilis*. (**a**, **b**) Plantation. (**c**) New shoot. PC: Prof. Yulong Ding

Fig. 1.8 *Dendrocalamus farinose*. (**a**, **b**) Plantation. (**c**) Shoot. PC: Prof. Yulong Ding

Fig. 1.9 *Bambusa rigida*. (**a**, **b**) Plantation. (**c**) New culms. PC: Prof. Yulong Ding

Fig. 1.10 *Bambusa emeiensis*. (**a**, **c**) Plantation. (**b**) New shoot. PC: Prof. Yulong Ding

Fig. 1.11 *Guadua angustifolia*. (**a**) Plantation. (**b**, **c**) New shoot. PC: Prof. Yulong Ding

Fig. 1.12 *Oxytenanthera abyssinica*. (**a**) Plantation. (**b**) Shoot. PC: Prof. Yulong Ding

1.8 Conclusions

Bamboo is one of the fast-growing and high-demand plants; however, its sustainability is at higher risk, and a critical examination for its renewability is a much-needed step to meet its traditional and modern demands. Its capacity to grow in almost any climate and ready to harvest in 3–5 years as compared to the other woody plants makes it a more versatile plant. For thousands of years, the bamboo plant has not been depleted despite its regular uses; however, the introduction of more advanced methods to make bamboo more resourceful is needed to be examined to avoid any consequences that claim its sustainability.

Acknowledgements This work was financially supported by the National Natural Science Foundation for Scholars of China (31870595) and the Priority Academic Program Development of Jiangsu Higher Education Institutions.

Conflict of Interest There is no conflict of interest.

Author Contribution ZA and AU wrote the manuscript. YD, AE and AS reviewed and added to the review. All authors read and approved the final manuscript.

References

Ahmed M, Kamke FA (2005) Analysis of Calcutta bamboo for structural composite materials: physical and mechanical properties. Wood Sci Technol 39:448–459

Akinlabi ET, Anane-Fenin K, Akwada DR (2017) Bamboo—the multipurpose plant. Springer, Switzerland

Amada S, Untao S (2001) Fracture properties of bamboo. Compos B 32:451–459

Banik RL (2000) Silviculture and field-guide to priority bamboos of Bangladesh and South Asia. BFRI, Chittagong, pp 1–187

Bamboo Phylogeny Group [BPG] (2012) An updated tribal and subtribal classification of the bamboos (Poaceae: Bambusoideae). In: Gielis J, Potters G (eds) Proceedings of the 9th world bamboo congress. Antwerp, Belgium, pp 3–27

Chaowana P (2013) Bamboo: an alternative raw material for wood and wood-based composites. J Mater Sci Res 2(2):90

Clayton WD, Renvoize SA (1986) Genera graminum, Grasses of the world, Vol XIII, Kew bulletin additional series. Her Majesty's Stationery Office, London, p 389

Dai L-M, Wang Y, Su D-K, Zhou L, Yu D-P, Lewis BJ, Qi L (2011) Major forest types and the evolution of sustainable forestry in China. Environ Manag 48:1066–1078

Dransfield S (1992) The bamboos of Sabah. Sabah forest records no. 14. Forestry Department, Sabah, Malaysia, p 94

Dransfield S (1994) The genus Hickelia (Gramineae-Bambusoideae). Kew Bull 49:429–443

Dransfield S, Widjaja EA (eds) (1995) Plant resources of South-East Asia No. 7: Bamboos. Backhuys, Leiden, p 189

Ferreira FM, Dorea MC, Leite KRB, Oliveira RP (2013) Eremitis afimbriata and E. magnifica (Poaceae, Bambusoideae, Olyreae): two remarkable new species from Brazil and a first record of blue iridescence in bamboo leaves. Phytotaxa 84:31–34

Fisher A, Triplett JK, Ho C-S, Schiller A, Oltrogge K, Schroder E, Kelchner SA, Clark LG (2009) Paraphyly in the Chusqueinae (Poaceae: Bambusoideae: Bambuseae). Syst Bot 34:673–683

Gadgil M, Prasad SN (1984) Ecological determinants of life history evolution of two Indian bamboo species. Biotropica 16:161–172

Gopakumar B, Motwani B (2013) Adaptive strategies of reed bamboos, Ochlandra spp., to the Western Ghat habitats of India. Bamboo Sci Cult 26:33–40

Grass Phylogeny Working Group [GPWG] (2001) Phylogeny and subfamilial classification of the grasses (Poaceae). Ann Missouri Bot Gard 88:373–457

Grass Phylogeny Working Group [GPWG II] (2012) New grass phylogeny resolves deep evolutionary relationships and discovers C4 origins. New Phytol 193:304–312

Janssen JJA (2000) Designing and building with bamboo. International Network for Bamboo and Rattan Technical Report No. 29. pp 207

Jiang Z (2007) Bamboo pulp and paper making (Chapter 11); bamboo charcoal. Vinegar and active carbon (Chapter 13). In: Bamboo and rattan in the world. China Forestry Publishing House, Beijing, pp 178–198. 214–226

Judziewicz EJ, Clark LG (2007) Classification and biogeography of New World grasses: Anomochlooideae, Pharoideae, Ehrhartoideae and Bambusoideae. Aliso 23:303–314

Judziewicz EJ, Clark LG, Londoño X, Stern MJ (1999) American bamboos. Smithsonian Institution, Washington, DC, p 392

Kelchner SA, Bamboo Phylogeny Group (2013) Higher level phylogenetic relationships within the bamboos (Poaceae: Bambusoideae) based on five plastid markers. Mol Phylogenet Evol 67:404–413

Li D-Z, Wang Z-P, Zhu Z-D, Xia N-H, Jia L-Z, Guo Z-H, Yang G-Y, Stapleton C (2006) Bambuseae. In: Wu Z-Y, Raven PH, Hong D-Y (eds) Flora of China: Poaceae, vol 22. Science Press, Missouri Botanical Garden Press, Beijing, St. Louis, pp 7–180

Li DZ, Xue JR (1997) The biodiversity and conservation of bamboos in Yunnan, China. In: Chapman GP (ed) The bamboos. Academic, London, pp 83–94

Li H, Shen S (2011) The mechanical properties of Bamboo and vascular bundles. J Mater Res 26 (21):2749–2756

Liese W, Tang TKH (2015) Morphology and growth. In: Liese W, Kohl M (eds) Bamboo: the plant and its uses. Springer International Publishing, Switzerland, pp 227–253

Linnaeus C (1753) "Species Plantarum," vol.1. London, pp 392

McClure FA (1966) The bamboos: a fresh perspective. Harvard University, Cambridge, MA, p 347

Nguyen HN (2006) Bamboos of Vietnam. Agricultural Publishing House

Noguchi M, Yoshida T (2005) Factors influencing the distribution of two co-occurring dwarf bamboo species (Sasa kurilensis and Sasa senanensis) in a conifer-mixed broadleaved stand in northern Hokkaido. Ecol Res 20:25–30

Ohrnberger D (1999) The bamboos of the world: annotated nomenclature and literature of the species and the higher and lower taxa. Elsevier, Amsterdam, p 585

Rao KS, Ramakrishnan PS (1988) Architectural plasticity of two bamboo species, Neohouzeaua dulloa Camus and Dendrocalamus hamiltonii Nees in successional environment in North-East India. Proc Indian Acad Sci (Plant Sci) 98:121–133

Ruiz-Sanchez E (2011) Biogeography and divergence time estimates of woody bamboos: insights in the evolution of Neotropical bamboos. Bol Soc Bot Me'x 88:67–75

Seethalakshmi KK, Kumar MSM (1998) Bamboos of India: a compendium. Kerala Forest Research Institute, Peechi. India and International Network for Bamboo and Rattan, New Delhi, p 342

Sharma B, Gatóo A, Bock M, Mulligan H, Ramage M (2014) Engineered bamboo: State of the art. Proc Inst Civil Eng. https://doi.org/10.1680/coma.14.00020

Singh O (2008) Bamboo for sustainable livelihood in India. Silviculture division, Forest Research Institute, Dehradun, pp 1193–1198

Soderstrom TR, Calderon CE (1979) V. Ecology and phytosociology of bamboo vegetation. In: Numata M (ed) Ecology of grasslands and bamboolands in the world. VEB Gustav Fischer Verlag, Jena, pp 223–236

Soderstrom TR, Ellis RP (1982) Taxonomic status of the endemic South African bamboo, Thamnocalamus tessellatus. Bothalia 14:53–67

Soderstrom TR, Ellis RP (1987) The position of bamboo genera and allies in a system of grass classification. In: Soderstrom TR, Hilu KW, Campbell CS, Barkworth ME (eds) Grass systematics and evolution. Smithsonian Institution, Washington, DC, pp 225–238

Soderstrom TR, Londoño X (1988) A morphological study of Alvimia (Poaceae: Bambuseae), a new Brazilian bamboo genus with fleshy fruits. Am J Bot 75:819–839

Soderstrom TR, Judziewicz EJ, Clark LG (1988) Distribution patterns of neotropical bamboos. In: Vanzolini PE, Heyer RE (eds) Proceedings of a workshop on neotropical distribution patterns, Academia Brasileira de Ciencias, Rio de Janeiro, pp 121–157

Stapleton CMA (1994a) The bamboos of Nepal and Bhutan Part I: Bambusa, Dendrocalamus, Melocanna, Cephalostachyum, Teinostachyum, and Pseudostachyum (Gramineae: Poaceae, Bambusoideae). Edinb J Bot 51:1–32

Stapleton CMA (1994b) The bamboos of Nepal and Bhutan Part II: Arundinaria, Thamnocalamus, Borinda and Yushania (Gramineae: Poaceae, Bambusoideae). Edinb J Bot 51:275–295

Stapleton CMA (1994c) The bamboos of Nepal and Bhutan Part III: Drepanostachyum, Himalayacalamus, Ampelocalamus, Neomicrocalamus and Chimonobambusa. Edinb J Bot 51:301–330

Stapleton CMA (1998) Form and function in the bamboo rhizome. J Am Bamb Soc 12(1):21–29

Steinfeld C (2001) A bamboo future. Environmental Design and Construction, pp 1–5
Sungkaew S, Stapleton CMA, Salamin N, Hodkinson TR (2009) Non-monophyly of the woody bamboos (Bambuseae; Poaceae): a multi-gene region phylogenetic analysis of Bambusoideae s.s. J Plant Res 122:95–108
Taylor AH, Qin Z-S (1997) The dynamics of temperate bamboo forests and panda conservation in China. In: Chapman GP (ed) The bamboos. Academic, London, pp 189–203
Triplett JK, Weakley A, Clark LG (2006) Hill cane (Arundinaria appalachiana), a new species of woody bamboo (Poaceae: Bambusoideae) from the Southern Appalachian Mountains. SIDA 22:79–95
Tsuyama I, Nakao K, Matsui T, Higa M, Horikawa M, Kominami Y, Tanaka N (2011) Climatic controls of a keystone understory species, Sasamorpha borealis, and an impact assessment of climate change in Japan. Ann For Sci 68:689–699
Viana P, Filgueiras T, Clark LG (2013) A new genus of woody bamboo (Poaceae: Bambusoideae: Bambuseae) endemic to Brazil. Syst Bot 38:97–103
Wong KM (1993) Four new genera of bamboos (Gramineae: Bambusoideae) from Malesia. Kew Bull 48:517–532
Wooldridge M (2012) Booming bamboo: The next super-material? BBC News Magazine 3
Wu R, Nan X, Yan H, Yang F, Shi Y, Bao Z (2020) A survey on the history of the introduction of bamboo from China to the United States (1898-2010) and the application of bamboo in urban greening. Urban For Urban Green 55:1–13
Yang Y-M, Xue J-R (1990) A preliminary study on the natural bamboo forests in the Dawei Mountain of southeastern Yunnan. J Southwest For Coll 10:22–29
Zhang W-P, Clark LG (2000) Phylogeny and classification of the Bambusoideae (Poaceae). In: Jacobs SWL, Everett J (eds) Grass systematics and evolution. CSIRO, Melbourne, pp 35–42

Chapter 2
Molecular Markers in Bamboos: Understanding Reproductive Biology, Genetic Structure, Interspecies Diversity, and Clonal Fidelity for Conservation and Breeding

Enéas Ricardo Konzen, Luciano Cesar Pozzobon, Denys Matheus Santana Costa Souza, Sérgio Bruno Fernandes, Wellington Ferreira Campos, Gilvano Ebling Brondani, Dulcinéia de Carvalho, and Siu Mui Tsai

Abstract Molecular markers have revolutionized analyses in population genetics, enabling precise estimates of the amount of genetic variability and how it is distributed within and among populations. The high diversity of bamboos, distributed throughout the world and of high economic relevance, has deserved several studies on molecular characterization. This chapter describes how distinct categories of molecular markers, such as isozymes, RAPD, AFLP, microsatellites, and SNP, have enabled the analysis of population genetic processes, assessments of the genetic diversity, and structure of natural populations and selected cultivars of bamboo species. One important application is their power of phylogenetic inference, enabling the distinction of the diverse set of bamboo species. With the genomic technologies, gene families have been characterized, mainly for *Phyllostachys edulis*, which has its genome sequenced and deposited to databases, enabling the detection of markers related with environmental constraints. As vegetative propagation is a common mechanism in bamboos and their cultivation relies on this strategy, molecular markers have been important for attesting genetic fidelity to their original source

---

E. R. Konzen (✉) · L. C. Pozzobon
Center for Coastal, Lymnological and Marine Studies, North Coastal Campus, Federal University of Rio Grande do Sul, Imbé, RS, Brazil

D. M. S. C. Souza · S. B. Fernandes · G. E. Brondani · D. de Carvalho
Department of Forestry Sciences, Federal University of Lavras, Lavras, MG, Brazil

W. F. Campos
Federal University of Jequitinhonha and Mucuri Valleys, Institute of Agricultural Sciences, Unaí, MG, Brazil

S. M. Tsai
Center for Nuclear Energy in Agriculture, University of Sao Paulo, Piracicaba, SP, Brazil

Z. Ahmad et al. (eds.), *Biotechnological Advances in Bamboo*,
https://doi.org/10.1007/978-981-16-1310-4_2

of propagules. Altogether, we provide a panorama of several applications of molecular markers to bamboo conservation and breeding.

Keywords Genetic diversity · Genetic fidelity · Microsatellites · Phylogenomics · Single-nucleotide polymorphisms

2.1 Introduction

Only a few decades back in the 1960s or even in the 1970s and 1980s, taxonomy involved almost exclusively the use of morphological traits to differentiate one species from another. Genetic variation of populations was also accounted for based on morphological variation. It was in the mid-1960s that Lewontin and Hubby revealed the application of protein electrophoresis to quantify genetic variation in populations of fruit flies (Lewontin and Hubby 1966). Soon enough, isozyme electrophoresis spread throughout the world. Numberless population genetic studies were produced, unraveling genetic systems of a diverse set of species, including plants. A revolution to that began with the use of markers directly at the level of DNA, in the 1980s. Restriction fragment length polymorphisms (RFLP) enabled not only population genetic structure but also the development of linkage maps (Tanksley et al. 1989). The development of polymerase chain reaction accelerated the discovery of molecular markers of various types in a few laboratory steps. Random amplified polymorphic DNA (RAPD) (Williams et al. 1990) and amplified fragment length polymorphisms (AFLP) (Vos et al. 1995) filled in countless issues of scientific journals in the 1990s and early 2000s. Simple sequence repeats (SSR) or microsatellites (Litt and Luty 1989) became more popular and are still being largely employed for population genetic studies that require the distinction of heterozygotes, such as in studies of genetic diversity, structure, paternity and system of mating. Currently, single-nucleotide polymorphisms (SNP) prevail, as those are the most abundant markers in organisms, accounting for the variation in a single base.

As members of the Poaceae family, bamboos comprise up to 1662 species and 121 genera described so far (Canavan et al. 2017), with broad distribution throughout the world. Extensive molecular marker-based surveys have been carried out to constantly improve the resolution of phylogenetic trees of such species (Bamboo Phylogeny Group 2012; Wysocki et al. 2015). Besides taxonomy and molecular phylogenetics, molecular markers have also been employed for the study of bamboos for diverse purposes such as genetic diversity and structure of populations, genetic fidelity assessment in clonal populations, candidate gene search for growth, and developmental processes, among others. The emerging field of epigenomics has also been explored in a few studies so far. It is the aim of this chapter to describe the fundamental applications of molecular markers to bamboos, briefly accounting for the main methods that have been employed, either PCR or gel-based, or obtained directly from sequencing. Up-to-date studies on genomics and perspectives into other "omics," such as epigenomics and phylogenomics, are also discussed.

2.2 Molecular Markers for Comprehensive Population Genetic Studies in Bamboos

The possibility of inferring genetic diversity and structure of populations with molecular markers allowed unprecedented discoveries in natural populations of several plants, including bamboos. Uncovering phylogenetic bases of the high species diversity of bamboos, in general, has been the target of various other studies employing molecular markers. An increasing number of studies, though, have also been devoted to deciphering intrapopulation genetic variation. Molecular surveys can explain the clonal structure of natural populations and the implications of the rare flowering events in various bamboos.

Bamboos are, in general, clonal plant species, possessing both sexual and asexual propagation systems. Asexual reproduction is advantageous as it enables the successful establishment in novel environments. On the other hand, it does not provide new genetic variation. This is achieved by sexual reproduction (Kitamura and Kawahara 2011). It is frequent that bamboos flower only once in their life cycle and after several decades (Kitamura and Kawahara 2009). A study of Isagi et al. (2004) investigated the flowering pattern of culms regenerated from seeds of *Phyllostachys pubescens* and that have naturally propagated their genotypes through leptomorphic rhizomes. AFLP screening allowed the identification of distinct genets distributed in the population. The genets had distinct flowering times, suggesting a genetic architecture involved in this trait. Further analysis with microsatellite markers identified the clonal structure of a population of *Sasa cernua* and revealed an overall synchronism of flowering culms of the same clone, but also that not all culms flowered at once. Culms that flowered died, leaving others of the same clone still able to flower in future events. This may enable novel opportunities for cross-pollination (Kitamura and Kawahara 2009). In *Sasamorpha borealis*, after vegetative propagation by rhizomes, almost all adults produce flowers, set seeds in large amounts, and die (Lee and Chung 1999). However, dying after producing flowers is not a general rule, as revealed for *S. pubiculmis*. Among distinct genets that were monitored, one had both flowering and nonflowering patches for 4 years. One genotype can maintain their rhizomes and nonflowering patches alive after mass flowering (Miyazaki et al. 2009).

The rare flowering events in bamboos have important implications in outcrossing rates. The synchronism of flowering within clones (Kitamura and Kawahara 2009) limits the possibility for outcrossing between distinct genotypes. Using microsatellite polymorphisms, Kitamura and Kawahara (2011) studied the mating system of the dwarf bamboo *S. cernua*. The authors found an overall low outcrossing rate ($t_m = 0.148$). The genotype pairs that presented the highest outcrossing rates also had between 2% and 17% of seeds with homozygote genotypes for the markers. Therefore, high inbreeding coefficients should be expected, but the authors observed some decline toward the stage of seedlings, suggesting some selection against inbred progenies. In *P. pubescens*, Lin et al. (2014) also detected low outcrossing rates ($t_m = 0.089$) and excess of homozygotes ($F_{IS} = 0.195$) after genotyping seeds with cDNA SSR markers from three locations separated by at least 100 km. This may be due to an auto compatibility system, such as shown in the Brazilian species *Merostachys riedeliana* (Guilherme and Ressel 2001).

The characteristics of bamboos are compatible with lower levels of genetic variation within populations and higher genetic differentiation among populations, as revealed by population genetic studies in some species (Table 2.1). However, various species still do not have population genetic surveys published, and high genetic variation is not that rare in data already available. It is not our goal to exhaust the literature on all the studies available for such a purpose, but to provide results and prospects of the use of molecular markers in bamboos.

Although with less polymorphism and as indirect products of gene expression, isozyme markers have been used in bamboo germplasm characterization. The advantage of such markers relies on the possibility of discriminating heterozygotes. Four enzyme systems were screened in accessions of hill bamboo, *Sinarundinaria anceps*. Out of 12 markers, approximately 75% were polymorphic, and similarities among accessions ranged from 54 to 100% (Tiwari et al. 2019). For isozymes, considerably high genetic variation, measured from the observed heterozygosity ($Ho = 0.219$), was found in populations of *S. borealis* from Korea. Relatively moderate genetic differentiation, however, was also identified, as measured by the proportion of genetic variation among populations ($G_{ST} = 0.310$) (Lee and Chung 1999). In 17 populations of *Pseudosasa japonica*, the mean genetic diversity encountered through 28 isozyme loci ($Ho = 0.099$) was lower than the expectation under Hardy-Weinberg equilibrium ($He = 0.167$). A high and significant deficit of heterozygotes was then detected ($F_{IS} = 0.457$), suggesting considerable inbreeding due to vegetative spread. In this species, most of the genetic variation was found within populations (84.8%) (Huh and Huh 2002).

AFLP, ISSR, and SRAP are dominant markers that, in general, are more suitable for genetic structure analysis rather than genetic diversity, as they do not allow the discrimination of heterozygotes. Sixteen cultivars of *P. violascens* were analyzed with these three types of markers. AFLP allowed the discrimination of 434 markers, while ISSR (209 bands) and SRAP (222) had a lower number. In general, AFLP showed the highest marker efficiency among other methods, although the polymorphism of AFLP (58.3%) was not higher than ISSR (65.1%) and SRAP (68.5%) (Lin et al. 2011). In the dwarf bamboo *Bashania fangiana*, AFLP markers showed considerable genetic polymorphism, and most of the genetic diversity was found within populations ($G_{ST} = 0.057$). The study indicated that the two populations investigated were multiclonal and diverse (Ma et al. 2013).

ISSR markers were used to analyze the genetic diversity and structure of seven populations of *Melocanna baccifera*, a non-clump and evergreen arborescent bamboo in India. Moderate values of Nei's genetic diversity ($H = 0.1639$) were detected through the analyses, with 88.4% polymorphic bands. From the hierarchical partition of the genetic diversity, most of it was found within populations ($G_{ST} = 0.1942$) (Nilkanta et al. 2017). Similar differentiation was also detected among populations of *Dendrocalamus membranaceus* from China. Among all the 155 bands, 153 were polymorphic, and the proportion of genetic differentiation among populations was 25.2% (from G_{ST}) and 21.1% based on AMOVA (Yang et al. 2012). Conversely, ISSR proved useful in detecting low genetic diversity ($H = 0.0418$) and high genetic

Table 2.1 Compilation of molecular marker studies with bamboo populations, with genetic diversity and structure estimates

				Genetic diversity estimates				Genetic divergence		
Species	Type of marker	Sampling	No. of markers[a]	*P* (%)	*A*	*He*[b]	*Ho*[c]	*Fst*	*Gst*	References
Sinarundinaria anceps	Isozymes	A	12	75.56						Tiwari et al. (2019)
Sasamorpha borealis	Isozymes	NP	14	51.43	2.23	0.219	0.317		0.3104	Lee and Chung (1999)
Pseudosasa japonica	Isozymes	NP	28	52.42	1.81	0.167	0.099		0.1520	Huh and Huh (2002)
Phyllostachys violascens	AFLP	C	434	58.29						Lin et al. (2011)
Bashania fangiana	AFLP	NP	202	54.95		0.1428			0.0571	Ma et al. (2013)
Phyllostachys violascens	ISSR	C	209	65.07						Lin et al. (2011)
Melocanna baccifera	ISSR	NP	-[4]	88.37	1.88	0.1639			0.1942	Nilkanta et al. (2017)
Dendrocalamus giganteus	ISSR	NP	140	88.57		0.0418			0.8474	Tian et al. (2012)
Dendrocalamus membranaceus	ISSR	NP	155	99		0.164			0.2520	Yang et al. (2012)
Oxytenanthera abyssinica	ISSR	NP	348	84.48		0.2702			0.2442	Oumer et al. (2020)
Aulonemia aristulata	Microsatellites	NP	13		2 to 5	0–0.753	0–0.200			Abreu et al. (2011)
Aulonemia aristulata	Microsatellites	NP	13		2 to 2.8	0.245–0.321	0.047–0.146			Abreu et al. (2014)
Kuruna debilis	Microsatellites	NP	12		7.83	0.708	0.758	0.113		Attigala et al. (2017)
Phyllostachys edulis	Microsatellites	NP	20		2 to 10	0.041–0.676	0–1			Jiang et al. (2013)
Bambusa arnhemica	Microsatellites	NP	9		6.8	0.69	0.36			Kaneko et al. (2008)
Dendrocalamus hamiltonii	Microsatellites	NP	17	53.72		0.132		0.165		Meena et al. (2019)

(continued)

Table 2.1 (continued)

				Genetic diversity estimates				Genetic divergence		
Species	Type of marker	Sampling	No. of markers[a]	*P* (%)	*A*	*He*[b]	*Ho*[c]	*Fst*	*Gst*	References
Dendrocalamus sinicus	Microsatellites	NP	16		2.6	0.311–0.754	0–1			Dong et al. (2012)
Fargesia denudata	Microsatellites	NP	14		2 to 19	0–0.87	0–1			Lv et al. (2016)
Dendrocalamus sinicus	Microsatellites	NP	8		18 to 37	0.541	0.488	0.306		Yang et al. (2018)
Sasa cernua	Microsatellites	SF	10		2 to 15	0.159–0.892	0.174–0.826			Kitamura et al. (2009)
Sasa kurilensis	Microsatellites	NP1	9		1 to 10	0–0.816	0–0.72			Kitamura et al. (2009)
Phyllostachys violascens	SRAP	C	222	68.47						Lin et al. (2011)

Sampling code: *A* accessions, *NP* natural populations, *NP1* 1 natural population only, *C* cultivars, *SF* secondary forest. *A* allele number, *He* expected heterozygosity, *Ho* observed heterozygosity, F_{ST} Wright's statistic of genetic differentiation, G_{ST} Nei's proportion of genetic variation among populations

[a]For dominant markers, the number of bands or *loci*; for codominant markers, the number of *loci*

[b]For codominant markers, the expected heterozygosity, in general. For dominant markers, Nei's genetic diversity, in general

[c]Observed heterozygosity, only for codominant markers

[c]Number of bands not revealed but based on five primers

differentiation ($G_{ST} = 0.8474$) in populations of *D. giganteus*, one of the largest bamboos in the world. A Mantel's test suggested that geographic and genetic distances were not significantly correlated, implicating in a high genetic differentiation across populations of the species. Differences in flowering times and limited pollen flow could explain the strong differentiation (Tian et al. 2012). Considerable genetic differentiation ($F_{ST} = 0.38949$) was also found among 13 populations of the Ethiopian lowland bamboo, through ISSR. In this case, considerably high genetic diversity was found at the species level ($H = 0.2702$) (Oumer et al. 2020).

The multiallelic nature and codominance of microsatellites markers, as well as their transferability to other species, make them important markers for more detailed and precise genetic variation and structure analyses (Table 2.1). Abreu et al. (2011) developed 13 novel microsatellite markers for a Brazilian species of bamboo, *Aulonemia aristulata*, native to the Atlantic rain forest. In a posterior study, those microsatellites were used to analyze the genetic diversity of two populations. The observed heterozygosity was calculated in saplings and seedlings of both populations, being higher in saplings of the populations (Table 2.1). The fixation indexes varied from 0.43 to 0.84, with higher values in the seedling stage. Inbreeding depression could be the main factor explaining a reduction in density of bamboos in the direction of the sapling stage (Abreu et al. 2014).

Inbreeding depression is a potential consequence of crossings between individuals identical by descent, as it happens among close relatives or by selfing (Frankham et al. 2010). As discussed before, the clonal structure of bamboos is favored by their limited flowering events, making opportunities from outcrossing scarcely available. Inbreeding depression can be manifested by several characteristics, such as low germination, mortality of seedlings due to factors such as abnormal germination, or lack of the ability to synthesize chlorophyll. Our group collected a source of seeds of *Dendrocalamus asper* that were obtained after a single flowering event of an individual in an urban area (Data unpublished). Significant mortality and abnormal germination were observed from such seeds. Moreover, several seedlings were not able to grow, as they lacked chlorophyll deposition in their leaves (Fig. 2.1a). Those that possessed partial green areas alternated with white surfaces on their leaves were able to grow normally (Fig. 2.1b). An ISSR profile with a few markers indicated low polymorphism on such materials, such as shown with the universal primer ISSR 827 (Fig. 2.1c).

With microsatellites, genetic diversity and structure of *Kuruna debilis*, a threatened bamboo species, were determined in populations located in Sri Lanka (Peng et al. 2013). High genetic diversity was found ($Ho = 0.708$ for all populations), but considerable deficiency of heterozygotes was also computed ($F_{IS} = 0.170$). The differentiation among populations ($F_{ST} = 0.113$) revealed that most of the diversity was also found within populations (Attigala et al. 2017). *P. edulis*, the model bamboo which has been sequenced (Peng et al. 2013), was also the object of a microsatellite study in populations of China. Considerable genetic variation, measured from heterozygosity, was measured from 20 novel microsatellite markers developed by Jiang et al. (2013). Nine microsatellites were developed for the Australian endemic species *Bambusa arnhemica*, revealing an average genetic variation of $Ho = 0.360$ in four populations. Once again, the deviation from the

Fig. 2.1 Plants of *Dendrocalamus asper* originated from seeds collected from a single plant. (**a**) Illustration of a few plants after approximately 4 months after germination. The plant circulated in yellow did not survive (white leaves). (**b**) Plants after 6 months of germination. Plants that have partial ability to accumulate chlorophyll survived. (**c**) ISSR profile of some of those plants, showing limited polymorphism among them

expected heterozygosity was considerably high (mean of $He = 0.690$) (Kaneko et al. 2008). In 19 populations of *D. hamiltonii* from the Himalayas, low genetic variation (Nei's $H = 0.175$ at the species level) and moderate genetic differentiation were found ($F_{ST} = 0.165$) (Meena et al. 2019).

From 16 microsatellite markers developed by Dong et al. (2012) for *D. sinicus*, 8 were screened in another study with natural populations of the species in its distribution range. The data on nuclear microsatellites enabled the discovery of high genetic differentiation among the populations, dividing them into two main subgroups that are consistent with different culm types. With an analysis of chloroplast DNA markers as well, one genetic group encompassed populations with straight culms, while the other genetic cluster grouped the sinuous-culm lineage. Based on both types of molecular markers, the authors concluded that the populations experienced dispersal and long-term isolation, with some dilution due to contemporary gene flow (Yang et al. 2018).

One of the most important food sources to the giant panda, in China, is the bamboo species *Fargesia denudata*. Therefore, it is straightforward to study the genetic resources available from natural populations of the species. The development of microsatellite markers for *F. denudata* is an important tool for exploring the genetic diversity of the species. According to a primer note, 14 markers were developed for further population genetic studies (Lv et al. 2016).

Due to the several related species of bamboos, microsatellites can usually be successfully transferred among species for genetic studies. Several of the microsatellites developed for the Brazilian species *Aulonemia aristulata* were successfully amplified in other species belonging to the genera *Bambusa*, *Dendrocalamus*,

Table 2.2 Compilation of a few molecular marker studies that were conducted to compare distinct taxa of bamboos

Molecular marker	Sampling	Objective	Citation
RFLP	USDA bamboo collection	Genetic relationships	Friar and Kochert (1991)
RAPD	5 bamboo genera from Indonesia	Molecular identification	Annisa et al. (2019)
RAPD	12 bamboo species	Genetic relationships among taxa	Nayak et al. (2003)
RAPD	9 bamboo species in Sri Lanka	Genetic relationships among taxa	Ramanayake et al. (2007)
RAPD-RFLP	13 bamboo taxa	Genetic relationships and validity of RAPD-RFLP	Konzen et al. (2017)
ISSR	15 species from India	Genetic relationships among species	Amom et al. (2018)
ISSR and EST primers	22 bamboo taxa	Genetic relationships among taxa	Mukherjee et al. (2010)
SRAP	13 bamboo species	Genetic relationships	Zhu et al. (2014)
AFLP	15 bamboo species	Phylogenetic analysis	Loh et al. (2000)
AFLP	12 bamboo species from India	Genetic relationships and phylogeny	Gosh et al. (2011)
AFLP and cpDNA	*Arundinaria* species	Genetic variation, hybridization and phylogeny	Triplett et al. (2010)
cDNA SSR	*Phyllostachys pubescens* and related taxa	Crosstransferability analysis and genetic characterization	Lin et al. (2014)
EST-SSR	USDA temperate bamboo collection	Genetic diversity and phylogenetic analysis	Barkley et al. (2005)
EST-SSR	*Phyllostachys violascens* and related species	Crosstransferability analysis	Cai et al. (2019)
EST-SSR	*Bambusa oldhamii* and other species	Crosstransferability analysis	Sharma et al. (2009)
SSR	*Dendrocalamus latiflorus* and related species	Crosstransferability analysis	Bhandawat et al. (2015)
Retrotransposon-based	58 bamboo accessions from distinct species from Asia	Development of markers and genetic characterization	Li et al. (2020a)

Gigantochloa, and *Guadua* (Abreu et al. 2011). Similarly, the microsatellites developed for *D. sinicus* were transferable to other species within the same genus (Dong et al. 2012). The microsatellites developed for *F. denudate* were transferable, in different numbers, to *F. scabrida*, *F. rufa*, *F. ferax*, *Arundinaria fargesii*, and *Yushania lineolata* (Lv et al. 2016). Bhandawat et al. (2015) dedicated a full study to analyze the crosstransferability between *D. latiflorus* and other bamboo species.

Another line of studies with molecular markers encompasses comparisons of distinct species and genera of bamboos (Table 2.2). RFLP markers were employed to characterize genetic relationships among several species of the genera *Phyllostachys*, *Bambusa*, *Arundinaria*, *Pseudosasa*, *Sinobambusa*, and *Sinocalamus* (Friar and Kochert 1991). RAPD markers were able to distinguish several bamboo species and genera (Nayak et al. 2003; Ramanayake et al. 2007; Konzen et al. 2017; Annisa et al. 2019). Moreover, RAPD markers were efficiently converted to RFLP markers by enzyme digestion. Those markers were able to distinguish four genera (*Bambusa*, *Dendrocalamus*, *Guadua*, and *Phyllostachys*) and species (Konzen et al. 2017). ISSR markers were also used in similar approaches (Mukherjee et al. 2010; Amom et al. 2018). Morphological variables and SRAP markers were also combined in a study with several bamboo taxa (Zhu et al. 2014). EST-derived microsatellites have been used in phylogenetic studies of bamboo species (Barkley et al. 2005; Sharma et al. 2009; Cai et al. 2019). EST-based random primers were also useful in the study of Mukherjee et al. (2010). cpDNA markers were the object of distinct studies as well, providing tools for examining genetic differentiation and haplotype diversity within populations (Abreu et al. 2014; Triplett et al. 2010; Yang et al. 2018) and among species (Triplett et al. 2010).

To detect natural hybridization among North American woody bamboos of the genus *Arundinaria*, AFLP and cpDNA markers were used (Triplett et al. 2010). The distinction of three species (*A. gigantea*, *A. appalachiana*, and *A. tecta*) that were previously recognized based on morphological variation was conducted. Those species were also analyzed for their intraspecies genetic diversity. In general, relatively low levels of genetic diversity were found within each of the species, based on the AFLP data (Triplett et al. 2010). The authors discussed the intergradation of molecular and phenotypic data among the parental species. cpDNA analyses suggested multiple and reciprocal hybridization events (Triplett et al. 2010).

Transposable elements are ubiquitous in eukaryotic genomes, being able to self-replication and insertion to other sites, affecting the stability of genomes. In bamboos, long terminal repeat (LTR) retrotransposons are highly abundant (more than 40% of the genome of *P. edulis*) (Zhou et al. 2017). Retrotransposon-based markers were used to screen 58 bamboo taxa, including 47 distinct species in the genus *Phyllostachys*. Those markers behaved as dominant, producing an average of 208.75 bands per primer and average polymorphic information content of 0.327 for the whole set of taxa. After AMOVA, 25% of the genetic variation was detected among the taxa (Li et al. 2020a).

The emerging field of epigenetics deals with heritable variation beyond the level of DNA, such as methylation and histone acetylation and patterns of gene expression. The considerable genomic and transcriptomic data available for bamboo is enabling the discovery of epigenomic variation among taxa. Using methylation-sensitive amplification polymorphism (MSAP), Yuan et al. (2014) encountered distinct patterns of methylation for chronological age of *P. edulis*. The increase in chronological age was accompanied by an augmented DNA methylation rate. Lu et al. (2012) also detected great differences of DNA methylation rates among developmental stages of *P. praecox*, based on AFLP using methylation-sensitive enzymes. The rapid increase in sequenced genomes will enable the discovery of methylation patterns across taxa of bamboo, assisting in the clarification of several processes that are accompanied by epigenetic changes.

2.3 SNP Markers and their Application to Conservation and Breeding

Genomic technologies accelerated the discovery of single-nucleotide polymorphisms (SNP), the most abundant type of molecular markers in all organisms. The publication of the genomes of *P. edulis* (Peng et al. 2013) has opened the field for genomic and transcriptomic studies. Restriction site-associated DNA sequencing (RAD-seq) is a method that enables the detection of thousands of SNP, by a strategy of reduced genomic representation (Baird et al. 2008). SNP markers detected through RAD-seq were used to evaluate phylogenetic relationships in the tribe Arundinarieae. In general, eight lineages were supported by the data, two of them in agreement with previous studies on nuclear markers (Wang et al. 2017a).

After comparing three moso bamboo samples through genome resequencing, 4,700,803 SNP markers were detected (Zhou et al. 2019a). The study also discriminated INDEL markers, accounting for 268,150 distinct positions throughout the genomes. Moreover, copy number variation (CNV = 65,935) and structural variations (SV = 215,297) were also analyzed. The variation detected was present in genes associated with ribosome genesis, caffeine metabolism, nucleotide binding, and ribonucleoside binding, among several other categories, probably involved in adaptation of moso bamboo to their environment (Zhou et al. 2019a).

To date, the number of studies involving SNP markers is yet insipient, but there will be certainly many more publications to come shortly. High-density SNP profiles have provided important information on genetic variation and structure of populations. They also have been largely employed in genome-wide association mapping in plants and other species. Such discoveries enable the design of proper conservation and breeding strategies, as phenotypes can be precisely correlated with SNP and genes that are located throughout genomes. Data on SNP can rapidly be obtained from genotyping by sequencing (GBS) (Elshire et al. 2011), even in species that do not have a reference genome for the alignment of reads (Poland et al. 2012).

2.4 Phylogenomics and its Application to Bamboo Taxonomy

Until a few decades ago, bamboo taxonomy relied exclusively on morphological variation. Moreover, the focus was mainly on vegetative characters. Due to the rare flowering events in bamboos, the morphology-based taxonomy of bamboo has been a challenge (Bhattacharya et al. 2006). The proper distinction between bamboo taxa and comprehension of phylogeographic and phylogenetic relationships are essential to guide conservation and development decisions. The development of molecular marker technologies has provided more consistent phylogenetic trees, improving the taxonomic resolution of bamboos.

Currently, high-throughput sequencing technologies (NGS) allow accessing genome sequences of virtually all species. Through genomic data, researchers can infer species relationships, understand mechanisms of molecular evolution, and control for stochastic events. This intersection between evolution and genomics has been called phylogenomics (Eisen and Fraser 2003; Delsuc et al. 2005; Philippe et al. 2005; Yu et al. 2018). In this topic, we provide a few case studies that used phylogenomics for taxonomical classification and a better understanding of the evolutionary relationships of bamboo species.

The tribe Arundinarieae belongs to the subfamily Bambusoideae (Poaceae), containing more than 1600 species, a highly complicated taxonomy (Bamboo Phylogeny Group 2012; Clark et al. 2015; Canavan et al. 2017). Plastid markers enabled the subdivision of bamboos into 12 major lineages (Zeng et al. 2010; Yang et al. 2013; Zhang et al. 2016). Plastid genome sequencing has also been used to resolve the phylogenetic relationships in Arundinarieae and in obtaining robust relationships among lineages (Ma et al. 2014; Wysocki et al. 2015). Recently, other phylogenomic approaches based on restriction site-associated DNA sequencing (RAD-seq) were used to estimate the phylogenetic relationships among Arundinarieae genera from a nuclear evolutionary trajectory perspective (Wang et al. 2017a).

The phylogenetic relationships of *Shibataea*, a genus of tribe Arundinarieae and endemic to China, have been reconstructed through a phylogenomic approach based on double digest restriction site-associated DNA sequencing (ddRAD-seq), and whole plastid genomes were generated using genome skimming (Guo et al. 2019). *Fargesia*, other genus of tribe Arundinarieae, has also been its phylogenetic relationships reconstructed based on complete plastid genome sequences of 26 species from *Fargesia* (Zhou et al. 2019b).

Phylogenomic approaches based on complete plastid genome sequences have also been used in evolutionary studies of bamboos. Through genome-wide comparative analyses of 76 chloroplast genomes of bamboos, extreme heterogeneity of the evolutionary rate within the bamboos was demonstrated, with the lowest value found in tribe Arundinarieae (Wang et al. 2020). Plastid genome sequences have also been used to estimate phylogenetic relationships among Bambuseae species (do Vieira et al. 2016). Bambuseae is another tribe of subfamily Bambusoideae (Poaceae), containing 66 genera and 812 species (Bamboo Phylogeny Group 2012; Clark et al. 2015).

Whole plastid genomes of Olyreae species were generated using genome-skimming approach and used to strongly support the phylogenetic positions of *Froesiochloa* and *Rehia* in the Olyreae (Wang et al. 2018). Olyreae is the smallest tribe of subfamily Bambusoideae (Poaceae), containing 22 genera and 124 species (Bamboo Phylogeny Group 2012; Clark et al. 2015). Therefore, there are few phylogenomic studies involving the Olyreae species.

Finally, all these studies and another show the potential of the phylogenomic approach based on NGS technologies combined with traditional morphological analyses for resolving difficult phylogenetic relationships in the intractable bamboo species, evolutionary, and biogeography studies.

2.5 Protein and Nucleotide Motifs for Evolutionary Studies with Gene Families

So far in this chapter, we have treated molecular markers as tools for comprehending population genetic processes as well as to understand phylogenetic relations among species. By a functional characterization of the allelic variation of genes, sequence motifs affecting phenotyping variation can be identified, those being referred to as functional markers (Andersen and Lübberstedt 2003). Genes that have been functionally characterized may be a reference to annotations of paralogs and orthologs, which enable their categorization in major families. Phylogenetic reconstruction with gene families provides insights into the evolution and adaptation of plants (Neale et al. 2017). Genome-wide searches for gene families use model plants to find homologs in the species of interest, but experiments are required to functionally confirm the inferred genes (Vaattovaara et al. 2019).

Various gene families of transcription factors are implicated in the response of plants to biotic and abiotic stresses. In *P. edulis*, genome-wide surveys of genes are available for *WRKY* (responsive to cold and drought) (Li et al. 2017); *no apical meristem* (*NAM*); *Arabidopsis* transcription activation factor (*ATAF*) and *cup-shape cotyledon* (*CUC*), collectively referred to as *NAC* genes (participate in lignin catabolic process and cellulose biosynthetic process) (Shan et al. 2019), heat shock transcription factors (*Hsfs*) (participate in shoot and flower development; expression is changed under cold, high temperature, drought and high salinity stresses) (Xie et al. 2019), dehydration responsive element-binding (*DREB*) (differentially expressed under drought, cold and high salinity) (Wu et al. 2015). Among various other categories of gene families studied in *P. edulis* are aquaporins (*AQP*) (Sun et al. 2017), late embryogenesis abundant (*LEA*) (Huang et al. 2016), amino acid/auxin permease (*AAAP*) (Liu et al. 2017), plant homeodomain zinc finger (*PHD-finger*) (Gao et al. 2018), *trihelix* genes (Gao et al. 2019), SQUAMOSA-promoter binding protein-like (*SBP-like*) transcription factors (Pan et al. 2017), IQ67-domain (*IQD*) genes (Wu et al. 2016), and basic leucine zipper domain (bZIP) transcription factors (Pan et al. 2019). In this chapter, we highlight the APETALA2/ethylene-responsive element binding protein (*AP2/EREBP*) superfamily and aquaporins as examples for the application of protein or nucleotide motifs for evolutionary analyses in bamboos.

The AP2/EREBP superfamily of proteins has a high conserved domain named APETALA2 (AP2), with 55 to 70 amino acids (Jofuku et al. 1994; Okamuro et al. 1997; Konzen et al. 2019). This superfamily is divided into three families that have different numbers of domains and participate in different physiological processes, which are RAV, AP2, and ERF (Sakuma et al. 2002). The RAV family has one AP2 domain and one B3 domain (Kagaya et al. 1999), the AP2 family has two AP2 domains (Okamuro et al. 1997), and the ERF family has only one AP2 domain (Sakuma et al. 2002).

Wu et al. (2016) identified 116 genes from the *AP2/EREBP* family in *P. edulis* and divided them into three subfamilies: 28 *AP2*, 7 *RAV*, 80 *ethylene response factors* (*ERF*), and 1 soloist; the *ERF* group is divided into two subgroups, *ERF*

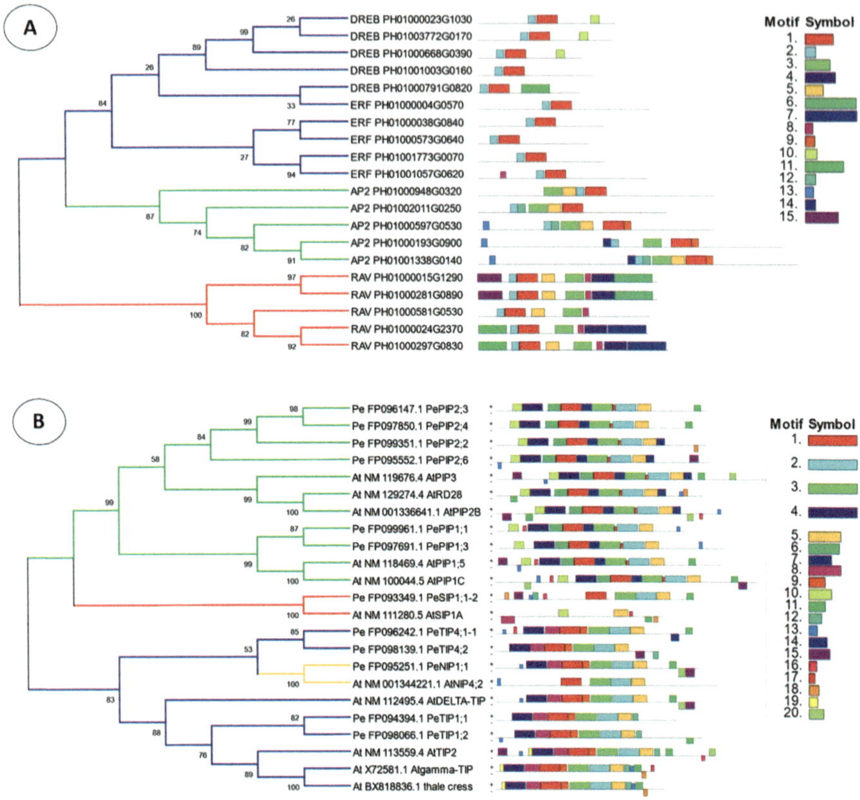

Fig. 2.2 Protein and nucleotide motifs as potential phylogenetic markers for the study of gene families in bamboos. (**a**) Protein motifs searched for some members of the AP2/EREBP superfamily. (**b**) Nucleotide motifs searched for 12 aquaporin genes These figures were elaborated based on analyses performed by the authors of the chapter, using MEME-suite396 website (http://meme-suite.org/) with sequences available from GenBank/NCBI (https://www.ncbi.nlm.nih.gov/) and the Plant Transcription Factor Database (http://planttfdb.gao-lab.org/)

and *DREB*. As an example in this chapter, we selected five amino acid sequences available in the Plant Transcription Factor Database (http://planttfdb.gao-lab.org/) from each group to analyze and identify motifs that can be used as molecular markers. First, we used the Clustal W algorithm for producing an alignment using BioEdit (Hall 1999). Afterward, we constructed a phylogenetic tree by maximum likelihood, using MEGA X (Kumar et al. 2018), and the amino acid substitution model WAG+I + G + F, as determined from Bayesian criterion information, with 500 bootstrap replications. The protein motifs were searched on the MEME-suite website (http://meme-suite.org/). From the phylogeny, two major groups were identified (Fig. 2.2a), one with RAV proteins and the other with AP2, DREB, and ERF, the last two being more similar between them (Data unpublished). Observing the protein motifs, similar motifs are shared within the major groups and according to the subdivision of the superfamily. The protein motifs one and two correspond to the domain AP2, a domain that is present in all sequences with certain differences among the subgroups. The protein motifs 8, 3, and 4 correspond to the domain B3, even though the motif 4 was not identified in the sequence RAV_PH01000581G0530.

We also performed a similar analysis of AQP proteins from sequences available in the public database. AQP proteins are channels to water transport along with CO_2 and nutrients (Maurel et al. 2015). *PeTIP4;1–1*, isolated from *P. edulis*, conferred tolerance to transgenic *Arabidopsis thaliana* in abiotic stresses caused by drought and high salinity (Sun et al. 2017). *PeTIP4;1* and *PeTIP4;2* were expressed in the parenchyma and epidermal cells and roots, indicating their role in water absorption and transport (Sun et al. 2018). Sun et al. (2018) classified the aquaporins from *P. edulis* into four groups: (a) plasma membrane intrinsic proteins 22 (PIP), (b) tonoplast intrinsic proteins 20 (TIP), (c) nodulin 26-like intrinsic proteins 17 (NIP), and (d) small basic intrinsic proteins 4 (SIP); the uncharacterized X intrinsic proteins (XIP) were not found.

We used the 12 sequences from the work of Sun et al. (2018), available from GenBank/NCBI (https://www.ncbi.nlm.nih.gov/) to search for the correspondent nucleotide sequences in *A. thaliana* with BLASTn. Sequence alignment and phylogenetic reconstruction were conducted similarly to the AP2/EREBP sequences. The maximum likelihood tree, this time, was constructed with the nucleotide substitution model K2 + G, as determined with MEGA X, after 500 bootstrap replications. The tree of aquaporins with the nucleotide sequences from bamboo and *A. thaliana* was divided into two major groups, the first with *SIP* and *PIP* genes and the second with *TIP* and *NIP* genes from both species (Fig. 2.2b). Motif 5 was identified in all nucleotide sequences. Motifs 6 and 7 were found only in *PIP* genes, while motifs 8 and 9 were found in *TIP* genes, in similar positions. Motifs 10 and 17 were found only in *PIP*. Motifs 1, 2, and 3 were found in all sequences except *SIP*. Motif 4 was detected in *TIP* and *PIP* genes (Data unpublished).

Therefore, protein or nucleotide motifs can be used as markers to identify and compare gene families, to better understand the evolution and adaptation of plants. With the sequences available for *P. edulis*, other bamboos may be searched for genes already annotated in this species.

2.6 Functional Gene Markers for Improving Abiotic Stress Tolerance in Bamboos

Drought, salinity, and low and high temperatures are the mains abiotic constraints that negatively affect bamboo growth and development and reduce its industrial productivity (Ramakrishnan et al. 2020). Therefore, understanding the stress responses will increase the ability to improve tolerance in bamboo species. However, the mechanisms of molecular responses in bamboo species under stress conditions are still not completely understood.

Table 2.3 Gene markers responsive to abiotic stresses in moso bamboo (*P. edulis*)

Gene	Protein function	Stress	Overexpression	Reference
PheWRKY72–2	Transcription factor	Drought, cold	*Arabidopsis*	Li et al. (2017)
PeWRKY83	Transcription factor	Drought, salinity	*Arabidopsis*	Wu et al. (2017)
PheMYB4–1	Transcription factor	Cold	*Arabidopsis*	Hou et al. (2018)
PheNAC3	Transcription factor	Drought, salinity	*Arabidopsis*	Xie et al. (2020)
PeSNAC-1	Transcription factor	Drought, salinity	Rice	Hou et al. (2020)
PheASR2	Transcription factor	Drought, salinity	Rice	Wu et al. (2020b)
PeTCP10	Transcription factor	Drought	*Arabidopsis*, rice	Liu et al. (2020)
PheDi19–8	Transcription factor	Drought	*Arabidopsis*, rice	Wu et al. (2020a)
PheDof12–1	Transcription factor	Drought, cold, salinity	*Arabidopsis*	Liu et al. (2019)
PheVQ28	Transcriptional regulator	Salinity	*Arabidopsis*	Cheng et al. (2020)
PeTIP4;1–1	Aquaporin	Drought, salinity	*Arabidopsis*	Sun et al. (2017)
PeLAC10	Oxidase	Drought, salinity	*Arabidopsis*	Li et al. (2020b)

The responses to drought, salinity, and low and high temperature stresses are under stronger biochemical and genetic control (Zhu 2016; Haak et al. 2017). As stated before, various gene families have been identified in moso bamboo. In general, no in-depth functional analyses were conducted for those genes. However, there are other gene families in that at least one member was functionally characterized. From these studies, we listed 12 genes that could be used as functional molecular markers in comprehending abiotic stress tolerance (Table 2.3). To select these stress-responsive genes, we considered their expression profile in moso bamboo plants under stress treatments. Moreover, we considered only those genes that were functionally analyzed in at least one model species such as *Arabidopsis*, rice, and yeast.

Several transcription factors have been identified and associated with drought, salinity, and cold-induced stresses in moso bamboo (Table 2.3). WRKY proteins are important transcription factors involved in different biological processes such as plant growth and abiotic stress responses (Rushton et al. 2010; Chen et al. 2012). A genome-wide survey identified 121 WRKY genes in *P. edulis* (Li et al. 2017). The authors showed that *PheWRKY72–2* is a drought- and cold-inducible gene. Furthermore, the overexpression of *PheWRKY72–2* in *Arabidopsis* enhanced plant tolerance to drought stress by promoting stomatal closure. In another study, *PeWRKY83* gene was highly upregulated in moso bamboo seedlings subjected to salinity and drought stress. After ectopic expression in *Arabidopsis*, it was demonstrated that *PeWRKY83* regulates the expression of genes involved in ABA biosynthesis, signaling, and responses to salt and improves salt tolerance (Wu et al. 2017).

The *MYB* gene family, first described as an oncogene from MYeloBlastosis virus (capital letters highlighting the attributed name of the gene family), also was identified in moso bamboo (Hou et al. 2018). Through in silico and expression analyses, these authors identified a *PheMYB4–1* stress-related gene. After ectopic overexpression, *PheMYB4–1* promoted improvement in tolerance to cold, drought, and salt stress in *Arabidopsis* seedlings. Two NAC transcription factors from moso bamboo were identified and functionally characterized. First, the expression levels of *PheNAC3* gene increased in leaves of moso bamboo after NaCl treatment. Overexpression in *Arabidopsis* improved drought and salt tolerance (Xie et al. 2020). Second, drought and salinity also induced the expression of *PeSNAC-1* in moso bamboo. Rice plants overexpressing *PeSNAC1* gene were more drought- and salt-tolerant (Hou et al. 2020). Moreover, these authors suggested that *PesSNAC-1* works as a positive stress regulator in moso bamboo.

The ABA-stress-ripening (*ASR*) gene family is a small group of chaperone-like proteins and plant-specific transcription factors involved in plant development, senescence, fruit ripening, and abiotic stresses (González and Iusem 2014). After being identified in moso bamboo, *ASR* genes were found upregulated under drought, NaCl, and ABA. Specifically, the overexpression of *PheASR2* in rice improved drought tolerance (Wu et al. 2020b). Stress response-related functions of Teosinte branched1/Cincinnata/proliferating cell factor (*TCP*) members have been investigated in different plant species including moso bamboo (Liu et al. 2018). A member of this family was characterized in rice and *Arabidopsis*, and the results showed that *PeTCP10* may have positive regulatory functions in drought tolerance (Liu et al. 2020). A drought-induced (Di19) protein, a zinc-finger transcription factor containing two conserved and atypical Cys2/His2 zinc-finger domains, isolated from moso bamboo, has been functionally characterized in *Arabidopsis* and rice (Wu et al. 2020a). Through transgenic lines, these authors concluded that *PheDi19–8* significantly increased drought tolerance. Moreover, complementation analyses showed that *PheDi19–8* works as a positive modulator of drought stress tolerance.

Twenty-six DNA binding with one finger (*Dof*) genes were identified in moso bamboo (Wang et al. 2016). A member of this transcription factor family has been functionally characterized by ectopic expression in *Arabidopsis* (Liu et al. 2019). These authors showed that the *PheDof12–1* gene is highly induced by cold, drought, and salt in moso bamboo and their overexpression promotes early flowering in *Arabidopsis*. The *Arabidopsis* lines overexpressing *PheDof12–1* gene were not analyzed under those abiotic stresses (Liu et al. 2019).

VQ motif-containing proteins are involved in responses to biotic and abiotic stresses by interaction with WRKY transcription factors (Lai et al. 2011; Hu et al. 2013). *VQ* genes were identified in moso bamboo genome, and the *PheVQ28* gene has been functionally characterized (Wang et al. 2017b; Cheng et al. 2020). *PheVQ28* was highly upregulated in moso bamboo seedlings under salt stress, and *Arabidopsis* plants overexpressing this gene were significantly more tolerant to salt stress (Cheng et al. 2020). Laccases (LAC) are oxidative enzymes involved in flavonoid and lignin biosynthesis (Pourcel et al. 2005; Berthet et al. 2011). Thus,

LAC genes are promising for woody, biofuel, and other biotechnological applications. Moreover, *LAC* genes have also been involved in plant responses to environmental stress (Wang et al. 2015). Laccase gene members have been identified in moso bamboo. *PeLAC10* gene was upregulated under ABA and NaCl, and its overexpression in *Arabidopsis* plants improved drought tolerance (Li et al. 2020b). These results suggest that laccase genes could be potential candidates for the molecular breeding of bamboo to increase the content of lignin and improve their adaptability to drought stresses.

Finally, although all genes list in Table 2.3 requires further functional characterization, they are a potential candidate for the breeding of stress tolerance in bamboo species and other crops through genetic engineering using approaches such as CRISPR (clustered regularly interspaced short palindromic repeats). Using molecular markers for deciphering genetic variation and its functional implications will certainly assist in improving stress tolerance and increase the productivity of bamboos in the climate change scenario that affects the whole planet.

2.7 Molecular Markers for Genetic Fidelity

Molecular markers are also a powerful tool to attest to the genetic fidelity of clonally produced plantlets of bamboos. The use of micropropagation has been a viable alternative for cloning superior genotypes, as well as for the production of seedlings with high quality and free of pathogens. There are strict precautions in micropropagation so that plants can express all their genetic potential, as on the course of in vitro cultivation regenerated plants may not maintain the genetic stability in relation to their mother plant (the donor of propagules) (Kaeppler et al. 2000; Ray and Ali 2017).

Changes in the genotype after micropropagation are referred to as somaclonal variation. Mutations may occur due to the use of growth regulators, the various subcultures of explants, the regeneration by callus induction or protoplasts, and somatic embryogenesis. Therefore, DNA can undergo irreversible and inheritable changes (Ray and Ali 2017). Genotypic changes can occur through chromosomal breaks and rearrangements, changes in the level of ploidy, activation of transposable elements, and DNA methylation (Kaeppler et al. 2000; Krishna et al. 2016). These are examples of genetic and epigenetic changes that may or may not cause phenotypic variations. In fact, the identification of possible phenotypic variation in early stages occurs by observing variations in leaf color or size (albinism, variegation, or dwarfism); however, in most cases, these variations do not become visible requiring analysis by molecular markers to identify somaclonal variants.

The most used markers for the identification of somaclonal variations in bamboo are AFLP, RAPD, ISSR, and SSR (Table 2.4). The principle of the analysis involves the comparison of DNA fingerprint among all propagules with their mother plant. Evidence of somaclonal variation occurs when the fingerprints differ from the original source (Fig. 2.3a). AFLP markers were used to verify the genetic stability

Table 2.4 Assessment of genetic fidelity in bamboos through molecular marker studies

Bamboo species	Protocol	Molecular markers	Results	Reference
Dendrocalamus strictus	In vitro regeneration through nodal culture in MS medium with BAP, NAA, and IBA in different stages	RAPD and ISSR	The RAPD decamers produced 58 amplicons, while nine ISSR primers generated a total of 66 bands All the bands were monomorphic	Goyal et al. (2015)
Bambusa arundinacea	Induction of high-frequency multiple shoots directly from the nodal explants in MS medium with BAP and kinetin	RAPD	Only 7 from 60 primers produced bands ranging from 350 to 1600 bp. The monomorphic pattern was observed in all plants compared to mother plant	Kalaiarasi et al. (2014)
Dendrocalamus asper	Micropropagation due to nodal explants in MS medium with BAP, subculturing in 2 years (30 passages) every 25 days	AFLP, RAPD, ISSR, and SSR	The RAPD showed 146 scorable bands; ISSR generated 170 bands; SSR showed 164 bands; AFLP showed 536 bands. All molecular markers showed monomorphic bands compared to matrix plant	Singh et al. (2013b)
Bambusa balcooa	The nodal explants were micropropagated in MS medium with kinetin, BAP, NAA, ascorbic acid, arginine, citric acid, and adenine sulfate.	ISSR and SCoT	The SCoT primers proved amplification of 48 scorable PCR bands. The ISSR showed 5 PCR bands per primer. This study showed that micropropagated plants under field conditions are anatomically and genetically similar to mother plant	Rajput et al. (2020)
Dendrocalamus hamiltonii	Axillary bud proliferation from nodal explant in MS with thidiazuron, ascorbic acid and IBA during 2 years (30 passages)	AFLP, RAPD, ISSR, and SSR	The RAPD produced 7 bands per primer and ISSR produced 181 bands. The SSR primers amplified 141 fragments and AFLP produced 46 bands. All the molecular markers showed monomorphism	Singh et al. (2013a)

(continued)

Table 2.4 (continued)

Bamboo species	Protocol	Molecular markers	Results	Reference
Guadua magna and *G. angustifolia*	Nodal segments were micropropagated in MS medium supplemented with plant preservative mixture and fungicide. After, BAP was used and five subcultures were conducted after every 30 days.	ISSR	The 20 primers allowed to amplify 223 loci for *G. magna* and 230 for *G. angustifolia* and showed no polymorphism compared to the mother plant	Nogueira et al. (2019)

Fig. 2.3 (**a**) Schematic illustration of the principle of the assessment of genetic fidelity after micropropagation of bamboo. A hypothetical fingerprint shows genetic fidelity vs evidence of somaclonal variation. Although no genetic variation may be present, novel technologies allow the identification of epigenetic mechanisms that might implicate in distinct phenotypic effects. Photo credit: Enéas Ricardo Konzen. (**b**) *Bambusa vulgaris* plant originated from micropropagation of mother plant. Photo credit: Siu Mui Tsai

of explants from a 2-year subculture of *D. hamiltonii* in a culture medium containing thidiazuron (TDZ) (Singh et al. 2013a). The combination of the enzymes *EcoRI* and *MseI* with 15 selective primers did not reveal any polymorphism between the fragments, which validated the protocol for the species. In another study, nodal segments of *Bambusa nutans* were used to obtain a somatic embryogenesis protocol in MS culture medium containing 2,4-D (dichlorophenoxyacetic acid) (Mehta et al.

2011). Six AFLP primer combinations produced 407 fragments, 98.8% of which were monomorphic. Therefore, low number of plants showed somaclonal variations. With this high percentage of monomorphism, the technique is efficient for verifying genetic fidelity.

There are also studies that report the success of RAPD markers for bamboo species as well as validating different types of protocols, such as the use of different growth regulators for the in vitro rooting of *D. strictus* (Goyal et al. 2015), regeneration of axillary buds of *B. arundinacea* (Kalaiarasi et al. 2014), and validation of commercial materials produced in vitro from *D. asper* (Singh et al. 2012). For bamboos, ISSR markers have become the most viable alternative for studies of genetic fidelity. The use of ISSR markers to verify the genetic fidelity of *B. balcooa*, a species of great importance in India, did not show somaclonal variation, even with variations in the concentrations of growth regulators (Rajput et al. 2020). Genetic stability using ISSR markers has also been shown in the in vitro cultivation of *G. magna* and *G. angustifolia* (Nogueira et al. 2019).

In *D. asper*, the use of four different markers, including SSR, did not show polymorphism for the use of nodal segments under different stages of in vitro development, with the use of cytokinin in the multiplication phase, indolebutyric acid, and acetic naphthalene for elongation (Singh et al. 2013b). The authors used 25 SSR primers based on the rice genome, obtaining 164 bands (all monomorphic) observing the absence of somaclonal variation, which shows the efficiency of these markers for studies of genetic fidelity.

The somaclonal variation observed in micropropagated bamboo species may be linked to the use of growth regulators, indirect organogenesis, as well as the number of subcultures. For the identification of variants, ISSR-type markers, which are based on the repetitive portion of eukaryotic DNA, are effective due to their low cost and repeatability. With the evolution of genomic studies in bamboo, other ways of identifying somaclonal variation will allow the selection of true-to-type planting material, such as studies of epigenetic variations, ploidy, and movement of transposable elements. Even if no mutations have occurred at the DNA level, epigenetic changes such as methylation of DNA, acetylation and methylation of histones, and altered gene expression are possible outcomes (Fig. 2.3a), which could implicate in distinct phenotypic effects and performances of the propagules. It is important to ascertain the genetic fidelity to obtain plants with similar performances to their sources (Fig. 2.3b).

2.8 Conclusion and Future Prospects

Molecular markers empowered the distinction of the diversity of bamboo species and enabled comprehensive analyses of their reproduction and population structure. The current genomic technologies will produce numberless novel reports on the genetic diversity and divergence within and among species. Phylogenomics offers novel methods for improving the resolution of phylogenetic trees, reaching a deeper understanding of the evolution of bamboo taxa. Moreover, in the era of systems biology, research is moving forward to integrative approaches for conservation and breeding of such important species. Further comprehension of bamboo DNA variation coupled with the modulation of methylation, acetylation, and gene expression in their environment is necessary to design proper management. From the classical to the ultimate technologies for their obtainment, molecular markers certainly have much more to contribute to future endeavors.

Acknowledgments We thank the National Council for Scientific and Technological Development (CNPq), Coordination for the Improvement of Higher Education Personnel (CAPES), Minas Gerais Research Funding Foundation (FAPEMIG), and São Paulo Research Foundation (FAPESP) for the financial support and scholarship.

Conflicts of Interest The authors declare that they have no conflicts of interest.

Author Contributions ERK conceived the general idea of the chapter. ERK, DC, and SMT described the application of molecular markers to bamboo genotyping and population genetic studies. LCP described and analyzed sequence data as potential phylogenetic markers. WFC contributed with the topics of phylogenomics and functional markers. DMSC, SBF, GEB, and DC described the application of molecular markers in studies of genetic fidelity. All authors read and approved the final version of the manuscript.

References

Abreu AG, Grombone-Guaratini MT, Monteiro M, Pinheiro JB, Tombolato AFC, Zucchi MI (2011) Development of microsatellite markers for *Aulonemia aristulata* (Poaceae) and cross-amplification in other bamboo species. Am J Bot 98(4):e90–e92. https://doi.org/10.3732/ajb.1000511

Abreu AG, Grombone-Guaratini MT, Moreira Val T, Zucchi MI (2014) Genetic diversity and age class structure of seedlings and saplings after a mast flowering of bamboo in the Brazilian Atlantic Forest. Int J Plant Sci 175(3):319–327. https://doi.org/10.1086/674448

Amom T, Tikendra L, Rahaman H, Potshangbam A, Nongdam P (2018) Evaluation of genetic relationship between 15 bamboo species of North-East India based on ISSR marker analysis. Mol Biol Res Commun 7(1):7–15. https://doi.org/10.22099/mbrc.2018.28378.1303

Andersen JR, Lübberstedt T (2003) Functional markers in plants. Trends Plant Sci 8(11):554–560. https://doi.org/10.1016/j.tplants.2003.09.010

Annisa HR, Setiawati T, Irawan B, Kusmoro J (2019) Evaluation of RAPD markers for molecular identification of five bamboo genera from Indonesia. Folia For Pol 61(4):255–266. https://doi.org/10.2478/ffp-2019-0025

Attigala L, Gallaher T, Nason J, Clark LG (2017) Genetic diversity and population structure of the threatened temperate woody bamboo *Kuruna debilis* (Poaceae: Bambusoideae: Arundinarieae) from Sri Lanka based on microsatellite analysis. J Natl Sci Found 45(1):53–65. https://doi.org/10.4038/jnsfsr.v45i1.8038

Baird NA, Etter PD, Atwood TS, Currey MC, Shiver AL, Lewis ZA, Selker EU, Cresko WA, Johnson EA (2008) Rapid SNP discovery and genetic mapping using sequenced RAD markers. PLoS One 3(10):e3376. https://doi.org/10.1371/journal.pone.0003376

Bamboo Phylogeny Group (2012) An updated tribal and subtribal classification of the bamboos (Poaceae: Bambusoideae). J Am Bamboo Soc 24:1–10

Barkley NA, Newman ML, Wang ML, Hotchkiss MW, Pederson GA (2005) Assessment of the genetic diversity and phylogenetic relationships of a temperate bamboo collection by using transferred EST-SSR markers. Genome 48(4):731–737. https://doi.org/10.1139/g05-022

Berthet S, Demont-Caulet N, Pollet B, Bidzinski P, Cézard L, Le Bris P, Borrega N, Hervé J, Blondet E, Balzergue S, Lapierre C, Jouanin L (2011) Disruption of *LACCASE4* and 17 results in tissue-specific alterations to lignification of *Arabidopsis thaliana* stems. Plant Cell 23 (3):1124. https://doi.org/10.1105/tpc.110.082792

Bhandawat A, Sharma V, Sharma H, Sood A, Sharma RK (2015) Development and crosstransferability of functionally relevant microsatellite markers in *Dendrocalamus latiflorus* and related bamboo species. J Genet 94(S1):48–55. https://doi.org/10.1007/s12041-014-0377-9

Bhattacharya S, Das M, Bar R, Pal A (2006) Morphological and molecular characterization of *Bambusa tulda* with a note on flowering. Ann Bot 98:529–535. https://doi.org/10.1093/aob/mcl143

Cai K, Zhu L, Zhang K, Li L, Zhao Z, Zeng W, Lin X (2019) Development and characterization of EST-SSR markers from RNA-Seq data in *Phyllostachys violascens*. Front Plant Sci 10:50. https://doi.org/10.3389/fpls.2019.00050

Canavan S, Richardson DM, Visser V, Roux JJL, Vorontsova MS, Wilson JRU (2017) The global distribution of bamboos: assessing correlates of introduction and invasion. AoB Plants 9: plw078. https://doi.org/10.1093/aobpla/plw078

Chen L, Song Y, Li S, Zhang L, Zou C, Yu D (2012) The role of WRKY transcription factors in plant abiotic stresses. BBA – Gene Regul Mech 819:120–128. https://doi.org/10.1016/j.bbagrm.2011.09.002

Cheng X, Wang Y, Xiong R, Gao Y, Yan H, Xiang Y (2020) A Moso bamboo gene *VQ28* confers salt tolerance to transgenic Arabidopsis plants. Planta 251:99. https://doi.org/10.1007/s00425-020-03391-5

Clark LG, Londoño X, Ruiz-Sanchez E (2015) Bamboo taxonomy and habitat BT - bamboo: the plant and its uses. In: Köhl M (ed) Liese W. Springer International Publishing, Cham, pp 1–30

Delsuc F, Brinkmann H, Philippe H (2005) Phylogenomics and the reconstruction of the tree of life. Nat Rev Genet 6:361–375

do Vieira LN, dos Anjos KG, Faoro H, de Fraga HPF, Greco TM, de Pedrosa FO, de Souza EM, Rogalski M, de Souza RF, Guerra MP (2016) Phylogenetic inference and SSR characterization of tropical woody bamboos tribe Bambuseae (Poaceae: Bambusoideae) based on complete plastid genome sequences. Curr Genet 62:443–453. https://doi.org/10.1007/s00294-015-0549-z

Dong Y-R, Zhang Z-R, Yang H-Q (2012) Sixteen novel microsatellite markers developed for *Dendrocalamus sinicus* (Poaceae), the strongest woody bamboo in the world. Am J Bot 99(9): e347–e349. https://doi.org/10.3732/ajb.1200029

Eisen JA, Fraser CM (2003) Phylogenomics: intersection of evolution and genomics. Science 300 (5323):1706–1707. https://doi.org/10.1126/science.1086292

Elshire RJ, Glaubitz JC, Sun Q, Poland JA, Kawamoto K, Buckler ES, Mitchell SE (2011) A robust, simple genotyping-by-sequencing (GBS) approach for high diversity species. PLoS One 6(5): e19379

Frankham R, Ballou J, Briscoe D (2010) Introduction to conservation genetics, 2nd edn. Cambridge University Press, Cambridge. https://doi.org/10.1017/CBO9780511809002

Friar E, Kochert G (1991) Bamboo germplasm screening with nuclear restriction fragment length polymorphisms. Theor Appl Genet 82(6):697–703. https://doi.org/10.1007/BF00227313

Gao H, Huang R, Liu J, Gao Z, Zhao H, Li X (2019) Genome-wide identification of *Trihelix* genes in Moso bamboo (*Phyllostachys edulis*) and their expression in response to abiotic stress. J Plant Growth Regul 38:1127–1140. https://doi.org/10.1007/s00344-019-09918-9

Gao Y, Liu H, Wang Y, Li F, Xiang Y (2018) Genome-wide identification of PHD-finger genes and expression pattern analysis under various treatments in moso bamboo (*Phyllostachys edulis*). Plant Physiol Biochem 123:378–391. https://doi.org/10.1016/j.plaphy.2017.12.034

González RM, Iusem ND (2014) Twenty years of research on *Asr* (ABA-stress-ripening) genes and proteins. Planta 239:941–949. https://doi.org/10.1007/s00425-014-2039-9

Gosh S, Waikhom SD, Sen Mandi S, Talukdar NC (2011) Amplified fragment length polymorphism based study of phylogenetic relationship and genetic variability among some edible bamboo species of North-East India. J Plant Mol Biol Biotechnol 2(2):8–15

Goyal AK, Pradhan S, Basistha BC, Sen A (2015) Micropropagation and assessment of genetic fidelity of *Dendrocalamus strictus* (Roxb.) nees using RAPD and ISSR markers. 3 Biotech 5 (4):473–482. https://doi.org/10.1007/s13205-014-0244-7

Guilherme FAG, Ressel K (2001) Biologia floral e sistema de reprodução de *Merostachys riedeliana* (Poaceae: Bambusoideae). Rev Bras Bot 24(2):205–211. https://doi.org/10.1590/S0100-84042001000200011

Guo C, Guo Z-H, Li D-Z (2019) Phylogenomic analyses reveal intractable evolutionary history of a temperate bamboo genus (Poaceae: Bambusoideae). Plant Divers 41:213–219. https://doi.org/10.1016/j.pld.2019.05.003

Haak DC, Fukao T, Grene R, Hua Z, Ivanov R, Perella G, Li S (2017) Multilevel regulation of abiotic stress responses in plants. Front Plant Sci 8:1564. https://doi.org/10.3389/fpls.2017.01564

Hall TA (1999) BioEdit: a user-friendly biological sequence alignment editor and analysis program for Windows 95/98/NT. Nucleic Acids Symp Ser 41:95–98

Hou D, Cheng Z, Xie L, Li X, Li J, Mu S, Gao J (2018) The R2R3MYB gene family in *Phyllostachys edulis*: genome-wide analysis and identification of stress or development-related *R2R3MYBs*. Front Plant Sci 9:738. https://doi.org/10.3389/fpls.2018.00738

Hou D, Zhao Z, Hu Q, Li L, Vasupalli N, Zhuo J, Zeng W, Wu A, Lin X (2020) *PeSNAC-1a* NAC transcription factor from moso bamboo (*Phyllostachys edulis*) confers tolerance to salinity and drought stress in transgenic rice. Tree Physiol 40:1792–1806. https://doi.org/10.1093/treephys/tpaa099

Hu Y, Chen L, Wang H, Zhang L, Wang F, Yu D (2013) Arabidopsis transcription factor WRKY8 functions antagonistically with its interacting partner VQ9 to modulate salinity stress tolerance. Plant J 74:730–745. https://doi.org/10.1111/tpj.12159

Huang Z, Zhong X-J, He J, Jin S-H, Guo H-D, Yu X-F, Zhou Y-J, Li X, Ma M-D, Chen Q-B, Long H (2016) Genome-wide identification, characterization, and stress-responsive expression profiling of genes encoding LEA (late embryogenesis abundant) proteins in Moso Bamboo (*Phyllostachys edulis*). PLoS One 11(11):e0165953. https://doi.org/10.1371/journal.pone.0165953

Huh HW, Huh MK (2002) Genetic diversity and population structure of *Pseudosasa japonica* (Bambusaceae) in Korea. Bamboo Sci Cult 16:9–17

Isagi Y, Shimada K, Kushima H, Tanaka N, Nagao A, Ishikawa T, OnoDera H, Watanabe S (2004) Clonal structure and flowering traits of a bamboo [*Phyllostachys pubescens* (Mazel) Ohwi] stand grown from a simultaneous flowering as revealed by AFLP analysis: clonal structure and flowering of bamboo. Mol Ecol 13(7):2017–2021. https://doi.org/10.1111/j.1365-294X.2004.02197.x

Jiang W-X, Zhang W-J, Ding Y-L (2013) Development of polymorphic microsatellite markers for *Phyllostachys edulis* (Poaceae), an important bamboo species in China. Appl Plant Sci 1 (7):1200012. https://doi.org/10.3732/apps.1200012

Jofuku KD, den Boer BGW, Montagu MV, Okamuro JK (1994) Control of Arabidopsis flower and seed development by the homeotic gene *APETALA2*. Plant Cell 6:1211–1225. https://doi.org/10.1105/tpc.6.9.1211

Kaeppler SM, Kaeppler HF, Rhee Y (2000) Epigenetic aspects of somaclonal variation in plants. In: Plant gene silencing. Springer, Dordrecht, pp 59–68

Kagaya Y, Ohmiya K, Hattori T (1999) RAV1, a novel DNA-binding protein, binds to bipartite recognition sequence through two distinct DNA-binding domains uniquely found in higher plants. Nucleic Acids Res 27(2):470–478. https://doi.org/10.1093/nar/27.2.470

Kalaiarasi K, Sangeetha P, Subramaniam S, Venkatachalam P (2014) Development of an efficient protocol for plant regeneration from nodal explants of recalcitrant bamboo (*Bambusa arundinacea* Retz. Willd) and assessment of genetic fidelity by DNA markers. Agrofor Syst 88 (3):527–537. https://doi.org/10.1007/s10457-014-9716-3

Kaneko S, Franklin DC, Yamasaki N, Isagi Y (2008) Development of microsatellite markers for *Bambusa arnhemica* (Poaceae: Bambuseae), a bamboo endemic to northern Australia. Conserv Genet 9(5):1311–1313. https://doi.org/10.1007/s10592-007-9467-z

Kitamura K, Kawahara T (2009) Clonal identification by microsatellite loci in sporadic flowering of a dwarf bamboo species, *Sasa cernua*. J Plant Res 122(3):299–304. https://doi.org/10.1007/s10265-009-0220-1

Kitamura K, Kawahara T (2011) Estimation of outcrossing rates at small-scale flowering sites of the dwarf bamboo species, *Sasa cernua*. J Plant Res 124(6):683–688. https://doi.org/10.1007/s10265-010-0398-2

Kitamura K, Saitoh T, Matsuo A, Suyama Y (2009) Development of microsatellite markers for the dwarf bamboo species *Sasa cernua* and *Sasa kurilensis* (Poaceae) in northern Japan. Mol Ecol Resour 9(6):1470–1472. https://doi.org/10.1111/j.1755-0998.2009.02675.x

Konzen E, Peron R, Ito M, Brondani G, Tsai S (2017) Molecular identification of bamboo genera and species based on RAPD-RFLP markers. Silva Fenn 51:4. https://doi.org/10.14214/sf.1691

Konzen ER, Recchia GH, Cassieri F, Caldas DGG, Berny Mier y Teran JC, Gepts P, Tsai SM (2019) *DREB* genes from common bean (*Phaseolus vulgaris* L.) show broad to specific abiotic stress responses and distinct levels of nucleotide diversity. Int J Genomics 2019:1–28. doi: https://doi.org/10.1155/2019/9520642

Krishna H, Alizadeh M, Singh D, Singh U, Chauhan N, Eftekhari M, Sadh RK (2016) Somaclonal variations and their applications in horticultural crops improvement. 3 Biotech 6(1):54. https://doi.org/10.1007/s13205-016-0389-7

Kumar S, Stecher G, Li M, Knyaz C, Tamura K (2018) MEGA X: molecular evolutionary genetics analysis across computing platforms. Mol Biol Evol 35(6):1547–1549. https://doi.org/10.1093/molbev/msy096

Lai Z, Li Y, Wang F, Cheng Y, Fan B, Yu J-Q, Chen Z (2011) Arabidopsis sigma factor binding proteins are activators of the WRKY33 transcription factor in plant defense. Plant Cell 23:3824–3841. https://doi.org/10.1105/tpc.111.090571

Lee NW, Chung MG (1999) High levels of genetic variation in Korean populations of *Sasamorpha borealis* (Poaceae). Bot Bull Acad Sin 40

Lewontin RC, Hubby JL (1966) A molecular approach to the study of genic heterozygosity in natural populations. II. Amount of variation and degree of heterozygosity in natural populations of *Drosophila pseudoobscura*. Genetics 54(2):595

Li L, Mu S, Cheng Z, Cheng Y, Zang Y, Miao Y, Hou C, Li X, Gao J (2017) Characterization and expression analysis of the WRKY gene family in moso bamboo. Sci Rep 7:6675. https://doi.org/10.1038/s41598-017-06701-2

Li S, Ramakrishnan M, Vinod KK, Kalendar R, Yrjälä K, Zhou M (2020a) Development and deployment of high-throughput retrotransposon-based markers reveal genetic diversity and population structure of Asian bamboo. Forests 11(1):31. https://doi.org/10.3390/f11010031

Li L, Yang K, Wang S, Lou Y, Zhu C, Gao Z (2020b) Genome-wide analysis of laccase genes in moso bamboo highlights PeLAC10 involved in lignin biosynthesis and in response to abiotic stresses. Plant Cell Rep 39:751–763. https://doi.org/10.1007/s00299-020-02528-w

Lin X, Lou Y, Zhang Y, Yuan X, He J, Fang W (2011) Identification of genetic diversity among cultivars of *Phyllostachys violascens* using ISSR, SRAP and AFLP markers. Bot Rev 77 (3):223–232. https://doi.org/10.1007/s12229-011-9078-8

Lin Y, Lu J-J, Wu M-D, Zhou M-B, Fang W, Ide Y, Tang D-Q (2014) Identification, cross-taxon transferability and application of full-length cDNA SSR markers in *Phyllostachys pubescens*. Springerplus 3(1):486. https://doi.org/10.1186/2193-1801-3-486

Litt M, Luty JA (1989) A hypervariable microsatellite revealed by in vitro amplification of a dinucleotide repeat within the cardiac muscle actin gene. Am J Hum Genet 44(3):397–401

Liu H, Gao Y, Wu M, Shi Y, Wang H, Wu L, Xiang Y (2020) TCP10, a TCP transcription factor in moso bamboo (*Phyllostachys edulis*), confers drought tolerance to transgenic plants. Environ Exp Bot 172:104002. https://doi.org/10.1016/j.envexpbot.2020.104002

Liu H, Wu M, Zhu D, Pan F, Wang Y, Wang Y, Xiang Y (2017) Genome-wide analysis of the AAAP gene family in moso bamboo (*Phyllostachys edulis*). BMC Plant Biol 17:29. https://doi.org/10.1186/s12870-017-0980-z

Liu H-L, Wu M, Li F, Gao Y-M, Chen F, Xiang Y (2018) TCP transcription factors in Moso bamboo (*Phyllostachys edulis*): genome-wide identification and expression analysis. Front Plant Sci 9:1263. https://doi.org/10.3389/fpls.2018.01263

Liu J, Cheng Z, Xie L, Li X, Gao J (2019) Multifaceted role of *PheDof12-1* in the regulation of flowering time and abiotic stress responses in moso bamboo (*Phyllostachys edulis*). Int J Mol Sci 20(2):424. https://doi.org/10.3390/ijms20020424

Loh J, Kiew R, Gan LH, Gan Y-Y (2000) A study of genetic variation and relationships within the bamboo subtribe Bambusinae using amplified fragment length polymorphism. Ann Bot 85 (5):607–612. https://doi.org/10.1006/anbo.2000.1109

Lu Y, Wang D, Li H, Jia Q, Wu Z, Lu W (2012) A comparison on DNA methylation levels in bamboo at five developmental stages. J For Res 23(1):157–159. https://doi.org/10.1007/s11676-012-0247-3

Lv Y, Yu T, Lu S, Tian C, Li J, Du FK (2016) Development of microsatellite markers for *Fargesia denudata* (Poaceae), the staple-food bamboo of the giant panda. Appl Plant Sci 4(6):1600005. https://doi.org/10.3732/apps.1600005

Ma P-F, Zhang Y-X, Zeng C-X, Guo Z-H, Li D-Z (2014) Chloroplast phylogenomic analyses resolve deep-level relationships of an intractable bamboo tribe Arundinarieae (Poaceae). Syst Biol 63:933–950. https://doi.org/10.1093/sysbio/syu054

Ma Q, Song H, Zhou S, Yang W, Li D, Chen J (2013) Genetic structure in dwarf bamboo (*Bashania fangiana*) clonal populations with different genet ages. PLoS One 8(11):e78784. https://doi.org/10.1371/journal.pone.0078784

Maurel C, Boursiac Y, Luu D-T, Santoni V, Shahzad Z, Verdoucq L (2015) Aquaporins in Plants. Physiol Rev 95:38

Meena RK, Bhandhari MS, Barhwal S, Ginwal HS (2019) Genetic diversity and structure of *Dendrocalamus hamiltonii* natural metapopulation: a commercially important bamboo species of Northeast Himalayas. 3 Biotech 9(2):60. https://doi.org/10.1007/s13205-019-1591-1

Mehta R, Sharma V, Sood A, Sharma M, Sharma RK (2011) Induction of somatic embryogenesis and analysis of genetic fidelity of in vitro-derived plantlets of *Bambusa nutans* Wall., using AFLP markers. Eur J For Res 130(5):729–736. https://doi.org/10.1007/s10342-010-0462-4

Miyazaki Y, Ohnishi N, Takafumi H, Hiura T (2009) Genets of dwarf bamboo do not die after one flowering event: evidence from genetic structure and flowering pattern. J Plant Res 122 (5):523–528. https://doi.org/10.1007/s10265-009-0241-9

Mukherjee AK, Ratha S, Dhar S, Debata AK, Acharya PK, Mandal S, Panda PC, Mahapatra AK (2010) Genetic relationships among 22 taxa of bamboo revealed by ISSR and EST-based random primers. Biochem Genet 48(11–12):1015–1025. https://doi.org/10.1007/s10528-010-9390-8

Nayak S, Rout GR, Das P (2003) Evaluation of the genetic variability in bamboo using RAPD markers. Plant Soil Environ 49(1):24–28. https://doi.org/10.17221/4085-PSE

Neale DB, Martínez-García PJ, De La Torre AR, Montanari S, Wei X-X (2017) Novel insights into tree biology and genome evolution as revealed through genomics. Annu Rev Plant Biol 68 (1):457–483. https://doi.org/10.1146/annurev-arplant-042916-041049

Nilkanta H, Amom T, Tikendra L, Rahaman H, Nongdam P (2017) ISSR marker based population genetic study of *Melocanna baccifera* (Roxb.) Kurz: a commercially important bamboo of Manipur, North-East India. Scientifica 2017:1–9. https://doi.org/10.1155/2017/3757238

Nogueira JS, Gomes HT, Scherwinski-Pereira JE (2019) Micropropagation, plantlets production estimation and ISSR marker-based genetic fidelity analysis of Guadua magna and

G. angustifolia. Pesqui Agropecu Trop 49:e53743. https://doi.org/10.1590/1983-40632019v4953743

Okamuro JK, Caster B, Villarroel R, Van Montagu M, Jofuku KD (1997) The AP2 domain of APETALA2 defines a large new family of DNA binding proteins in *Arabidopsis*. Proc Natl Acad Sci U S A 94(13):7076–7081. https://doi.org/10.1073/pnas.94.13.7076

Oumer OA, Dagne K, Feyissa T, Tesfaye K, Durai J, Hyder MZ (2020) Genetic diversity, population structure, and gene flow analysis of lowland bamboo [*Oxytenanthera abyssinica* (A. Rich.) Munro] in Ethiopia. Ecol Evol 2020:1–20. https://doi.org/10.1002/ece3.6762

Pan F, Wang Y, Liu H, Wu M, Chu W, Chen D, Xiang Y (2017) Genome-wide identification and expression analysis of SBP-like transcription factor genes in Moso bamboo (*Phyllostachys edulis*). BMC Genomics 18:486. https://doi.org/10.1186/s12864-017-3882-4

Pan F, Wu M, Hu W, Liu R, Yan H, Xiang Y (2019) Genome-wide identification and expression analyses of the bZIP transcription factor genes in moso bamboo (*Phyllostachys edulis*). Int J Mol Sci:20. https://doi.org/10.3390/ijms20092203

Peng Z, Lu Y, Li L, Zhao Q, Feng Q, Gao Z et al (2013) The draft genome of the fast-growing non-timber forest species moso bamboo (*Phyllostachys heterocycla*). Nat Genet 45(4):456–461

Philippe H, Delsuc F, Brinkmann H, Lartillot N (2005) Phylogenomics. Annu Rev Ecol Evol Syst 36:541–562

Poland JA, Brown PJ, Sorrells ME, Jannink JL (2012) Development of high-density genetic maps for barley and wheat using a novel two-enzyme genotyping-by-sequencing approach. PLoS One 7(2):e32253

Pourcel L, Routaboul J-M, Kerhoas L, Caboche M, Lepiniec L, Debeaujon I (2005) *TRANSPARENT TESTA10* encodes a laccase-like enzyme involved in oxidative polymerization of flavonoids in *Arabidopsis* seed coat. Plant Cell 17:2966–2980. https://doi.org/10.1105/tpc.105.035154

Rajput BS, Jani M, Ramesh K, Manokari M, Jogam P, Allini VR, Kher MM, Shekhawat MS (2020) Large-scale clonal propagation of *Bambusa balcooa* Roxb.: An industrially important bamboo species. Ind Crop Prod 157:112905. https://doi.org/10.1016/j.indcrop.2020.112905

Ramakrishnan M, Yrjälä K, Vinod KK, Sharma A, Cho J, Satheesh V, Zhou M (2020) Genetics and genomics of moso bamboo (*Phyllostachys edulis*): current status, future challenges, and biotechnological opportunities toward a sustainable bamboo industry. Food Energy Secur 2020 (00):e229. https://doi.org/10.1002/fes3.229

Ramanayake SMSD, Meemaduma VN, Weerawardene TE (2007) Genetic diversity and relationships between nine species of bamboo in Sri Lanka, using random amplified polymorphic DNA. Plant Syst Evol 269:55–61. https://doi.org/10.1007/s00606-007-0587-1

Ray SS, Ali MN (2017) Factors affecting macropropagation of bamboo with special reference to culm cuttings: a review update. N Z J For Sci 47(1):17. https://doi.org/10.1186/s40490-017-0097-z

Rushton PJ, Somssich IE, Ringler P, Shen QJ (2010) WRKY transcription factors. Trends Plant Sci 15:247–258. https://doi.org/10.1016/j.tplants.2010.02.006

Sakuma Y, Liu Q, Dubouzet JG, Abe H, Shinozaki K, Yamaguchi-Shinozaki K (2002) DNA-binding specificity of the ERF/AP2 domain of *Arabidopsis* DREBs, transcription factors involved in dehydration- and cold-inducible gene expression. Biochem Biophys Res Commun 290(3):998–1009. https://doi.org/10.1006/bbrc.2001.6299

Shan, Yang, Xu, Zhu, Gao (2019) Genome-wide investigation of the *NAC* gene family and its potential association with the secondary cell wall in moso bamboo. Biomol Ther 9(10):609. https://doi.org/10.3390/biom9100609

Sharma V, Bhardwaj P, Kumar R, Sharma RK, Sood A, Ahuja PS (2009) Identification and cross-species amplification of EST derived SSR markers in different bamboo species. Conserv Genet 10(3):721–724. https://doi.org/10.1007/s10592-008-9630-1

Singh SR, Dalal S, Singh R, Dhawan AK, Kalia RK (2012) Micropropagation of *Dendrocalamus asper* {Schult. & Schult. F.} Backer ex k. Heyne: an exotic edible bamboo. J Plant Biochem Biotechnol 21(2):220–228

Singh SR, Dalal S, Singh R, Dhawan AK, Kalia RK (2013a) Ascertaining clonal fidelity of micropropagated plants of *Dendrocalamus hamiltonii* Nees et Arn. ex Munro using molecular markers. In Vitro Cell Dev Biol Plant 49(5):572–583. https://doi.org/10.1007/s11627-013-9520-1

Singh SR, Dalal S, Singh R, Dhawan AK, Kalia RK (2013b) Evaluation of genetic fidelity of in vitro raised plants of *Dendrocalamus asper* (Schult. & Schult. F.) Backer ex K. Heyne using DNA-based markers. Acta Physiol Plant 35(2):419–430. https://doi.org/10.1007/s11738-012-1084-x

Sun H, Li L, Lou Y (2017) The bamboo aquaporin gene *PeTIP4;1–1* confers drought and salinity tolerance in transgenic *Arabidopsis*. Plant Cell Rep 36:597–609. https://doi.org/10.1007/s00299-017-2106-3

Sun H, Wang S, Lou Y, Zhu C, Zhao H, Li Y, Li X, Gao Z (2018) Whole-genome and expression analyses of bamboo aquaporin genes reveal their functions involved in maintaining diurnal water balance in bamboo shoots. Cell 7:195. https://doi.org/10.3390/cells7110195

Tanksley SD, Young ND, Paterson AH, Bonierbale MW (1989) RFLP mapping in plant breeding: new tools for an old science. Nat Biotechnol 7(3):257–264. https://doi.org/10.1038/nbt0389-257

Tian B, Yang H-Q, Wong K-M, Liu A-Z, Ruan Z-Y (2012) ISSR analysis shows low genetic diversity versus high genetic differentiation for giant bamboo, *Dendrocalamus giganteus* (Poaceae: Bambusoideae), in China populations. Genet Resour Crop Evol 59(5):901–908. https://doi.org/10.1007/s10722-011-9732-3

Tiwari C, Bakshi M, Gupta D, Razjput B (2019) Genetic diversity assessment of Sinarundinaria anceps (Mitf.) Chao and Renvoize accessions through isozyme markers. J Pharmacogn Phytochem 2019:515–519

Triplett JK, Oltrogge KA, Clark LG (2010) Phylogenetic relationships and natural hybridization among the North American woody bamboos (Poaceae: Bambusoideae: *Arundinaria*). Am J Bot 97(3):471–492

Vaattovaara A, Leppälä J, Salojärvi J, Wrzaczek M (2019) High-throughput sequencing data and the impact of plant gene annotation quality. J Exp Bot 70(4):1069–1076. https://doi.org/10.1093/jxb/ery434

Vos P, Hogers R, Bleeker M, Reijans M, van de Lee T, Hornes M, Friters A, Pot J, Paleman J, Kuiper M, Zabeau M (1995) AFLP: a new technique for DNA fingerprinting. Nucleic Acids Res 23(21):4407–4414. https://doi.org/10.1093/nar/23.21.4407

Wang J, Feng J, Jia W, Chang S, Li S, Li Y (2015) Lignin engineering through laccase modification: a promising field for energy plant improvement. Biotechnol Biofuels 8:145. https://doi.org/10.1186/s13068-015-0331-y

Wang T, Yue J-J, Wang X-J, Xu L, Li L-B, Gu X-P (2016) Genome-wide identification and characterization of the *Dof* gene family in moso bamboo (*Phyllostachys heterocycla* var. *pubescens*). Genes Genom 38:733–745. https://doi.org/10.1007/s13258-016-0418-2

Wang W, Chen S, Guo W, Li ZX (2020) Tropical plants evolve faster than their temperate relatives: a case from the bamboos (Poaceae: Bambusoideae) based on chloroplast genome data. Biotechnol Biotechnol Equip 34:482–493. https://doi.org/10.1080/13102818.2020.1773312

Wang W, Chen S, Zhang X (2018) Whole-genome comparison reveals divergent IR borders and mutation hotspots in chloroplast genomes of herbaceous bamboos (Bambusoideae: Olyreae). Molecules 23(7):1537. https://doi.org/10.3390/molecules23071537

Wang X, Ye X, Zhao L, Li D, Guo Z, Zhuang H (2017a) Genome-wide RAD sequencing data provide unprecedented resolution of the phylogeny of temperate bamboos (Poaceae: Bambusoideae). Sci Rep 7(1):11546. https://doi.org/10.1038/s41598-017-11367-x

Wang Y, Liu H, Zhu D, Gao Y, Yan H, Xiang Y (2017b) Genome-wide analysis of VQ motif-containing proteins in Moso bamboo (*Phyllostachys edulis*). Planta 246:165–181. https://doi.org/10.1007/s00425-017-2693-9

Williams JGK, Kubelik AR, Livak KJ, Rafalski JA, Tingey SV (1990) DNA polymorphisms amplified by arbitrary primers are useful as genetic markers. Nucleic Acids Res 18 (22):6531–6535. https://doi.org/10.1093/nar/18.22.6531

Wu H, Lv H, Li L, Liu J, Mu S, Li X, Gao J (2015) Genome-wide analysis of the AP2/ERF transcription factors family and the expression patterns of DREB genes in Moso Bamboo (*Phyllostachys edulis*). PLoS One 10(5):e0126657. https://doi.org/10.1371/journal.pone.0126657

Wu M, Li Y, Chen D, Liu H, Zhu D, Xiang Y (2016) Genome-wide identification and expression analysis of the IQD gene family in moso bamboo (*Phyllostachys edulis*). Sci Rep 6:24520. https://doi.org/10.1038/srep24520

Wu M, Liu H, Gao Y, Shi Y, Pan F, Xiang Y (2020a) The moso bamboo drought-induced 19 protein PheDi19-8 functions oppositely to its interacting partner, *PheCDPK22*, to modulate drought stress tolerance. Plant Sci 299:110605. https://doi.org/10.1016/j.plantsci.2020.110605

Wu M, Liu H, Han G, Cai R, Pan F, Xiang Y (2017) A moso bamboo *WRKY* gene *PeWRKY83* confers salinity tolerance in transgenic *Arabidopsis* plants. Sci Rep 7(1):11721. https://doi.org/10.1038/s41598-017-10795-z

Wu M, Liu R, Gao Y, Xiong R, Shi Y, Xiang Y (2020b) PheASR2, a novel stress-responsive transcription factor from moso bamboo (*Phyllostachys edulis*), enhances drought tolerance in transgenic rice via increased sensitivity to abscisic acid. Plant Physiol Biochem 154:184–194. https://doi.org/10.1016/j.plaphy.2020.06.014

Wysocki WP, Clark LG, Attigala L et al (2015) Evolution of the bamboos (Bambusoideae; Poaceae): a full plastome phylogenomic analysis. BMC Evol Biol 15:50. https://doi.org/10.1186/s12862-015-0321-5

Xie L, Cai M, Li X, Zheng H, Xie Y, Cheng Z, Bai Y, Li J, Mu S, Gao J (2020) Overexpression of PheNAC3 from moso bamboo promotes leaf senescence and enhances abiotic stress tolerance in *Arabidopsis*. PeerJ 8:e8716. https://doi.org/10.7717/peerj.8716

Xie L, Li X, Hou D, Cheng Z, Liu J, Li J, Mu S, Gao J (2019) Genome-wide analysis and expression profiling of the heat shock factor gene family in *Phyllostachys edulis* during development and in response to abiotic stresses. Forests 10(2):100. https://doi.org/10.3390/f10020100

Yang H-M, Zhang Y-X, Yang J-B, Li D-Z (2013) The monophyly of *Chimonocalamus* and conflicting gene trees in Arundinarieae (Poaceae: Bambusoideae) inferred from four plastid and two nuclear markers. Mol Phylogenet Evol 68:340–356. https://doi.org/10.1016/j.ympev.2013.04.002

Yang H-Q, An M-Y, Gu Z-J, Tian B (2012) Genetic diversity and differentiation of *Dendrocalamus membranaceus* (Poaceae: Bambusoideae), a declining bamboo species in Yunnan, China, as based on inter-simple sequence repeat (ISSR) analysis. Int J Mol Sci 13(4):4446–4457. https://doi.org/10.3390/ijms13044446

Yang JB, Dong YR, Wong KM, Gu ZJ, Yang HQ, Li DZ (2018) Genetic structure and differentiation in *Dendrocalamus sinicus* (Poaceae: Bambusoideae) populations provide insight into evolutionary history and speciation of woody bamboos. Sci Rep 8(1):16933. https://doi.org/10.1038/s41598-018-35269-8

Yu X, Yang D, Guo C, Gao L (2018) Plant phylogenomics based on genome-partitioning strategies: Progress and prospects. Plant Divers 40:158–164. https://doi.org/10.1016/j.pld.2018.06.005

Yuan J-L, Sun H-M, Guo G-P, Yue J-J, Gu X-P (2014) Correlation between DNA methylation and chronological age of Moso bamboo (*Phyllostachys heterocycla* var. *pubescens*). Bot Stud 55 (1):4. https://doi.org/10.1186/1999-3110-55-4

Zeng C-X, Zhang Y-X, Triplett JK, Yang J-B, Li D-Z (2010) Large multi-locus plastid phylogeny of the tribe Arundinarieae (Poaceae: Bambusoideae) reveals ten major lineages and low rate of molecular divergence. Mol Phylogenet Evol 56:821–839. https://doi.org/10.1016/j.ympev.2010.03.041

Zhang X-Z, Zeng C-X, Ma P-F, Haevermans T, Zhang Y-X, Zhang L-N, Guo Z-H, Li D-Z (2016) Multi-locus plastid phylogenetic biogeography supports the Asian hypothesis of the temperate woody bamboos (Poaceae: Bambusoideae). Mol Phylogenet Evol 96:118–129. https://doi.org/10.1016/j.ympev.2015.11.025

Zhou M, Hu B, Zhu Y (2017) Genome-wide characterization and evolution analysis of long terminal repeat retroelements in moso bamboo (*Phyllostachys edulis*). Tree Genet Genomes 13(2):43. https://doi.org/10.1007/s11295-017-1114-3

Zhou M, Wu J, Ramakrishnan M, Meng X, Vinod KK (2019a) Prospects for the study of genetic variation among Moso bamboo wild-type and variants through genome resequencing. Trees (Berl West) 33(2):371–381. https://doi.org/10.1007/s00468-018-1783-z

Zhou Y, Zhang Y-Q, Xing X-C, Zhang J-Q, Ren Y (2019b) Straight from the Plastome: molecular phylogeny and morphological evolution of *Fargesia* (Bambusoideae: Poaceae). Front Plant Sci 10:981. https://doi.org/10.3389/fpls.2019.00981

Zhu J-K (2016) Abiotic stress signaling and responses in plants. Cell 167:313–324. https://doi.org/10.1016/j.cell.2016.08.029

Zhu S, Liu T, Tang Q, Fu L, Tang S (2014) Evaluation of bamboo genetic diversity using morphological and SRAP analyses. Russ J Genet 50(3):267–273. https://doi.org/10.1134/S1022795414030132

Chapter 3
Standard Protocols for in Vitro Propagation of Bamboo with Emphasis on Axillary Shoot Proliferation

Víctor M. Jiménez, Andrea Holst, Paula Carvajal-Campos, and Eric Guevara

Abstract Bamboo is a group of plants in which in vitro propagation truly offers an option to overcome the reported constraints of conventional propagation, allowing to potentially obtain large amounts of uniform selected plants. Plant in vitro propagation can occur following different pathways, i.e., activation of preexisting buds, organogenesis, and somatic embryogenesis, with the involvement of de novo formation of proliferating structures in the latter two. The utilization of preformed vegetative meristems to develop bamboo plants in vitro will be the subject of this chapter. Bamboo plants produce thin lateral branches in addition to the straight and comparatively massive culms. These side branches are the source of the single-node segments used as explants to start tissue cultures. After lateral bud sprouting, growth, and multiplication, rooting and acclimatization steps are necessary to obtain plants to be grown in the field. Considering the existence of recent and comprehensive review papers on the use of axillary shoot proliferation for bamboo mass propagation, we will try to set in this chapter common approaches to fulfill the requirements of the distinct taxa, in addition to emphasizing the most critical problems, and on how to counteract them to increase the chances of success.

Keywords Acclimatization · Bud sprouting · Clonal propagation · Rooting

V. M. Jiménez (✉) · A. Holst · P. Carvajal-Campos · E. Guevara
Centro para Investigaciones en Granos y Semillas (CIGRAS), Universidad de Costa Rica, San Pedro, Costa Rica
e-mail: victor.jimenez@ucr.ac.cr

Z. Ahmad et al. (eds.), *Biotechnological Advances in Bamboo*,
https://doi.org/10.1007/978-981-16-1310-4_3

3.1 Why Is in Vitro Culture an Option for Bamboo Propagation?

In spite of the considerable large areas still covered with natural bamboo forests in some tropical and subtropical regions of the world, there is an increasing necessity to generate a constant supply of abundant and adequate planting material to replant deforested stands and establish new managed plantations. This is important due to the relevance that particularly woody bamboos have acquired for several industrial purposes and environmental services (Ramakrishnan et al. 2020).

Propagation of bamboo plants by using seeds is not feasible in most cases because many species flower gregariously and very sporadically, and the recalcitrant seeds lose their viability very quickly. In addition, conventional vegetative propagation by using cuttings or rhizome sections is rather limitedly used in large scale because of its very low efficiency and erratic multiplication rates, while employed propagules are usually very bulky and prone to disperse soil-borne pests and systemic diseases (Ray and Ali 2017). These constraints to sexual and asexual standard propagation of most bamboo species require the development of novel and reliable methods. In vitro propagation offers the potential to overcome most of these problems and has been the subject of continuous work during the past decades, trying to develop particular procedures, mainly for commercially important species (Nogueira et al. 2017; Sandhu et al. 2018). Implementation of in vitro protocols for the mass propagation of bamboo plants allows the possibility of selecting elite individuals as explant donors and to design the forthcoming commercial plantation. It might be desirable to choose a combination of different source plants with desirable characteristics within the same species to avoid having a plantation 100% uniform, which might be disadvantageous in case biotic or abiotic constraints arise, such as has been observed in other forest species (Rosvall et al. 2019).

3.2 Developmental Pathways for Bamboo Regeneration

Obtaining a complete and functional plant from an explant exposed to particular in vitro conditions can occur through one of the three possible pathways. One implies further development of preformed structures that have this potential, such as sprouting and further growth of axillary or apical buds. The other two, organogenesis and somatic embryogenesis, involve the establishment of newly (de novo) formed structures, which can occur directly or indirectly, the latter after a cell dedifferentiation phase usually evident through callus formation (Ikeuchi et al. 2016). All three pathways have been explored in bamboo attempting to regenerate whole plants from excised explants (Thapa et al. 2018). After the first successful report on somatic embryogenesis in *Bambusa arundinacea* (Mehta et al. 1982), additional research articles involving this developmental pathway in bamboo have been published (Mudoi et al. 2013; Thapa et al. 2018). The first report on bamboo organogenesis describes successful callus development from the excised shoot apices of *Bambusa oldhamii* and *Phyllostachys aurea* on 2,4-dichlorophenoxyacetic acid-supplied medium and subsequent development of adventitious shoots (Huang et al. 1989). Although relevant reports were subsequently published (Ali et al. 2009; e.g., Ye et al. 2017; Zang et al. 2019), the literature on bamboo organogenesis is less abundant than that of somatic embryogenesis and axillary shoot proliferation (Mudoi et al. 2013), with a much larger prevalence of the latter (Sandhu et al. 2018; Suwal et al. 2020), certainly the preferred way nowadays because of the high response rates obtained in different genera and species and the low levels of somaclonal variation reported (Sandhu et al. 2018).

Though initial reports on the use of tissue culture for bamboo propagation employed zygotic embryos or seedlings as explant sources (Alexander and Rao 1968; Nadgir et al. 1984; Nadgauda et al. 1990; Ravikumar et al. 1998; Mudoi et al. 2013), the breakthrough for the potential use of this methodology for commercial purposes arose when explants from adult plants, whose performance has been already evaluated in the field, could be used as starting material (Vongvijitra 1988). Axillary shoot proliferation allows researchers to fully take advantage of the genetic background of selected plants, because explants from adult individuals are normally used and the plantlet development occurs without cell dedifferentiation steps, which could negatively affect (epi)genetic fidelity (Bednarek and Orłowska 2020).

3.3 Use of Axillary Buds for Bamboo Micropropagation

In vitro propagation of bamboo through axillary shoot proliferation starts by selecting the proper source of explants, i.e., thin lateral branches with prominent buds in the nodes. Explants should be subsequently successfully disinfected to achieve manageable contamination rates and then placed on culture medium appropriately supplemented to induce bud break and shoot growth, without showing symptoms of tissue darkening. Afterward, the growing shoots should produce new leaves and then lateral shoots that grow to form small shoot clumps that, spontaneously or after the proper treatment, start to build roots. The last steps could occur in a single phase without adapting the composition of the culture media or resulting from clear steps aiming at inducing the desired outcomes by modifying medium constitution or culture conditions. Sometimes, the development of lateral shoots is needed for rooting to occur, while in other cases, rooting might be necessary to induce lateral branching and to avoid tissue browning.

Since micropropagation aims at obtaining a large number of plants from few original explants, the shoot clumps that developed from the initial segments should be divided to obtain more clumps. This is a repetitive process that keeps running until some of the rooted clumps are hardened or acclimatized to greenhouse conditions. Greenhouse plants are then grown until transplanted in the field.

Due to the very large genetic diversity in the Bambusoideae subfamily, with at least 1641 species (Soreng et al. 2015), significant variation in terms of identifying the proper conditions to successfully propagate bamboo through axillary shoot proliferation in different species is to be expected. Gielis and Oprins (2002) claimed that it is possible to propagate many diverse bamboo species with a single method; however, the vast literature available that reports differences in the requirements according to species, clones, age, physiological stage, time of the year, etc. points out to the necessity of adapting the protocols frequently by "guided" trial and error (Goyal and Sen 2016; Sandhu et al. 2018; Suwal et al. 2020).

3.4 Relevant Steps during Bamboo Axillary Bud Micropropagation

3.4.1 Explant Selection

In vitro propagation of bamboo through sprouting of axillary buds takes advantage of the particular architecture of the bamboo plant, which has a more complicated branching system than the other members of the Poaceae. In woody bamboos, once the main culms reach their final height, some lateral buds become active. This can occur in the same year of culm development or during the next year and can be observed in the basal part of the culms, like in the genus *Guadua* and some individual species in *Bambusa*, or only at a higher level, such as in several *Dendrocalamus* species. Further branching of these stout branches produces thinner branches in subsequent orders (Banik 2015). It is from these thin branches, coming from either field- or greenhouse-growing plants, that the single-node segments are usually taken to be used as initial explants for axillary shoot proliferation (Fig. 3.1) (Jiménez et al. 2006; Silveira et al. 2020).

Additional criteria for selecting the most adequate branches include the thickness and presence of an active bud underneath the sheath. There are diverse examples in the literature related to the effect of explant diameter conditioning the bud sprouting rate in different bamboo species (Bag et al. 2000; Sanjaya et al. 2005; Jiménez et al. 2006; Somashekar et al. 2008; Kabade 2009; Sandhu et al. 2018). According to our personal experience, thicker explants are more prone to release exudates that turn brown in the culture medium and can subsequently induce explant necrosis, while thinner ones are more affected by the disinfection process. Therefore, the most adequate diameter for a particular system (the combination of species, plant age, position, season, disinfection treatment, etc.) should probably be determined case by case (Jiménez and Guevara 2007). Nodal sections displaying a prominent (dormant) bud should be preferred because the chances of these buds to become activated and continue growing are much higher than when smaller buds are present. Nodes that do not show the presence of a conspicuous bud should be discarded at this point. Therefore, it is recommended to check one or two nodes in a branch, by removing

Fig. 3.1 Thin bamboo lateral branch showing conspicuous axillary buds suitable for in vitro culture (pointed by arrows). *Bar = 5 cm*

the leaf sheath, before selecting the branch for further processing (disinfection and dissection) (Jiménez and Guevara 2007).

The use of branches or branch sections with sprouted buds should be avoided because these growing tissues are more sensitive to damage during disinfection and also because contaminating microorganisms can escape the effect of sterilants when located in the space between the stem and the bud, increasing the chances of contamination (Fig. 3.2a). Furthermore, explants showing immature buds, generally with green or white colorations, should be avoided because they tend to oxidize during the disinfection process and may not have the ability to sprout (Fig. 3.2b). Mature buds showing a yellow coloration should be preferred. These buds should be healthy, without evident insect or fungus damage (Fig. 3.2c).

Because of the inherent and sometimes severe contamination problems associated with bamboo plants growing in the field, especially in (sub)tropical climates, where most economically important species develop (Ray and Ali 2016), a good alternative is to maintain the explant donor plants under greenhouse conditions, where they can be managed to reduce fungal and bacterial populations more easily (Niedz and Bausher 2002). To accomplish this, the selected field plants can be propagated by conventional vegetative methods and then established in the greenhouse in manageable-sized pots. To produce the abovementioned thin lateral branches more profusely, the plants can be intensely pruned a few weeks before starting an experiment (Jiménez and Guevara 2007).

There is also the option of using plantlets derived from in vitro germinated seeds or isolated embryos. While explants taken from younger tissues respond quicker and

Fig. 3.2 Selection of axillary buds to be used as explants for bamboo in vitro culture. Explants showing sprouted (**a**) or immature buds (**b**) should be avoided. Explants with healthy and mature buds are to be selected (**c**). *Bar = 1 cm*

at higher rates (Saxena and Bhojwani 1993; Ravikumar et al. 1998; George et al. 2008; dos Santos et al. 2019), the same concerns described above for sexual seeds apply to them, and this hampers selecting elite genotypes for propagation. Nevertheless, for bamboo species with gregarious flowering and consequent plant death, this approach has the advantage of resetting the ontogenetic plant age extending the productive life of the micropropagated plants when compared to plants raised from mature explant donors (Zhang et al. 2020).

To facilitate the subsequent disinfection process, nodal sections should be individually split at this moment. Care should be taken to leave at least 2–4 cm above and below the node because the disinfecting agents damage the exposed tissues and could also harm the bud if it is too close to the cutting site. It is also advisable to identify the top/bottom of the individual cuttings at this time (e.g., by leaving a longer internode section systematically in one extreme). This will enable recognizing explant's orientation in further steps.

3.4.2 Disinfection

Like most plants, the two explant-related contamination sources affecting bamboo tissue culture are the superficial contaminants and the internal microorganisms (Ray and Ali 2016). Though much effort is sometimes required to clean the explants of the former ones, there are several decontaminating options (by applying surface disinfectants and additives at various concentrations and for diverse time intervals) that have shown their efficiency in bamboo (Jiménez and Guevara 2007; Ray and Ali 2016). More difficult to deal with are the internal contaminants, which could be systemic (endophytic) or nonsystemic, because they can become evident only after the explants have been in culture for a certain time and can be associated with the

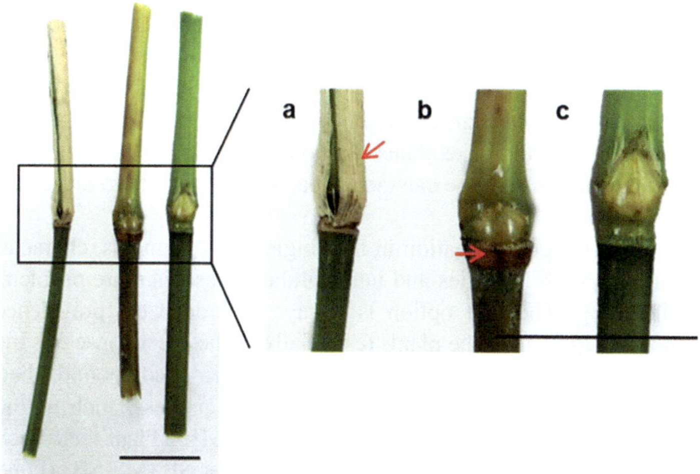

Fig. 3.3 Removal of the bud sheath (**a**) and remnants with a sharp scalpel (**b**) before disinfection of the bamboo explants. Explant ready to disinfect (**c**). *Arrows* show the bud sheath and remnants to be removed. *Bar = 1 cm*

protection that the explant tissues offer to the disinfection procedure (Orlikowska et al. 2017; Quambusch and Winkelmann 2018).

To reduce the microbial load present in the donor plants grown in a greenhouse, it is important to carefully irrigate the plants only to the base to reduce water sprinkling and humidity accumulation in the aerial parts of the plant. The sustained application of systemic products (e.g., the fungicides Benomyl and Bavistin and the bactericides Agri-Mycin and Kilol), starting several weeks before collecting the axillary buds, is advisable (Jiménez and Guevara 2007; Ray and Ali 2016; Sandhu et al. 2018).

Mechanical and chemical pretreatments have been reported to be helpful to eliminate superficial microorganisms and dirt present in the axillary buds. In our experience, contamination can be reduced by scrubbing the explants with antibacterial soap and properly cutting off the bud sheath (Fig. 3.3a), trying to remove all remnants of the bud sheath with the help of a sharp scalpel (Fig. 3.3b) (Furlan et al. 2018), because spores and bacteria can accumulate in these sites. Furthermore, the axillary buds without the bud sheath (Fig. 3.3c) can be rubbed with a cotton ball soaked in ethanol 70% (w/v), and the nodal sections are then immersed in an alkaline detergent, like Extran® MA 01 (Merck, Darmstadt, Germany), and subsequently shaken in a mixture of fungicide and bactericide, as used for treating the donor plants (Jiménez and Guevara 2007).

Superficial disinfection per se relies in most bamboo reports on the use of mercuric chloride, in spite of its toxicity (Cappelletti et al. 2019), because it has delivered better results than less toxic options, such as sodium and calcium hypochlorite (Ray and Ali 2016). Nevertheless, sodium hypochlorite has been successfully employed in bamboo as well (Jiménez et al. 2006; Ornellas et al. 2019; Torres et al. 2019). However, in many cases, superficial disinfection with the abovementioned compounds needed to be complemented with additional actions to reduce in vitro contamination in bamboo. An option to eliminate or reduce bacterial surface contamination is to wash the explants with antibiotics. It is possible to use broad-spectrum antibiotics or combinations of narrow-spectrum ones. The latter are preferred because lower concentrations can be used, which reduces the damage that these substances cause to the explants. Although antibiotics can help in reducing contamination, their constant use can induce bacterial resistance to these compounds (Ray and Ali 2016).

Managing internal contamination in the single-node segments, characterized by having significant vessel cavities and intercellular spaces, is more problematic and can be challenging. The first option is always to reduce the prevalence of the microorganisms by handling the plants (especially in the greenhouse), as mentioned above. This can be complemented with the use of biocidal compounds that translocate within the plant tissues during the disinfection process, such as the broad-spectrum products Plant Preservative Mixture™ (PPM™, Plant Cell Technology, Washington DC, USA) and Microbiological Suppressor SB (MS-SB) (Laboratorios Químicos ARVI S.A., San José, Costa Rica), and several antibiotics (Jiménez et al. 2006; dos Santos et al. 2019).

Another option to handle contamination is to include in the culture medium compounds that can exert a more enduring effect over the internal contaminants, such as Benomyl, PPM™, or MS-SB (Ramanayake et al. 1995; Jiménez et al. 2006; Ornellas et al. 2019; Pasqualini et al. 2019). However, this approach can also mask prevalence of internal contaminants, usually observed in the form of exudates caused by microorganisms growing out of the explants and into the culture medium. Care should be then taken to avoid further propagation of asymptomatic (due to the supplemented compounds) but contaminated in vitro developed plants. One option is to subculture part of the stock plants to medium devoid of these biocidal compounds during the multiplication phase, to evidence any contamination and continue selecting only contaminant-free plants, but without the risk of losing the whole batch. If antibiotics are used as a component of the culture medium to control endophytic bacteria, it is very important to molecularly identify the species that are affecting the culture. In this way, it is possible to determine which is the most suitable antibiotic and the right concentration to use (Ray and Ali 2016).

3.4.3 Dissection and Positioning

Once disinfection has been conducted, individual explants have to be further dissected. In the laminar flow chamber, both ends of the explants, damaged by the disinfection process, must be removed with a sharp scalpel. It is advisable to make a straight cut in the upper part of the explant and a clean oblique cut in the lower part, to avoid placing the bud upside down when culturing (Fig. 3.4a). The most suitable explant size (comprising the node and the two internode sections) depends largely on the species and can range from 2 to 3.5 cm in length for successful bamboo

Fig. 3.4 Bamboo explant with an axillary bud after disinfection. A straight cut in the upper part and an oblique clean cut in the lower part, to remove damaged tissue after disinfection, are preferred (**a**). Explant vertical positioning in the culture medium with the bud at the culture medium edge level (**b**). *Bar = 1 cm*

micropropagation (Sandhu et al. 2018). The explants should be placed vertically into the culture medium. The oblique cut in the bottom end helps the explant to enter into the medium without significant rupture, thus maintaining good contact to avoid explant dehydration. Placing the bud right at the culture medium edge, and not above or below, allows a suitable supply of medium components and humidity while impedes anoxia conditions (Fig. 3.4b) (Jiménez and Guevara 2007).

3.4.4 *Bud Sprouting*

Sprouting of the dormant axillary buds to develop into shoots with vigorous growth is the first achievement that needs to be fulfilled to use this technique for the clonal propagation of bamboos. Murashige and Skoog (1962) is the most commonly referenced mineral salt composition for the induction of bud sprouting in bamboo. According to Sandhu et al. (2018), over 95% of all publications using nodal explants to regenerate bamboo plants relied on this medium for culture initiation and bud break. In the few reports in which different culture media were compared vis-à-vis, however, MS medium was superior to other commonly used formulations [e.g., B5 (Gamborg et al. 1968), WPM (McCown and Lloyd 1981), SH (Schenk and Hildebrandt 1972)] to induce shoots. MS medium composition has been modified in some cases, by reducing the concentration of micronutrients or changing individual compounds, to improve responses. Carbohydrate supplementation of the culture medium mostly relies on the use of sucrose (2.0–3.0% w/w) (Sandhu et al. 2018).

Although most reports describe the utilization of semisolid medium for the initial phase, mainly because of the advantage of offering support to maintain the initial explant straight and the bud in contact with the edge of the culture medium, without immersion (as described above), liquid media have been used as well (Sandhu et al. 2018). The main advantages of the liquid medium include increased nutrient diffusion to the explant surroundings to replenish uptaken minerals and dilution of exudates leaching from the explants that rapidly oxidate and exert negative effects to the plant tissues (Choudhary et al. 2017). However, several reports refer to increased hyperhydricity rates when the liquid medium is used in bamboo tissue cultures (Saxena and Bhojwani 1993; Ramanayake and Yakandawala 1997).

Most published reports on bamboo relate including a cytokinin in the culture medium with increased bud sprouting. Among representatives of this group of plant growth regulators, benzyladenine (BA) is the most prevalent and effective (Sandhu et al. 2018), with a response clearly related to concentration (Jiménez et al. 2006; Ornellas et al. 2019). However, there are some reports on the synergistic effect of BA with other cytokinins and auxins (Sanjaya et al. 2005; Negi and Saxena 2011) and few examples of the positive impact of an auxin alone on shoot induction (Chaturvedi et al. 1993). The addition of other compounds, such as ascorbic and citric acid, cysteine, glutamine, and coconut milk among others, to the culture medium is less widespread. Information about specific examples can be found elsewhere (Sandhu et al. 2018).

Early stages of bud sprouting are noticeable by an increase in the bud volume, followed by a separation (opening) of the buds' most internal sheaths, which can look dried out due to the disinfection process. Sometimes, when spores or bacterial cells escape disinfection because they are allocated beneath external tissues, contamination signs become evident at this point when they are exposed to the culture medium and high humidity. Therefore, a second disinfection process may be necessary. Since the growing shoots are now exposed and thus more susceptible to the sterilants, a milder procedure should be preferred (e.g., with lower concentrations of sodium hypochlorite, such as 0.5–1.0 mg/l).

After buds sprout, shoots start growing usually continuously. Subculture frequency needs to be defined according to the growth rate. On the one side, it is important to consider the size increase of the explants, because it is desirable to avoid that newly formed plant structures (leaves, shoots, etc.) reach the vessel walls because the simple permanent contact may induce damage to the tissues. Therefore, subculturing to larger vessels might be soon necessary depending on the growth rate (Fig. 3.5). In addition, very active growth can deplete medium components (inorganic and organic compounds) very quickly. According to published reports, subculture rate in bamboo varies from every 2 to 6 weeks, depending on the species, physiological stage, culture system, etc. (Sandhu et al. 2018). It is advisable to recut, in each subculture, the oblique bottom end with a sharp scalpel to facilitate adequate absorption of nutrients. Once the clump is formed, remnants of the mother explant can be removed.

One of the most recurring problems in this stage of in vitro bamboo cultivation is browning. This phenomenon can occur both in the original explant and in the growing shoots. In the case of shoots, this darkening can occur from the tip of the shoot toward its base (Fig. 3.6) or vice versa. This has been associated with the oxidation of phenols or lack of nutrient mobility such as that of calcium, due to

Fig. 3.5 Subculturing of bamboo in different size culture vessels depending on the growth rate. Culture of the sprouted shoot in a 2.5-cm-wide test tube (**a**). Subculture of the newly forming clump in a 5.8-cm-wide culture vessel (**b**). Subculture of the formed rooted clumps in a 9.5-cm-wide culture jar (**c**). *Bar = 1 cm*

Fig. 3.6 Development of shoot-browning in a newly sprouted bamboo shoot. Evolution of symptoms is shown by *thick arrows*. Direction of the necrosis advancement in the shoot is represented by thin arrows. *Bar = 1 cm*

limited transpiration related to the high humidity that exists in the culture vessel (Teixeira da Silva et al. 2020). Strategies to reduce this problem, in different bamboo species, include the use of antioxidants in the culture medium and also transferring the explants more frequently to fresh medium, the latter being the most widely used (Mudoi et al. 2013; Sandhu et al. 2018).

3.4.5 Culture Conditions

Bamboo axillary buds are usually initially placed in the darkness to reduce oxidation. As soon as the buds begin to sprout, usually after 2 weeks, they are moved into light conditions. Light intensity requirements are variable, with literature reporting PAR values between 10 and 140 μmol m^{-2} s^{-1}, mostly using white light (Jiménez and Guevara 2007). In most cases, a 16-h photoperiod is reported, except for a few that used 12 h, for example, in *G. angustifolia* (Marulanda et al. 2005; Jiménez et al. 2006). Mean culture temperature conditions used in most cases are around 25 °C. Culture requirements do not show variations between tropical and temperate climate bamboos.

3.4.6 Multiplication

In contrast to conventional bamboo vegetative propagation methods, one explant (bud) does not only result in a single plant or a limited number of plants, but it has the potential to produce several thousands in few months. Two methods have been explored for in vitro multiplication of bamboo plantlets developing from growing axes. One follows the same procedure used to establish in vitro cultures, by sectioning single nodal segments from the in vitro growing shoots. In the second strategy, the shoot clumps that form after activation of lateral or axillary buds in the newly formed structures of the growing axis, usually with cytokinins, are divided into smaller clumps (Jiménez and Guevara 2007).

Although there are reports in the literature about the use of both systems (Sandhu et al. 2018), in some cases, the division of clumps has proved to generate much higher multiplication rates (e.g., Jiménez et al. 2006; Ornellas et al. 2019). Therefore, this will be the method of choice for further description in this review (Fig. 3.7). When dividing the clumps to produce new individual in vitro plantlets, care should be taken to avoid dissecting structures that only comprise few shoots. According to several reports, less crowded clumps tend to reestablish growth at a slower pace, and therefore, the multiplication rate is lower than when larger clumps are generated (Ravikumar et al. 1998). Depending on the species, it has been reported that clumps

Fig. 3.7 Division of clumps for multiplication of in vitro bamboo plants. Newly formed clump ready to be divided, with cut sites denoted by *dotted lines* (**a**). Recently separated clumps (**b**). Bamboo clumps after 3–6 weeks, ready to be divided again in the following subculture (**c**). *Bar = 1 cm*

ranging from two to ten shoots are suitable for bamboo in vitro multiplication (Sandhu et al. 2018).

Likewise, MS mineral composition predominates over other formulations for this multiplication phase (Sandhu et al. 2018). Regarding the potential effect of the physical consistency of the culture medium, numerous reports recommend the semisolid (agar- or gellan gum-gelled) over the liquid preparations, not only because of increased growth and shoot multiplication rate but also due to the lower hyperhydricity frequency observed in the former (Arya and Sharma 1998; Goyal and Sen 2016; Sandhu et al. 2018; Suwal et al. 2020). However, several publications describe better multiplication rates and superior general condition of the plantlets when they were multiplied in temporary immersion systems, including the absence of physiological disorders like hyperhydricity (Nogueira et al. 2019). Another strategy includes the use of a combination of semisolid medium during the sprouting stage, and then these shoots are transferred to a liquid medium for their multiplication. In some cases, the medium composition remains the same with the removal of the gelling agent as a single change (Gantait et al. 2018), while in others the plant growth regulators are modified as well (Ray et al. 2018). Similar to what was described for the bud sprouting phase, during shoot multiplication sucrose also predominates over other carbohydrate sources (Sandhu et al. 2018).

3.4.7 Rooting

Rooting is a relevant step during the in vitro development of plants. The promotion of functional roots is a determining factor for subsequent acclimatization. Usually, younger tissues tend to root easier than those taken from adult plants. This has been observed in bamboos when seedling shoots are employed for micropropagation protocols (Ramanayake et al. 2006). In addition, an effect of the in vitro ready-to-root plantlet height and the number of shoots in the clump has been associated with the number of in vitro newly developed roots (do Vale et al. 2019). In some cases, rooting has occurred spontaneously during the multiplication phase (i.e., under the same conditions used for lateral bud sprouting and shoot growth, which means on culture medium containing cytokinins) (Jiménez et al. 2006; Ramanayake et al. 2008; Furlan et al. 2018), while in most cases it had to be induced in a separate phase, usually devoid of cytokinins (Ornellas et al. 2019) and with supplemented auxins (Ye et al. 2017; Sandhu et al. 2018; Obsuwan et al. 2019; Rajput et al. 2020).

In vitro plant roots are generally greenish on the surface due to exposure to light (Fig. 3.8a). However, once the plants are acclimatized in the greenhouse, the new roots are whitish (Fig. 3.8b). The functionality of the in vitro greenish roots has not been systematically explored. However, in an experiment conducted in our

Fig. 3.8 Radical development of an in vitro versus an acclimatized bamboo plant (*Guadua angustifolia*). In vitro plant at the multiplication stage (**a**.1) showing greenish roots (**a**.2). Two-month-old acclimatized plant (**b**.1) showing a whitish root system (**b**.2). *Bar = 1 cm*

laboratory, uptake and transport of acid fuchsin from the roots to the leaves were confirmed in both root types in *Guadua angustifolia* (Carvajal-Campos and Holst, unpublished). In addition, histological sectioning revealed they share a similar structure, supporting their functionality (Fig. 3.9).

MS medium continues to be the most recommended mineral salt composition for this phase. Nevertheless, there are several reports that favor reducing the salt concentration of all elements or of only part of them (like nitrogen) to increase the rooting rate (Saxena 1990; Somashekar et al. 2008; Singh et al. 2012; Sandhu et al. 2018). Two to three percent sucrose has been reported adequate to induce rooting in several bamboo publications, with very few reports in which different compounds and/or concentrations were compared vis-à-vis. Moreover, a positive effect of a pre-conditioning phase without sugar has been reported. García-Ramírez et al. (2019) related reduced rooting with the presence of sucrose in the culture medium when acclimatizing *Bambusa vulgaris* in vitro developed plants. They observed a higher rooting frequency after 30 days in the greenhouse when sucrose was not included in the last stages during in vitro culture.

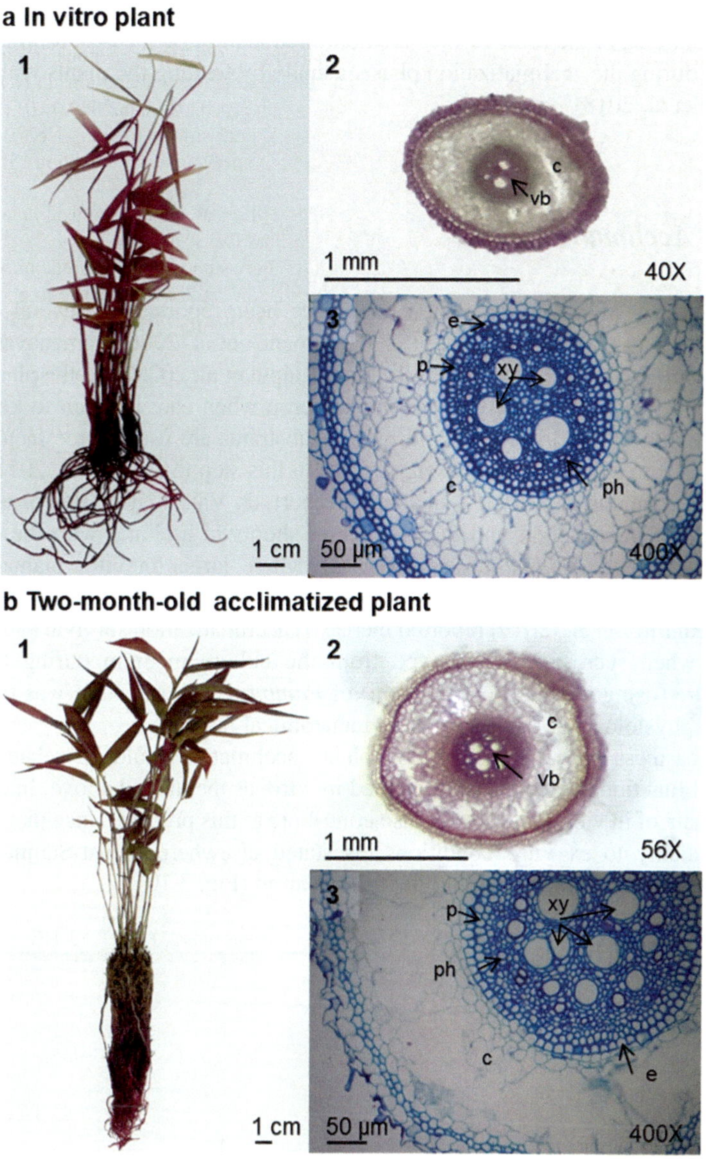

Fig. 3.9 Root functionality and root histology of an in vitro and an acclimatized bamboo plant (*Guadua angustifolia*). In vitro plant at the multiplication stage (**a**.1) and a 2- month-old acclimatized plant (**b**.1) showing stained roots, leaves, and stems after dipping the root distal section in acid fuchsin. In vitro (**a**.2) and acclimatized (**b**.2) root cross-section view after staining in acid fuchsin. In vitro (**a**.3) and acclimatized (**b**.3) microscopic view of the root main structures stained with toluidine blue. *Vb*: *vascular bundle. Xy*: *xylem. E*: *endodermis. C*: *cortex. P*: *pericycle. Ph*: *phloem*

Sometimes it is not possible or extremely hard to induce in vitro root formation at a sufficient rate, or the roots formed are nonfunctional. In such cases, rooting can be induced during the acclimatization phase, usually by treating the plants with auxins (Sandhu et al. 2018).

3.4.8 Acclimatization

Although very high acclimatization rates have been reported for several bamboo species (Bag et al. 2000; Arya et al. 2001; Jiménez et al. 2006; Ramanayake et al. 2006; Singh et al. 2012; do Vale et al. 2019; Rajput et al. 2020) and the plant group has been accepted as an easy-to-acclimatize crop when care is taken to keep high humidity during the first weeks and proper substrates are used, some factors have been related to a lesser or increasing success in this step (Mishra et al. 2011; Singh et al. 2012; Sandhu et al. 2018). In a recent report, do Vale et al. (2019) observed an effect of the plant height and the number of shoots in a clump with the ex vitro plantlet vigor of *Guadua* aff. *chaparensis*, where larger in vitro plants do not necessarily conduct to taller plants after the acclimatization phase. In addition, García-Ramírez et al. (2019) reported increased acclimatization survival and plantlet growth when sucrose was removed from the culture medium during the final subculture (using temporary immersion) of *Bambusa vulgaris*. This was related to morpho-physiological, biochemical, and anatomical factors.

The easiness with which bamboo plants acclimatize could be related to the potential functionality of the roots formed in vitro, as mentioned above. In addition, the stomata of in vitro plants could also contribute to this process, since they seem to adapt quickly to ex vitro conditions, as stated elsewhere (Holst-Sanjuán 2014; Rajput et al. 2020), thus avoiding their desiccation (Fig. 3.10).

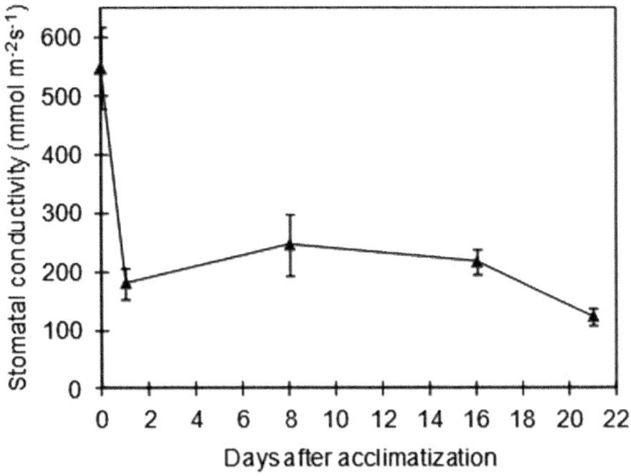

Fig. 3.10 Stomatal conductivity of in vitro *Guadua angustifolia* bamboo plants (day 0) during the acclimatization process (day 1 to 21). Each value shows the mean ± standard error

3.5 Conclusion and Future Perspective

The entire process for bamboo in vitro propagation through axillary buds is shown in Fig. 3.11. The steps involved have been described individually in this chapter and comprises of selection of the explant and its disinfection, induction of bud sprouting

Fig. 3.11 In vitro propagation process of bamboo through axillary buds. From potted plants grown under greenhouse conditions (**a**), nodal sections are selected and disinfected and placed on the culture medium (**b**). Axillary buds sprout in dark conditions (**c**), and the explants are subsequently moved to light conditions (**d**). Several lateral shoots develop from an explant (**e**), which multiply until they form clumps with roots (**f**). Plants are acclimatized under greenhouse conditions (**g**), and they are then transferred to bags to be transplanted in the field (**h**). Six-year-old plant growing in the field (**i**). Direction of the process is indicated by arrows

under dark conditions, growth of the shoots under the light regime, the formation of the clumps and sequential subculture to larger vessels, multiplication of the clumps, and their rooting and acclimatization. Further steps include the sustained plant growth in bags in the greenhouse and their subsequent development in the field.

With the global trend into bioeconomy, renewable resources are becoming more relevant. Bamboos fulfill the conditions of a multipurpose plant with industrial, food, and environmental applications. Therefore, large-scale availability of bamboo stands is of the utmost relevance, and in vitro propagation of selected individuals offers an option to provide planting material in the required numbers. Recent developments include an increase in the number of species for which propagation methodologies have been established and refined, as well as a better understanding of the physiological and developmental processes involved. This progress points out a direction for a wider commercial application of bamboo plant tissue culture in the near future.

Acknowledgments We would like to thank the very active undergraduate and graduate students who contributed during the past years to generate data and discussion around bamboo tissue culture in our lab.

Author Contributions V.M.J. structured the chapter layout. A.H. and P.C.-C. conducted pending experiments to be included in this work. V.M.J., A.H., and P.C.-C. revised literature and wrote the manuscript drafts. All authors revised and corrected subsequent versions. All authors read and approved the final manuscript.

Funding Sources Funding to conduct specific experiments included in this chapter was provided by the University of Costa Rica under the research project VI-734-B7-077.

Conflict of Interest The authors declare that preparation of this manuscript was conducted in the absence of any commercial or financial relationships that could be construed as a potential conflict of interest.

References

Alexander MP, Rao TCR (1968) In vitro culture of bamboo embryos. Curr Sci 37:415

Ali AH, Nirmala C, Badal T, Sharma ML (2009) In-vitro organogenesis and simultaneous formation of shoots and roots from callus in *Dendrocalamus asper*. In: 8th World Bamboo Congress. Thailand, pp 31–40

Arya ID, Satsangi R, Arya S (2001) Rapid micropropagation of edible bamboo *Dendrocalamus asper*. J Sustain For 14:103–114

Arya S, Sharma S (1998) Micropropagation technology of *Bambusa bambos* through shoot proliferation. Indian For 124:725–731

Bag N, Chandra S, Palni LMS, Nandi SK (2000) Micropropagation of Dev-ringal [*Thamnocalamus spathiflorus* (Trin.) Munro]—a temperate bamboo, and comparison between in vitro propagated plants and seedlings. Plant Sci 156:125–135

Banik RL (2015) Morphology and growth. In: Liese W, Köhl M (eds) Bamboo: the plant and its uses. Springer International Publishing, Cham, pp 43–89

Bednarek PT, Orłowska R (2020) Plant tissue culture environment as a switch-key of (epi)genetic changes. Plant Cell Tissue Organ Cult 140:245–257. https://doi.org/10.1007/s11240-019-01724-1

Cappelletti S, Piacentino D, Fineschi V, Frati P, D'Errico S, Aromatario M (2019) Mercuric chloride poisoning: symptoms, analysis, therapies, and autoptic findings. A review of the literature. Crit Rev Toxicol 49:329–341. https://doi.org/10.1080/10408444.2019.1621262

Chaturvedi HC, Sharma M, Sharma AK (1993) In vitro regeneration of *Dendrocalamus strictus* Nees through nodal segments taken from field-grown culms. Plant Sci 91:97–101

Choudhary A, Priyanka K, Ashish R (2017) Refinement of protocol for rapid clonal regeneration of economical bamboo, *Bambusa balcooa* in the agroclimatic conditions of Bihar, India. Afr J Biotechnol 16:450–462

do Vale PAA, de Oliveira Júnior JB, da Costa FHS, Scherwinski-Pereira JE (2019) Height and number of shoots on the survival and development of micropropagated bamboo plantlets during pre-acclimatization. Pesqui Agropecuária Trop 49:e53751. https://doi.org/10.1590/1983-40632019v4953751

dos Santos DWR, Rocker TP, Ornellas TS, Guerra MP (2019) Effects of a commercial biocide, casugamycin and consistency of the culture medium in the in vitro establishment of bamboo. Pesqui Agropecuária Trop 49:e55435. https://doi.org/10.1590/1983-40632019v4955435

Furlan FC, Gavilan NH, Zorz AZ, de Oliveira LS, Konzen ER, Brondani GE (2018) Active chlorine and charcoal affect the in vitro culture of *Bambusa vulgaris*. Bosque Valdivia 39:61–70. https://doi.org/10.4067/S0717-92002018000100061

Gamborg OL, Miller RA, Ojima K (1968) Nutrient requirements of suspension cultures of soybean root cells. Exp Cell Res 50:151–158

Gantait S, Pramanik BR, Banerjee M (2018) Optimization of planting materials for large scale plantation of *Bambusa balcooa* Roxb.: influence of propagation methods. J Saudi Soc Agric Sci 17:79–87. https://doi.org/10.1016/j.jssas.2015.11.008

García-Ramírez Y, Barrera GP, Freire-Seijo M, Barbón R, Concepción-Hernández M, Mendoza-Rodríguez MF, Torres-García S (2019) Effect of sucrose on physiological and biochemical changes of proliferated shoots of *Bambusa vulgaris* Schrad. Ex Wendl in temporary immersion. Plant Cell Tissue Organ Cult 137:239–247. https://doi.org/10.1007/s11240-019-01564-z

George EF, Hall MA, Klerk G-JD (2008) Stock plant physiological factors affecting growth and morphogenesis. In: George EF, Hall MA, Klerk G-JD (eds) Plant propagation by tissue culture, vol 1. The Background. Springer Netherlands, Dordrecht, pp 403–422

Gielis J, Oprins J (2002) Micropropagation of temperate and tropical woody bamboos—from biotechnological dream to commercial reality. In: Bamboo for Sustainable Development. Proceedings of the Vth International Bamboo Congress and the VIth International Bamboo Workshop. San José, Costa Rica, pp 333–344

Goyal AK, Sen A (2016) In vitro regeneration of bamboos, the "green gold": an overview. Indian J Biotechnol 15:9–16

Holst-Sanjuán A (2014) Aislamiento, purificación y cultivo de protoplastos en *Guadua angustifolia* Kunth (Poaceae: Bambusoideae). Thesis M.Sc., Universidad de Costa Rica

Huang L-C, Huang B-L, Chen W-L (1989) Tissue culture investigations of bamboo—IV. Organogenesis leading to adventitious shoots and plants in excised shoot apices. Environ Exp Bot 29:307–315. https://doi.org/10.1016/0098-8472(89)90004-X

Ikeuchi M, Ogawa Y, Iwase A, Sugimoto K (2016) Plant regeneration: cellular origins and molecular mechanisms. Development 143:1442–1451. https://doi.org/10.1242/dev.134668

Jiménez VM, Castillo J, Tavares E, Guevara E, Montiel M (2006) In vitro propagation of the neotropical giant bamboo, *Guadua angustifolia* Kunth, through axillary shoot proliferation. Plant Cell Tissue Organ Cult 86:389–395

Jiménez VM, Guevara E (2007) Micropropagation of bamboo species through axillary shoot proliferation. In: Jain SM, Haggman H (eds) Protocols for micropropagation of Woody trees and fruits. Springer, pp 465–476

Kabade AU (2009) Studies on refinement of protocols for rapid and mass in vitro clonal propagation, evaluation of genetic fidelity and growth performance of bamboo species—*Bambusa bambos* (L.) Voss and *Dendrocalamus strictus* (Roxb.) Nees. Forest Research Institute Dehradun

Marulanda ML, Gutiérrez LG, del Márquez MP (2005) Micropropagación de *Guadua angustifolia* Kunth. Actual Biol 27:5–15

McCown BH, Lloyd G (1981) Woody plant medium (WPM)-a mineral nutrient formulation for microculture of woody plant-species. HortScience 16:453–453

Mehta U, Rao IVR, Ram HYM (1982) Somatic embryogenesis in bamboo. In: Proceedings 5th International Congress Plant Tissue and Cell Culture. Japanese Association for Plant Tissue Culture, Tokyo and Lake Yamanaka, pp 109–110

Mishra Y, Patel P, Ansari SA (2011) Acclimatization and macroproliferation of micropropagated plants of *Bambusa tulda* Roxb. Asian J Exp Biol Sci 2:498–501

Mudoi KD, Saikia SP, Goswami A, Gogoi A, Bora D, Borthakur M (2013) Micropropagation of important bamboos: a review. Afr J Biotechnol 12:2770–2785

Murashige T, Skoog F (1962) A revised medium for rapid growth and bio assays with tobacco tissue cultures. Physiol Plant 15:473–497

Nadgauda RS, Parasharami VA, Mascarenhas AF (1990) Precocious flowering and seeding behaviour in tissue-cultured bamboos. Nature 344:335–336

Nadgir AL, Phadke CH, Gupta PK, Parsharami VA, Nair S, Mascarenhas AF (1984) Rapid multiplication of bamboo by tissue culture. Silvae Genet 33:219–233

Negi D, Saxena S (2011) In vitro propagation of *Bambusa nutans* Wall. ex Munro through axillary shoot proliferation. Plant Biotechnol Rep 5:35–43. https://doi.org/10.1007/s11816-010-0154-z

Niedz RP, Bausher MG (2002) Control of in vitro contamination of explants from greenhouse- and field-grown trees. In Vitro Cell Dev Biol Plant 38:468–471. https://doi.org/10.1079/IVP2002316

Nogueira JS, da Costa FS, do Vale PAA, Luis ZG, Scherwinski-Pereira JE (2017) Micropropagação de bambu em largaescala: princípios, estratégias e desafios. In: Drumond PM, Wiedman G (eds) Bambus no Brasil: da biologia à tecnologia. Embrapa, Rio Branceo, pp 103–129

Nogueira JS, Gomes HT, Scherwinski-Pereira JE (2019) Micropropagation, plantlets production estimation and ISSR marker-based genetic fidelity analysis of *Guadua magna* and *G. angustifolia*. Pesqui Agropecuária Trop 49:e53743

Obsuwan K, Duangmanee A, Thepsithar C (2019) In vitro propagation of a useful tropical bamboo, *Thyrsostachys siamensis* Gamble, through shoot-derived callus. Hortic Environ Biotechnol 60:261–267. https://doi.org/10.1007/s13580-018-00119-z

Orlikowska T, Nowak K, Reed B (2017) Bacteria in the plant tissue culture environment. Plant Cell Tissue Organ Cult 128:487–508. https://doi.org/10.1007/s11240-016-1144-9

Ornellas TS, Marchetti CK, de Oliveira GH, Fritsche Y, Guerra MP (2019) Micropropagation of *Guadua chacoensis* (Rojas) Londoño & P. M. Peterson. Pesqui Agropecuária Trop 49:e55450. https://doi.org/10.1590/1983-40632019v4955450

Pasqualini APA, Schneider GX, Fraga HPF, Biasi LA, Quoirin M (2019) In vitro establishment of *Bambusa oldhamii* Munro from field-grown matrices and molecular identification of endophytic bacteria. Pesqui Agropecuária Trop 49:e53673. https://doi.org/10.1590/1983-40632019v4953673

Quambusch M, Winkelmann T (2018) Bacterial endophytes in plant tissue culture: mode of action, detection, and control. In: Loyola-Vargas VM, Ochoa-Alejo N (eds) Plant cell culture protocols. Springer, New York, pp 69–88

Rajput BS, Jani M, Ramesh K, Manokari M, Jogam P, Allini VR, Kher MM, Shekhawat MS (2020) Large-scale clonal propagation of *Bambusa balcooa* Roxb.: an industrially important bamboo species. Ind Crop Prod 157:112905. https://doi.org/10.1016/j.indcrop.2020.112905

Ramakrishnan M, Yrjälä K, Vinod KK, Sharma A, Cho J, Satheesh V, Zhou M (2020) Genetics and genomics of moso bamboo (*Phyllostachys edulis*): current status, future challenges, and biotechnological opportunities toward a sustainable bamboo industry. Food Energy Secur 9: e229. https://doi.org/10.1002/fes3.229

Ramanayake SMSD, Maddegoda KMMN, Vitharana MC, Chaturani GDG (2008) Root induction in three species of bamboo with different rooting abilities. Sci Hortic 118:270–273

Ramanayake SMSD, Meemaduma VN, Weerawardene TE (2006) In vitro shoot proliferation and enhancement of rooting for the large-scale propagation of yellow bamboo (*Bambusa vulgaris* 'Striata'). Sci Hortic 110:109–113

Ramanayake SMSD, Yakandawala K (1997) Micropropagation of the giant bamboo (*Dendrocalamus giganteus* Munro) from nodal explants of field grown culms. Plant Sci 129:213–223

Ramanayake SMSD, Yakandawala K, Nilmini Deepika PKD, Ikbal MCM (1995) Studies on micropropagation of *Dendrocalamus giganteus* and *Bambusa vulgaris* var. striata. In: Bamboo, People and the Environment. INBAR, pp 75–85

Ravikumar R, Ananthakrishnan G, Kathiravan K, Ganapathi A (1998) In vitro shoot propagation of *Dendrocalamus strictus* Nees. Plant Cell Tissue Organ Cult 52:189–192

Ray SS, Ali MN (2017) Factors affecting macropropagation of bamboo with special reference to culm cuttings: a review update. N Z J For Sci 47:17. https://doi.org/10.1186/s40490-017-0097-z

Ray SS, Ali N (2016) Biotic contamination and possible ways of sterilization: a review with reference to bamboo micropropagation. Braz Arch Biol Technol 59:e160485. https://doi.org/10.1590/1678-4324-2016160485

Ray SS, Ali N, Banerjee M, Yeasmin L (2018) In vitro and in vivo assessment of thidiazuron mediated micro-clones of *Dendrocalamus asper*, an ornamental bamboo species. J Crop Weed 14:150–157

Rosvall O, Bradshaw RHW, Egertsdotter U, Ingvarsson PK, Mullin TJ, Wu H (2019) Using Norway spruce clones in Swedish forestry: implications of clones for management. Scand J For Res 34:390–404. https://doi.org/10.1080/02827581.2019.1590631

Sandhu M, Wani SH, Jiménez VM (2018) In vitro propagation of bamboo species through axillary shoot proliferation: a review. Plant Cell Tissue Organ Cult 132:27–53. https://doi.org/10.1007/s11240-017-1325-1

Sanjaya, Rathore TS, Rai VR (2005) Micropropagation of *Pseudoxytenanthera stocksii* Munro. In Vitro Cell Dev Biol-Plant 41:333–337

Saxena S (1990) In vitro propagation of the bamboo (*Bambusa tulda* Roxb.) through shoot proliferation. Plant Cell Rep 9:431–434

Saxena S, Bhojwani SS (1993) In vitro clonal multiplication of 4-year-old plants of the bamboo, *Dendrocalamus longispathus* Kurz. In Vitro Cell Dev Biol Plant 29:135–142

Schenk RU, Hildebrandt AC (1972) Medium and techniques for induction and growth of monocotyledonous and dicotyledonous plant cell cultures. Can J Bot 50:199–204

Silveira AAC, Lopes FJF, Sibov ST (2020) Micropropagation of *Bambusa oldhamii* Munro in heterotrophic, mixotrophic and photomixotrophic systems. Plant Cell Tissue Organ Cult 141:315–326. https://doi.org/10.1007/s11240-020-01788-4

Singh SR, Dalal S, Singh R, Dhawan AK, Kalia RK (2012) Micropropagation of *Dendrocalamus asper* {Schult. & Schult. F.} Backer ex k. Heyne: an exotic edible bamboo. J Plant Biochem Biotechnol 21:220–228

Somashekar PV, Rathore TS, Shashidhar KS (2008) Rapid and simplified method of micropropagation of *Pseudoxytenanthera stocksii*. In: Ansari SA, Narayanan C, Mandal AK (eds) Forest biotechnology in India. Satish Serial Publishing House, Delhi, pp 165–182

Soreng RJ, Peterson PM, Romaschenko K, Davidse G, Zuloaga FO, Judziewicz EJ, Filgueiras TS, Davis JI, Morrone O (2015) A worldwide phylogenetic classification of the Poaceae (Gramineae). J Syst Evol 53:117–137. https://doi.org/10.1111/jse.12150

Suwal MM, Lamichhane J, Gauchan DP (2020) Regeneration technique of bamboo species through nodal segments: a review. Nepal J Biotechnol 8:54–68

Teixeira da Silva JA, Nezami-Alanagh E, Barreal ME, Kher MM, Wicaksono A, Gulyás A, Hidvégi N, Magyar-Tábori K, Mendler-Drienyovszki N, Márton L, Landín M, Gallego PP,

Driver JA, Dobránszki J (2020) Shoot tip necrosis of in vitro plant cultures: a reappraisal of possible causes and solutions. Planta 252:47. https://doi.org/10.1007/s00425-020-03449-4

Thapa P, Bhattacharya A, Sood P, Devi K, Sood A (2018) Advances in bamboo biotechnology: present status and future perspective. In: Gosal SS, Wani SH (eds) Biotechnologies of crop improvement, vol 1.: Cellular Approaches. Springer International Publishing, Cham, pp 243–265

Torres GRDC, Coutinho FP, Araújo BGP, Abreu GPD, Sales Júnior R (2019) Thermotherapy as a microbial contaminant-reducing agent in micropropagation of bamboo. Rev Caatinga 32:690–697. https://doi.org/10.1590/1983-21252019v32n313rc

Vongvijitra R (1988) Traditional vegetative propagation and tissue culture of some Thai bamboos. In: Proceedings of the International Bamboo Workshop. The Kerala Forest Research Institute, Cochin, pp 148–150

Ye S, Cai C, Ren H, Wang W, Xiang M, Tang X, Zhu C, Yin T, Zhang L, Zhu Q (2017) An efficient plant regeneration and transformation system of Ma bamboo (*Dendrocalamus latiflorus* Munro) started from young shoot as explant. Front Plant Sci 8:1298. https://doi.org/10.3389/fpls.2017.01298

Zang Q, Liu Q, Zhuge F, Wang X, Lin X (2019) In vitro regeneration via callus induction in *Dendrocalamus asper* (Schult.) Backer. Propag Ornam Plants 19:66–71

Zhang Z, Sun Y, Li Y (2020) Plant rejuvenation: from phenotypes to mechanisms. Plant Cell Rep 39:1249–1262. https://doi.org/10.1007/s00299-020-02577-1

Chapter 4
Somatic Embryogenesis in Bamboos: Advances and Prospects

Luiza Giacomolli Polesi, Leila do Nascimento Vieira, Miguel Pedro Guerra, and Hugo Pacheco de Freitas Fraga

Abstract Bamboos are one of the largest Poaceae groups and comprise about 1600 species distributed worldwide. Many of these species are sources of food, handicrafts, construction, charcoal, paper, and fiber. However, the obtainment of financially viable and high-quality seedlings of bamboo is not an easy task. Micropropagation has been successfully used to obtain large-scale bamboo seedlings and is one of the most promising techniques to overcome the bottlenecks in the propagation of these species. Two morphogenetic routes are being used in bamboo micropropagation: organogenesis and somatic embryogenesis (SE). In SE, one or more somatic cells differentiate into somatic embryos, resembling the structure of the zygotic embryo, with bipolar axes that are not vascular connected with the source tissue. Historically, SE has been neglected in comparison to organogenesis, possibly due to its greater complexity and difficulty in control and modulation. So far, ~30 bamboo species have established SE protocols, most of which from genera *Bambusa* and *Dendrocalamus*. This chapter discusses the principal advances in bamboo SE, with emphasis on complementary approaches as morpho-anatomy, biochemistry, and molecular biology. Also, we outline the main constraints and limitations of this promising technique and new paths to develop more efficient protocols.

Keywords Bamboo biotechnology · Genetic fidelity · Micropropagation · Morpho-anatomy · Somatic embryogenesis

---

L. G. Polesi
Graduate Program in Plant Genetic Resources, Laboratory of Plant Developmental Physiology and Genetics, Federal University of Santa Catarina, Florianópolis, SC, Brazil

L. do Nascimento Vieira · H. Pacheco de Freitas Fraga (✉)
Department of Botany, Federal University of Paraná, Curitiba, PR, Brazil
e-mail: hugofraga@ufpr.br

M. P. Guerra
Graduate Program in Plant Genetic Resources, Laboratory of Plant Developmental Physiology and Genetics, Federal University of Santa Catarina, Florianópolis, SC, Brazil

Graduate Program in Agricultural and Natural Ecosystems, Federal University of Santa Catarina, Curitibanos Campus, Curitibanos, SC, Brazil

Z. Ahmad et al. (eds.), *Biotechnological Advances in Bamboo*,
https://doi.org/10.1007/978-981-16-1310-4_4

4.1 Introduction

Due to the sessile condition of plants, they need to adjust to the different variations of the surrounding environment. The impossibility of migration and the need for continuous organogenesis during its development lead to the maintenance of undifferentiated cell niches with morphogenetic potential, the meristems. Thus, the ontogenetic program of plants is highly flexible and is closely linked to the reversibility of somatic cell differentiation (Fehér 2019). Somatic embryogenesis (SE) is a complete model of cell totipotency in which a somatic cell development is restructured toward an embryogenic route (Karami et al. 2009). This morphogenetic route is dependent on the reprogramming of gene expression and the action of a complex signaling network (Méndez-Hernández et al. 2019). Somatic embryogenesis follows a morphogenetic program analogous to zygotic embryos development after egg cell fertilization, presenting similar stages of embryonic development (Zimmerman 1993). However, unlike their zygotic analogous, somatic embryos can be easily manipulated and culture conditions controlled.

Somatic embryogenesis is associated with a myriad of applications: (1) obtaining a model system for fundamental studies in morpho-anatomy, physiology, biochemistry, and molecular biology; (2) large-scale propagation, synthetic seeds, and secondary metabolite production; (3) integration to breeding programs and genetic engineering; and (4) in vitro germplasm conservation by slow growth and cryopreservation (Karami et al. 2009; Guzmán-García et al. 2013). A summary of these applications is in Fig. 4.1.

To present and discuss the main achievements obtained for bamboo SE, we indicate the strengths, weaknesses, and opportunities that this morphogenetic route can provide to the advancement of knowledge in bamboo biotechnology.

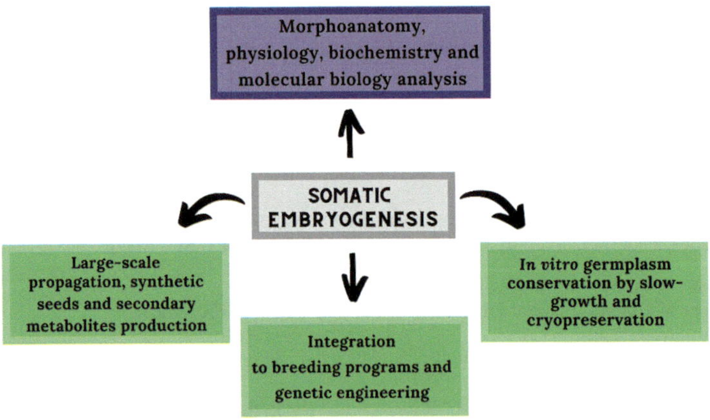

Fig. 4.1 Somatic embryogenesis main applications. The purple box indicates the application of somatic embryogenesis for fundamental studies. The green boxes illustrate the possibilities that somatic embryogenesis provides for applied studies

4.2 Timeline of Bamboo Somatic Embryogenesis

Since the first SE report for bamboo species (Mehta et al. 1982), 38 years have passed, and a lot of effort has been made, but the number of species with established protocols is still limited. Nowadays, there are 29 species of bamboo with published SE protocols (Fig. 4.2, Table 4.1). Among them, 65% used bamboo species belonging to the genera *Bambusa* (41%) and *Dendrocalamus* (24%).

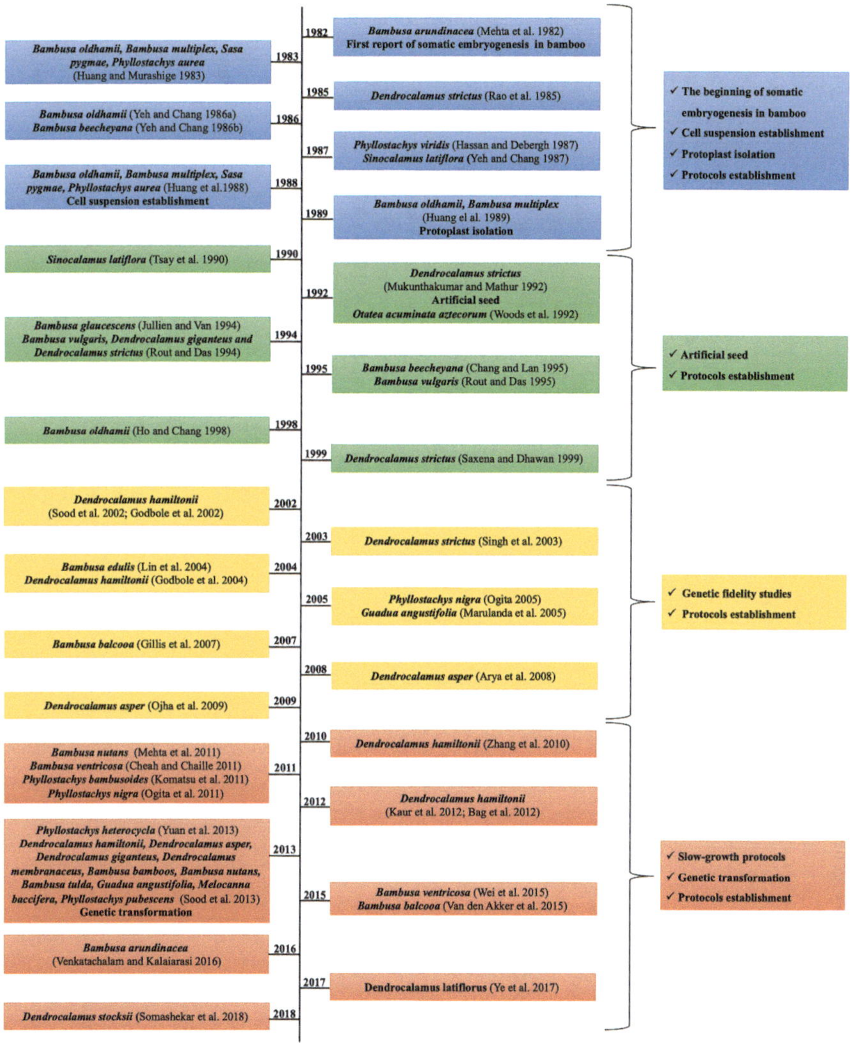

Fig. 4.2 Timeline of bamboo somatic embryogenesis studies from the first report to the present

Table 4.1 Somatic embryogenesis (SE) protocols published for bamboo species

Specie	Explant used	Medium	Response	References
Bambusa arundinacea	Zygotic embryo	N6 + 2,4-D 2,4-D + BA + PVP	Somatic embryogenesis	Mehta et al. (1982)
	Seed	MS + 1.0 mg. l^{-1} of 2,4-D and BAP MS + 1.0 mg l^{-1} of BAP, 2,4-D, NAA + 3% sucrose MS + 1.0 mg l^{-1} of BAP, 2,4-D, NAA + 3% sucrose	Embryogenic callus Embryo maturation Germination	Venkatachalam and Kalaiarasi (2016)
Bambusa balcooa	Pseudospikelets	MS + 1.0 mg l^{-1} 2,4-D + 3% sucrose MS + 5.0 mg l^{-1} BAP + 3% sucrose	Embryogenic callus Germination	Gillis et al. (2007)
Bambusa beecheyana	Inflorescence	MS + 3.0 mg l^{-1} 2,4-D + 2 mg l^{-1} KIN +6% sucrose	Somatic embryogenesis	Yeh and Chang (1986b)
	Roots of in vitro plantlets	MS + 3.0 mg l^{-1} 2,4-D + 2 mg l^{-1} KIN +6% sucrose	Somatic embryogenesis	Chang and Lan (1995)
Bambusa Edulis	Nodal and internodal segments of in vitro plantlets	MS + 13.6 μM 2,4-D + 9.2 μM KIN +0.1% coconut milk +6% sucrose MS + 13.6 μM 2,4-D + 0.046 μM TDZ + 6% sucrose MS + 0.455 μM TDZ + 3% sucrose	Embryogenic callus Embryo maturation Germination	Lin et al. (2004)
Bambusa glaucescens	Young leaves of in vitro shoots	MS + 18 μM 2,4-D + 3% sucrose MS + 9.0 μM 2,4-D + 1.14 μM IAA + 0.22 μM BA + 3% sucrose MS + 9.0 μM 2,4-D + 2.85 μM IAA + 2.22 μM BA +0.6 μM ABA +3% sucrose	Callus induction Callus multiplication Embryo formation	Jullien and Van (1994)
Bambusa nutans	Sprouted buds	MS + 5.0 mg l^{-1} 2,4-D + 3% sucrose MS + 1.0 mg l^{-1} of BAP and 2,4-D + 20 mg l^{-1} ascorbic acid +2% glucose	Embryogenic callus Embryo maturation and germination	Mehta et al. (2011)

Bambusa oldhamii	Young inflorescence	MS + 3.0 mg l^{-1} 2,4-D + 2.0 mg. l^{-1} KIN +6% sucrose	Somatic embryogenesis	Yeh and Chang (1986a)
B. oldhamii, B. multiplex, Phyllostachys aurea, Sasa pygmaea	Shoot tips	MS + 3.0 mg l^{-1} 2,4-D + 3% sucrose	Callus induction	Huang and Murashige (1983)
Bambusa ventricosa	Elongated in vitro shoots	MS + 3.0 mg l^{-1} 2,4-D + 2.0 mg. l^{-1} KIN +3% sucrose MS + 1.0 mg l^{-1} 2,4-D + 1.0 mg. l^{-1} KIN +3% sucrose MS + 1,5 mg l^{-1} BAP + 0.2 mg l^{-1} IBA + 3% sucrose WPM + 0.5 mg l^{-1} BAP + 3% sucrose	Callus induction Callus multiplication Embryo formation/ maturation Germination	Cheah and Chaille (2011)
	Sterile buds	MS + 27.0 μM 2,4-D + 2.7 μM NAA + 0.0045 μM TDZ + 3% sucrose MS + 22.6 μM 2,4-D + 2.2 μM 6-BA +5.4 μM NAA + 3% sucrose MS + 13.3 μM 6-BA +2.7 μM NAA + 3% sucrose	Callus induction Embryo formation/ maturation Germination	Wei et al. (2015)
Bambusa vulgaris, Dendrocalamus giganteus, and *D. strictus*	Nodal tissues and excised zygotic embryo	MS + 3.0 mg l^{-1} 2,4-D + 0.25 mg. l^{-1} KIN +3% sucrose ½ MS + 2.0 mg. l^{-1} 2,4-D + 0.5 mg l^{-1} KIN +10 mg l^{-1} adenine sulfate +3% sucrose	Callus induction Embryo formation/ maturation/ germination	Rout and Das (1994)
Bambusa vulgaris	Zygotic embryos	MS + 3.0 mg l^{-1} 2,4-D + 0.25 mg. l^{-1} KIN +3% sucrose ½ MS + 2.0 mg. l^{-1} 2,4-D + 0.5 mg l^{-1} KIN +10.0 mg. l^{-}adenine sulfate +3% sucrose MS + 0.25 mg l^{-1} IBA + 0.5 mg l^{-1}	Callus induction Embryo formation/ maturation Germination	Rout and Das (1995)

(continued)

Table 4.1 (continued)

Specie	Explant used	Medium	Response	References
		adenine sulfate + 0.5 mg l^{-1} GA_3		
Dendrocalamus asper	Nodal tissues and basal parts of leaves	MS + 30.0 μM 2,4-D + 2% sucrose MS + 9.0 μM 2,4-D + 2,85 μM IAA + 0.88 μM BAP + 2% sucrose MS + 5.0 μM ABA +6% sucrose MS + 4.4 μM BAP + 2.8 μM GA_3 + 2% sucrose	Callus induction Embryo formation Maturation Germination	Arya et al. (2008)
	Roots, leaves, and nodal segments	MS + 30.0 μM 2,4-D + 3% sucrose MS + 20.0 μM BAP + 3% sucrose	Embryogenic callus Germination	Ojha et al. (2009)
Dendrocalamus hamiltonii	Sprouted buds from selected nodal explants	MS + 1.0 mg l^{-1} 2,4-D + 1.0 mg. l^{-1} BA +3% sucrose MS + 1.0 mg l^{-1} 2,4-D + 1.0 mg. l^{-1} BA +0.5 mg. l^{-1} GA_3 + 3% sucrose MS + 3% sucrose	Callus induction Embryo formation/ maturation Germination	Sood et al. (2002)
	New sprouts from nodal segments	MS + 1.0 mg l^{-1} 2,4-D + 1.0 mg. l^{-1} BA +3% sucrose ½ MS + 2.5 mg. l^{-1} BA +3% sucrose MS + 8% sucrose	Callus induction Embryo formation Maturation/ germination	Godbole et al. (2002)
	New sprouts from nodal segments	½ MS + 1.0 mg l^{-1} 2,4-D + 1.0 mg l^{-1} BA +1.0 mg l^{-1} NAA + 3% sucrose ½ MS + 1.0 mg. l^{-1} 2,4-D + 1.0 mg l^{-1} BA +3% sucrose ½ MS + 2.5 mg. l^{-1} BA+ 3% sucrose MS + 8% sucrose	Callus induction Callus multiplication Embryo formation/ maturation Germination	Godbole et al. (2004)
	Mature zygotic embryos	MS + 1.0–3.0 mg. l^{-1} 2,4-D + 3% sucrose MS + 3.0 mg l^{-1} BA +3.0 mg l^{-1} KIN +3% sucrose	Callus induction Embryo formation,	Zhang et al. (2010)

			maturation, and germination	
	Nodal segments	MS + 1.0 mg l^{-1} 2,4-D + 1.0 mg. l^{-1} BAP + 2% sucrose MS + 1.0 mg l^{-1} BAP + 2% sucrose	Embryogenic callus Maturation and germination	Kaur et al. (2012)
	Sprouted buds	MS + 5.0 μM 2,4-D + 5.0 μM BAP + 2% sucrose	Somatic embryogenesis	Bag et al. (2012)
Dendrocalamus hamiltonii, *D. asper*, *D. giganteus*, *D. membranaceus*, *Guadua angustifolia*, *Melocanna baccifera*, *Phyllostachys pubescens*, *Bambusa bambos*, *B. nutans*, *B. tulda*	Nodal segments	MS + 2,4-D + BAP + 3% sucrose MS + BAP + 3% sucrose MS + GA_3 + 3% sucrose MS + 8% sucrose	Callus induction Embryo formation Maturation Germination	Sood et al. (2013)
Dendrocalamus latiflorus	Young shoots	MS + 8.0 mg l^{-1} 2,4-D + 0.5 mg. l^{-1} IBA + 3% sucrose ¾ MS + 2.0 mg l^{-1} 2,4-D + 250 mg l^{-1} PVP + 3 g l^{-1} sorbitol +3% sucrose MS + 2.0 mg l^{-1} BAP + 0.5 mg l^{-1} NAA + 3% sucrose	Callus induction Embryogenic callus Maturation/ germination	Ye et al. (2017)
Dendrocalamus stocksii	Nodal segments	MS + 2.1 μM 2,4-D + 0.10 μM KIN +10% coconut milk +3% sucrose MS + 2.1 μM 2,4-D + 10% coconut milk +3% sucrose MS + 0.25 μM NAA + 0.49 μM BAP + 3% sucrose	Callus induction Embryo formation/ maturation Germination	Somashekar et al. (2018)
Dendrocalamus strictus	Seeds	B5 + 10^{-5} M or 3x10 $^{-5}$ M of 2,4-D + 2% sucrose B5 + 5 × 10^{-7} M IBA + 10^{-7} M NAA + 2% sucrose	Embryogenic callus Germination	Rao et al. (1985)
	Seeds	MS + 3.0 mg l^{-1} 2,4-D + 0.5 mg. l^{-1} KIN +3% sucrose	Embryogenic callus Germination	Mukunthakumar and Mathur (1992)

(continued)

Table 4.1 (continued)

Specie	Explant used	Medium	Response	References
		MS + 1.0 mg l^{-1} NAA + 0.5 mg l^{-1} KIN +3% sucrose		
	Seeds	MS + 3 × 10^{-5} M 2,4-D + 3% sucrose MS + 1 × 10^{-5} M 2,4-D + 5 × 10^{-6} M KIN +2x10^{-6} M IBA + 250 mg l^{-1} PVP + 3% sucrose MS + 5 × 10^{-6} M NAA + 5 × 10^{-6} M KIN +250 mg l^{-1} PVP + 3% sucrose	Embryogenic callus Multiplication/maturation Germination	Saxena and Dhawan (1999)
Guadua angustifolia	Axillary buds	MS + 6.0 mg l^{-1} 2,4-D + 3% sucrose MS + 0.5 mg l^{-1} AIA + 0.5 mg l^{-1} GA_3 + 3% sucrose	Embryogenic callus Embryo maturation	Marulanda et al. (2005)
Otatea acuminata aztecorum	Seed	MS + 3.0 mg l^{-1} 2,4-D + 0.5 mg. l^{-1} BAP + 2% sucrose	Somatic embryogenesis	Woods et al. (1992)
Phyllostachys bambusoides	In vitro leaf sheath from nodal segments	MS + 8.0 mg l^{-1} Picloram +2% sucrose	Callus induction and embryo formation	Komatsu et al. (2011)
Phyllostachys heterocycla	Zygotic embryo	MS + 4.0 mg l^{-1} 2,4-D + 0.1 mg. l^{-1} ZT + 3% sucrose MS + 0.5–2.0 mg. l^{-1} 2,4-D + 3% sucrose MS + 5.0–7.0 mg. l^{-1} ZT + 3% sucrose	Callus induction Embryogenic callus Maturation/ germination	Yuan et al. (2013)
Phyllostachys nigra	Shoots	½ MS + 3.0 μM 2,4-D + 3% sucrose	Callus induction and embryo formation	Ogita (2005)
Phyllostachys viridis	Leaf explants	MS + 9 × 10^{-6} M 2,4-D + 2% sucrose MS + 2% sucrose	Embryogenic callus Germination	Hassan and Debergh (1987)
Sasa pygmaea	Shoot tips	MS + 3.0 mg l^{-1} 2,4-D + 3% sucrose	Callus induction	Huang and Murashige (1983)

Sinocalamus latiflora	Anthers	N6 + 1.0 mg l^{-1} 2,4-D + 1.0 mg. l^{-1} BA +2.0 g l^{-1} activated charcoal +9% sucrose N6 + 1.0 mg l^{-1} BA +2.0 g l^{-1} charcoal +9% sucrose	Embryogenic callus Germination	Tsay et al. (1990)
	Mature zygotic embryo	MS + 6.0 mg l^{-1} 2,4-D + 3.0 mg. l^{-1} KIN +250 mg. l^{-1} PVP + 5% sucrose MS + 3.0 mg l^{-1} 2,4-D + 2.0 mg. l^{-1} KIN +5% sucrose	Embryogenic callus Germination	Yeh and Chang (1987)

N6 Chu et al. (1975), *MS* Murashige and Skoog (1962), *WPM* woody plant medium (Lloyd and McCown 1981), *B5* Gamborg O.L (1968), *2,4-D* 2,4-dichlorophenoxyacetic acid, *BA* benzyladenine, *PVP* polyvinylpyrrolidone, *BAP* 6-benzylaminopurine, *NAA* 1-naphthaleneacetic acid, *KIN* Kinetin, *TDZ* Thidiazuron, *IAA* 3-indoleacetic acid, *ABA* abscisic acid, *IBA* 3-butyric acid, *6-BA* 6-benzyladenine, GA_3 gibberellic acid, *ZT* zeatin

The timeline of the SE in bamboo initiated when Mehta et al. (1982) described an SE protocol for *Bambusa arundinacea* (synonym of *Bambusa bambos*) using zygotic embryos as explant and N6 culture medium (Chu et al. 1975) supplemented with 2,4-dichlorophenoxyacetic acid (2,4-D), 6-benzylaminopurine (BA), and polyvinylpyrrolidone (PVP). After that, most of the protocols used the following explant sources: (1) seeds (Rao et al. 1985), (2) zygotic embryos (Yeh and Chang 1987), (3) inflorescences (Yeh and Chang 1986a, 1986b), and (4) leaves (Hassan and Debergh 1987). At the end of the 1980s, nine species of bamboo had established SE protocols, e.g., *Bambusa arundinacea*, *B. beecheyana*, *B. multiplex*, *B. oldhamii*, *Dendrocalamus strictus*, *Phyllostachys aurea*, *P. viridis*, *Sasa pygmaea* (synonym of *Pleioblastus fortunei*), and *Sinocalamus latiflora* (synonym of *Dendrocalamus latiflorus*). Additionally, during this decade, the establishment of cell suspension (Huang et al. 1988) and protoplast isolation (Huang et al. 1989) from bamboo cells (Huang and Murashige 1983) represented relevant progress for further studies.

In the 1990s, protocols for four other species were reported: *B. glaucescens* (synonym of *B. multiplex*), *B. vulgaris*, *Dendrocalamus giganteus*, and *Otatea acuminata*. In 1992, it was reported the first protocol for synthetic seed obtainment, through the encapsulation of *Dendrocalamus strictus* somatic embryos in a matrix of MS culture medium with sodium alginate, calcium chloride, and mineral oil (Mukunthakumar and Mathur 1992). After that, Jullien and Van (1994) reported, for the first time, the use of young leaves as explants for SE induction. The authors used semi-thin sectioned leaves of *B. glaucescens* and MS culture medium with different concentrations of 2,4-D (Jullien and Van (1994). Their results showed the existence of two gradient responses: the leaf position (1) from the plant base to the apex and (2) from the basis to the apical portion of the leaf. Embryogenic callus and embryo-like structures were obtained, but no plant regeneration could be achieved (Jullien and Van 1994).

In 1995, a protocol using the roots of in vitro plants of *B. beecheyana* in an MS culture medium (Murashige and Skoog 1962) supplemented with 2,4-D, kinetin (KIN) and high sucrose concentration (6%) was described, with successfully somatic embryo induction and rooted somatic plants (Chang and Lan 1995). In an intriguing study, Ho and Chang (1998) reported that embryogenic callus of *B. oldhamii* obtained in 1986 from inflorescence tissue in an MS culture medium with 2,4-D, KIN, and 6% sucrose retained the embryonic potential after 11 years in culture. These authors observed that after 4 years of culture, 75% of the callus maintained their embryogenic potential; however, the number of abnormal plants increased. After 7 years, only 25% of the regenerated plants were normal, and after 11 years, the embryogenic potential was maintained, but all the regenerated plants were abnormal. Moreover, they observed that 5% of the abnormal plants flowered spontaneously and produced viable pollen (Ho and Chang 1998).

In the 2000s, protocols for six other species were published: *Bambusa edulis* (synonym of *Sinoarundinaria edulis*), *B. balcooa*, *Dendrocalamus asper*, *D. hamiltonii*, *Phyllostachys nigra*, and *Guadua angustifolia*. In 2002, an efficient SE protocol was established for *D. hamiltonii* using nodal segments in a half-strength MS culture medium supplemented with 2,4-D. BAP, and sucrose 8%

(Godbole et al. 2002). In this protocol, the authors reported a conversion rate of somatic embryos in plants of 80% (Godbole et al. 2002). Shortly after, Godbole et al. (2004) reported starch deposition and amylase accumulation during the SE of *D. hamiltonii*. In the same year, Lin et al. (2004) showed that the use of thidiazuron (TDZ) in proliferation culture medium resulted in higher conversion rates (more than 80%) of somatic embryos for *B. edulis*, when compared to other plant growth regulators evaluated (kinetin, BAP, naphthaleneacetic acid, and zeatin). These authors also showed that the TDZ supplementation promoted somatic embryo formation and reduced callus browning in proliferation medium (Lin et al. 2004).

The first SE protocol for the genus *Guadua* was described by Marulanda et al. (2005). In this study, *Guadua angustifolia* calluses were obtained after the inoculation of nodal segments in MS culture medium supplemented with 2,4-D. During the multiplication, nodular and compact callus with pro-embryos was observed, and the maturation process could be achieved in a plant growth regulator-free culture medium (Marulanda et al. 2005). Soon after, Gillis et al. (2007) reported an SE protocol for *B. balcooa*, which proved to be an efficient and low-cost alternative for mass-scale propagation in comparison with organogenesis through axillary branching. Similarly, Arya et al. (2008) proposed, for the first time, an effective protocol for *D. asper* SE that allowed the production of more than 10,000 plants in 1 single year.

In the 2010s, new protocols were proposed for 11 bamboo species, e.g., *Bambusa nutans*, *B. bambos*, *B. tulda*, *B. ventricosa* (synonym of *Bambusa tuldoides*), *Dendrocalamus latiflorus*, *D. membranaceus*, *D. stocksii* (a synonym of *Pseudoxytenanthera stocksii*), *Melocanna baccifera*, *Phyllostachys bambusoides*, *P. pubescens*, and *P. heterocycla* (synonym of *Phyllostachys edulis*).

Mehta et al. (2011) proposed a complete SE protocol for *B. nutans* initiated from immature buds in an MS culture medium supplemented with 2,4-D, obtaining callus, globular and mature somatic embryos, and regenerated plants. In the same year, another study reported a protocol of particle bombardment for genetic transformation in cell suspension cultures of *Phyllostachys nigra* (Ogita et al. 2011). This protocol was obtained by the optimization of the growth efficiency of the target cells, an important step that is directly associated with the success of genetic transformation. Additional details regarding this study can be found in the topic "Molecular Biology Approaches" in this chapter.

In 2012, it was proposed an SE protocol combined with a slow-growth method for *D. hamiltonii* (Kaur et al. 2012). In this study, the authors observed that even after 365 days in culture, the somatic embryos had a 59.9% conversion rate. In 2013, an in vitro slow-growth study with *D. hamiltonii* evaluated varying sucrose concentrations, basal salts, and retardant substances and proposed a protocol for genetic transformation (Sood et al. 2013).

In 2016, a protocol for indirect SE in *B. arundinacea* using mature seeds as explants was reported (Venkatachalam and Kalaiarasi 2016). In this study, the embryo maturation reached 94%, but the conversion rate was only 25.1%. Recently, Somashekar et al. (2018) reported a protocol for *D. stocksii* SE. In this study, the authors evaluated several explant sources (leaf segment and sheath, nodal segments, and shoot tips) in an MS culture medium, and the best results found were using nodal segments (Somashekar et al. 2018). Also, the most suitable medium composition was MS supplemented with 2,4-D, KIN and coconut milk for callus induction, 2,4-D and coconut milk for embryo formation and maturation, and 1-naphthaleneacetic acid (NAA) and BAP for germination (Somashekar et al. 2018).

Low regeneration rates (Rao et al. 1985; Venkatachalam and Kalaiarasi 2016) or even somatic embryo formation without plant regeneration has been reported (Jullien and Van 1994; Marulanda et al. 2005; Daquinta et al. 2007; Komatsu et al. 2011) showing that they are recurrent issues in some bamboo SE protocols. Further studies should focus on the establishment of more effective protocols for bamboo SE, also aiming to expand the species spectra. There are more than 1600 bamboo species and only 24 species with SE protocols described and published. Additional efforts to expand the knowledge of this morphogenetic route need to be done.

4.3 Morpho-Anatomy of Bamboo Somatic Embryogenesis

The morpho-anatomy analysis represents a valuable tool for SE characterization in bamboos. The evaluation of the callus morphology, cell structure, and ultrastructure is essential for the understanding of the morphogenetic features underlying bamboo SE. In bamboo SE, there are mainly three types of callus morphology: (1) compact, nodular, yellow-white colored presenting small, rounded, and isodiametric cells with a thick cell wall, dense cytoplasm, prominent nucleus, and starch grains; (2) friable and soft, yellow-white colored, exhibiting large and vacuolated cells with a thin cell wall and few starch grains; and (3) mucilaginous or gelatinous translucent callus varying between the compact and friable features (Yeh and Chang 1986a, 1986b; Jullien and Van 1994; Arya et al. 2008). Some studies reported the presence of only two morphologies: compact or friable callus (Rao et al. 1985; Tsay et al. 1990; Saxena and Dhawan 1999; Godbole et al. 2002; Mehta et al. 2011).

Godbole et al. (2002) reported that friable callus presented fast-growing features, while the compact nodular was a slow-growing callus. Similarly, Jullien and Van (1994) described that compact embryogenic callus showed meristematic structures, epidermal cells, and tracheids. In contrast, the friable non-embryogenic callus showed undifferentiated cells. Regarding the origin of the callus, cell proliferation was initiated in the cells surrounding vascular bundles and formed a cambium-like layer that spreads faster, allowing callus formation (Jullien and Van 1994; Godbole et al. 2002; Venkatachalam and Kalaiarasi 2016).

Somatic embryos or embryoid formation occurs either spontaneously (Yeh and Chang 1986b; Tsay et al. 1990; Chang and Lan 1995) or after the transference of the compact callus to the maturation culture medium (Godbole et al. 2002; Arya et al. 2008; Venkatachalam and Kalaiarasi 2016). Secondary embryogenesis events were noticed in bamboo SE, whereas new embryos were usually formed from the surface of the compact globular callus or somatic embryos (Rao et al. 1985; Yeh and Chang

1986b; Arya et al. 2008; Komatsu et al. 2011). Arya et al. (2008) reported the formation of pro-embryos during the induction phase of *D. asper* SE and globular somatic embryos during the multiplication phase, followed by scutellar and coleoptilar stages. The secondary embryogenesis was also observed on the surface of the scutellar somatic embryos.

Yeh and Chang (1986a, b) reported meristematic structures with the scutellum, coleoptile, coleorhiza, and shoot meristem, during *B. beecheyana* SE. According to Godbole et al. (2002), somatic embryogenesis of monocotyledons occurs through the following developmental stages: globular, elongated, scutellar, and coleoptilar. However, the characterization of the developmental stages in bamboo SE is still a challenge since it is rarely clearly demonstrated in the published studies.

In *D. hamiltonii*, the amylase activity changes during SE developmental stages, being absent in initial stages (induction, multiplication, and globular embryos), increased in scutellum formation and embryo conversion, and finally, decreased in further developmental stages (Godbole et al. 2004; Sood et al. 2013). Scanning electron microscopy and light microscopy demonstrated the presence of starch grains in the scutellum of *D. hamiltonii* mature embryos (Godbole et al. 2004; Sood et al. 2013).

Although some works have well described the morpho-anatomy of calluses in the initial steps of SE, anatomical characterizations of further developmental stages, i.e., embryo formation, maturation, and conversion, were insufficient or even not mentioned. The most challenging point for the morpho-anatomical characterization of bamboo SE is the scarce information available in the literature and also how laborious is to perform such a description. Thus, the anatomical characterization of all developmental stages should be strongly encouraged to understand this process in bamboos.

4.4 Biochemical Approaches

Somatic embryogenesis is a model for investigating the dynamics of several biochemical parameters, such as amino acids/proteins, plant hormones, or other essential molecules for plant metabolism. The study of biochemical aspects of SE can help to elucidate this morphogenetic route, especially in recalcitrant and neglected species. The first biochemical analysis carried out in bamboo SE investigated isozyme pattern. The legendary Toshio Murashige, in addition to his valuable contribution to the development of the MS salt composition, performed studies with callus cultures of four bamboo species to compare their isozyme banding pattern (Huang and Murashige 1983). The isozyme bands of glutamate-oxaloacetate transaminase (GOT) were different among the evaluated bamboo species. Only one GOT isozyme was common to all species.

The isozyme profile was also investigated in *Bambusa vulgaris* SE (Rout and Das 1995). This study compared the transition of non-embryogenic to embryogenic calli, somatic embryo development, and somatic embryo conversion. The authors reported marked differences in the isozyme profile among the SE stages: e.g., the transition of non-embryogenic to embryogenic callus was marked by the appearance of five new peroxidase bands. In general, during SE, some peroxidase and esterase bands are suppressed, while some new isoforms appear. Isozyme analyses are a valuable technique but have relevant limitations, as those studies suffered from a contemplative effect of the results.

The analysis of protein profiles by SDS-PAGE is also a suitable approach to study bamboo SE. Jullien and Van (1994) reported an SE protocol for *Bambusa glaucescens*, in which it was possible to obtain embryogenic and non-embryogenic calluses and the production of "embryo-like" structures. In this study, the authors compared callus protein profiles with distinct morphogenic features: i.e., friable, compact, and embryogenic. The protein profiles between compact and embryogenic calluses were quantitative identical, with minor qualitative differences. Friable calluses showed a different pattern, which coincided with its non-embryogenic features. The correlation between protein profiles and embryogenic potential is a major point to be investigated.

With the advancement of technologies for quantification and analysis of metabolites, studies involving more sensitive techniques were carried out in bamboo SE. Analysis of the amino acid composition during SE can also provide useful information for understanding the morphogenetic events. Besides reporting a protocol for callus induction and cell suspension establishment of *Phyllostachys nigra*, Ogita (2005) evaluated the free amino acid composition in bamboo shoots and callus tissues. The author reported that glutamine (41.9%), ɣ-aminobutyric acid (21.9%), and alanine (14.3%) were the major amino acids in the callus tissues. These results are in agreement with several other studies that report the high importance of glutamine and ɣ-aminobutyric acid in SE development (Ogita et al. 2001; Pescador et al. 2012).

In *Dendrocalamus strictus*, amino acids and sugar were differentially accumulated in NaCl-tolerant callus (Singh et al. 2003). In this study, the authors selected cell lineages tolerant of NaCl (100 mM) and compared them with callus induced and maintained in NaCl-free culture medium. The results indicated a higher content of total sugar and free amino acids in the tolerant callus than in the NaCl-free callus, which may be related to the capacity of plant tissues to tolerate salt stress (Singh et al. 2003). These results illustrate the importance of SE as a model system to investigate abiotic stresses.

The importance of plant hormones or plant growth regulators in regulating plant morphogenesis is widespread, especially in plant tissue culture. The endogenous concentrations of these hormones can vary greatly depending on the supplementation of plant regulators in the culture medium, affecting the stages of SE. In a more recent study, Van den Akker et al. (2015) reported the use of UPLC-MS/MS (ultraperformance/pressure liquid chromatography) for the cytokinin quantification in cell suspension cultures of *B. balcooa* culture on 2,4-D-supplemented culture medium without exogenous cytokinins. The predominant forms of cytokinins in cell cultures were trans-zeatin-O-glucoside and trans-zeatin-riboside-O-glucoside, i.e., the predominant forms were glucosides and ribosides. Free bases were not detected in cell suspension cultures, indicating that the pool of endogenous cytokinins was mostly conjugated.

Despite the existence of the abovementioned studies, very few reports regarding the available dynamics of biochemical parameters. Therefore, much effort still needs to be done to better understand the SE in bamboo. Strategies based on obtaining large-scale data, such as proteomics, metabolomics, and hormonomics, can bring more comprehensive information that allows a clearer understanding of this complex morphogenetic route.

4.5 Molecular Biology Approaches

During the long-term tissue culture process, morphological and (epi)genetic variations are highly frequent. These variations are, generally, induced by its long-time maintenance under artificial conditions and the exposure of propagated tissue to high concentrations of plant growth regulators, leading to disorganized cell division (Li et al. 2019). The variant plants regenerated from tissue culture have been termed somaclones (soma = vegetative; clone = identical copy; Larkin and Scowcroft 1981). Somaclonal variation is a phenotypic variation, either genetic or epigenetic in origin, displayed among somaclones. The genetic events include alteration in chromosome numbers (i.e., polyploidy and aneuploidy), alteration in chromosome structure (i.e., translocation, deletion, duplication), and mutation in the DNA sequence. The epigenetic changes can occur as a result of alterations in DNA methylation patterns, changes in histone modifications, or a combination of these mechanisms (Vroh-Bi et al. 2011).

Among the epigenetic changes, DNA methylation is a major factor causing differential gene expression and gene silencing (Vroh-Bi et al. 2011). The methylation of gene promoters may be associated with alteration of gene transcription that leads to morphological changes. Thus, the methylation of cytosine in both CG and CNG sites has received considerable attention in recent years.

The analysis of both genetic and epigenetic fidelities in bamboo somatic plantlets was described. Gillis et al. (2007) analyzed the ploidy level and assessed the genetic and epigenetic stability (by methylation-sensitive amplified polymorphism (MSAP)) of somatic plantlets of *B. balcooa* up to 18 months after transplanting. The analysis indicated no alteration in the ploidy levels and very low or inexistent epigenetic changes, demonstrating the feasibility of this protocol *B. balcooa* propagation. Similarly, Mehta et al. (2011) reported no genetic variation during SE in *Bambusa nutans*. The authors analyzed the whole SE process: donor plant, the globular and mature somatic embryos, and regenerated plantlets by amplified fragment length polymorphism (AFLP) fingerprinting. Thus, both studies reported no somaclonal variation during SE of these species. However, a larger spectrum of bamboo species should be analyzed to confirm this tendency.

Somatic embryogenesis offers another relevant possibility through molecular biology approaches, genetic transformation. Somatic embryos are ideal explants for carrying out transformation studies and regeneration of transgenic plants. The compact organized structures can be visibly screened, multiplied, and converted into plantlets, representing a relevant advantage of SE (Sood et al. 2013). Ogita et al. (2011) proposed an efficient transformation protocol using the particle bombardment method for *Phyllostachys nigra* suspension cells, being first detailed information on the transformation of cultured bamboo cells. Sood et al. (2013) also successfully obtained transgenic plants through SE of *Dendrocalamus hamiltonii* both by *Agrobacterium tumefaciens* and microprojectile bombardment-mediated approach.

Recently, Ye et al. (2017) reported an efficient and reproducible protocol for *Dendrocalamus latiflorus* regeneration through SE and *Agrobacterium*-mediated transformation. These authors achieved the callus induction from young shoots as explants and could regenerate bamboo somatic plantlets by optimizing the culture medium composition. In this study, a heterologous expression of the maize leaf color (Lc) gene in *D. latiflorus* somatic embryos was obtained, which indicated the efficiency of the method for further studies on genetic transformation in this species and other related bamboos.

All these reports represent a relevant achievement due to the challenging application of the genetic transformation in monocots, especially in bamboos. The development of efficient genetic transformation systems in bamboo is greatly dependent on the development of reliable and efficient regeneration SE protocols, which implicates in the need for better elucidation of this morphogenetic route.

4.6 In Vitro Conservation Strategies

An effective germplasm conservation method is essential to securely and efficiently preserve endangered species, plant genetic variability, and elite planting materials. Somatic embryogenesis is a suitable technique for ex situ conservation of plant species, despite being underreported for bamboos. This technique can provide appropriate plant materials to be conserved in vitro, due to the low cell differentiation status and the possibility to manipulate the culture conditions. Culture callus or somatic embryos on a slow-growth condition is one of the possible strategies for maintaining propagules and in vitro conservation of plant species. Storage under slow-growth conditions allows the maintenance of multiplying cultures and ensures constant availability for basic or applied studies of planting propagules.

Kaur et al. (2012) reported an effective method to maintain cultures of *D. hamiltonii* under slow-growth conditions. They used a liquid paraffin overlay (LPO) to induce slow growth through the imposition of oxygen stress, which subsequently affected carbon metabolism (Kaur et al. 2012). The maintenance of active growth somatic embryos of *D. hamiltonii* requires regular subculture at 30-day intervals, whereas the use of LPO eliminated the need for subculture for 365 days (Kaur et al. 2012). Soon after, Sood et al. (2013) used sucrose 8% and half strength MS culture medium to slow down the somatic embryonic growth of *D. hamiltonii*. The use of an inhibitor of gibberellin synthesis, such as paclobutrazol, retarded the development of the somatic embryo (Sood et al. 2013).

Another valuable technique compatible with in vitro conservation is the synthetic seeds (synseeds). This technique is especially attractive for most bamboo species due to its very long intervals of seed production (long vegetative periods) and a low pollen and seed viability. The synseeds are a potential propagative material and can be used for cryopreservation procedures. Mukunthakumar and Mathur (1992) assessed the in vitro and in vivo germination and development of *D. strictus* synseeds. The synseeds reached 56% of germination in soil and ~ 95% in culture medium. The plantlets were successfully acclimatized, indicating the feasibility of the technique. Unfortunately, the authors did not evaluate the synseed performance for cryopreservation purposes.

There is an important knowledge gap in the use of cryopreservation for the in vitro conservation of bamboo through SE. Cryopreservation, which involves the conservation of propagules under ultralow temperatures (usually −196 °C), is a technique widely used for numerous plant species. Somatic embryogenesis is a very valuable system for propagule generation (e.g., calluses and somatic embryos) and cryopreservation. However, until now, there are no clear and defined published protocols that allow conservation via cryopreservation of these materials for bamboo species. The possible reasons for this scenario are the complexity of SE route in bamboos, the recalcitrance of some explant sources, the low number of SE protocols defined, and the inherent complexity of the cryopreservation technique. Therefore, the best exploitation of this technique can contribute enormously to the in vitro conservation of many bamboo species.

The main advances and prospects for bamboo SE were summarized in Fig. 4.3.

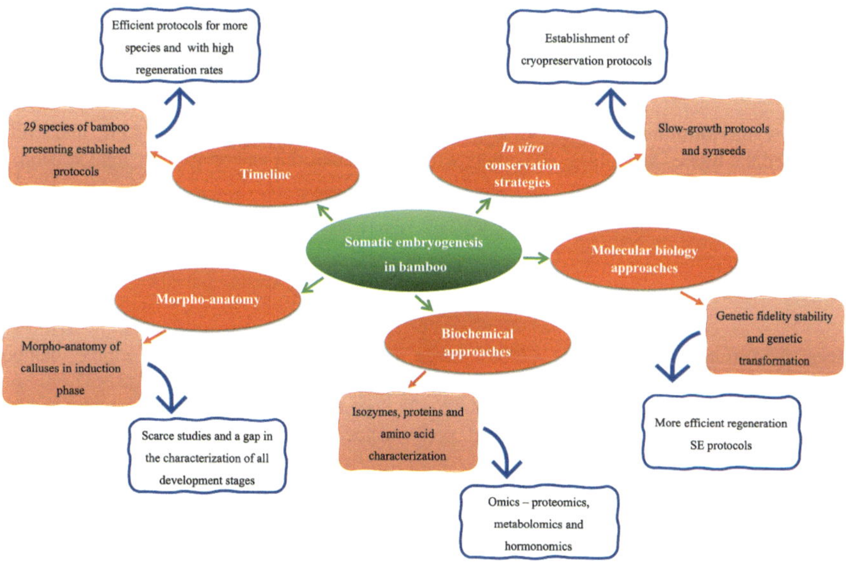

Fig. 4.3 Schematic figure summarizing the main advances and prospects in bamboo somatic embryogenesis. Orange and blue rectangles indicate the main advances and the main limitations, respectively

4.7 Conclusion and Future Perspectives

So far, we do have major advances, such as the establishment of protocols for 29 bamboo species, the characterization of embryogenic and non-embryogenic calluses, studies using biochemical and molecular biology approaches (i.e., isozymes, proteins, amino acids, genetic and epigenetic stability), the establishment of genetic transformation protocols, and in vitro conservation strategies (slow growth). However, much remains to be done: (1) to establish more efficient SE regeneration protocols, (2) to increase the number of relevant bamboo species with described SE protocols, (3) to characterize the morpho-anatomy and biochemical studies of the developmental stages of somatic embryos, (4) to increase the omics data in bamboo SE, and (5) to establish cryopreservation protocols.

Bamboos are a unique group of plants, which have several peculiarities and complexities. We firmly believe that efforts aimed at elucidating bamboo SE can revolutionize the production of seedlings of these species, which are increasingly required worldwide. Within the context of bamboo micropropagation techniques, organogenesis is the most used technique today. However, serious problems related to the control of microorganisms associated with plants limit the in vitro propagation. SE has the potential to circumvent this chronic problem of in vitro contamination, with the appropriate selection of explants and greater control of this morphogenetic route. In this sense, future advances in the knowledge of bamboo SE can improve both fundamental studies and the production chain of bamboo.

Acknowledgments This work was supported by the Brazilian funding agencies CAPES (Coordenação de Aperfeiçoamento de Pessoal de Nível Superior), CNPq (Conselho Nacional de Desenvolvimento Científico e Tecnológico), and FAPESC (Fundação de Amparo à Pesquisa e Inovação de Santa Catarina).

Author Contributions LGP, LNV, and HPFF conceptualized the chapter; LGP and HPFF elaborated the figures and tables; all authors wrote the manuscript and revised the final version.

References

Arya S, Satsangi R, Arya ID (2008) Large scale plant propagation of edible bamboo *Dendrocalamus asper* through somatic embryogenesis. J Am Bamboo Soc 21:21–31

Bag N, Palni LMS, Chandra S, Nandi SK (2012) Somatic embryogenesis in 'maggar' bamboo (*Dendrocalamus hamiltonii*) and field performance of regenerated plants. Curr Sci 102 (9):1279–1287

Chang WC, Lan TH (1995) Somatic embryogenesis and plant regeneration from roots of bamboo (*Bambusa beecheyana* Munro var *beecheyana*). J Plant Physiol 145:535–538

Cheah KT, Chaille LC (2011) Somatic embryogenesis from mature *Bambusa ventricosa*. In: Biotechnology. Coll Trop Agr Hum Res, Honolulu, pp 1–5

Chu CC, WANG CC, Sun CS, HSU C, YIN KC, BI FY (1975). Establishment of an efficient medium for anther culture of rice through comparative experiments on the nitrogen sources. Sci Sin 18(5):659–668

Daquinta M, Gregori A, Cid M, Lezcano Y, Sagarra F (2007) Formación de callos e inducción de brotes a partir de tejido intercalar de ramas de plantas adultas de Guadua angustifolia Kunth. Biotecnología vegetal, 7(2)

Fehér A (2019) Callus, dedifferentiation, totipotency, somatic embryogenesis: what these terms mean in the era of molecular plant biology? Front Plant Sci 10:536

Gamborg OL, Miller R, Ojima K (1968) Nutrient requirements of suspension cultures of soybean root cells. Exp Cell Res 50(1):151–158

Gillis K, Gielis J, Peeters H, Dhooghe E, Oprins J (2007) Somatic embryogenesis from mature *Bambusa balcooa* Roxburgh as basis for mass production of elite forestry bamboos. Plant Cell Tissue Organ Cult 91(2):115–123

Godbole S, Sood A, Sharma M, Nagar PK, Ahuja PS (2004) Starch deposition and amylase accumulation during somatic embryogenesis in bamboo (*Dendrocalamus asper*). J Plant Physiol 161:245–248

Godbole S, Sood A, Thakur R, Sharma M, Ahuja PS (2002) Somatic embryogenesis and its conversion into plantlets in a multipurpose bamboo, *Dendrocalamus hamiltonii* Nees et Arn. Ex Munro. Curr Sci 83:885–889

Guzmán-García E, Sánchez-Romero C, Panis B, Carpentier SC (2013) The use of 2D-DIGE to understand the regeneration of somatic embryos in avocado. Proteomics 13:3498–3507

Hassan AAE, Debergh P (1987) Embryogenesis and plantlet development in the bamboo *Phyllostachys viridis* (Young) McClure. Plant Cell Tissue Organ Cult 10:73–77

Ho CW, Chang WC (1998) *In vitro* flowering of albino bamboo (*Bambusa oldhamnii* Munro) regenerants derived from an eleven-year old embryogenic cell line. Acta Hortic 461:433–438

Huang LC, Chen WL, Huang BL (1988) Tissue culture investigations of bamboo II. Liquid suspension cultures of *Bambusa*, *Phyllostachys*, and *Sasa* cells. Bot Bull Acad Sin 29:177–182

Huang LC, Chen WL, Huang BL (1989) Tissue culture investigations of bamboo. III: a method for viable protoplast isolation from *Bambusa* cells of liquid suspension culture. Bot Bull Acad Sin 30(1):49–57

Huang LC, Murashige T (1983) Tissue culture investigations of bamboo, I. Callus cultures of *Bambusa*, *Phyllostachys* and *Sasa*. Bot Bull Acad Sin 24:31–52

Jullien F, Van KTT (1994) Micropropagation and embryoid formation from young leaves of *Bambusa glaucescens* 'Golden goddess'. Plant Sci 98(2):199–207

Karami O, Aghavaisi B, Pour AM (2009) Molecular aspects of somatic-to-embryogenic transition in plants. J Chem Biol 2(4):177–190

Kaur D, Ogra RK, Bhattacharya A, Sood A (2012) Changes in sugar levels during slow growth of *Dendrocalamus hamiltonii* somatic embryos due to liquid paraffin overlay. Vitro Cell Dev Biol Plant 48(1):120–126

Komatsu YH, Piotto KDB, Brondani GE, Gonçalves AN, Almeid M (2011) *In vitro* morphogenic response of leaf sheath of *Phyllostachys bambusoides*. J For Res 22(2):209–215

Larkin PJ, Scowcroft WR (1981) Somaclonal variation—a novel source of variability from cell cultures for plant improvement. Theor Appl Genet 60(4):197–214

Li J, Manghwar H, Sun L, Wang P, Wang G et al (2019) Whole genome sequencing reveals rare off-target mutations and considerable inherent genetic or/and somaclonal variations in CRISPR/Cas9-edited cotton plants. Plant Biotech J 17(5):858–868

Lin CS, Lin CC, Chang WC (2004) Effect of thidiazuron on vegetative tissue derived somatic embryogenesis and flowering of bamboo *Bambusa edulis*. Plant Cell Tissue Organ Cult 76:75–82

Lloyd G, McCown B (1981) Commercially-feasible micropropagation of mountain laurel, Kalmia latifolia, by use of shoot-tip culture. Proc Intl Plant Prop Soc 30:421–427

Marulanda M, Gutiérrez L, Márquez M (2005) Micropropagación de *Guadua angustifolia* Kunt. Rev Col Biotech Veg 87(27):5–15

Mehta R, Sharma V, Sood A, Sharma M, Sharma RK (2011) Induction of somatic embryogenesis and analysis of genetic fidelity of *in vitro*-derived plantlets of *Bambusa nutans* Wall., using AFLP markers. Eur J For Res 130(5):729–736

Mehta U, Rao IVR, Ram HYM (1982) Somatic embryogenesis in bamboo. Plant Tissue Culture. In: Fujiwara A (ed) Jpn. Assoc. Proc 5th Intl Cong Plant Tiss Cell Cult pp 109–110

Méndez-Hernández HA, Ledezma-Rodríguez M, Avilez-Montalvo RN, Juárez-Gómez YL, Skeete A, Avilez-Montalvo J, Loyola-Vargas VM (2019) Signaling overview of plant somatic embryogenesis. Front Plant Sci 10:77

Mukunthakumar S, Mathur J (1992) Artificial seed production in the male bamboo *Dendrocalamus strictus* L. Plant Sci 87(1):109–113

Murashige T, Skoog F (1962) A revised medium for rapid growth and bio assays with tobacco tissue cultures. Physiol Plantarum 15(3):473–497

Ogita S (2005) Callus and cell suspension culture of bamboo plant, *Phyllostachys nigra*. Plant Biotech 22(2):119–125

Ogita S, Kikuchi N, Nomura T, Kato Y (2011) A practical protocol for particle bombardment-mediated transformation of *Phyllostachys* bamboo suspension cells. Plant Biotech 28(1):43–50

Ogita S, Sasamoto H, Yeung EC, Thorpe TA (2001) The effects of glutamine of the maintenance of embryogenic cultures of *Cryptomeria japonica*. In Vitro Cell Dev Biol Plant 37(2):268–273

Ojha A, Verma N, Kumar A (2009) *In vitro* micropropagation of economically important edible bamboo (*Dendrocalamus asper*) through somatic embryos from root, leaves and nodal segments explants. Res Crops 10(2):430–436

Pescador R, Kerbauy GB, Fraga HPF, Hamasaki RM, Tavares LBB, Guerra MP (2012) Dynamics of free and 3H-labelled glutamine concentrations during zygotic and somatic embryogenesis of Feijoa [*Acca sellowiana* (O. berg.) Burret]. J Hortic Sci Biotechnol 87(6):583–587

Rao IU, Rao IR, Narang V (1985) Somatic embryogenesis and regeneration of plants in the bamboo *Dendrocalamus strictus*. Plant Cell Rep 4(4):191–194

Rout GR, Das P (1994) Somatic embryogenesis and *in vitro* flowering of 3 species of bamboo. Plant Cell Rep 13:683–686

Rout GR, Das P (1995) Isozyme profile during somatic embryogenesis and *in vitro* flowering of *Bambusa vulgaris*. J Plant Biochem Biotechnol 4(1):43–45

Saxena S, Dhawan V (1999) Regeneration and large-scale propagation of bamboo (*Dendrocalamus strictus* Nees) through somatic embryogenesis. Plant Cell Rep 18:438–443

Singh M, Jaiswal U, Jaiswal VS (2003) *In vitro* selection of NaCl-tolerant callus lines and regeneration of plantlets in a bamboo (*Dendrocalamus strictus* Nees). In Vitro Cell Dev Biol Plant 39(2):229–233

Somashekar PV, Rathore TS, Fatima T (2018) *In vitro* plant regeneration of *Dendrocalamus stocksii* (Munro) M. Kumar, Remesh & Unnikrisnan, through somatic embryogenesis. Am J Plant Sci 9(12):2429–2445

Sood A, Ahuja PS, Sharma M, Sharma OP, Godbole S (2002) *In vitro* protocols and field performance of elites of an important bamboo *Dendrocalamus hamiltonii* Nees et Arn. Ex Munro. Plant Cell Tissue Organ Cult 71:55–63

Sood A, Bhattacharya A, Sharma M, Sharma RK, Nadha HK, Sood P, Mehta R, Kaur D, Brar J, Ahuja PS (2013) Somatic embryogenesis and *Agrobacterium* mediated genetic transformation in bamboos. In: Aslam J, Srivastava PS, Sharma MP (eds) Somatic embryogenesis and genetic transformation in plants. Narosa Publishing House, New Delhi, pp 166–178

Tsay HS, Yeh CC, Hsu JY (1990) Embryogenesis and plant regeneration from another culture of bamboo (*Sinocalamus latiflora* (Munro) McClure). Plant Cell Rep 9:349–351

Van den Akker S, Bormans P, Peeters H, Prinsen E, Gielis J (2015) Cytokinin dynamics in cell suspension cultures of *Bambusa balcooa* Roxburgh using UPLC-ESI/MS/MS. J Am Bamboo Soc 32

Venkatachalam P, Kalaiarasi K (2016) Indirect somatic embryogenesis and plantlet development from mature seed embryo explants of *Bambusa arundinacea* (Retz.) wild. In: Anis M, Ahmad N (eds) Plant tissue culture: propagation, conservation and crop improvement. Springer, Singapore, pp 509–519

Vroh-Bi I, Anagbogu C, Nnadi S, Tenkouano A (2011) Genomic characterization of natural and somaclonal variations in bananas (*Musa* spp.). Plant Mol Biol Rep 29(2):440–448

Wei Q, Cao J, Qian W, Xu M, Li Z, Ding Y (2015) Establishment of an efficient micropropagation and callus regeneration system from the axillary buds of *Bambusa ventricosa*. Plant Cell Tissue Organ Cult 122(1):1–8

Woods SH, Philips GC, Woods JE, Collins GB (1992) Somatic embryogenesis and plant regeneration from zygotic embryo explants in Mexican weeping bamboo, *Otatea acuminata* Aztecorum. Plant Cell Rep 11:257–261

Ye S, Cai C, Ren H, Wang W, Xiang M, Tang X, Zhu C, Yin T, Zhang L, Zhu Q (2017) An efficient plant regeneration and transformation system of ma bamboo (*Dendrocalamus latiflorus* Munro) started from young shoot as explant. Front Plant Sci 8:1298

Yeh ML, Chang WC (1986a) Plant regeneration through somatic embryogenesis in callus culture of green bamboo (*Bambusa oldhamii* Munro). Theor Appl Genet 73:161–163

Yeh ML, Chang WC (1986b) Somatic embryogenesis and subsequent plant regeneration from inflorescence callus of *Bambusa beecheyana* Munro var. beecheyana. Plant Cell Rep 5:409–411

Yeh ML, Chang WC (1987) Plant regeneration via somatic embryogenesis in mature embryo-derived callus culture of *Sinocalamus latiflora* (Munro) McClure. Plant Sci 51:93–96

Yuan JL, Yue JJ, Wu XL, Gu XP (2013) Protocol for callus induction and somatic embryogenesis in Moso bamboo. PLoS One 8(12):e81954

Zhang N, Fang W, Shi Y, Liu Q, Yang H, Gui R, Lin X (2010) Somatic embryogenesis and organogenesis in *Dendrocalamus hamiltonii*. Plant Cell Tissue Organ Cult 103(3):325–332

Zimmerman JL (1993) Somatic embryogenesis: a model for early development in higher plants. Plant Cell 5(10):1411

Chapter 5
Initiation and Establishment of Cell Suspension Cultures in Bamboo

Fernanda Furlan Queiroz, Inaê Mariê de Araújo Silva-Cardoso, and Jonny Everson Scherwinski-Pereira

Abstract Bamboos have peculiar reproductive characteristics, such as irregular flowering, low seed viability, and monocarpic nature, which make it difficult to introduce characteristics of interest by classical breeding. In addition, conventional breeding methods are considered inefficient when aiming at mass production. In this sense, biotechnological tools, such as protoplast culture, somatic hybridization, somatic embryogenesis, and genetic transformation, in addition to vegetative propagation, are configured as important alternative methods to subsidize programs of genetic improvement of bamboo and boost mass-scale production of interest varieties and superior genotypes. However, for the application of the aforementioned techniques, the establishment and the ability to control the multiplication of cell suspensions, with a high growth rate and with interest characteristics, seem to be of fundamental importance. Although the direction of research for the establishment of cell suspensions in bamboo has been reported since the 1980s, currently there are relatively few advances obtained, with limited work on the number of species used. In this sense, this chapter compiles the available works on the establishment of cell suspensions in bamboo, with emphasis on the two species of *Guadua*, *G. magna*, and *G.* aff. *chaparensis*, and describes a pioneering and detailed methodology of easy application for the establishment and maintenance of cell suspensions with embryogenic potential.

Keywords Callus culture · Cell suspension culture · Micropropagation · Secondary metabolites · Somatic embryogenesis

F. F. Queiroz
Universidade de Brasília, Brasília, Distrito Federal, Brazil

I. M. de Araújo Silva-Cardoso
CNPq/Embrapa Postdoctoral Fellowship Program, Embrapa Recursos Genéticos e Biotecnologia, Brasília, Brazil

J. E. Scherwinski-Pereira (✉)
Embrapa Recursos Genéticos e Biotecnologia, Brasília, Distrito Federal, Brazil
e-mail: jonny.pereira@embrapa.br

Z. Ahmad et al. (eds.), *Biotechnological Advances in Bamboo*,
https://doi.org/10.1007/978-981-16-1310-4_5

5.1 Introduction

Bamboo is a plant of arborescent habit that belongs to the family Poaceae and subfamily Bambusoideae, which covers about 120 genera (Soreng et al. 2017) and more than 1600 species (Vorontsova et al. 2016). Considered the only grass that develops in a wide variety of climates, it is widely distributed on different continents, such as Asia Pacific, the Americas, and Africa (Ramakrishnan et al. 2020), and is identified as one of the most abundant providers of natural materials (Shi et al. 2019).

Worldwide, bamboos stand out due to the multipurpose profile, ensuring several socioeconomic and environmental benefits. Among its various applications, cellulose is obtained for paper production (Rao and Rao 1990; Guan et al. 2019) and tissue (Alvarez et al. 2020); in construction (Hong et al. 2019; Fatimah et al. 2019; Sun et al. 2020), which includes the production of cement materials (Moraes et al. 2019; Rodier et al. 2019); biomass supply for fuel production (Yuan et al. 2018; Marafon et al. 2019), production of water pipelines (Shi et al. 2019); furniture manufacturing (Chang et al. 2018; Fatimah et al. 2019), crafts (Singh et al. 2013; Chang et al. 2018; Fatimah et al. 2019); food supply (Felisberto et al. 2017; Chang et al. 2018; Nirmala et al. 2018; Thapa et al. 2019); drug production; and medicinal applications (Singh et al. 2013; Nirmala et al. 2018).

Directly or indirectly, the products originating from bamboo also mitigate various environmental problems. According to Shi et al. (2019), the use of bamboo as an alternative to wood in its various applications can help reduce global warming by slowing down the destruction of forests. In addition, bamboos absorb carbon dioxide during their growth, and when used as structural materials, they are considered more economical and more effective in mitigating the greenhouse effect (Sun et al. 2020).

Considered to be of great agricultural potential and competitive power in relation to other types of raw materials (Mudoi et al. 2013; Sandhu et al. 2018), bamboo species stand out due to their rapid growth, in addition to reproducing in the field without the need for replanting (Pereira and Beraldo 2008; Komatsu et al. 2011; Singh et al. 2013). Due to their wide diversity of applications and agricultural versatility, different species of bamboos are cultivated around the world (Carnegie 1997; Arya et al. 1999; Singh et al. 2013), which requires a continuous and effective supply of plantlets.

However, bamboo species have limitations of propagation by traditional methods, sexed and asexual (e.g., dismemberment of clumps and cutting). The sexed pathway is characterized by low availability and viability of seeds (Rao et al. 1985; Arya et al. 2008; Mudoi et al. 2013; Singh et al. 2013), the occurrence of irregular flowering (Tsay et al. 1990), and quite long (up to 120 years), high consumption of seeds by fauna, in addition to the resulting genetic heterogeneity (Singh et al. 2013).

The conventional asexual pathway is considered slow (Woods et al. 1995), marked by low availability of propagules and high costs (Mudoi et al. 2013), in addition to low income and need for a lot of labor and allocation of space for production, therefore, not being an adequate strategy for commercial scale (Singh et al. 2013). Moreover, the vegetative bamboo propagation from propagules from plants of unknown age is configured as a risk due to the peculiar behavior of mass flowering and deaths of flowering groups (Bag et al. 2012), since vegetatively propagated plants tend to bloom at the same time as the mother clump (Rao and Rao 1990; Woods et al. 1995).

It should also be noted that due to the aforementioned characteristics inherent to sexual reproduction associated with the perennial nature and monocarpic behavior of bamboo plants, the genetic improvement of this species by conventional methods is a practice of high difficulty (Rao and Rao 1990; Singh et al. 2013; Wei et al. 2015; Ye et al. 2017; Guo et al. 2018; Ramakrishnan et al. 2020). In this sense, aiming to boost the mass propagation of plantlets and subsidize genetic improvement in bamboo species, including, through the use of genetic transformation strategies, in vitro cultivation techniques, are seen as viable for these purposes, enabling the supply of a large number of plants with phytosanitary quality and in a short period of time, in addition to, depending on the route used, making it possible to reset the internal calendar that controls flowering (Arya et al. 2008; Sood et al. 2013).

In bamboos, tissue culture from micropropagation by bud proliferation is already well documented for different species (Singh et al. 2012; Bakshi et al. 2015; Bhadrawale et al. 2017; Furlan et al. 2018; Nogueira et al. 2019; Vale et al. 2019; Silveira et al. 2020). However, it is noteworthy that despite the relatively high number of publications related to in vitro cultivation of bamboo, most of the available methods are insufficient and/or not applicable to all species (Mudoi et al. 2013; Suwal et al. 2020). Therefore, alternative and more efficient in vitro routes need to be investigated and developed. In this sense, the culture of embryogenic cells in suspension and subsequent regeneration of plants via somatic embryogenesis emerge as techniques capable of providing considerable increases in mass production of plantlets. However, studies related to the establishment of cell suspensions in bamboo are limited.

Cell cultures in suspension can allow the propagation and maintenance of cells in a liquid medium, favoring a higher rate of cell division and, thus, exponentially expanding the clonal propagation capacity (Litz and Gray 1995; Cid 1998; Teixeira et al. 2004; Vanisree et al. 2004; Matsumoto 2006; Kong et al. 2020). In addition to being an intermediate step of protocols aimed at the clonal regeneration of plants via somatic embryogenesis, the establishment of cell suspensions guarantees other advantages, such as favoring the investigation of biological processes of interest (Mustafa et al. 2011; Ogita et al. 2012b, c; Nomura et al. 2013), making it possible to obtain secondary metabolites (Taghizadeh et al. 2019; Vijendra et al. 2020), allowing for mutagenesis studies (Jeoung et al. 1998), consisting of a source of protoplasts (Woods et al. 1995; Van den Akker et al. 2015), and being an alternative to enable genetic transformation (Woods et al. 1995; Ogita et al. 2011; Ogita et al. 2012a) and conservation via cryopreservation (Al-Bahrany and Al-Khayri 2012) (Fig. 5.1).

In view of this situation and due to the limited availability of literature and the direction of research on the subject, this chapter compiles the available works on the establishment of cell suspensions in bamboo and describes a pioneering and detailed

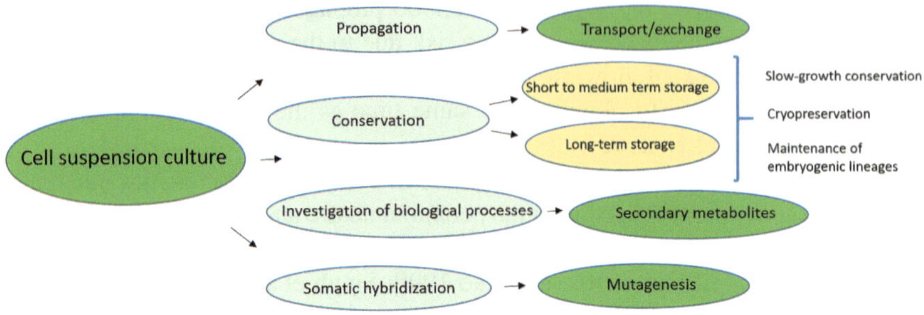

Fig. 5.1 Advantages of cell suspension culture

methodology of easy application for the successful establishment and maintenance of cell suspensions with embryogenic potential in bamboo.

5.2 Cell Suspensions in Bamboo

Cell cultures in suspension, or cell suspension, are the method that allows the induction, propagation, and maintenance of cells in a liquid medium, that is, without the addition of a gelling agent (Table 5.1). Generally, for the culture of suspended cells, friable calluses that have been induced in vitro are selected and inoculated in a liquid culture medium (Bhatia 2015). In these conditions, materials are usually kept under agitation and in the dark, in order to form a suspended content in which cells acquire competence to develop and form cellular aggregates (Mazareia et al. 2011; Kshirsagar et al. 2015).

Due to this direct contact of the cells with the nutrients of the culture medium, the cell cultures in suspension may favor a higher rate of cell division, allowing an accelerated and exponential growth of cells (Cid 1998; Vanisree et al. 2004; Mohd et al. 2012). Due to the characteristics inherent in the liquid medium, such as better availability of nutrients (Gawel and Robacker 1990; Preil 2005), gradual pH change during cultivation (Gawel and Robacker 1990), reduction of toxin effect, and removal of polarity, among others, the use of this system favors the attainment of a high number of somatic embryos (Ascough and Fennell 2004) and, consequently, potentiates the clonal propagation and production of synthetic seeds of species of interest. Generally, in systems aimed at clonal regeneration via somatic embryogenesis, the cultivation in liquid medium consists of an intermediate step aiming at multiplication concomitantly with the maintenance of embryogenic cell potential (Ascough and Fennell 2004). In addition, the use of liquid media has the advantage of reducing costs during plant production, in addition to enabling automation (Teixeira et al. 1995).

In bamboos, still in the 1990s, Rao and Rao (1990) already warned about the need to direct research for the development of embryogenic cell suspensions, aiming at the production of plantlets on a mass scale with minimal costs. These same authors

Table 5.1 Protocols for the establishment of cell suspensions in different bamboo species

Species	Purpose	Explant	Cultivation medium	References
*	To evaluate the affinity of the *Agrobacterium tumefaciens* to bamboo cells for the purpose of genetic transformation	*	MS + 45.2 μM 2,4-D	Douglas et al. (1985)
*	Somatic embryogenesis	*	Adaptation of Finer and Nagasawa (1988)s methodology	Woods et al. (1995)
Bambusa multiplex, *B. oldhamii*, *Phyllostachys aurea*, and *Sasa pygmaea*	*	Shoot apices	MS + 13.5 μM 2,4-D	Huang et al. (1988) cited by Chang (1991)
Bambusa multiplex and *B. oldhamii*	Isolate protoplasts	Shoot apices	*	Huang et al. (1989) cited by Chang (1991)
Bambusa multiplex and *B. oldhamii*	Isolate protoplasts	Shoot apices	*	Huang et al. (1990) cited by Chang (1991)
Dendrocalamus giganteus	Somatic embryogenesis	Shoots	MS + 33.9 μM 2,4-D with or without 16.2 μM NAA	Ramanayake and Wanniarachchi (2003)
Bambusa edulis	Investigate sucrose metabolism	*	MS + 13.5 μM 2,4-D	Liu et al. (2005)
Phyllostachys nigra	Subsidize subsequent manipulations	Shoots	MS/2 + 3 μM 2,4-D	Ogita (2005)
Phyllostachys spp.	Carbohydrate absorption by cells in bamboo suspension	*	*	Ogita et al. (2008)
Bambusa oldhamii	Isolate chitinases and investigate their biochemical characteristics	*	MS + 13.5 μM 2,4-D	Kuo et al. (2009)
Phyllostachys nigra	Genetic transformation	*	MS/2 + 3 μM 2,4-D	Ogita et al. (2011)
Phyllostachys bambusoides	Evaluate the processes of division, differentiation,	*	MS/2 + 10 μM 2,4-D	Ogita et al. (2012b)

(continued)

Table 5.1 (continued)

Species	Purpose	Explant	Cultivation medium	References
	and lignification of cells			
Phyllostachys nigra	Investigate the process of cell lignification	*	MS/2 + 3 μM 2,4-D	Ogita et al. (2012c)
Phyllostachys nigra	Investigate the composition of secondary metabolites involved with the process of xylogenesis	*	MS + 680 mg L^{-1} KH_2PO_4 + 10 μM picloram	Nomura et al. (2013)
Bambusa balcooa	Analyze the dynamics of endogenous cytokinins	Somatic embryos induced from pseudospikelets	MS + 6.7 or 2.2 μM 2,4-D	Van den Akker et al. (2015)
Bambusa multiplex, Phyllostachys bambusoides, P. nigra, Sasa kurilensis	Isolate protoplasts	Node e young shoots	MS + 680 mg L^{-1} KH_2PO_4 + 10 μM picloram	Ogita and Sasamoto (2017)
Guadua magna e *G.* aff. *Chaparensis*	Mass propagation	Nodal segments	MS + 4.4 μM picloram or 2,4-D	Queiroz (2020)

*Information not available. *KH_2PO_4* monopotassium phosphate, *MS* Murashige and Skoog (1962), *NAA* naphthaleneacetic acid, *Picloram* 4-amino-3,5,6-trichloropicolinic acid, *2,4-D* 2,4-dichlorophenoxyacetic acid

reported success in establishing suspended crops. Rao and Rao (1990) also highlighted the importance of the establishment of bamboo cell suspensions aiming at advances in protoplast isolation, which would open opportunities for induction of somatic embryogenesis, isolation of somaclonal variants, and somatic hybridization.

A few years later, Woods et al. (1995) mentioned the regeneration of bamboo plants from somatic embryos from embryogenic cell suspensions by adapting a suspension culture system previously developed for *Glycine max* by Finer and Nagasawa (1988). Chang (1991), while reviewing the literature on the establishment of cell suspensions in bamboo, reported the pioneering work on the establishment of cell suspension protocols by Huang et al. (1988) in *Bambusa multiplex*, *B. oldhamii*, *Phyllostachys aurea*, and *Sasa pygmaea* and Huang et al. (1989, 1990) in *B. multiplex* and *B. oldhamii*.

Despite the recognized importance of directing research to establish cell suspensions in bamboo, from the 1980s to the present day, few advances have been obtained, with studies limited to a few species, such as *Dendrocalamus giganteus*. Aiming at the in vitro regeneration of clonal plants of this species, Ramanayake and Wanniarachchi (2003) established cell suspensions in culture medium MS (Murashige and Skoog 1962) supplemented with 33.9 μM of 2,4-dichlorophenoxyacetic acid (2,4-D) alone or in combination with 40.3 μM of naphthaleneacetic acid (NAA), from friable calluses from shoots. According to the authors, it took 2 months for the formation of cell suspensions, which reached continuous growth after subsequent transfers to larger bottles with fresh media. Despite the finding of pro-embryos in established cell suspensions, the authors did not verify further development of these structures, classifying them as transient. Despite the nonsuccess in obtaining plants via pro-embryos, Ramanayake and Wanniarachchi (2003) managed to keep *D. giganteus* cells in suspension for a period of more than 2 years, which indicates the potential use of this methodology for in vitro conservation in germplasm banks.

Ogita (2005), with the objective of investigating the morphoanatomic and physiological characteristics of *Phyllostachys nigra* cells in order to subsidize subsequent manipulations, established cell suspensions using culture medium MS/2 plus 3.0 μM of 2,4-D, from calluses from shoots. Suspensions established after 3–6 weeks of culture were morphologically characterized by the presence of aggregates composed of cells with shapes ranging from round to oval and with 200–500 μm of diameter. According to the author, the high proliferative potential of *P. nigra* suspension cells with expressive deposition of cell wall compounds (callose, β-1,3-glucans) is the main information provided by this work.

According to Ogita et al. (2012c), the optimization of the protocols of cell cultures in suspension is an important and useful approach for a better understanding of biological processes in plant species of interest. These authors, for example, investigated the lignification process in living bamboo cells (*Phyllostachys* genus) using cell suspensions. According to Liu et al. (2005), cell suspension cultures develop under strictly controlled growth conditions, thus avoiding that experiments are affected by differentiation processes. According to these authors, it is an ideal system to investigate sucrose metabolism in *B. edulis* cells. This system was also used in *B. oldhamii* for the isolation of chitinases and investigation of their biochemical characteristics (Kuo et al. 2009).

Cell suspensions were also established in order to evaluate the processes of division, differentiation, and lignification of *P. bambusoides* cells (Ogita et al. 2012b). Similarly, Ogita et al. (2008) used suspended cell cultures to analyze carbohydrate metabolism related to the growth and development of the bamboo cell wall. Nomura et al. (2013), in turn, used *P. nigra* suspension cells to investigate the composition of secondary metabolites involved with the xylogenesis process.

In this context of various applications, Van den Akker et al. (2015) highlighted the importance of optimizing cultures of suspended cells in order to use them as a model system for various studies, such as those related to obtaining protoplasts and genetic transformation. These authors also established cell suspensions from somatic embryos of *B. balcooa*, in the medium of MS supplemented with 2,4-D, and analyzed the dynamics of endogenous cytokinins.

Protocols for the establishment of cell suspension, developed specifically for the isolation of bamboo protoplasts, were reported by Huang et al. (1989, 1990) in *B. multiplex* and *B. oldhamii* and Ogita and Sasamoto (2017) in *B. multiplex*, *P. bambusoides*, *P. nigra*, and *Sasa kurilensis*. To obtain cell suspensions, these last authors used node and young shoots grown in culture medium MS supplemented with 10 μM of 4-amino-3,5,6-trichloropicolinic acid (picloram). The suspensions were kept under dark, at 27 °C in a rotary agitator at 100 rpm of speed.

Ogita et al. (2011) obtained stable transgenic bamboo cells through a genetic transformation protocol via bombardment of *P. nigra* suspension cells. In addition to the success in the transformation, Ogita et al. (2011) discuss technical details related to maintaining high-efficiency growth of calluses/bamboo suspension cells. According to the authors, the efficient establishment of a suspension cell culture system is one of the main bottlenecks that need to be superimposed to obtain bamboo transgenic cells, in addition to the need for improvement in transformation procedures. The authors established a culture system of cells in bamboo suspension with high efficiency by supplementation of the medium MS with 680 mg L^{-1} of monopotassium phosphate (KH_2PO_4) and addition of 10 μM picloram. As a result, Ogita et al. (2011) reported suspended cells with morphological uniformity and synchrony in the cell division process, which were used to develop the transformation protocol.

It is important to mention that already in the 1980s, Douglas et al. (1985) successfully established bamboo suspension cell culture in MS medium with 10 mg L^{-1} of 2,4-D under constant light, in order to evaluate the binding ability of *Agrobacterium tumefaciens* bacteria to bamboo cells for the purpose of genetic transformation into monocotyledonous. It is emphasized that an efficient tissue regeneration system is a prerequisite for genetic transformation in any plant species. However, despite efforts, genetic transformation is still quite limited in bamboo (Wei et al. 2015; Ye et al. 2017; Ramakrishnan et al. 2020), due to the low efficiency and slowness of the in vitro regeneration process (Ramakrishnan et al. 2020), which requires continuous investigation of the various factors and phases involved in the process and optimization of available protocols. According to Ramakrishnan et al. (2020), improvement reports of interest agronomic characteristics in bamboos through genetic engineering remain limited to a single species, *D. latiflorus*.

5.3 Stages of Callogenesis and Establishment of Cell Suspensions

A brief outline of the different steps for establishing cell suspension in bamboo is illustrated in Fig. 5.2. For experimental installations and, consequently, obtaining calluses in bamboo, mother plants found in natural regions are used as plant materials, from which seeds are collected and subjected to in vitro germination. However, this material is difficult to obtain, due to the reproductive peculiarities already presented earlier. Therefore, field collection of developing seedlings can also be carried out, which are transplanted into pots and kept in greenhouses. From this material, vegetative parts of plants can be used as alternatives for the in vitro establishment of bamboo species, such as nodal segments that correspond to efficient

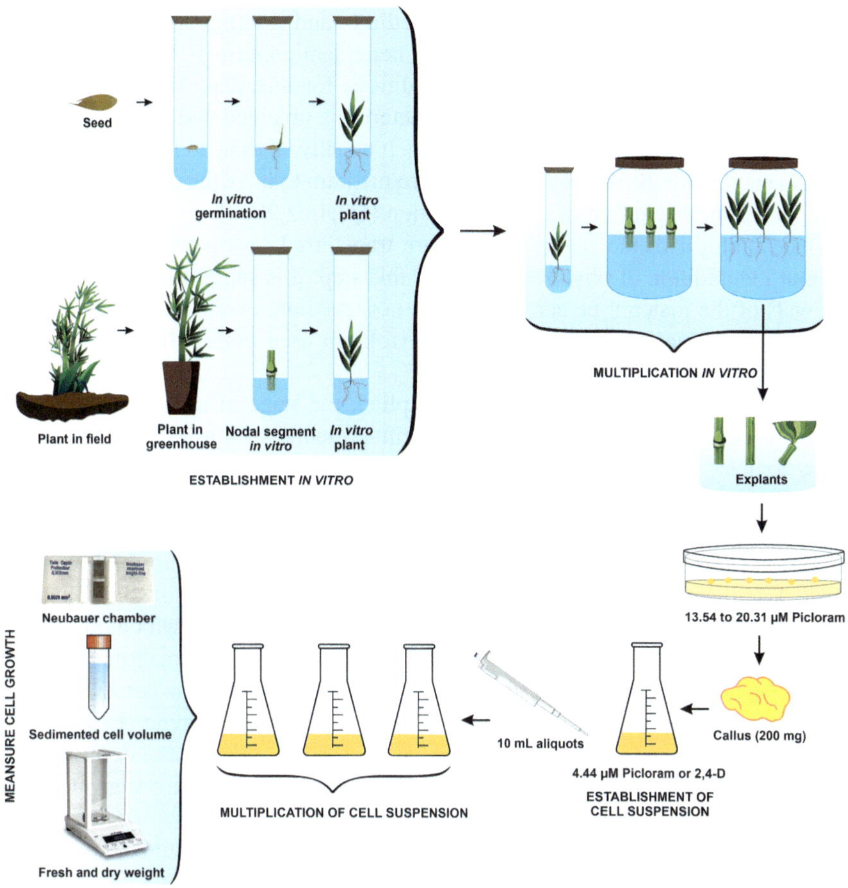

Fig. 5.2 Illustrative summary of the steps involved in establishing cell suspensions in bamboo. First, plants are established in vitro (seed germination or multiplication by nodal segments of plants collected in the field and previously kept in a greenhouse). The established plants are then multiplied, and from these, different explants can be obtained (nodal and internodal segments and leaf sheaths), which are inoculated in a callus-inducing medium composed of salts and vitamins of Murashige and Skoog (1962) (MS) supplemented with 13.54–20.31 µM picloram. For the establishment of cell suspensions, 200–500 mg of the calli (friable) are inoculated in Erlenmeyer flasks containing MS medium added of 4.4 µM picloram or 4.44 µM 2,4-D. Subsequently, the material is filtered through a 100-µm sieve, and 10-mL aliquots of the dense phase of the cell suspensions are induced to multiply in new flasks containing 20 mL of culture medium. To measure the growth of the cell suspension, different methodologies may be used, such as cell counting in the Neubauer chamber, measurement by cell volume after sedimentation (CVS), and measurement of fresh and dry mass

explants (Queiroz et al. 2019; Nogueira et al. 2019). It is emphasized that for this, efficient disinfestation methods should be applied (Nogueira et al. 2019).

Once established, the plants are subjected to the in vitro multiplication step. For the multiplication of *G. magna* and *G.* aff. *chaparensis*, using 650-mL transparent glass vials containing 70 mL of MS medium and adding 20 g L^{-1} sucrose, 1.0–3.0 mg L^{-1} metaTopoline (mT), or 6-benzylaminopurine (BAP), without the addition of gelling agent, are suitable conditions for this step. The objective is to maintain such conditions until sufficient material is obtained and without contamination for the installation of the experiments. It usually takes up to five subcultures of 60 days each, with the renewal of the culture medium to have the optimal amount of material for the beginning of callus induction (Queiroz 2020).

Once multiplied, the plants in vitro are transferred to a new culture medium, without the addition of phytoregulators. In this step, it is suggested that up to three renewals of the material be carried out in a new medium every 30 days, in order to reduce the residual impact of exogenous cytokinin used during the plant multiplication step (Queiroz 2020).

With the material established and multiplied, the stage of induction of calluses begins, which can be obtained from different explants, such as leaves, roots, stem parts, axillary shoots, anthers, stilettos, nodal segments (Fig. 5.3a), internodal segments (Fig. 5.3b), and leaf sheath (Fig. 5.3c).

One of the explants with the greatest potential for callogenesis in bamboo is nodal segment, where portions of 1–2 cm containing an immature axillary bud protected by the leaf sheath are subjected to callus induction treatments (Fig. 5.3a). This type of material in particular is formed by young and tender tissues, which, in general, facilitate the process of cell de-differentiation and consequent formation of calluses (Fig. 5.3d–i), characterized as clusters of cells with heterogeneous constitution and different development potentials, consequent to cell reprogramming (Fehér 2019). Additionally, the presence of meristematic areas, which are common in this type of explant, is a factor that determines the success of in vitro cultivation in monocotyledons (Benson 2000).

Nodal segments respond satisfactorily when inoculated in a culture medium of MS supplemented with sucrose (30 mg L^{-1}), hydrolyzed casein (500 mg L^{-1}), glutamine and cysteine (100 mg L^{-1}), Phytagel (2.2 g L^{-1}), and between 13.54 and 20.31 μM of picloram or 2,4-D, although picloram in some species may present even better results than 2,4-D (Queiroz 2020).

The formation of calluses in nodal segments occurs between 60 and 120 days in a nonluminous environment. The formed calluses may present a differentiated morphological aspect, such as those with a color ranging between beige and yellow (Fig. 5.3d, e) and friable ones with a color varying between white and yellow (Fig. 5.4a). Compact callus is usually anatomically characterized by zones with cells with meristematic characteristics (cells with smaller diameter, isodiametric, high nucleus/cytoplasm ratio, with small starch grains) (Verdeil et al. 2007) and areas with vacuolated cells, with no visible nuclei (Fig. 5.3f). Anatomical analyses reveal that these calluses are possibly originated from the multiplication of perivascular meristematic cells (Fig. 5.3g–i), also known as procambial, *stem cells* or *stem cell-like cells*, which naturally recruit vascular tissues (Fukuda 2004; Liu et al. 2014; Jiang et al. 2015).

On the other hand, friable calluses (Fig. 5.4a) are usually morphologically characterized by ease of fragmentation (Fransz and Schel 1991) and, anatomically,

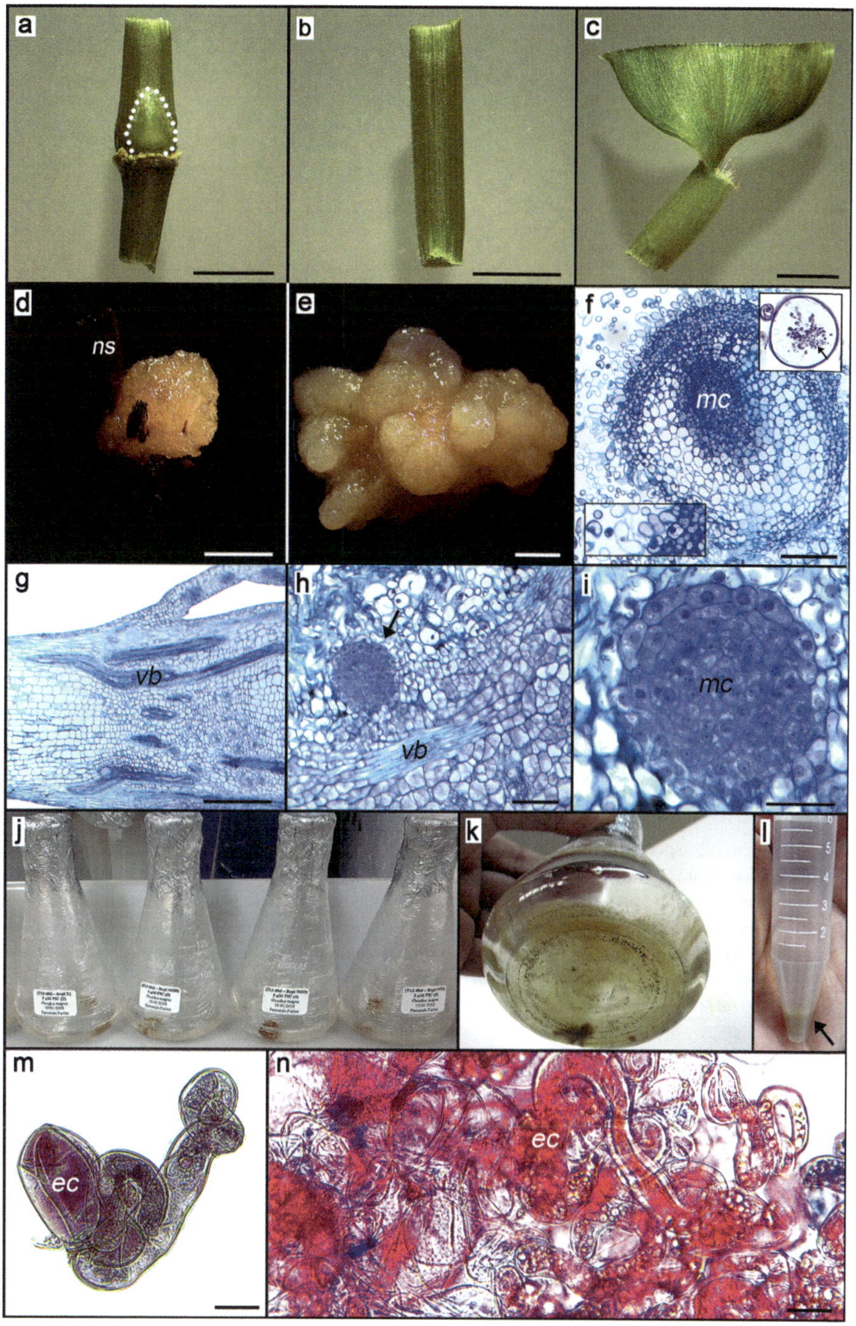

Fig. 5.3 General aspects of the establishment of cell suspensions in *Guada magna*. (**a–c**) Explant nodal segment (dotted axillary gem), internodal segment, and leaf sheath, respectively, used for callus induction. (**d, e**) Compact-type calluses originating from nodal segments. (**f**) Anatomical section of compact callus derived from nodal segment; note desquamation of peripheral cells with meristematic characteristics (cells in greater magnification – lower detail) and cell with

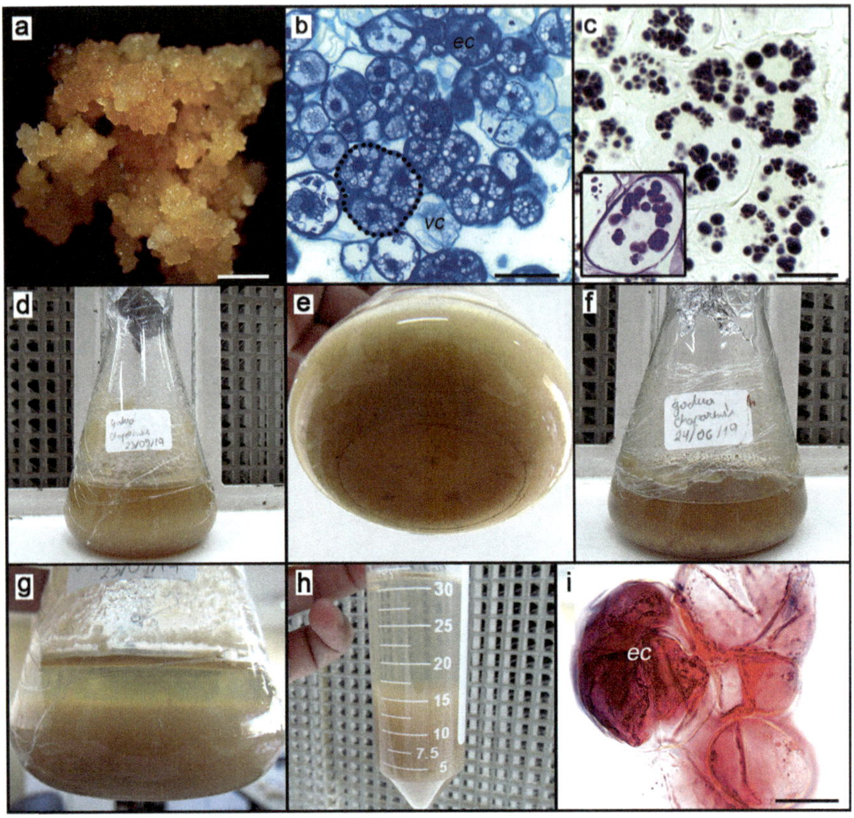

Fig. 5.4 General aspects of the establishment of cell suspensions in *Guadua* aff. *chaparensis*. (**a**) Friable callus from nodal segment. (**b**) Anatomical section of friable callus; note cells with embryogenic potential, pro-embryos (dotted), and vacuolated cells. (**c**) Friable callus cells with starch grains stained with Lugol and *Periodic Acid Schiff* (lateral detail). (**d–g**) Cell suspensions established. (**h**) Sedimented cell volume. (**i**) Cytochemical analysis with carmine acetic and Evans blue to verify embryogenic viability of suspended cells. *ec* embryogenic cell, *vc* vacuolated cell. Bars: 2 mm, **a**; 50 μm, **b**; 200 μm, **c**; 20 μm, **i**

Fig. 5.3 (continued) polysaccharides stained with *Periodic Acid Schiff* (upper detail – arrow). (**g**) Anatomical section of nodal segment after 7 days in the middle of callus induction; observe multiplication of perivascular meristematic cells (lateral detail – dotted). (**h, i**) Anatomical section of nodal segment with multiplication of perivascular meristematic cells after 60 days in callus induction medium. (**j**) Cultivation of calluses in liquid medium for establishment of cell suspension. (**k**) Cell suspension established. (**l**) Sedimented cell volume. (**m, n**) Cytochemical analysis with carmine acetic and Evans blue to verify embryogenic viability of suspended cells. *ec* embryogenic cell, *mc* meristem cells, *ns* nodal segment, *vb* vascular bundle. Bars: 2 mm, **a–e**; 200 μm, **f**; 500 μm, **g**; 100 μm, **h**; 50 μm, **i**; 25 μm, **m–n**

are formed by isolated cells with embryogenic characteristics (isodiametric cells, with evident nuclei, usually centralized, and with a single nucleolus, with high nucleus/cytoplasm ratio, fragmented vacuoles, and starch grains) (Verdeil et al. 2007; Silva-Cardoso et al. 2019), distributed among vacuolated cells in apparent degeneration (Figs. 5.4b, c).

Such cells with embryogenic characteristics usually divide and form structures known as proembrions (Fig. 5.4b), which precede the formation of somatic embryos. These calluses are often used to establish cell cultures in suspension, due to their rarefied structure that facilitates dissociation.

It should be noted that although friable calluses are preferred for the establishment of cell suspensions, since less force is needed for cell separation (Bhatia 2015), calluses with compact centers, originating from nodal segments, are also able to generate cell suspensions successfully (Figs. 5.3j–l), probably, by desquamation of peripheral cells with meristematic characteristics (Fig. 5.3f).

For the establishment of cell suspensions, approximately 200–500 mg of the formed calluses are transferred to Erlenmeyer flasks (125 mL) (Fig. 5.1j), containing 20 mL of liquid MS medium, supplemented with 30 g L^{-1} sucrose, in addition to the amino acid hydrolyzed casein, glutamine, proline, and cysteine (100 mg L^{-1}), with 4.4 μM picloram or 2,4-D.

The material is kept in an orbital agitator at 100 rpm at low luminosity incidence and at 25 ± 2 °C. In a few days of cultivation, it is possible to verify a growing cell content coming from the inoculated calluses. In general, both compact calluses (Fig. 5.3d, e) and friable calluses (Fig. 5.4a), but preferably friable, form cultures of dense consistency, with yellowish-white coloration and homogeneous and milky appearance (Figs. 5.3k and 5.4d, e).

After 30–40 days of cultivation, there is an intense growth of cell culture in suspension, which can double cultivated content (Fig. 5.4f, g). It is emphasized that this behavior is more characteristic of cell suspensions established from friable calluses. Thus, periodically, it is necessary to renew the nutrient medium, because the material presents intense proliferation of cell content, which can cause oxidation of the material or even death of the culture due to the low availability of essential nutrients.

Once obtained, the material must be filtered in a metal sieve of 100 μm, and aliquots of 10 mL of the dense phase of the cell suspensions are induced to multiply in new vials of the same composition containing 20 mL of culture medium.

To measure the cell suspension growth, there are several methodologies, being the most used cell count in Neubauer chamber, measurement by packed cell volume (PCV), and weight measurement of fresh mass and dry table (Khanpour-Ardestani et al. 2015).

For cell count in Neubauer chamber, homogeneous samples between 1 and 5 mL (depending on sampling) are taken from the liquid phase of the cell suspension. This material is arranged in microtubes or conical flasks containing equal volume of Trypan Blue solution (0.4%). This dilution should be resuspended at least three times, and after up to 60 minutes, aliquots of 10 μL are obtained from the samples and carefully introduced into the space between the laminula and the Neubauer chamber, filling it completely by capillarity. Then, with the aid of an optical microscope, the quantification of the results obtained is carried out, which can provide important information about cell culture, such as the kinetics of cell growth and the total and percentage of viable (non-stained) and nonviable (stained) cells, in addition to the concentration by volume (mL) of the cells of the culture. Once it is known that the volume of each of the four corners is 0.1 mm^3 (1 mm^2 area × 0.1 mm^2 depth) or 1×10^{-4} mL for quantification by the Neubauer chamber, the following formula is applied:

$$\text{NC} = \frac{(A + B + C + D)}{4} \times 10^4 \times 2$$

where NC is the number of cells (viable and/or non-viable), resulting from counting the four quadrants of the chamber (*A*, *B*, *C*, *D*), divided by four to obtain the mean. This result is multiplied (10^4) to obtain the number of cells per mL of the sample applied in the Neubauer chamber and further multiplied by two, taking into account the dilution of 1:1 of the cell sample and the Trypan Blue (Louis and Siegel 2011). If there is a need for additional dilutions in the cell volume, because, for example, the cell concentration is very high, this should also be taken into account in the final calculation.

Cell counting is laborious and requires care and attention, as well as adjustments in cell dilution so that at the time of reading the cells do not overlap. In this sense, to know the estimate of the percentage of viable cells of the cell suspension, the following formula is applied:

$$\%\text{viable cells} = \frac{\text{number of viable cells (not stained)}}{\text{total number of cells (viable + dead)}} \times 100$$

Additionally, measurement of suspension multiplication can be performed by sedimented volume package (PVS). For this, the cell material grown in Erlenmeyer flasks (125 mL) is poured into graduated *Falcon* tubes that are previously autoclaved. Once the amount of medium has been deposited, it is recommended that the suspensions be allowed to stand for at least 30 minutes before the evaluation. After this period, the separation of the volumes of the liquid phase corresponds to the culture medium (supernatant), and the volume of the dense phase corresponds to the sedimented cells (Figs. 5.11 and 5.2h). Then the values of sedimented cell volumes are noted. It is also recommended that the procedure be repeated after 30 minutes for taking the second volume reading, which gives greater reliability to the data collected. Thus, this volume of cells obtained is known as packed cell volume (PCV) (Teixeira et al. 2004; Loyola-Vargas and Váquez-Flota 2006).

As for cultures with less expressive cell growth, usually coming from calluses with compact aspect (Fig. 5.1d, e) and which, consequently, are difficult to dissociate, the determination of the growth curve of the cultures is verified by means of fresh and dry mass. For this purpose, it is possible to use the methodology where the vials containing the cell suspension are homogenized manually. Then, 2 mL of the material is obtained with the help of an automatic pipette. This volume is transferred to microtubes of 2 mL capacity, which are subjected to centrifugation at 14,000 rpm for 20 minutes, when then carefully collect 1 mL of the liquid supernatant. Then the microtubes are again subjected to centrifugation, and then the supernatants from the liquid phase are carefully removed, leaving a fresh mass *pellet* (mg) that is measured by precision analytical balance.

Microtubes containing the fresh mass of cell cultures in suspension are stored in a greenhouse at 50 °C until the dry mass is obtained. After this period, the dry mass (mg) is measured with the aid of a precision analytical balance.

Finally, the TCM (average growth rate) is calculated, according to the following formula:

$$\text{TCM } (\%) = \left[\left(\sum X \right) / t \right] \times 100$$

where $\sum X$ is the sum of the mean of the fresh mass and/or dry mass over the cultivation time and t is the cultivation time.

It is emphasized here the importance of identifying cells with embryogenic potential in cell cultures through histological studies and histochemical analyses, such as the technique of double staining with carmine acetic and Evans blue (Khanpour-Ardestani et al. 2015).

In this context, 1 mL aliquots of cell suspension cultured in Erlenmeyer flasks are collected. The collected material is then transferred to microtubes (2 mL). After approximately 1 h of decanting, the supernatant is carefully removed with the aid of a pipette, and then 100 μL of Evans blue dye (0.1%) is added, which remains in contact with the cell formations for 2 minutes. After this period, the excess Evans blue is removed, and 100 μL of carmine acetic dye (2%) is added for the same period of time (Durzan 1988).

Finally, 200 μL of glycerinated water (50%) are added to each microtube, of which 20 μL aliquots are distributed in histological slide. The samples are then covered with foil, subjecting the ready blades to light beatings with a glass stick, for spreading the material. The samples can then be viewed and photographed under a light microscope (Figs. 5.1m, n and 5.2i). Carmine acetic and Evans blue dyes reflect on the integrity of cellular structures such as the nucleus and plasma membrane (Báez et al. 2002; Pline et al. 2002), and potentially embryogenic cells form aggregates that are reactive to carmine acetic and blush red (Steiner et al. 2005).

5.4 Conclusion and Future Prospects

In this chapter, the detailed methodology for the successful establishment of cell suspensions with embryogenic potential in bamboo is first described, based on the species *Guada magna* and *G.* aff. *chaparensis*, which can support the development of efficient protocols in other species for different purposes. It should be noted that despite the high use potential of cell cultures in bamboo suspension and considering the high number of existing species, there is a considerable deficit of research aimed at the establishment of these systems in bamboo, aiming, above all, at mass-scale propagation via somatic embryogenesis. In this sense, future studies covering other bamboo species are necessary, as well as the development of somatic embryogenesis protocols that encompass the establishment and regeneration of suspended cell cultures as an intermediate phase of the process, in order to potentialize them.

Acknowledgments The authors thank the Conselho Nacional de Desenvolvimento Científico e Tecnológico (CNPq Grant 458151/2013-0), Coordenação de Aperfeiçoamento de Pessoal de Nível Superior (Capes/Embrapa 001-2011/Grant 39), and Centro de Pesquisa e Aplicação de Bambu e Fibras Naturais (CPAB) for their financial support and fellowships.

Conflict of Interest The authors declare that they have no conflict of interest.

References

Al-Bahrany AM, Al-Khayri JM (2012) Optimizing *in vitro* cryopreservation of date palm (*Phoenix dactylifera* L.). Biotechnology 11:59–66. https://doi.org/10.3923/biotech.2012.59.66

Alvarez J, Ławińska K, Falkiewicz-Dulik M (2020) Quality assessment of viscose bamboo fabrics intended for use inside children's and special footwear. Fibres Text East Eur 4(142):82–88. https://doi.org/10.5604/01.3001.0014.0940

Arya S, Satsangi R, Arya ID (2008) Large scale production of edible bamboo *Dendrocalamus asper* through somatic embryogenesis. J Am Bamboo Soc 21(1):21–31

Arya S, Sharma S, Kaur R, Arya ID (1999) Micropropagation of *Dendrocalamus asper* by shoot proliferation using seeds. Plant Cell Rep 18(10):879–882

Ascough GD, Fennell CW (2004) The regulation of plant growth and development in liquid culture. S Afr J Bot 70(2):181–190. https://doi.org/10.1016/S0254-6299(15)30234-9

Báez P, Riveros M, Lehnebach C (2002) Viability and longevity of pollen of *Nothofagus* species in South Chile. N Z J Bot 40(4):671–678

Bag N, Palni LMS, Chandra S, Nandi SK (2012) Somatic embryogenesis in 'Maggar' bamboo (*Dendrocalamus hamiltonii*) and field performance of regenerated plants. Curr Sci 102 (9):1279–1287. https://doi.org/10.13140/RG.2.2.13665.81763

Bakshi M, Tiwari C, Razvi S (2015) Conservation of an important montane bamboo *Thamnocalamus falconeri*, Hook.F. ex Munro through axillary bud proliferation. J For Res 26 (1):179–185. https://doi.org/10.1007/s11676-015-0022-3

Benson EE (2000) Special symposium: *In vitro* plant recalcitrance—*In vitro* plant recalcitrance: an introduction. In Vitro Cell Dev Biol-Plant 36:141–148. https://doi.org/10.1007/s11627-000-0029-z

Bhadrawale D, Mishra JP, Mishra Y (2017) An improvised *in vitro* vegetative propagation technique for *Bambusa tulda*: influence of season, sterilization and hormones. J For Res 29:1069–1074. https://doi.org/10.1007/s11676-017-0569-2

Bhatia S (2015) Plant tissue culture. In: Bhatia S, Sharma K, Dahiya R, Bera T (eds) Modern applications of plant biotechnology in pharmaceutical sciences. Academic Press, Boston, pp 31–107

Carnegie AJM (1997) Sugarcane and bamboo. In: Ben-Dov Y, Hodgson CJ (eds) Soft scale insects—their biology, natural enemies and control, vol 7. Elsevier Science, pp 333–341. https://doi.org/10.1016/s1572-4379(97)x8073-x

Chang F-C, Chen K-S, Yang P-Y, Ko C-H (2018) Environmental benefit of utilizing bamboo material based on life cycle assessment. J Clean Prod 204:60–69. https://doi.org/10.1016/j.jclepro.2018.08.248

Chang W-C (1991) Bamboos. In: Bajaj YPS (ed) Biotechnology in agriculture and forestry 16 trees III. Springer-Verlag, Berlin, pp 221–237. https://doi.org/10.1007/978-3-662-13231-9

Cid LPB (1998) Suspensão celular. In: Torres AC, Caldas LS, Buso JA (eds) Cultura de tecidos e transformação genética de plantas, vol 1. EMBRAPA-SPI/EMBRAPA-CNPH, Brasília, pp 331–353

Douglas C, Halperin W, Gordon M, Nester E (1985) Specific attachment of *Agrobacterium tumefaciens* to bamboo cells in suspension cultures. J Bacteriol 161(2):764–766

Durzan DJ (1988) Applications of cell and tissue culture in tree improvement. In: Applications of plant cell and tissue culture. Ciba foundation symposium 137. Wiley, Chichester, pp 36–58

Fatimah I, Rubiyanto D, Taushiyah A, Najah FB, Azmi U, Sim Y-L (2019) Use of ZrO_2 supported on bamboo leaf ash as a heterogeneous catalyst in microwave-assisted biodiesel conversion. Sustain Chem Pharm 12:100129. https://doi.org/10.1016/j.scp.2019.100129

Fehér A (2019) Callus, dedifferentiation, totipotency, somatic embryogenesis: what these terms mean in the era of molecular plant biology? Front Plant Sci 10:536. https://doi.org/10.3389/fpls.2019.00536

Felisberto MHF, Miyake PSE, Beraldo AL, Clerici MTPS (2017) Young bamboo culm: potential food as source of fiber and starch. Food Res Int 101:96–102. https://doi.org/10.1016/j.foodres.2017.08.058

Finer JJ, Nagasawa A (1988) Development of an embryogenic suspension culture of soybean (*Glycine max* Merrill.). Plant Cell Tissue Organ Cult 15:125–136. https://doi.org/10.1007/BF00035754

Fransz PF, Schel JHN (1991) Cytodifferentiation during the development of friable embryogenic callus of maize (*Zea mays*). Can J Bot 69(1):26–33. https://doi.org/10.1139/b91-005

Fukuda H (2004) Signals that control plant vascular cell differentiation. Nat Rev Mol Cell Biol 5:379–391. https://doi.org/10.1038/nrm1364

Furlan FC, Gavilan NH, Zorz AZ, Oliveira LS, Konzen ER, Brondani GE (2018) Active chlorine and charcoal affect the *in vitro* culture of *Bambusa vulgaris*. Bosque 39(1):61–70. https://doi.org/10.4067/S0717-92002018000100061

Gawel NJ, Robacker CD (1990) Somatic embryogenesis in two *Gossypium hirsutum* genotypes on semisolid versus liquid proliferation media. Plant Cell Tissue Organ Cult 23:201–204. https://doi.org/10.1007/BF00034432

Guan M, An X, Liu H (2019) Cellulose nanofiber (CNF) as a versatile filler for the preparation of bamboo pulp based tissue paper handsheets. Cellulose 26:2613–2624. https://doi.org/10.1007/s10570-018-2212-6

Guo X, Ji L, Zhu L, Song L, Liu S, Zang Q, Lin X (2018) Bamboo NiR gene is associated with regeneration capacity. J Hortic Sci Biotechnol 93(2):142–149. https://doi.org/10.1080/14620316.2017.1353894

Hong C, Li H, Lorenzo R, Wu G, Corbi I, Corbi O, Xiong Z, Yang D, Zhang H (2019) Review on connections for original bamboo structures. J Renew Mater 7(8):713–730. https://doi.org/10.32604/jrm.2019.07647

Huang LC, Chen W, Huan B (1989) Tissue culture investigations of bamboo III. A method for viable protoplast isolation from *Bambusa* cells of liquid suspension culture. Bot Bull Acad Sin 30:49–57

Huang LC, Chen WL, Huang BL (1988) Tissue culture investigations of bamboo II. Liquid suspension cultures of *Bambusa, Phyllostachys* and *Sasa* cells. Bot Bull Acad Sin 29 (3):117–182

Huang LC, Huang BL, Chen WL (1990) Tissue culture investigations of bamboo V. recovery of callus from protoplasts of suspension-cultured *Bambusa* cells. Bot Bull Acad Sin 31(1):29–34

Jeoung J-M, Fritz JO, Liang GH (1998) Callus induction, plant regeneration and mutagenesis of Eastern Gamagrass. Cytologia 63(4):381–386. https://doi.org/10.1508/cytologia.63.381

Jiang F, Feng Z, Liu H, Zhu J (2015) Involvement of plant stem cells or stem cell-like cells in dedifferentiation. Involvement of plant stem cells or stem cell-like cells in dedifferentiation. Front Plant Sci 6:1028. https://doi.org/10.3389/fpls.2015.01028

Khanpour-Ardestani N, Sharifi M, Behmanesh M (2015) Establishment of callus and cell suspension culture of *Scrophularia striata* Boiss.: an *in vitro* approach for acteoside production. Cytotechnology 67:475–485. https://doi.org/10.1007/s10616-014-9705-4

Komatsu YH, Batagin-Piotto KD, Brondani GE, Gonclaves AN, Almeida MD (2011) *In vitro* morphogenic response of leaf sheath of *Phyllostachys bambusoides*. J For Res 22:209–215. https://doi.org/10.1007/s11676-011-0152-1

Kong EYY, Biddle J, Foale ASW (2020) Cell suspension culture: a potential *in vitro* culture method for clonal propagation of coconut plantlets via somatic embryogenesis. Ind Crop Prod 147:112125. https://doi.org/10.1016/j.indcrop.2020.112125

Kshirsagar PR, Chavan JJ, Umdale SD, Nimbalkar MS, Dixit GB, Gaikwad NB (2015) Highly efficient *in vitro* regeneration, establishment of callus and cell suspension cultures and RAPD analysis of regenerants of *Swertia lawii* Burkill. Biotechnol Rep 6:79–84. https://doi.org/10.1016/j.btre.2015.03.003

Kuo C-J, Huang L-C, Liao Y-C, Chang C-T, Sung H-Y (2009) Biochemical characterization of a novel chitotriosidase from suspension-cultured bamboo (*Bambusa oldhamii*) cells. Bot Stud 50 (3):281–289

Litz RE, Gray DJ (1995) Somatic embryogenesis for agricultural improvement. World J Microbiol Biotechnol 11:416–425. https://doi.org/10.1007/BF00364617

Liu C-C, Huang L-C, Chang C-T, Sung H-Y (2005) Purification and characterization of soluble invertases from suspension-cultured bamboo (*Bambusa edulis*) cells. Food Chem 96 (4):621–631. https://doi.org/10.1016/j.foodchem.2005.02.044

Liu J, Sheng L, Xu Y, Li J, Yang Z, Huang H, Xu L (2014) WOX11 and 12 WOX11 and 12 are involved in the first-step cell fate transition during de novo root organogenesis in *Arabidopsis*. Plant Cell 26(3):1081–1093. https://doi.org/10.1105/tpc.114.122887

Louis KS, Siegel AC (2011) Cell viability analysis using trypan blue: manual and automated methods. Methods Mol Biol 740:7–12. https://doi.org/10.1007/978-1-61779-108-6_2

Loyola-Vargas V, Váquez-Flota F (2006) Plant cell culture protocols, 2rd edn. Humana, Totowa

Marafon AC, Amaral AFC, Lemos EEP (2019) Characterization of bamboo species and other biomasses with potential for thermal energy generation. Pesqui Agropecu Trop 49:e55282. https://doi.org/10.1590/1983-40632019v4955282

Matsumoto K (2006) Cultura de células em suspensão: focalizando a bananeira. Embrapa, Brasília

Mazareia M, Al-Ahmada H, Rudis MR, Blake LJ, Stewart CN Jr (2011) Switchgrass (*Panicum virgatum* L.) cell suspension cultures: establishment, characterization, and application. Plant Sci 181:712–715. https://doi.org/10.1016/j.plantsci.2010.12.010

Mohd RZ, Kukreja AK, Bisht NS (2012) Plant regeneration in *Chlorophytum borivilianum* Sant. Et Fernand. From embryogenic callus and cell suspension culture and assessment of genetic fidelity of plants derived through somatic embryogenesis. Physiol Mol Biol Plants 18 (3):253–263. https://doi.org/10.1007/s12298-012-0113-y

Moraes MJB, Moraes JCB, Tashima MM, Akasaki J, Soriano L, Borrachero MV, Payá J (2019) Production of bamboo leaf ash by auto-combustion for pozzolanic and sustainable use in cementitious matrices. Constr Build Mater 208:369–380. https://doi.org/10.1016/j.conbuildmat.2019.03.007

Mudoi KD, Saikia SP, Goswami A, Gogoi A, Bora D, Borthakur M (2013) Micropropagation of important bamboos: a review. Afr J Biotechnol 12(20):2770–2785. https://doi.org/10.5897/AJB12.2122

Murashige T, Skoog F (1962) A revised medium for rapid growth and bioassays with tobacco tissue cultures. Physiol Plant 15(3):473–497. https://doi.org/10.1111/j.1399-3054.1962.tb08052.x

Mustafa NR, Winter W, van Iren F, Verpoorte R (2011) Initiation, growth and cryopreservation of plant cell suspension cultures. Nat Protoc 6:715–742. https://doi.org/10.1038/nprot.2010.144

Nirmala C, Bisht MS, Bajwa HK, Santosh O (2018) Bamboo: a rich source of natural antioxidants and its applications in the food and pharmaceutical industry. Trends Food Sci Technol 77:91–99. https://doi.org/10.1016/j.tifs.2018.05.003

Nogueira JS, Gomes HT, Scherwinski-Pereira JE (2019) Microprapagation, plantlets production estimation and ISSR marker-based genetic fidelity analysis of *Guadua magna* and *G. angustifolia*. Pesqui Agropecu Trop 49:e53743. https://doi.org/10.1590/1983-40632019v4953743

Nomura T, Shiozawa M, Ogita S, Kato Y (2013) Occurrence of hydroxycinnamoylputrescines in xylogenic bamboo suspension cells. Plant Biotechnol 30(5):447–453. https://doi.org/10.5511/plantbiotechnology.13.0704a

Ogita S (2005) Callus and cell suspension culture of bamboo plant, *Phyllostachys nigra*. Plant Biotechnol 22:119–125. https://doi.org/10.5511/plantbiotechnology.22.119

Ogita S, Kikuchi N, Nomura T, Kato Y (2011) A practical protocol for particle bombardment-mediated transformation of *Phyllostachys* bamboo suspension cells. Plant Biotechnol 28:43–50. https://doi.org/10.5511/plantbiotechnology.10.1101a

Ogita S, Kikuchi N, Nomura T, Kato Y (2012a) The mutated acetolactate synthase gene from rice as a non-antibiotic selection marker for transformation of bamboo cells. Am J Plant Sci 3 (3):368–372. https://doi.org/10.4236/ajps.2012.33044

Ogita S, Nomatu T, Kishimoto T, Kato Y (2012c) A novel xylogenic suspension culture model for exploring lignification in *Phyllostachys* bamboo. Plant Methods 8(40):1–9. https://doi.org/10.1186/1746-4811-8-40

Ogita S, Ohki S, Kato Y (2008) Uptake of carbohydrates by suspension cultured cells of bamboo plants. In: Teixeira SJA (ed) Floriculture, ornamental and plant biotechnology, vol 5. Global Science Books, Islewort, pp 240–244

Ogita S, Ohki S, Nomura T, Kato Y (2012b) A β-glucosidase activity potentially involved in cell division and wall development of *Phyllostachys* bamboo suspension cells. Am J Plant Sci 3 (8):1066–1072. https://doi.org/10.4236/ajps.2012.38127

Ogita S, Sasamoto H (2017) *In vitro* bioassay of allelopathy in four bamboo species; *Bambusa multiplex*, *Phyllostachys bambusoides*, *P. nigra*, *Sasa kurilensis*, using sandwich method and protoplast co-culture method with digital image analysis. Am J Plant Sci 8(7):1699–1710. https://doi.org/10.4236/ajps.2017.87117

Pereira MA, Beraldo AL (2008) Bambu de corpo e alma. Canal 6 Editora, Bauru

Pline WA, Edmisten KL, Oliver T, Wilcut JW, Wells R, Allen NS (2002) Use of digital image analyses, viability stains, and germination assays to estimate conventional and glyphosate-resistant cotton pollen viability. Crop Sci 42(6):2193–2200. https://doi.org/10.2135/cropsci2002.2193

Preil W (2005) General introduction: a personal reflection on the use of liquid media for *in vitro* culture. In: Preil W, Hvoslef-Eide AK (eds) Liquid culture systems for *in vitro* plant propagation. Springer, Berlin, pp 1–18. https://doi.org/10.1007/1-4020-3200-5

Queiroz FCF (2020) Estabelecimento e cultivo de células em suspensão e uso de biorreatores como estratégias de propagação de bambus do gênero *Guadua*. PhD thesis, Universidade de Brasília

Queiroz FF, Gomes HT, Scherwinski-Pereira JE (2019) Clonagem *in vitro* de um importante bambu brasileiro e avaliação da fidelidade genética das plantas produzidas utilizando marcadores moleculares ISSR. In: Anais da Semana da Biologia de Tangará da Serra e Workshop de Instrumentação para o Ensino de Ciências e Biologia. Universidade do Estado de Mato Grosso, Tangará da Serra, pp 1

Ramakrishnan M, Yrjälä K, Vinod KK, Sharma A, Cho J, Satheesh V, Zhou M (2020) Genetics and genomics of moso bamboo (*Phyllostachys edulis*): current status, future challenges, and biotechnological opportunities toward a sustainable bamboo industry. Food Energy Secur 00 (e229):1–36. https://doi.org/10.1002/fes3.229

Ramanayake SMSD, Wanniarachchi WAVR (2003) Organogenesis in callus derived from an adult giant bamboo (*Dendrocalamus giganteus* wall. Ex Munro). Sci Hortic 98:195–200. https://doi.org/10.1016/S0304-4238(02)00204-2

Rao IU, Rao IR, Narang V (1985) Somatic embryogenesis and regeneration of plants in the bamboo *Dendrocalamus strictus*. Plant Cell Rep 4(4):191–194. https://doi.org/10.1007/BF00269286

Rao IVR, Rao IU (1990) Tissue culture approaches to the mass-propagation and genetic improvement of bamboos. In: Rao IVR, Gnanaharan R, Sastry CB (eds) , pp 151–158Proceedings of the Int. Bamboo Workshop. Kerala Forest Research Institute, India, International Development Research Centre, Canada

Rodier L, Villar-Cocina E, Ballesteros JM, Junior HS (2019) Potential use of sugarcane bagasse and bamboo leaf ashes for elaboration of green cementitious materials. J Clean Prod 231:54–63. https://doi.org/10.1016/j.jclepro.2019.05.208

Sandhu M, Wani SH, Jiménez VM (2018) *In vitro* propagation of bamboo species through axillary shoot proliferation: a review. Plant Cell Tissue Organ Cult 132:27–53. https://doi.org/10.1007/s11240-017-1325-1

Shi SQ, Cai L, Weng Y, Wang D, Sun Y (2019) Comparative life-cycle assessment of water supply pipes made from bamboo vs. polyvinyl chloride. J Clean Prod 240:118172. https://doi.org/10.1016/j.jclepro.2019.118172

Silva-Cardoso IMA, Meira FS, Gomes ACMM, Scherwinski-Pereira JE (2019) Histology, histochemistry and ultrastructure of pre-embryogenic cells determined for direct somatic embryogenesis in the palm tree *Syagrus oleracea*. Physiol Plant 168(4):845–875. https://doi.org/10.1111/ppl.13026

Silveira AAC, Lopes FJF, Sibov ST (2020) Micropropagation of *Bambusa oldhamii* Munro in heterotrophic, mixotrophic and photomixotrophic systems. Plant Cell Tissue Organ Cult 141:315–326. https://doi.org/10.1007/s11240-020-01788-4

Singh SR, Dalal S, Singh R, Dhawan AK, Kalia RK (2012) Micropropagation of *Dendrocalamus asper* {Schult. &Schult. F.} Backer ex k. Heyne: an exotic edible bamboo. J Plant Biochem Biotechnol 21(2):220–228. https://doi.org/10.1007/s13562-011-0095-9

Singh SR, Singh R, Kalia S, Dalal S, Dhawan AK, Kalia RK (2013) Limitations, progress and prospects of application of biotechnological tools in improvement of bamboo—a plant with extraordinary qualities. Physiol Mol Biol Plants 19:21–41. https://doi.org/10.1007/s12298-012-0147-1

Sood A, Bhattacharya A, Sharma M, Sharma RK, Nadha HK, Sood P, Mehta R, Kaur D, Brar J, Ahuja PS (2013) Somatic embryogenesis and *Agrobacterium* mediated genetic transformation in bamboos. In: Aslam J, Srivastava PS, Sharma MP (eds) Somatic embryogenesis and genetic transformation in plants. Narosa Publishing House, New Delhi, pp 166–178

Soreng RJ, Peterson PM, Romaschenko K, Davidse G, Teisher JK, Clark LG, Barberá P, Gillespie LJ, Zuloaga FO (2017) A worldwide phylogenetic classification of the Poaceae (Gramineae) II: An update and a comparison of two 2015 classifications. J Syst Evol 55(4):259–290. https://doi.org/10.1111/jse.12262

Steiner N, Vieira FN, Maldonado S, Guerra MP (2005) Effect of carbon source on morphology and histodifferentation of *Araucaria angustifolia* embryogenic cultures. Braz Arch Biol Technol 48 (6):895–903. https://doi.org/10.1590/S1516-89132005000800005

Sun X, He M, Li Z (2020) Novel engineered wood and bamboo composites for structural applications: state-of-art of manufacturing technology and mechanical performance evaluation. Constr Build Mater 249:118751. https://doi.org/10.1016/j.conbuildmat.2020.118751

Suwal MM, Lamichhane J, Gauchan DP (2020) Regeneration technique of bamboo species through nodal segments: a review. Nepal J Biotechnol 8(1):54–68. https://doi.org/10.3126/njb.v8i1.30209

Taghizadeh M, Nasibi F, Kalantari KM, Ghanati F (2019) Evaluation of secondary metabolites and antioxidant activity in *Dracocephalum polychaetum* Bornm. Cell suspension culture under magnetite nanoparticles and static magnetic field elicitation. Plant Cell Tissue Organ Cult 136:489–498. https://doi.org/10.1007/s11240-018-01530-1

Teixeira JB, Junqueira CS, Pereira AJPC, Melo RIS, Silva APD, Mundim DA (2004) Multiplicação clonal de café (*Coffea arabica* L.) via embriogênese somática. Embrapa Recursos Genéticos e Biotecnologia, Brasília

Teixeira JB, Sondahl MR, Nakamura T, Kirby EG (1995) Establishment of oil palm cell suspensions and plant regeneration. Plant Cell Tissue Organ Cult 40:105–111. https://doi.org/10.1007/BF00037662

Thapa P, Kaur D, Sood P, Mehta R, Brar J, Naddha H, Ogra RK, Prakash O, Bhattacharya A, Sood A (2019) Biotechnology of bamboos. In: Kaushik S, Singh YP, Kumar D, Thapliyal M, Barthwal S (eds) Bamboos in India. ENVIS Centre on Forestry, Dehradun, pp 147–186

Tsay HS, Yeh CC, Hu JY (1990) Embryogenesis and plant regeneration from anther culture of bamboo (*Sinocalamus latiflora* (Munro) McClure). Plant Cell Rep 9:349–351. https://doi.org/10.1007/BF00232396

Vale PAA, Oliveira Júnior JB, Costa FHS, Scherwinski-Pereira JE (2019) Height and number of shoots on the survival and development of micropropagated bamboo plantlets during pre-acclimatization. Pesqui Agropecu Trop 49:e53751. https://doi.org/10.1590/1983-40632019v4953751

Van den Akker S, Bormans P, Peeters H, Prinsen E, Gielis J (2015) Cytokinin dynamics in cell suspension cultures of *Bambusa balcooa* Roxburgh using UPLC-ESI/MS/MS. Bamboo Sci Cult: J Am Bamboo Soc 28(1):32–39

Vanisree M, Lee CY, Lo SF, Nalawade SM, Lin CY, Tsay HS (2004) Studies on the production of some important secondary metabolites from medicinal plants by tissue cultures. Bot Bull Acad Sinica 45:1–22

Verdeil JL, Alemanno L, Niemenak N, Trambarger TJ (2007) Pluripotent versus totipotent plant stem cells: dependence versus autonomy? Trends Plant Sci 12(6):245–252. https://doi.org/10.1016/j.tplants.2007.04.002

Vijendra PD, Jayanna SG, Kumar V, Sannabommaji T, Rajashekar J, Gajula H (2020) Product enhancement of triterpenoid saponins in cell suspension cultures of *Leucas aspera* Spreng. Ind Crop Prod 156:112857. https://doi.org/10.1016/j.indcrop.2020.112857

Vorontsova MS, Clark LG, Dransfeld J, Govaert RS, Baker WJ (2016) World checklist of bamboos and rattans. INBAR Technical Report, Kew

Wei Q, Cao J, Qian W, Xu M, Li Z, Ding Y (2015) Establishment of an efficient micropropagation and callus regeneration system from the axillary buds of *Bambusa ventricosa*. Plant Cell Tissue Organ Cult 122:1–8. https://doi.org/10.1007/s11240-015-0743-1

Woods SH, Woods JE, Collins GB (1995) Somatic embryogenesis in bamboo. In: Jain SM, Gupta PK, Newton RJ (eds) Somatic embryogenesis in Woody plants. Forestry sciences, vol 44-46. Springer, Dordrecht, pp 41–51. https://doi.org/10.1007/978-94-011-0491-3_3

Ye S, Cai C, Ren H, Wang W, Xiang M, Tang X, Zhu C, Yin T, Zhang L, Zhu Q (2017) An efficient plant regeneration and transformation system of ma bamboo (*Dendrocalamus latiflorus* Munro) started from young shoot as explant. Front Plant Sci 8:1298. https://doi.org/10.3389/fpls.2017.01298

Yuan Z, Wen Y, Kapu NS (2018) Ethanol production from bamboo using mild alkaline pre-extraction followed by alkaline hydrogen peroxide pretreatment. Bioresour Technol 247:242–249. https://doi.org/10.1016/j.biortech.2017.09.080

Chapter 6
Foliar-Anatomical Adaptations of Micropropagated Plants of *Dendrocalamus strictus* (Roxb.) Nees Towards Photoautotrophic Conditions

Mahipal S. Shekhawat, Bharat S. Rajput, and M. Manokari

Abstract *Dendrocalamus strictus* (Roxb.) Nees (family: Poaceae) is an economically important bamboo species explored worldwide for its wider adaptability. In vitro propagation is an alternate platform and has astonishing benefits in large-scale production of disease-free clones of the important plant species. Despite several efforts implicated to improve the in vitro propagation of *D. strictus*, the protocols were hindered by certain morpho-anatomical and physiological disorders of in vitro system that is directly connected with the physiology and plantlet development. Hence, this study is centred on the analysis of foliar-anatomical alterations persuaded in the foliar apparatus regenerated under controlled conditions and after acclimatization. The structural, functional, and enhanced level of photosynthetic pigments has evolved with subsequent stages of plantlet developments from lab to the greenhouse acclimatization, and the alteration of necessary structural changes allowed successful plantlet survival in the field. The parameters evaluated revealed the gradual adaptational developments through the acclimatization process in the greenhouse which could help in the prediction of the survival percentage of plantlets in the field. The present protocol has potential applications in the propagation and supply of structurally developed and physiologically active tissue culture-raised plantlets of *D. strictus* to meet the commercial demand.

M. S. Shekhawat (✉)
Biotechnology Unit, Department of Botany, Kanchi Mamunivar Government Institute for Postgraduate Studies and Research, Puducherry, India

B. S. Rajput
Tissue Culture Lab, Silviculture and Forest Utilization Circle, Research Range, Rajpipla, Gujarat, India

M. Manokari
Biotechnology Unit, Department of Botany, Kanchi Mamunivar Government Institute for Postgraduate Studies and Research, Puducherry, India

Siddha Clinical Research Unit, Central Council for Research in Siddha (M/o AYUSH), Palayamkottai, TN, India

Z. Ahmad et al. (eds.), *Biotechnological Advances in Bamboo*,
https://doi.org/10.1007/978-981-16-1310-4_6

Keywords Acclimatization · Bulliform cells · *Dendrocalamus strictus* · Foliar anatomy · Micropropagation

Abbreviations

BAP	6-benzylaminopurine
$HgCl_2$	Mercuric chloride
NAA	α-Naphthalene acetic acid
NaOCl	Sodium hypochlorite
PGRs	Plant growth regulators
RH	Relative humidity
SPFD	Spectral photon flux density

6.1 Introduction

Bamboo species are popularly termed as 'green gold' and have been well acknowledged for improving the world economy (Owen 2015; Fernandes 2017). Globally, bamboo cultivation is promoted by intergovernmental organizations due to its market demand and industrial potential (Londono 2002; National Bamboo Mission 2020). The versatile applications of bamboo also provide sustainable opportunities to support rural livelihood (Basumatary et al. 2015; Musau 2016).

Dendrocalamus strictus (Roxb.) Nees (Poaceae: Bambusoideae) is commonly called 'giant bamboo', 'Calcutta bamboo', and 'male bamboo', which is predominantly distributed in India particularly in semidry zones and hilly tracts (Saxena and Dhawan 1999). It can withstand the temperature ranges between −5 and 45 °C; therefore, the commercial cultivation is recorded from tropical Asia, India, Taiwan, and China (Das et al. 2017). In India, it is cultivated in the Western Ghats, Andhra Pradesh, Madhya Pradesh, Orissa, Uttar Pradesh, and Himachal Pradesh for its multipurpose industrial applications (Amit and Uniyal 2008; Das et al. 2017). It provides raw materials for furniture, agricultural implements, musical instruments, paper mills, baskets, raft household products, etc. (Reddy 2006).

The phytochemical characterization of *D. strictus* reveals the presence of carbohydrates, fats, fibres, vitamins C and E, essential minerals, magnesium, calcium and iron, flavonoids, polysaccharides, phenolic acid, and amino acids (Wangawar et al. 2017). Due to the nutritional properties of young shoots and seeds, male bamboo is cultivated and consumed in northeast India (Nongdam and Tikendra 2014). The leaves are endowed with antioxidant potential, and the combination of *D. strictus* and *Curcuma longa* leaves is traditionally used to cure wounds, cough, cold, fever, etc. (Kamble et al. 2010; Goyal et al. 2011). Moreover, it also possesses antifungal, antibacterial, anti-inflammatory, anti-hyperglycaemic, and antioxidant properties (Nongdam and Tikendra 2014; Wangawar et al. 2017).

Conventionally, *D. strictus* is propagated through seeds and vegetative methods (rhizome and culm cuttings). But the natural propagation methods are besieged by poor seed set and short viability of seeds, and the vegetative organ carries fungal consortium and other pests, which hinder the successful propagation to meet the demand (Singh et al. 2012; Sarkar et al. 2020). Therefore, in vitro propagation only could satisfy the global demand of propagules that can also offer large-scale production of elite clones within a short period irrespective of the season. Nevertheless, the in vitro propagation technique is a promising alternative, but it is besieged by in vitro persuaded morphological, anatomical, physiological, and biochemical disorders (Bairu et al. 2011; Ruffoni and Savona 2013). The developed plantlets with such deformities find difficulties while transferring to the in vivo conditions; hence, the widespread application of in vitro regeneration is restricted (Chandra et al. 2010). The prime success of in vitro propagation and crop improvement lies in the production of physiologically active plantlets for successful survival in the field (Pospisilova et al. 2009).

Several physicochemical factors, such as nutrient medium, plant growth regulators, growth conditions, vessel types, etc., cause variations in the micro-morpho-anatomy and the physiology of in vitro regenerated plants (Kozai and Kubota 2001; Hazarika 2006). The poorly differentiated photosynthetic tissues in unorganized mesophyll cells, limited palisade and spongy tissues, non-functional stomata, absence of cuticular wax, underdeveloped epidermal and subepidermal layers, and irregular cortical cells were reported under in vitro conditions. These abnormalities led to a low rate of physiological activities (Lamhanedi et al. 2003; Lebedev and Schestibratov 2013). Such anatomically and physiologically inefficient plantlets fail to withstand out of culture vessels, i.e. in greenhouse as well as in field (Bairu and Kane 2011; Kumar and Rao 2012). The available evidence and theories of in vitro developmental processes via histological approaches using microscopic techniques open a new way to address these problems in recent times (Trigiano et al. 2004; Motte et al. 2014).

Therefore, the assessment of the anatomical uniformity of regenerated plants is important for the growth and development of tissue culture-raised plants in the natural habitat. Macro- and micro-morpho-anatomical features of leaves are extensively used to assess structural uniformity with physiological functionality of micropropagated plantlets before field transfer (Afreen 2005; Costa et al. 2009; Shekhawat and Manokari 2018), and they have been successfully applied in the survival success of *Morinda coreia* (Shekhawat et al. 2017), *Bauhinia racemosa* (Sharma et al. 2017), *Bambusa balcooa* (Rajput et al. 2020), *Vitex negundo* (Manokari et al. 2020a), etc.

Considering the economic importance of this species and the problems faced during acclimatization stages of tissue culture-raised plantlets, the objective of this study was to elucidate the comparative foliar-anatomical transitions of in vitro regenerated *D. strictus* at subsequent phases of micropropagation to demonstrate the developmental adaptations and to compare the foliar structural stability with that of the mother plant. The present study is the foremost report to determine the survival rate of micropropagated plants of *D. strictus* using foliar anatomy.

6.2 Materials and Methods

6.2.1 Explant Source and Sterilization Procedures

The freshly sprouted nodal shoot segments from the superior genotype of *Dendrocalamus strictus* have been collected from the coastal areas of Puducherry (south India) and washed using tap water for 30 min. Surface sterilization of explants was done using series of disinfectants for 5 min each such as sodium hypochlorite (1%, v/v), which is followed by an aqueous solution of Bavistin (0.1%, w/v) and mercuric chloride (0.1%, w/v) and finally with 70% (v/v) ethanol for 30 s. The explants were then washed with sterile distilled water several times and inoculated under aseptic conditions.

6.2.2 Establishment of Aseptic Cultures and Development of Plantlets

The nodal shoot segments were inoculated on full-strength MS (Murashige and Skoog) medium (Murashige and Skoog 1962), fortified with 4.0 mg L^{-1} 6-benzylaminopurine (BAP) as per Rajput et al. (2019). The regenerated shoots from nodal cultures were proliferated using MS medium fortified with 3.0 mg L^{-1} BAP and 0.5 mg L^{-1} α-naphthalene acetic acid (NAA). Healthy shoots (≥ 5.0 cm in length) were dissected and transferred to half-strength MS medium augmented with 3.0 mg L^{-1} NAA for root induction (Rajput et al. 2019). The culture room condition was maintained at 25 ± 2 °C temperature and 65–70% relative humidity (RH) and 40–45 μmol m^{-2} s^{-1} spectral photon flux density (SPFD) irradiance for 16/8-h day^{-1} light/dark period. The percentage of axillary bud break, proliferation, and rooting rate were recorded after 4–5-week interval.

6.2.3 Acclimatization of Plantlets

After 4 weeks, the plantlets with healthy, well-developed roots (≥ 4.0 cm lengthy roots) were transferred into eco-friendly paper cups filled with sterile cocopeat and Soilrite® mixture (1:1 ratio) and maintained in a greenhouse for 4 weeks. Thereafter, the plantlets were irrigated with 1/4 MS macro-salt solution and placed under a shade net with ventilation. The percentage of survival was recorded after 12 weeks.

6.2.4 Foliar-Anatomical Evaluation

The foliar-anatomical transitions were determined through subsequent anatomical evaluation of regenerated plantlets to understand the structural transition in plants from in vitro to field environments. The leaves from five stages and three environments [in vitro, the fourth subculture of multiplication stage (1) and 4 weeks old in vitro rooted plantlets (2); ex vitro, 4-week-old greenhouse hardened (3) and 12-week-old shade net acclimatized plantlets (4); in vivo, mother plant (5)] were subjected to foliar-anatomical evaluation. Ten samples from each stage were evaluated and repeated thrice to conclude the developmental changes in structure and functioning of foliar constants such as cuticle, epidermal layers, stomata, mesophyll, vascular tissues, bulliform, and transfusion cells, and the thickness (μm) of adaxial (ADE), abaxial (ABE), and mesophyll (MP) area were measured. Regardless of the phases or stages and environments, completely developed, fully opened leaves from the third node towards the base were selected. In the case of the mother plant, the branches were pruned, and newly emerged leaves (near the base) after 4 weeks were evaluated.

Initially, the leaves were pretreated with xylene solution for 2 h to remove the microscopic debris and fixed in FAA (formalin, acetic acid, and ethyl alcohol) solution in the ratio of 1:1:3. To study the gradual developmental adaptations, thin transverse sections of leaves from the above-mentioned five phases were stained with safranin (1%, w/v) (Johansen 1940). The photomicrographs were taken using a digital camera (Leica MC170 HD) at 40× and 100× magnifications fixed on bright light field Leica trinocular microscope (Leica DM 750). These photomicrographs were used to study the developments in dermal, ground, and vascular tissue systems such as ribs and furrows, bulliform and fusoid cells, mesophyll tissues, marginal sclerenchyma, vascular elements, etc. which were characteristics to *D. strictus*.

6.2.5 Analysis of Photosynthetic Pigments

The concentrations of chlorophyll *a* and *b* and total carotenoids (C_{x+c}) from the five phases (multiplication and rooting stage, 4 weeks old hardened, 12 weeks old acclimatized and mother plant) were evaluated following the report of Saini et al. (2001). Fresh leaves from third to seventh nodes were excised, and about 500 mg of leaf tissues was ground using 1.0 mL of 80% cold acetone. The mixture was centrifuged at 10000 rpm for 15 min at 4 °C, and the pellet was used for evaluation of photosynthetic pigments. UV-visible double beam spectrophotometer (Systronics, Model 2202, Ahmadabad, Gujarat, India) was used to detect chlorophyll *a* (at 663 nm), chlorophyll *b* (646 nm), and total carotenoids (470 nm).

6.2.6 *Statistical Analysis*

The experimental design to evaluate foliar anatomy was consisted of five treatments [(a) in vitro multiplied shoots, (b) in vitro rooted shoots, (c) 4-week-old hardened plantlets, (d) 12-week-old acclimatized plantlets, and (e) mother plant] with ten replications and repeated thrice. The analysis of variance (ANOVA) was performed using SPSS software (v.17.0, SPSS Inc., Chicago, USA) with $P < 0.05$ significance of using Duncan's Multiple Range Test (DMRT). The morphometric values were presented as mean ± SD of triplicates.

6.3 Results and Discussion

The foliar-anatomical analysis revealed the qualitative anatomical and physiological developmental phases of micropropagated plantlets of *D. strictus*.

6.3.1 *Micropropagation of* D. strictus

The nodal shoot segments were used for the development of cultures by following our earlier report (Rajput et al. 2019). The optimized concentrations of chemical disinfectants resulted in the establishment of 98% contaminant-free cultures. Nodes cultured on full-strength MS medium fortified with 4.0 mg L^{-1} BAP produced 5.2 ± 0.23 shoots with 5.0 ± 0.14 cm length within 4 weeks of inoculation (Fig. 6.1a). The simultaneous subculturing of regenerated shoots on MS medium fortified with 3.0 mg L^{-1} BAP and 0.5 mg L^{-1} NAA yielded 33.8 ± 0.30 shoots with 6.5 cm length after the fourth subculture (Fig. 6.1b). Half-strength MS medium containing 3.0 mg L^{-1} NAA resulted in regeneration of the maximum number of roots (13.0 ± 0.22 roots with 4.5 cm length) from the robust shoots (Fig. 6.1c) (Table 6.1). The number of shoots that proliferated in the present study was comparatively higher as compared to the existing reports (Reddy 2006; Pandey and Singh 2012; Goyal et al. 2015; Bora and Borah 2019).

6.3.2 *Hardening and Acclimatization of* D. strictus

The rooted shoots were placed on cocopeat and Soilrite® mixture (1:1 ratio) and maintained under the semi-controlled conditions of the greenhouse for 4 weeks (Fig. 6.1d) and irrigated with one-fourth-strength MS nutrients twice a day to support growth and development. Then the plantlets were gradually shifted to

Fig. 6.1 Micropropagation of *D. strictus* using nodal explants. (**a**) Axillary bud breaks and shoot regeneration from nodal explants. (**b**) Proliferated shoots of *D. strictus*. (**c**) In vitro rhizogenesis on half-strength MS + 3.0 mg L^{-1} NAA. (**d**) 4-week-old greenhouse hardened plantlets. (**e**) 12-week-old acclimatized plantlet. (**f**) Completely acclimatized *D. strictus* plantlets under shade net

Table 6.1 Micropropagation studies of *D. strictus*

Stages of micropropagation	Culture medium and concentration of PGRs	Responses	Shoot morphometry (mean ± SD)	Root morphometry (mean ± SD)
Initiation	Full-strength MS + BAP 4.0 mg L^{-1}	98% of axillary bud break	5.2 ± 0.23 shoots with 5.0 ± 0.14 cm length	–
Shoot proliferation	Full-strength MS + BAP 3.0 mg L^{-1} + 0.5 mg L^{-1} NAA	–	33.8 ± 0.30 shoots with 6.5 cm length	–
In vitro rooting	Half-strength MS + 3.0 mg L^{-1}	100%	–	13.0 ± 0.22 roots with 4.5 cm length

The mean values were analysed statistically at $P < 0.05$ by the Duncan's multiple range test

shade net and maintained further for 8 weeks (Fig. 6.1e). The maximum percentage of survival (100%) was recorded after 12 weeks of acclimatization (Fig. 6.1f).

The in vitro conditions are characterized by the aseptic environments, restricted gas exchanges, higher humidity, lower irradiance by artificial light sources and plenty of nutrients, readymade carbon source, and growth regulators, which are essential for large-scale propagation of plants (Bhatia 2015; Espinosa-Leal et al. 2018). However, these optimized in vitro conditions may not serve better in ex vitro growth response and acclimatization, during which the plantlets have to transform from heterotrophic to photoautotrophic growth. The connecting link of in vitro growth and ex vitro adaptation is largely deficient in the previous micropropagation studies. Hence, careful observation and gradual shifting from one phase to another phase aid in the successful acclimatization of tissue culture-raised plantlets. The initial maintenance of plantlets under higher humidity and lower irradiance could assist the plant to attain structural differentiation. Therefore, the plantlets will be physiologically active and maintain their survival in the field.

6.3.3 Foliar-Anatomical Evaluation at Subsequent Stages of Micropropagation

The evaluation of foliar anatomy from in vitro and ex vitro conditions illuminates the development of structural plasticity in the micropropagated plantlets of *D. strictus*. The transverse section of leaves of the multiplication stage possessed underdeveloped structural elements. In the frontal view, the epidermal layers of the lamina consisted of single-layered cells and numerous papillae and are covered with a thin layer of the cuticle (Fig. 6.2a). The stomata were underdeveloped, paracytic, and more on the!!abaxial surface than the adaxial surface. Trichomes were silicified and pointed unicellular projections which were emerged out from the epidermis as prickle hairs (Fig. 6.2b). The adaxial epidermis was less undulated and consisted of fan-shaped arrays formed by three to four large bulliform cells in between the vascular bundles (Fig. 6.2b). The mesophyll was poorly differentiated and comprised of asymmetrically invaginated large fusoid cells (Figs. 6.2a, b). The vascular bundles were collateral, surrounded by a thin single-layered parenchymatous sheath and consisted of underdeveloped xylem and phloem elements (Fig. 6.2b).

The leaves of rooted shoots were characterized by the development of structural parameters to a certain extent only under in vitro conditions. The epidermal cells were single layered with highly pointed papillae and thin cuticle. Stomata and prickly hairs were differentiated. The epidermal cells showed highly sinuous anticlinal walls, and the abaxial epidermis possessed four to five bulliform cells (Fig. 6.2c). Mesophyll was comprised of differentiated fusoid cells with intense chlorophylls. Vascular bundles with few lignified tracheary elements and a small patch of phloem were surrounded by a parenchymatous sheath (Fig. 6.2d). There was no differentiation observed between middle and lateral bundles under in vitro conditions. Plicate or arm cells and rosette cells were absent at both multiplication and rooting stages.

The leaves of 4-week-old hardened plantlets established anatomical developments to adapt to the slightly harsh ex vitro environment. The epidermal cells

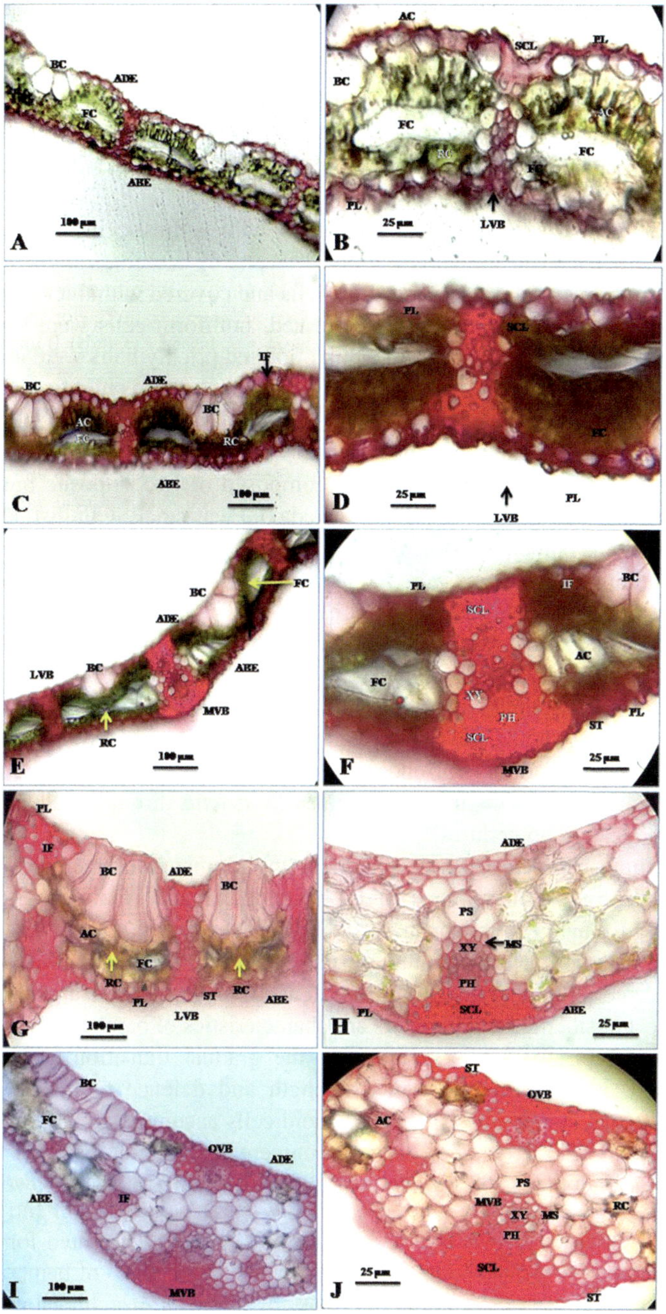

Fig. 6.2 Comparative foliar-anatomical adaptation at subsequent stages of micropropagation. (**a**, **b**) Transverse section of leaf from in vitro multiplication stage with underdeveloped structural parameters. (**c**, **d**) Foliar-anatomical developments in the in vitro rooting stage. (**e**, **f**) – Structural transition of foliar anatomy in the 4-week-old greenhouse hardened plantlets. (**g**, **h**) Completely adapted foliar anatomy in the 12-week-old acclimatized plantlets. (**i**, **j**) Foliar anatomy of the mother

were highly wavy, bulliform cells were more than six, and prickly hairs were abundant in the greenhouse grown plants (Fig. 6.2e). The mesophyll was highly differentiated, fusoid cells were small in size, and arm cells were also detected (Figs. 6.2e, f). The middle vascular bundles consisted of a two-layered bundle sheath (inner mestome and outer parenchymatous), and the bundles of lamina were covered by parenchymatous sheath alone. Differentiation of protoxylem and metaxylem was observed after 4 weeks of hardening (Fig. 6.2f).

Completely grown adaptational characters with a drastic increase in internal tissues were detected in the leaves after 12 weeks of acclimatization of plantlets. The epidermis consisted of long and short cells and covered with thick cuticle layers (Fig. 6.2g). Functional stomata were observed, bulliform cells were 6–12 on the adaxial surface, and the silicified unicellular pointed prickly hairs were abundant on both the epidermal surfaces (Figs. 6.2g, h). The mesophyll was highly organized and invaginated by the arm and fusoid cells. The vascular bundles were collateral with highly differentiated and functional vascular elements. The midrib complex was characterized by the first-order vein and composed of two opposite bundles. The abaxial vascular bundle was large, and the adaxial was small with distinct protoxylem, metaxylem, phloem, double sheath, and marginal sclerenchyma (Fig. 6.2h). Second- and third-order veins were represented by lateral vascular bundles with few lignified tracheary elements (Figs. 6.2g, h).

The foliar-anatomical evaluation of four stages of micropropagation has revealed that the plantlets gradually developed adaptational traits resembling the donor plant and the leaves of acclimatized plants showed structural similarity with the mother plant (Figs. 6.2i, j). Hence, the regenerated plantlets exhibited 100% survival success in the field. These findings corroborate the internal development and adaptational mechanism achieved by the micropropagated *D. strictus* through gradual hardening and acclimatization procedures.

The optimized in vitro physicochemical environments adopted for large-scale propagation promote anatomical and physiological abnormalities in the tissue culture-raised plants (Soundappan et al. 2018; Rajput et al. 2020). The structural changes in micropropagated plants are used to detect the adaptational characters towards altered environments (Rodrigues et al. 2014; Pita-Barbosa et al. 2015).

Papillae and silicified prickly hairs are characteristics of bamboo species, and the developmental plasticity is promoted by the gradual transition of the external environment; it provides mechanical strength and defence to the aerial organs (Hodson 2016; Liana et al. 2017). The fusoid cells accumulate CO_2 during photorespiration and also depend on the environmental conditions (March and Clark 2011). The development of a waxy coat in the acclimatized plants lowers the rate

Fig. 6.2 (continued) plant (*ABE* abaxial epidermis, *AC* arm cell, *ADE* adaxial epidermis, *BC* bulliform cells, *FC* fusoid cell, *IF* intercostal fibre, *LVB* lateral vascular bundle, *MS* mestome sheath, *MVB* middle vascular bundle, *OVB* opposite vascular bundle to the middle bundle, *PH* phloem, *PL* papillae, *PS* parenchymatous sheath, *RC* rosette cell, *SCL* sclerenchyma, *ST* stomata, *XY* xylem) of transpiration and protects the internal tissues from high light intensity (Souza et al. 2010). The completely developed vascular elements substantiate the effective transport of water and minerals to withstand under harsh in vivo environment (Sack et al. 2012; Rathore et al. 2013).

6.3.4 Analysis of Photosynthetic Pigments

The leaves regenerated in multiplication and rooting stages under in vitro conditions resulted in a lower level of photosynthetic pigments. Instead, leaves developed during hardening and acclimatization stages were visibly greener and similar to that of the mother plant. The spectroscopic evaluation of photosynthetic pigments revealed the gradual increase in the pigment contents in the leaves from multiplication to acclimatized plants (Table 6.2). The quantification of photosynthetic pigments visualized that the leaves developed in the in vitro multiplication stage possessed lower content of chlorophylls (Chl *a*, 120.0; Chl *b*, 143.0; $C_{x + c}$, 37.0 μg g^{-1} FW). The level of pigments was slightly enhanced during rooting stage (Chl *a*, 142.0; Chl *b*, 160.5; $C_{x + c}$, 39.0 ± 0.15 μg g^{-1} FW), whereas the hardened and acclimatized plantlets showed linear and double-fold increase of pigments after 4 weeks (Chl *a*: 198.0; Chl *b*: 210.5; $C_{x + c}$: 49.0 μg g^{-1} FW) and 12 weeks (Chl *a*, 229.0; Chl *b*, 240.0; $C_{x + c}$, 55.0 μg g^{-1} FW) under the greenhouse conditions (Table 6.2). The completely acclimatized plantlets (after 12 weeks) resembled the mother plants in pigment quantification experiments (Table 6.2).

The readymade source of carbon, constraints of gas exchange, low water availability, and irradiance of in vitro conditions results in the least synthesis of photosynthetic pigments (Ramirez-Mosqueda et al. 2017). Similarly, the reduced photosynthetic pigments were also detected in the in vitro regenerated leaves of *Carica papaya* (Schmildt et al. 2015), *Liriodendron tulipifera* (Chan et al. 2016), *Vitex negundo* (Manokari et al. 2020b), and *Hemidesmus indicus* (Shekhawat et al. 2020). Alterations in the in vitro environment by reduced sucrose in the nutrient medium can improve the photosynthetic parameters in the leaves (Mosaleeyanon et al. 2004; Rybczynski et al. 2007; Eckstein et al. 2012). This statement is in agreement with the present report that the gradual increase in photosynthetic pigments resulted in the rooting stage, where the shoots were rooted on reduced (half-strength) MS nutrients and sucrose. The exemption of exogenous carbohydrates promotes photoautotrophic growth in the micropropagated plantlets (Nicoloso et al. 2003). Hence, the sugar-free hardening medium, relatively higher light intensity, lower humidity, and gaseous circulation promoted the development of photosynthetic pigments in the 4-week and 12-week acclimatized plantlets of *D. strictus*.

Table 6.2 Comparative photosynthetic pigment analysis (μg g^{-1} fresh weight) in leaves at subsequent stages of micropropagation

Pigments	In vitro conditions (μg g^{-1} FW) (mean ± SD)		Acclimatized plantlets (ex vitro) (μg g^{-1} FW) (mean ± SD)		Mother plant (μg g^{-1} FW) (mean ± SD)
	Multiplication	Rooting	4 week old	12 week old	
Chlorophyll *a*	120.0 ± 0.27^{e}	142.0 ± 0.33^{d}	198.0 ± 0.30^{c}	229.0 ± 0.26^{b}	240.0 ± 0.35^{a}
Chlorophyll *b*	143.0 ± 0.20^{e}	160.5 ± 0.26^{d}	210.5 ± 0.41^{c}	240.0 ± 0.29^{b}	252.0 ± 0.23^{a}
Total chlorophyll (*a* + *b*)	263.0 ± 0.31^{e}	282.5 ± 0.28^{d}	408.5 ± 0.20^{c}	469.0 ± 0.43^{b}	492.0 ± 0.20^{a}
Ratio of chlorophyll (Chl *a*/Chl *b*)	0.83 ± 0.22^{e}	0.88 ± 0.30^{d}	0.94 ± 0.35^{c}	0.95 ± 0.40^{b}	0.95 ± 0.32^{a}
Total carotenoid (C_{x+c})	37.0 ± 0.17^{e}	39.0 ± 0.15^{d}	49.0 ± 0.25^{c}	55.0 ± 0.20^{b}	57.0 ± 0.22^{a}
Ratio of chlorophyll and carotenoid (Chl (*a* + *b*)/C_{x+c})	7.1 ± 0.15^{e}	7.24 ± 0.27^{d}	8.33 ± 0.20^{c}	8.5 ± 0.36^{b}	8.63 ± 0.29^{a}

The means followed by the same letter within rows do not differ statistically at $P < 0.05$ by the Duncan's multiple range test

6.4 Conclusions

The present study emphasized the application of anatomical diagnosis and photosynthetic pigment analysis for effective survival percentage of *D. strictus* during field transfer. Foliar anatomy revealed that the plantlets developed in this protocol were anatomically true to type with the donor plant. Furthermore, the comparative evaluation of photosynthetic pigments of micropropagated plants could help in identifying the physiologically active plantlets for field transfer. It is concluded that the successful survival of plantlets in the field was accomplished by the selection of anatomically and physiologically stable micropropagated plantlets.

Acknowledgements Author MSS is grateful to the National Medicinal Plants Board, Ministry of AYUSH, Government of India, for providing financial support to his laboratory (grant number NMPB/IFD/GIA/NR/PL/2018-19/187). BR is thankful to the Silviculture and Forest Utilization Circle, Gujarat Forest Department, for the infrastructure support. MM acknowledges the Central Council for Research in Siddha (Ministry of AYUSH), Government of India.

Author Contribution MSS, MM, and BR designed the research and conducted the experiments; MM and MSS drafted the manuscript. All authors have read, revised, and approved the final manuscript.

Conflict of Interest The authors declare that they have no conflicts of interest.

Human and Animal Rights This research did not involve experiments with human or animal participants.

Informed Consent Informed consent was obtained from all individual participants included in the study. Additional informed consent was obtained from all individual participants for whom identifying information is included in this article.

References

Afreen F (2005) Physiological and anatomical characteristics of *in vitro* photoautotrophic plants. In: Kozai T, Afreen F, Zobayed S (eds) Photoautotrophic (sugar-free medium) micropropagation as a new micropropagation and transplant production system. Springer, Dordrecht, pp 61–90. https://doi.org/10.1007/1-4020-3126-2_6

Amit K, Uniyal SK (2008) Distribution pattern of *Dendrocalamus strictus* in Kangra district of Himachal Pradesh, India. J Bamboo Rattan 7:211–217

Bairu MW, Aremu AO, van Staden J (2011) Somaclonal variation in plants: causes and detection methods. Plant Growth Regul 63:147–173

Bairu MW, Kane ME (2011) Physiological and developmental problems encountered by *in vitro* cultured plants. Plant Growth Regul 63:101–103

Basumatary A, Middha SK, Usha T, Brahma BK, Goyal AK (2015) Bamboo, as potential sources of food security, economic prosperity and ecological security in north-East India: an overview. Res Plant Biol 5:17–23

Bhatia S (2015) Plant tissue culture. Modern application of plant biotechnology in pharmaceutical sciences, pp. 31–107, https://doi.org/10.1016/B978-0-12-802221-4.00002-9

Bora A, Borah S (2019) Micro-propagation of *Dendrocalamus strictus* and its importance. Int J Sci Technol Res 8:430–432

Chan HA, Yong WK, Heung KM, Jae SY (2016) Effects of *in vitro* culture types on regeneration and acclimatization of yellow poplar (*Liriodendron tulipifera* L.) from somatic embryos. J Plant Biotechnol 43:110–118. https://doi.org/10.5010/JPB.2016.43.1.110

Chandra S, Bandopadhyay R, Kumar V, Chandra R (2010) Acclimatization of tissue cultured plants: from laboratory to land. Biotechnol Lett 32:1199–1205

Costa FHS, Pasqual M, Pereira JES, de Castro EM (2009) Anatomical and physiological modifications of micropropagated 'Caipira' banana plants under natural light. Sci Agric 66:323–330. https://doi.org/10.1590/S0103-90162009000300007

Das S, Singh YP, Negi YK, Shrivastav PC (2017) Genetic variability in different growth forms of *Dendrocalamus strictus*: Deogun revisited. N Z J For Sci 47:43. https://doi.org/10.1186/s40490-017-0104-4

Eckstein A, Zieba P, Gabrys H (2012) Sugar and light effects on the condition of the photosynthetic apparatus of *Arabidopsis thaliana* cultured *in vitro*. J Plant Growth Regul 31:90–101. https://doi.org/10.1007/s00344-011-9222-z

Espinosa-Leal CA, Puente-Garza CA, García-Lara S (2018) *In vitro* plant tissue culture: means for production of biological active compounds. Planta 248:1–18. https://doi.org/10.1007/s00425-018-2910-1

Fernandes V (2017) Bamboo can be more profitable than sugarcane and rice! Check out how [Website]. Financ. Express. URL https://www.financialexpress.com/opinion/the-bamboo-boon/974764/ (accessed 13 September, 2020)

Goyal AK, Middha SK, Sen A (2011) *In vitro* antioxidative profiling of different fractions of *Dendrocalamus strictus* (Roxb.) Nees leaf extracts. Free Radicals Antioxid 1:42–48. https://doi.org/10.5530/ax.2011.2.9

Goyal AK, Pradhan S, Basistha BC, Sen A (2015) Micropropagation and assessment of genetic fidelity of *Dendrocalamus strictus* (Roxb.) Nees using RAPD and ISSR markers. 3 Biotech 5:473–482. https://doi.org/10.1007/s13205-014-0244-7

Hazarika BN (2006) Morpho-physiological disorders in *in vitro* cultured plants. Sci Hortic 108:105–120

Hodson MJ (2016) The development of phytoliths in plants and its influence on their chemistry and isotopic composition. Implications for palaeoecology and archaeology. J Archaeol Sci 68:62–69. https://doi.org/10.1016/j.jas.2015.09.002

Johansen DA (1940) Plant Microtechnique. Ed.1. McGraw Hill Book Co., New York, London, pp 182–197

Kamble SY, Patil SR, Sawant PS, Sawant S, Pawar SG, Singh EA (2010) Studies on plant used in traditional medicine by Bhilla tribe of Maharashtra. Indian J Trad Knowledge 9:591–598

Kozai T, Kubota C (2001) Developing a photoautotrophic micropropagation system for woody plants. J Plant Res 114:525–537

Kumar K, Rao IU (2012) Morphophysiologicals problems in acclimatization of micropropagated plants in *ex vitro* conditions—a reviews. J Ornam Hortic Plants 2:271–283

Lamhanedi MS, Chamberland H, Tremblay FM (2003) Epidermal transpiration, ultrastructural characteristics and net photosynthesis of white spruce somatic seedlings in response to *in vitro* acclimatization. Physiol Plant 118:554–561

Lebedev V, Schestibratov K (2013) Effect of natural and synthetic growth stimulators on *in vitro* rooting and acclimatization of common ash (*Fraxinus excelsior* L.) microplants. Nat Sci 5:1–7

Liana A, Purnomo P, Sumardi I, Daryono B (2017) The classification of *Bambusa* spp. from celebes based on the micromorphological characters of leaf epidermis. J Trop Life Sci 7:197–207. https://doi.org/10.11594/jtls.07.03.02

Londono X (2002) Evaluation of bamboo resources in Latin America. In: Kumar A, Ramanuja Rao IV, Sastry C (eds) Bamboo for sustainable development. VSP Publications, Utrecht, pp 49–78

Manokari M, Cokul Raj M, Priyadharshini S, Phulwaria M, Shekhawat MS (2020a) Foliar micro-morphology—a promising tool to improve survival percentage of tissue culture raised plantlets with special reference to *in vitro* propagation of *Vitex negundo* L. Vegetos 33:504–515. https://doi.org/10.1007/s42535-020-00134-95

Manokari M, Priyadharshini S, Shekhawat MS (2020b) Mitigating *in vitro* induced macro- and micro-morphological anomalies using seismic stress in *Vitex negundo* L. Plant Cell Tissue Organ Cult 143:411–420. https://doi.org/10.1007/s11240-020-01927-x

March RH, Clark LG (2011) Sun-shade variation in bamboo (Poaceae: Bambusoideae) leaves. Telopea 13:93–104. https://doi.org/10.7751/telopea20116007

Mosaleeyanon K, Cha-um S, Kirdmanee C (2004) Enhanced growth and photosynthesis of rain tree (*Samanea saman* Merr.) plantlets *in vitro* under a CO_2-enriched condition with decreased sucrose concentrations in the medium. Sci Hortic 103:51–63

Motte H, Vereecke D, Geelen D, Werbrouck S (2014) The molecular path to *in vitro* shoot regeneration. Biotechnol Adv 32:107–121

Murashige T, Skoog F (1962) A revised medium for rapid growth and bioassay with tobacco tissue cultures. Plant Physiol 15:473–497. https://doi.org/10.1111/j.1399-3054.1962.tb08052.x

Musau Z (2016) Bamboo: Africa's untapped potential [Website]. Africa Renew. URL https://www.un.org/africarenewal/magazine/april-2016/bamboo-africa's-untapped-potential (Accessed 10 September, 2020)

National Bamboo Mission (2020) National Bamboo Mission Ministry of Agriculture & Farmers Welfare Government of India [Website]. URL https://www.nbm.nic.in/ (Accessed 13 September, 2020)

Nicoloso FT, Erig AC, Russowski D, Martins CF (2003) Effect of carbohydrate concentration and source on growth of brazilian ginseng [*Pfaffia glomerata* (spreng.) pedersen] *in vitro* cultured plants. Cienc Agrotec 27:84–90

Nongdam P, Tikendra L (2014) The nutritional facts of bamboo shoots and their usage as important traditional foods of Northeast India. Int Sch Res Notices. https://doi.org/10.1155/2014/679073

Owen A (2015) Bamboo!! Improving island economy and resilience with Guam College students. J Mar Island Cult 4:65–75

Pandey BN, Singh NB (2012) Micropropagation of *Dendrocalamus strictus* Nees from mature nodal explants. J Appl Nat Sci 4:5–9

Pita-Barbosa A, Goncalves EC, Azevedo AA (2015) Morpho-anatomical and growth alterations induced by arsenic in *Cajanus cajan* (L.) DC (Fabaceae). Environ Sci Pollut Res 22:11265–11274. https://doi.org/10.1007/s11356-015-4342-9

Pospisilova J, Synkova H, Haisel D, Batkova-Spoustova P (2009) Improvement of *ex vitro* transfer of tobacco plantlets by addition of abscisic acid to the last subculture. Biol Plant 53:617–624

Rajput BS, Jani M, Ramesh K, Manokari M, Jogam P, Allini VR, Kher MM, Shekhawat MS (2020) Large-scale propagation of *Bambusa balcooa* Roxb.: an industrially important bamboo species. Ind Crops Prod:157. https://doi.org/10.1016/j.indcrop.2020.112905

Rajput BS, Jani MD, Gujjar MR, Shekhawat MS (2019) Effective and large scale *in vitro* propagation of *Dendrocalamus strictus* (Roxb.) Nees using Nodal segments as explants. World Sci News 130:238–249

Ramirez-Mosqueda MA, Iglesias-Andreu LG, Luna-Sanchez IJ (2017) Light quality affects growth and development of *in vitro* plantlet of *Vanilla planifolia* Jacks. S Afr J Bot 109:288–293

Rathore NS, Rathore N, Shekhawat NS (2013) *In vitro* propagation and micromorphological studies of *Cleome gynandra*: a C_4 model plant closely related to *Arabidopsis thaliana*. Acta Physiol Plant 35:2691–2698. https://doi.org/10.1007/s11738-013-1301-2

Reddy GM (2006) Clonal propagation of bamboo (*Dendrocalamus strictus*). Curr Sci 91:1462–1464

Rodrigues SP, Picoli EAT, Oliveira DC, Carneiro RGS, Isaias RMS (2014) The effects of *in vitro* culture on the leaf anatomy of *Jatropha curcas* L. (Euphorbiaceae). Biosci J Uberlandia 30:1933–1941

Ruffoni B, Savona M (2013) Physiological and biochemical analysis of growth abnormalities associated with plant tissue culture. Hortic Environ Biotechnol 54:191–205

Rybczynski JJ, Borkowska B, Fiuk A, Gawronska H, Sliwinska E, Mikula A (2007) Effect of sucrose concentration on photosynthetic activity of *in vitro* cultures *Gentiana kurroo* (Royle) germlings. Acta Physiol Plant 29:445–453. https://doi.org/10.1007/s11738-007-0054-1

Sack L, Scoffoni C, McKown AD, Frole K, Rawls M, Havran JC, Tran H, Tran T (2012) Developmentally based scaling of leaf venation architecture explains global ecological patterns. Nat Commun 3:1–6

Saini RS, Sharma KD, Dhankhar OP (2001) Laboratory manual of analytical in horticulture. Agrobios, India

Sarkar PK, Kumar PR, Singh AK, Bhatt BP (2020) Effect of priming treatments on seed germination and seedling growth in bamboo [*Dendrocalamus strictus* (Roxb.) Nees]. Acta Ecol Sin 40:128–133

Saxena S, Dhawan V (1999) Regeneration and large-scale propagation of bamboo (*Dendrocalamus strictus* Nees.) through somatic embryogenesis. Plant Cell Rep 18:438–443. https://doi.org/10.1007/s002990050600

Schmildt O, Netto AT, Schmildt ER, Calvalho VS, Otoni WC, Campostrini E (2015) Photosynthetic capacity, growth and water relations in 'Golden' papaya cultivated *in vitro* with modifications in light quality, sucrose concentration and ventilation. Theor Exp Plant Physiol 27:7–18. https://doi.org/10.1007/s40626-014-0026-y

Sharma U, Kataria V, Shekhawat NS (2017) *In vitro* propagation, *ex vitro* rooting and leaf micromorphology of *Bauhinia racemosa* lam.: a leguminous tree with medicinal values. Physiol Mol Biol Plants 23:969–977

Shekhawat MS, Manokari M (2018) Micromorphological and anatomical evaluation of *in vitro* and field transferred plants of *Coccinia indica*. Agric Res 7:135–144. https://doi.org/10.1007/s40003-018-0326-6

Shekhawat MS, Manokari M, Kannan N (2017) Micromorphological response towards altered environmental conditions in subsequent stages of *in vitro* propagation of *Morinda coreia*. Environ Exp Biol 15:37–46. https://doi.org/10.22364/eeb.15.06

Shekhawat MS, Manokari M, Kannan N, Priyadharshini S (2020) Exogenous implication of seismic stress in attenuation of *in vitro* induced morpho-anatomical aberrations in *Hemidesmus indicus* (L.) R. Br. Ex Schult. S Afr J Bot. https://doi.org/10.1016/j.sajb.2020.09.011

Singh S, Kumar A, Pandey A, Palni LMS (2012) Dendrocalamus strictus ((Roxb.) Nees): a suitable host for the maintenance and propagation of AM fungi under temperate conditions. Int Scholar Res Not. https://doi.org/10.5402/2012/607213

Soundappan K, Bajaj R, Vaidya BN, Joshee N (2018) Anatomical and biochemical investigations on medicinal tree *Murraya koenigii* Spreng. J Biotechnol Res 9:70–78

Souza TC, de Magalhães PC, Pereira FJ, Castro EM, de Silva Junior JM, da Parentoni SN (2010) Leaf plasticity in successive selection cycles of "Saracura" maize in response to periodic soil flooding. Pesqui Agropecuária Bras 45:16–24. https://doi.org/10.1590/S0100-204X2010000100003

Trigiano RN, Windham MT, Windham AS (2004) Plant pathology. Concepts and laboratory exercises. CRC Press, Boca Raton, Florida, p 413

Wangawar SN, Shendarkar GR, Shelke DP, Daswad AK, Pohare JG, Roge AB (2017) Phytochemical screening and antimicrobial activity of *Dendrocalamus strictus* leaves extracts. World J Pharma Res 6:1029–1041. https://doi.org/10.20959/wjpr20174-8128

Chapter 7
Micropropagation of Bamboos and Clonal Fidelity Assessment Using Molecular Markers

Leimapokpam Tikendra, Ravish Choudhary, Rajkumari Sanayaima Devi, Abhijit Dey, Angamba Meetei Potshangbam, and Potshangbam Nongdam

Abstract Bamboos, the non-timber forest trees, are economically important plants with over 4000 traditional uses and 1500 commercial applications. Even though bamboos are fast-growing plants, their populations are diminishing at an alarming rate due to extensive forest habitat destruction, rampant illegal collection, and mismanagement of bamboo resources. Conservation of bamboo natural resources through rapid and mass propagation is the need of the hour. However, conventional propagation through seeds and other vegetative methods is besotted with several limitations, and bamboo production will not be sufficient to meet the demands of bamboo stocks. Micropropagation of bamboos provides an excellent alternative to ineffective classical propagation methods by in vitro propagating bamboos rapidly on a large scale. However, somaclonal variation may appear among the in vitro bamboo clones as they are constantly confronted with diverse culture conditions. Assessing the clonal fidelity remains one of the most important prerequisites as heterogeneity can severely limit the purpose of bamboo micropropagation. Several molecular markers have been efficaciously employed to evaluate the clonal fidelity of the bamboo clones so that only the elite, genetically identical plants are propagated. This chapter highlights the recent advancement in bamboo micropropagation and the application of DNA molecular markers in clonal fidelity assessment of the in vitro propagated bamboos.

Keywords Bamboos · Clonal fidelity · DNA markers · Genetic variation · Micropropagation · Somaclonal variation

---

L. Tikendra · A. M. Potshangbam · P. Nongdam (✉)
Department of Biotechnology, Manipur University, Canchipur, Manipur, India

R. Choudhary
DSST, Indian Agricultural Research Institute, New Delhi, India

R. Sanayaima Devi
Department of Botany, Deen Dayal Upadhyaya College, University of Delhi, New Delhi, India

A. Dey
Department of Life Sciences, Presidency University, Kolkata, India

Z. Ahmad et al. (eds.), *Biotechnological Advances in Bamboo*,
https://doi.org/10.1007/978-981-16-1310-4_7

7.1 Introduction

Bamboos, the multipurpose forest tree grasses belonging to the family Poaceae, are one of the most economically important plants in the world (Clark et al. 2015). They are long-lived, woody perennial monocotyledonous grasses having considerable economic, social, and ecological importance. The woody plant family comprises 9–10 subtribes with 59 to 111 genera and approximately 1642 species (Choudhury et al. 2012; Goyal and Brahma 2014). They are very fast-growing evergreen plants with sporadic flowering patterns. Every bamboo species has a unique duration of the flowering cycle, ranging from 15 to 60 years or even more than 100 years for some temperate bamboos (Banik 2015). They are distributed widely across the globe existing mostly as natural vegetation in tropical, subtropical, and temperate regions (Soreng et al. 2015). They account for about 1% of the total forest area of the world, with an estimated bamboo forest area of about 31.5 Mha (Wang et al. 2020). India produces the largest bamboo forest area of 9.5 Mha followed by China with a bamboo area of 6.01 Mha (Wang et al. 2020).

Bamboos attach significant importance to the everyday life of many people around the globe because of their enormous utility (Bisht et al. 2012). They are known to have more than 4000 traditional uses and 1500 commercial applications (Sandhu et al. 2018). The highly diverse utilities of bamboos are found extensively in the paper, textile, and handicraft industries, house construction, and making furniture, water pipes, storage vessels, ladders, water, and food containers, and many other household items (Sawarkar et al. 2020). Bamboos are also attractive from ecology viewpoints as they have not only carbon-sequestering potential but also high nitrogen-fixing capability (Gu et al. 2019). Juvenile bamboo shoots are also consumed as popular traditional health foods in many Asian countries (Nongdam 2015).

Bamboo genetic resources are depleting rapidly due to widespread forest destruction, rampant exploitation as timber substitutes, and increasing demand for industrial and household use (Nongdam and Tikendra 2014). Innovative conservation strategies have to be formulated to preserve and expand the rapidly reducing bamboo resources to fulfill global demands. Conventional approaches of bamboo propagation through seeds, rhizome division, and culm cutting are often limited due to nonavailability of plant materials, handling of large-size propagules, erratic flowering pattern, short seed viability, and low seed set (Gantait et al. 2018). The classical propagation techniques can yield insufficient plants (10,000 plants per year), which are not viable for large-scale bamboo propagation (more than 500,000 plants per year) (Goyal and Sen 2016). The increasing bamboo demand can be met through micropropagation techniques, which provide an alternative to conventional methods for mass production.

Micropropagation of several bamboo species has been successfully performed using different explants by many researchers (Mudoi et al. 2014; Venkatachalam et al. 2015; Wei et al. 2015; Kahsay et al. 2017; Ray et al. 2017). However, the main objective of tissue culture is to propagate true-to-type plants without any molecular and phenotypic defects to preserve and commercialize the elite genotypes (Tikendra et al. 2019a; Dey et al. 2020a). Genetic fidelity of in vitro propagated bamboos must be tested as somaclonal variation may occur among the clones due to different factors associated with culture conditions. Retaining the clonal uniformity for a prolonged culture period also has immense commercial significance as the initiation of bamboo culture is always a daunting task due to excessive phenol exudation, persistent contamination risk, and problem in explant selection (Singh et al. 2013b). There are reports of clonal fidelity assessment of several micropropagated bamboos using different DNA markers (Agnihotri et al. 2009; Singh et al. 2013b; Goyal et al. 2015; Desai et al. 2019). The present chapter features the crucial aspects of bamboo micropropagation using different explants and the application of DNA markers in ascertaining the clonal fidelity of different in vitro regenerated bamboos.

7.2 Bamboo Micropropagation

Conventionally, bamboos are propagated using seeds, macroproliferation, layering, rhizome, and culm cuttings. However, inadequate bamboo production through conventional approaches is due to low seed viability, sporadic flowering with seed production after a long interval, lack of vegetative propagules and difficulty in long-distance transport, seasonal dependence, low survival rate, and lack of rooting of propagules (Singh et al. 2013a). Micropropagation of bamboos is an important alternative to classical propagation methods for rapid and large-scale production of different bamboos. Several bamboo species have been successfully micropropagated by employing different explants. Micropropagation involves explant selection, surface sterilization, and inoculation to appropriate nutrient media for culture initiation, shoot and root developments, and subsequent acclimatization and hardening of the in vitro regenerated plants.

7.2.1 Factors Affecting Bamboo Micropropagation

Although micropropagation helps generate disease-free and genetically identical plants (Jiménez and Guevara 2007), it is associated with various factors that cause hindrances at different stages of growth and development.

7.2.1.1 Nature of Explant

The choice of explant is a major determining factor in successful micropropagation of bamboos because different plant parts vary in their regenerative potential (Evans et al. 1981; Chaturvedi 1984). While in vitro propagation via seed culture is comparatively easier in terms of regeneration and low rate of contamination than the nodal explants, its limited availability and loss of germination capability are major setbacks (Mudoi et al. 2013). The leaf and nodal explants are readily available throughout the year, but the seed explant does not. Therefore, mature tissue explants such as leaf and nodal segment provide an alternative from bamboo flowering dependency (for seeds) and are a good source of explant for large-scale propagation of valuable bamboo plants (Mudoi et al. 2013). However, successful regeneration of plantlets using leaf explant requires callus induction, an event that increases the occurrence of somaclonal variation (Hassan and Debergh 1987; Jullien and Van 1994; Komatsu et al. 2011; Saravanan et al. 2011). In vitro regeneration through the nodal explant is also dependent on the size, age, and position of the nodal segments (Mudoi et al. 2014). Mid-culm nodes of secondary branches are reported as the best part of nodal explant for axillary shoot initiation (Saxena and Bhojwani 1993). But the nature of dormancy and nodal bud break varied with its position in the plant and also with the season and the species involved (McClure 1966; Hu and Wang 1983; Singh et al. 2012b).

7.2.1.2 Microbial Contamination

Media used for plant propagation are also favorable for the growth of various microbial contaminants. They compete with the growing plant tissues for nutrients from the media (Omamor et al. 2007). There is also a strong correlation between the level of contamination and the seasons of sample collection. Though the explants collected during the rainy season generally have high-frequency bud break, they are associated with greater chances of contamination. The explants collected during the rainy season produced high endogenous contamination in *Bambusa nutans* (Mehta et al. 2011), *Arundinaria callosa* (Devi and Sharma 2009), *D. asper* (Singh et al. 2012c; Nadha et al. 2013), *D. hamiltonii* (Singh et al. 2012a), and *D. giganteus* (Ramanayake and Yakandawala 1997). Fungal and bacterial contaminants are the most common biotic contaminants causing difficulties during bamboo micropropagation. Fungal contamination can either be endogenous or through contacts during the culture processes (Babaolu et al. 2001; Ankur et al. 2014). There are certain reasons which make removal of contaminants through surface sterilization a challenge. The cut ends of nodal explant have large intercellular spaces and vessel cavities which were exposed to bacterial and fungal contaminants during the sample collection. This allows the microbial pathogens to settle inside the nodal tissues avoiding the effect of sterilants (Thakur and Sood 2006). They induce latent contamination during micropropagation, causing necrosis, which results in high plant mortality (Kane 2003; Abdulminam et al. 2009). Fungicide and antibiotic treatment of explant tissues are usually performed to manage such problems (George 1993; Herman 1996; Niedz 1998). Though various antibiotics are useful at different stages of bacterial cellular activities, broad-ranged and stable bactericide or fungicide having low phytotoxicity must be chosen for the sterilization of the explants (Shields et al. 1984; Luna et al. 2008).

7.2.1.3 Browning of Plantlet

Browning due to phenolic exudation is another major constraint faced during in vitro propagation of bamboos (Fig. 7.1a). Injury to the tissues during explant preparation or subculturing increases the release of polyphenol oxidase (PPO). The oxidation

Fig. 7.1 In vitro propagation of *Dendrocalamus hamiltonii* using nodal explant. (**a**) Browning and necrosis in the regenerated shoot. (**b**) Nodal explant preparation. (**c**) Sprouting of bud (bud break) from the node. (**d**) Shoot induction. (**e**) Multiple shoot proliferation. (**f**) Transplantation of the well-grown bamboo plantlets

products can cause necrosis, which leads to the death of the culture. Huang et al. (2002) affirmed the occurrence of shoot browning due to PPO activity to be pH-dependent. The browning was low at pH 5.7, but a high rate was witnessed with the medium at pH 7 and 8. The enzymatic activity of PPO was optimum between pH 9 and 11, but its action declined above and below this range. Incorporating a high concentration of BAP in the medium escalated the browning of bamboos (Huang et al. 1989). Several authors reportedly tried to control browning by supplementing various additives like polyvinylpyrrolidone (PVP), activated charcoal, ascorbic acid, cysteine, ferulic acid, kojic acid, and thiourea to culture media (Saxena and Dhawan 1999; Huang et al. 2002; Singh et al. 2012c). While PVP in the medium controlled browning in *D. strictus* (Saxena and Dhawan 1999), its failure to prevent it was evident in *Phyllostachys nigra* (Ogita 2005) and *D. hamiltonii* (Singh et al. 2012a). The ineffectiveness of these additives indicates their inability to counter the activity of PPO, which suggests that the exogenous substances are not readily taken up by the growing cells or transported to the effective sites of action (Mayer and Harel 1979; Murata et al. 1997). Transferring of culture to a freshly prepared medium at regular intervals is the most dependable way to control browning, as was reported in *D. giganteus* (Mudoi et al. 2014).

7.2.1.4 Hyperhydricity

The bamboo tissues during in vitro culture exhibited a physiological disorder of hyperhydricity. This phenomenon can lead to irreversible loss of the regenerative ability of plant tissues leading, finally, to plant death (Gaspar et al. 2000). The main factors which cause hyperhydricity are prolonged culture and delayed subculturing, which triggered culture stress due to decreasing nutritional and moisture content in the culture vessel (Mudoi and Borthakur 2012). Franclet (1991) and Lin and Chang (1998) indicated that the explant (clump) number, culture incubation time, and culture vessel size exerted a significant effect on the shoot hyperhydricity of *Bambusa balcooa*. There is also the likelihood that growth hormone type and concentration may influence the development of a morphological symptom of hyperhydricity. The application of liquid medium for in vitro plant culture is the most common cause of hyperhydricity (Kevers et al. 2004; Scheidt et al. 2011). Some of the measures suggested for avoiding this disorder include the addition of low ammonium ion concentration in the medium (Brand 1993), increasing agar concentration (Bornman and Vogelmann 1984), regular subculturing, alternate use of liquid and solid media (Arshad et al. 2005; Mudoi and Borthakur 2012), and using of luffa pieces as supporting medium for the shoots (Mudoi et al. 2014).

7.2.2 *Sterilization of Culture Explants*

Removal of microbial contaminants that can produce immediate or latent infection through proper explant sterilization is vital for successful culture initiation and plantlet development of bamboos. The possible surface sterilization protocols for different explants of bamboos are described in Fig. 7.2. Explants are dehusked (seeds) or have culm sheath detached (if any, in case of nodal explant) and cut into nodal segments before washing with detergents such as Tween 20 or Teepol (Fig. 7.1b). To improve sterilization, Thakur and Sood (2006) utilized the multi-nodal explant to reduce the permeation of contaminants and disinfectants into the intercellular spaces and vascular cavities at the cut end. The use of multi-nodal explant lowered the level of contaminants and phytotoxic effect of the disinfectants to the explant tissues. In another approach, rinsing the explants with acetone for 3–4 s before treatment with bleaching solution enhanced the success rate of raising aseptic cultures (Oprins et al. 2004).

Jha et al. (2013) reported the usefulness of ethanol dip treatment as a part of the sterilization process in controlling culture contamination. For effective surface sterilization, the choice of chemicals and treatment duration may vary depending on the explant source and types (Reed et al. 1998). Sodium hypochlorite, calcium hypochlorite, and mercuric chloride are the main sterilizing agents used to eliminate microbial contaminants in bamboo explants (Jiménez and Guevara 2007). However, depending on the severity of contamination, antibiotics and fungicides are utilized prior to mercuric chloride treatment to remove endophytic contaminants residing in the internal explant tissues that cause latent infection (Torres et al. 2016). Some of the fungicides which are used at different concentration and treatment duration in

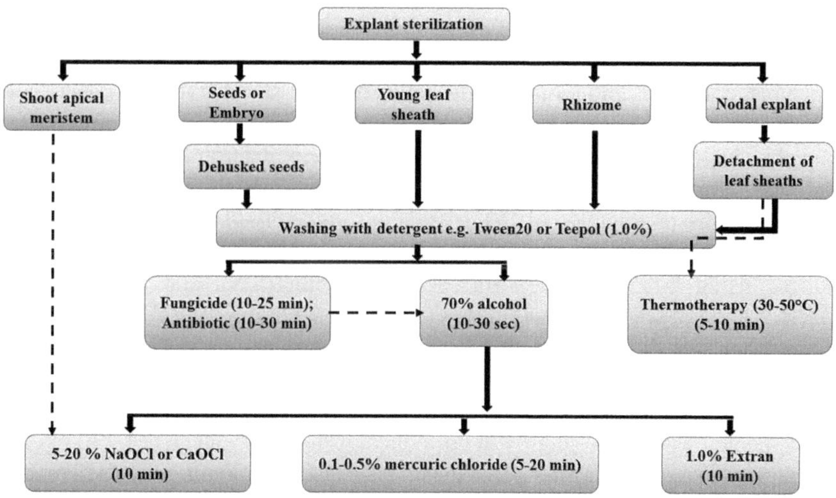

Fig. 7.2 Surface sterilization protocols for different bamboo explants involving various experimental steps

explant sterilization are Benomyl (Ahmadi et al. 2012), mancozeb (Mudoi et al. 2014), and bavistin (Venkatachalam et al. 2015). Although Benomyl and bavistin are composed of the same active ingredient of benzimidazole, Benomyl is reported to be more effective in controlling fungal contamination (Ahmadi et al. 2012). Several authors also recommended the combined use of fungicides with antibiotics such as streptomycin (Kant et al. 2009), streptomycin sulfate and tetracycline hydrochloride (Singh et al. 2012a), bacteriomycin (Sharma and Sarma 2011), and gentamycin (Sharma and Sarma 2013). In *Guadua angustifolia*, Jiménez et al. (2006) performed a combined treatment of the explants with Extran, Agri-Mycin, and Benomyl followed by treatment with sodium hypochlorite and PPM (Plant Preservative Mixture), which is a heat-stable biocide effective against a broad spectrum of contaminants. The sterilization process that they performed helped in reducing the contaminants to 11%. Ali et al. (2009) used antibiotics (ciprofloxacin, streptomycin, and rifampicin) and bavistin fungicide to successfully eliminate endophytic contaminants in the nodal explants of *Bambusa tulda*, *B. wamin*, *B. balcooa*, *B. bambos*, and *D. asper*. However, bactericidal (that kills bacteria) antibiotics were reported to be more beneficial as compared to bacteriostatic ones (that prevents bacterial growth) (Young et al. 1984; Fisse et al. 1987; Leifert et al. 1992). Alternatively, thermotherapy of explant was also performed, which exposed the explants to a certain temperature followed by cooling at room temperature and subsequent treatment with chemical disinfectants (Torres and De Lemos 2017).

7.2.3 Culture Initiation

Following the first propagation of bamboo from a mature embryo in an artificial medium (Alexander and Rao 1968), several efficient in vitro culture protocols of various bamboo species have been successfully developed. Commonly used bamboo explants for culture initiation include embryos, leaf sheath, young branch node, rhizome segments, and roots of young seedlings. The possible in vitro developmental pathways for different explants leading to bamboo plant production after proper acclimatization and hardening are depicted in Fig. 7.3. Waikhom et al. (2012) initiated in vitro propagation of *Dendrocalamus giganteus* using seeds in Murashige and Skoog (MS) medium (Murashige and Skoog 1962). Seed germination was induced using 6-benzylaminopurine (BAP), kinetin (KN), and gibberellic acid (GA_3). Although seed germination was also observed in the control medium devoid of any plant growth regulators, the percentage of germination was low compared to the other medium incorporated with BAP, KN, or GA_3. Ojha et al. (2009) used leaf sheaths of *Dendrocalamus asper* to initiate in vitro culture on MS medium supplemented with different growth hormones. A high percentage (77.7%) of callus formation was reported from the leaf sheath base of *D. asper* cultured in MS medium appended with 30μM 2,4-D (2,4-dichlorophenoxyacetic acid). Komatsu et al. (2011) also attempted callus induction from the in vitro nodal segment-derived leaf sheath of *Phyllostachys bambusoides*. The leaf sheaths were inoculated in MS medium

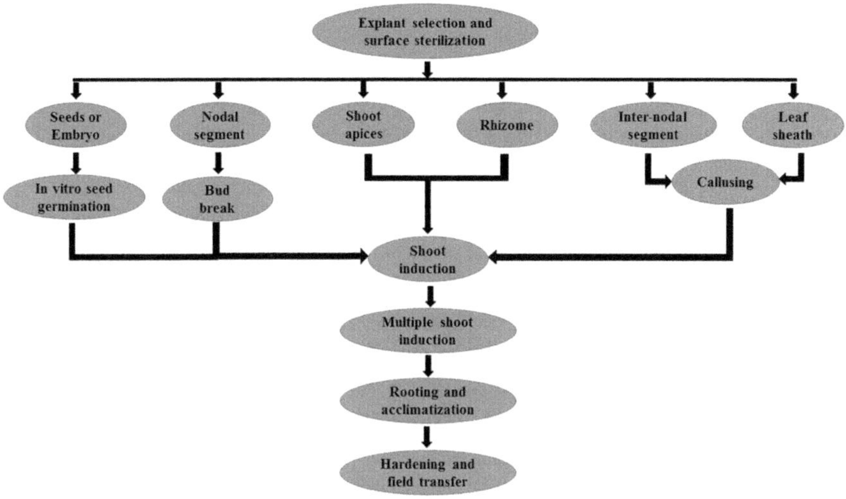

Fig. 7.3 The possible in vitro developmental pathways of different explants for bamboo plant production

incorporated with 8.0 mg L^{-1} picloram. White-friable and yellow-globular calli were observed, of which the latter was unable to undergo in vitro morphogenesis even after consecutive subculture. The addition of carbohydrate sources (glucose, fructose, sucrose) in the medium favored the morphogenetic ability of the callus to induce meristematic activities and root formation. Lin et al. (2007) exploited the property of bamboo's shoot apical meristem (SAM) that could generate organs in a very regular arrangement. To propagate the disease-free plant, shoot tip meristems free from bamboo mosaic virus (BaMV) were used to initiate in vitro propagation of *Bambusa oldhamii*. By inoculating SAM into MS medium with plant growth hormones, multiple shoots were developed within 2 weeks of culture.

Nodal segments, because of their easy availability, are the most widely used explants for in vitro propagation and clonal fidelity assessment of bamboos (George and Debergh 2008). This propagation technique involves breaking of dormant bud under the influence of artificial medium and growth hormones (Fig. 7.1C). Similar to the ontogenetic development of axillary meristem along the branches, the dormant buds are induced to shooting initiation and further multiplication under the impact of cytokinins (Figs. 7.1D and E). The culture upon transferring to the auxin-enriched rooting medium generates roots, and plantlets with well-developed leaves and roots are subsequently hardened in the greenhouse after proper acclimatization (Fig. 7.1F). Bisht et al. (2010) employed nodal segments having a single axillary bud for propagation of *Gigantochloa atroviolacea* in liquid MS medium. Bud break was observed in 2–3 weeks of culture; however, the sprouting bud that developed into shoots failed to survive when cultured in the control medium devoid of growth hormones. Seasonal factors influence the bud break of the nodal explant. Initiating the in vitro culture of *Dendrocalamus hamiltonii* using nodal explants collected in the first week of every month, Singh et al. (2012a) suggested that the explants gathered during spring and early summer took a shorter period for bud break. Contamination that occurred during the period was relatively low, with optimal bud break frequency and early shoot initiation. This effect may be due to auxin production in the young buds during the spring season. Funada et al. (2001) also opined seasonal dependence on the role of endogenous hormones in stimulating plant cell division and growth.

Somashekar et al. (2018) described various explant potential for somatic embryogenesis of *Dendrocalamus stocksii* via callus formation. Leaf segments, leaf sheath, shoot tips, and nodal and internodal segments were used to induce callusing. The frequency and intensity of callus induction were best observed in cultures derived from nodal explants. Callusing was also mainly dependent on the culture conditions and its components. Of the various culture compositions experimented for every explant, nodal segments inoculated in MS + additives (49.9 μM L^{-1} ascorbic acid, 22.8 μM L^{-1} citric acid, 24.9 μM L^{-1} cysteine, and 100 μM L^{-1} glutamine) + 2.1μM 2,4-D + 0.22μM KN (kinetin) + 10% coconut water exhibited the highest callus induction frequency (70%) and intensity of embryogenesis when maintained under dark condition (Somashekar et al. 2018). Rhizome, the underground part of bamboo stem (culm), is also an important propagule for initiating culture. Nirmala et al. (2011) inoculated rhizome segments of *Dendrocalamus asper* and *D. membranaceus* on MS medium supplemented with different concentrations of BAP. The rhizome-induced shoots increased their multiplication rate with repeated subculturing, but propagules with less than three to four shoots showed a decrease in multiplication.

7.2.4 *In Vitro Shooting and Rooting Responses*

While successful in vitro bamboo propagation mainly depends on the composition of growth media and hormone combinations, other factors also determine culture development depending on a particular species. The pH of the medium is one such factor whose values ranging from 5.0 to 5.5 are generally optimum for the of bamboo micropropagation (Butenko et al. 1984). However, due to the variability of plant tissues, the pH of the medium varying from 4.5 to 5.8 was optimum for induction of shoot growth and development in *D. giganteus* (Arya et al. 2006). The recommended sucrose content of the medium is 3.0%, but there was a report of ideal shoot development at lower concentration (2.0% w/v) in *B. tulda* (Saxena 1990). The higher concentration (6.0% w/v) of sucrose in the medium also induced callusing in *Bambusa edulis* (Lin et al. 2004). Singh et al. (2012c) observed deleterious effects on shoot multiplication when sucrose was substituted by glucose as the carbon source, while table sugar had little or no impact on the rate of shoot multiplication of *D. asper*.

Kapoor and Rao (2006) used caryopses of *Bambusa bamboos* var. *gigantea* to induce in vitro rhizome development via multiple shoot formation. High-shoot proliferation was observed in the MS medium enriched with 2.5μM BAP. The incorporation of a low concentration of 0.1μM GA_3 increased shoot length but inhibited shoot multiplication rate. The same growth response was not evident when 0.1μM GA_3 was added to the medium, along with a higher BAP (5.0μM). Rhizome induction was not witnessed when the medium was supplemented with either individual growth regulators (BAP, GA_3, and NAA) or sucrose alone. High rhizome induction (85%) was achieved in MS + 2.5μM BAP + 50.0μM NAA + 5% sucrose. The importance of sucrose for culture growth was also identified when rhizome length and growth were significantly reduced in its absence in the medium. Moreover, the shoots did not survive after 5 weeks of culture in the non-sucrose medium. GA_3 in the medium enhanced node formation in the rhizomes, and thick roots were generated from the nodes of the rhizome. Lin et al. (2004) cultured nodal and internodal tissues of in vitro plantlets of *Bambusa edulis* on MS + 9.2μM kinetin +13.6μM 2,4-D + coconut milk (0.1% v/v) + sucrose (6% w/v) for obtaining embryogenic callus. The effect of sucrose and thidiazuron (TDZ) was also examined for callus initiation and proliferation. MS medium supplemented with 0.046μM TDZ, 13.6μM 2,4-D, and sucrose (3% w/v) was found most suitable for embryogenic callus proliferation. Medium with TDZ (0.455μM) produced a maximum rate of somatic embryo germination (80%), while the presence of naphthaleneacetic acid (NAA) reduced it.

Waikhom and Louis (2014) propagated *Bambusa tulda* and *Melocanna baccifera* using different strengths of MS medium. Shoot induction and multiplication from the nodal explant were achieved in full-strength MS medium appended with either benzylaminopurine (BAP) or KN alone or in combination. The rooting initiation, however, was achieved in half-strength MS medium containing 3.0% sucrose; various plant growth regulators, viz., BAP and indole-3-butyric acid (IBA); and coumarin. Rajput et al. (2019a) similarly reported shoot induction and multiplication in *Dendrocalamus strictus* from the nodal explant in full-strength MS medium containing either BAP or KN singly or a combination of BAP and NAA at different concentrations. Maximum rooting response (93.0%) was observed in a half-strength MS medium supplemented with 3.0 mg L^{-1} NAA. Wei et al. (2015) performed successful callus regeneration and micropropagation of *Bambusa ventricosa* using axillary buds. Robust bud sprouting and multiplication were witnessed in MS media appended with either 22.2μM BAP or 26.6μM BAP. While the highest bud proliferation was achieved in MS + 22.2μM BAP + 0.23μM thidiazuron (TDZ) + 0.27μM NAA, maximum frequency of callus formation was observed in MS + 27μM 2,4-D + 2.7μM NAA + 0.0045μM TDZ. Rooting from the proliferated buds was optimum in MS + 2.7μM NAA + 4.9μM IBA + 4.4μM BAP.

Chambers et al. (1991) used nodal explants from young seedlings of *D. hamiltonii* to produce multiple shoots and flowers on MS medium enriched with varying concentrations of BAP. The maximum number of shoots (8 per explant) was produced after 12 weeks on MS medium containing 4.4μM BAP. In vitro flowering was first observed after 13 weeks with the highest flowering (47% of nodal explants) attained in 8 weeks after the seedlings were transferred from 22.2μM BA incorporated medium to a basal MS medium. In an attempt to prevent potential loss due to unpredictable flowering in bamboos and post-flowering consequences, Arya et al. (2008) initiated direct shoot regeneration from immature inflorescence explants of *D. asper*. Though shoot generation was witnessed in MS +7.0 mg L^{-1} BAP, the best shoot induction was achieved with explant size ranging between 0.5 and 1.0 cm. A 12- to 15-fold increase in shoot multiplication was observed in medium appended with a lower concentration of BAP (3.0 mg L^{-1}). Root induction in 90–95% of the regenerated shoots was obtained in MS medium enriched with 10 mg L^{-1} IBA. Waikhom et al. (2012) reported successful propagation of *D. giganteus* in full-strength liquid MS medium using seed explant. In vitro seed germination was enhanced in the medium incorporated with GA_3. Shoot induction and multiplication were accomplished in MS medium supplemented with 3.0 mg L^{-1} BAP. Callus induction and proliferation were prominent in the medium augmented with 3.0 mg L^{-1} 2,4-D and 0.5 mg L^{-1} KN. De novo production of the shoot was achieved from the white, compact, and nodular calli after being transferred to MS medium containing 1.0 mg L^{-1} NAA and 0.5 mg L^{-1} KN. A high percentage (86.0%) of root induction from the regenerated shoots was also obtained when the shoots were relocated to a half-strength MS medium containing 5.0 mg L^{-1} IBA. Micropropagation work on several bamboo species by different researchers in the past 10 years are listed in Table 7.1.

7.2.5 Acclimatization and Hardening

The soft in vitro bamboo tissues accustomed to low irradiance and high humidity experienced unfavorably when transferred to the natural environment with high irradiance and low humidity. Moreover, the in vitro regenerated bamboos usually have leaves with little or no development of cuticular wax, distressed stomatal mechanism, low chlorophyll pigments, insufficient photosynthetic activity, and underdeveloped vascular and connective tissues, which are not conducive for self-sustenance. Thus, effective standardized protocols for hardening and acclimatization should be established to overcome the bottleneck of fortuitous transfer of in vitro propagated bamboos from the glasshouse to the field. The regenerants with poorly developed physiological and metabolic pathways are toughened by culturing them in a medium whose component and other additives are gradually lowered or withdrawn during regular passages to adapt themselves to the natural environment (Singh et al. 2013b). The well-rooted plants are hardened using potting mixtures like soil, sand, perlite, Soilrite, vermicompost, or farmyard manure (FYM), individually or in combination at various proportions (Mishra et al. 2011; Rajput et al. 2019a). Waikhom and Louis (2014) hardened the plantlets of *B. tulda* and *M. baccifera* in bottle jars that contained sterilized rice straw-vermicompost and sand (4:1% w/v) for 25 days. The acclimatization was successfully maintained at 30 ± 2 °C and 84% humidity under greenhouse conditions with a survival rate of 81.81% and 70.31% for *B. tulda* and *M. baccifera*, respectively. Rajput et al. (2019a) transplanted the 3–4-wee-old well-rooted shoots to a hardening medium that contained Soilrite

Table 7.1 Micropropagation of different bamboo species in the last 10 years

Species	Explant used	Optimal growth response	References
Bambusa arundinacea	Nodal explants	• MS + 3.0 mg L^{-1} BAP + 0.5 mg L^{-1} IBA was the best medium for shoot bud initiation (87.2%) with 24.2 shoots per explant • MS + 3.0 mg L^{-1} BAP + 4.0% CW + 4.0% sucrose recorded the highest frequency (95.2%) of shoot bud multiplication • MS + 3.0 mg L^{-1} IBA + 2.0 mg L^{-1} $AgNO_3$ produced maximum rooting (85%)	Venkatachalam et al. (2015)
B. balcooa	Nodal	• Bud break achieved on MS (basal) + 2.0% sucrose in about 15 days • Shoot multiplication was best observed on MS + 4.4 μM BAP + 0.53 μM NAA with 19.8 ± 1.4 shoots per explant • Rooting was highest on MS + 16.11 μM NAA + 2.0% sucrose with 11.5 ± 1.58 number of roots	Brar et al. (2014)
		• Axillary shoot sprouting (97.22%) was best in liquid MS + 7.0 mg L^{-1} gentamicin with the lowest contamination • Shoot multiplication was highest (12–15 shoots) in MS + 3.0 mg L^{-1} BAP + 0.5 mg L^{-1} NAA • Root induction was optimum (83% average 9.6 roots) in MS + 4.0 mg L^{-1} NAA	Patel et al. (2015)
	Culm	• The highest bud break (92%) was achieved from the explant collected during early may with low contamination (18%) • Shoot multiplication (35 shoots) and bud sprouting (96%) were recorded highest on the MS (liquid) + 2.5 mg L^{-1} BAP + 0.1 mg L^{-1} NAA + 50 mg L^{-1} ascorbic acid +25 mg L^{-1} citric acid	Choudhary et al. (2017)

(continued)

Table 7.1 (continued)

Species	Explant used	Optimal growth response	References
		+25 mg L^{-1} cysteine • 100% rooting was evident on ½ MS + 2.5 mg L^{-1} NAA	
	Nodal	• MS + 0.8% agar +3.0% sucrose +100 mg L^{-1} inositol +5.0 mg L^{-1} BAP was the most suitable medium for bud break • MS + 3.0 mg L^{-1} BAP recorded the highest shoot multiplication (95.67 ± 12.89) • MS + 4.5 mg L^{-1} NAA produced the highest root number (236 ± 11.53)	Thapa et al. (2018)
B. bambos	Nodal	• The highest percentage (90 ± 14) of bud sprouting was observed in MS + 1.07 μM NAA, MS + 8.87 μM BAP, and MS + 21.48 μM BAP • The maximum number of shoots (7.10 ± 0.26) and leaves (7.20 ± 0.47) was produced in MS + 8.87 μM BAP • Rooting response was highest (100%) in ½ MS + 53.71 μM choline chloride +10.74 μM IBA	Desai et al. (2019)
		• Explant in MS (liquid) + 2.0 mg L^{-1} BAP + 1.0 mg L^{-1} TDZ induced the highest (90%) shooting response and shoot number (3.14 ± 0.06) per explant • Shoot proliferation was highest (16.58 ± 0.24) in ½ MS + 2.0 mg L^{-1} BAP + 1.0 mg L^{-1} TDZ + 10% CW + 4% sucrose	Raju and Roy (2017)
B. nutans	Nodal	• Maximum shoot number (11.0) was recorded in MS + 2.0 mg L^{-1} BAP • Callus induction (30–40%) from the in vitro sprouted buds was best achieved in	Mehta et al. (2011)

(continued)

Table 7.1 (continued)

Species	Explant used	Optimal growth response	References
		MS + 5.0 mg L^{-1} 2,4-D • Maturation and germination of well-developed somatic embryos were achieved in MS +1.0 mg L^{-1} BAP + 20.0 mg L^{-1} 2,4-D	
		• MS + 4.4 μM BAP + 2.32 μM KN + 0.2% Gelrite yielded 80% aseptic cultures with 100% bud break • MS (liquid) + 13.2 μM BAP + 0.98 μM IBA recorded the best shoot proliferation (3.5-fold by fifth passage) • Rooting was 100% successful when cluster of 3–65 shoots were cultured on MS (liquid) + 9.8 μM IBA + 2.85 μM IAA + 2.68 μM NAA + 3.0% sucrose	Negi and Saxena (2011)
		• Explant treated with 5.0% tween 20 + 0.1% Mancozeb +0.1% gentamicin +70% OH produced the highest survivability rate (45%) with lowest rate of contamination (55%) • ½ MS (basal) medium generated the highest (63%) seed germination in a 16-hour light condition • The shoot number was highest (11.33) when proliferated using 4 shoots per cluster on MS + 0.5 mg L^{-1} BAP + 0.1 mg L^{-1} NAA • Rooting of 85% was recorded on MS + 2.0 mg L^{-1} NAA with the highest root number (2.60)	Mudoi et al. (2014)
B. oldhamii	Nodal	• The highest control of contamination was achieved in MS + BAP + 4 mlL^{-1} PPM • Shoot formation was highest (40%) in MS + 4 mlL^{-1} PPM	Pasqualini et al. (2019)

(continued)

Table 7.1 (continued)

Species	Explant used	Optimal growth response	References
		• The best growth response was observed for the explants collected from June to July • An increase in the leaf number (81.79%) was witnessed under the light condition of 30% blue +70% red LED in comparison to the control medium • MS + 2.27 μM TDZ + 100% blue light showed an increase in shoot number (78.45%) • Relatively high level of Chl a was observed in (MS + PBZ) + (30% blue +70% red light) or (30% red +70% blue light) • Chl b content increased on MS + (TDZ or PBZ) + (100% blue light) or (30% red +70% blue light)	Silveira et al. (2020)
B. pallida	Nodal	• Shoot induction were recorded high (98.57% and 98.52%) on MS (liquid) + 22.2 μM BAP + 1.34 μM NAA and MS (liquid) + 1.125 μM TDZ + 1.34 μM NAA, respectively • Shoot number per explant was maximum (4.90) on MS (liquid) + 1.125 μM TDZ + 1.34 μM NAA • Shoot multiplication from the clump was highest (8.05 shoots/clump) on MS (liquid) + 22.2 μM TDZ + 1.34 μM NAA • Maximum rooting (67.5%) and root number (5.36) were recorded on MS + 0.5 $mgml^{-1}$ IBA	Beena and Rathore (2012)
B. tulda	Nodal	• The highest bud break frequency was observed in MS + 3.0 mg L^{-1} BAP • Maximum number of shoots (17.67 ± 0.56) was recorded in MS + 2.0 mg L^{-1} KN + 3.0 mg L^{-1} BAP	Waikhom and Louis (2014)

(continued)

Table 7.1 (continued)

Species	Explant used	Optimal growth response	References
		• Highest rooting (86.67 ± 3.33%) was witnessed in MS + 3.0 mg L^{-1} IBA + 3.0 mg L^{-1} IAA + 10 mg L^{-1} coumarin	
B. ventricosa	Nodal	• MS + 22.2 μM BAP was optimal for bud initiation (90.4 ± 2.77%), while MS + 26.6 μM BAP provided multiple bud induction (up to 84%) • MS + 22.2 μM BAP + 0.23 μM TDZ + 0.27 μM NAA recorded the highest bud proliferation (>1.5-fold) • MS + 2.7 μM NAA + 4.9 μM IBA + 4.4 μM BAP was optimal for rooting (>70%) • MS + 27 μM 2,4-D + 2.7 μM NAA + 0.0045 μM TDZ was the most effective for callus formation (>60%), while MS + 22.6 μM 2,4-D + 5.4 μM NAA + 2.2 μM BAP produced the highest rate of proliferation • The regeneration efficiency of embryogenic callus was highest in MS + 13.3 μM BAP + 2.7 μM NAA	Wei et al. (2015)
B. vulgaris 'wamin'	Nodal	• Bud sprouting was highest (90 ± 14%) in MS + 0.54 μM NAA, MS + 1.07 μM NAA, MS + 8.87 μM BAP + 0.54 μM NAA, and MS + 8.87 μM BAP + 2.69 μM NAA • The maximum number of the shoot (6.20 ± 0.32) and leaf (7.50 ± 0.61) formation was achieved in MS + 8.87 μM BAP + 0.54 μM NAA • Rooting multiplication was best (4.4 ± 0.18) in ½	Desai et al. (2019)

(continued)

Table 7.1 (continued)

Species	Explant used	Optimal growth response	References
		MS + 53.71 μM choline chloride +21.48 μM IBA	
Dendrocalamus asper	Nodal	• Maximum shoot multiplication (14) was observed on MS (liquid) + 5.0 mg L^{-1} BAP + 40 mg L^{-1} adenine sulfate • 93.33% of shoots were successfully rooted when transferred to MS (liquid) + 1.0 mg L^{-1} IBA with the highest number of roots (7.33 ± 0.33)	Banerjee et al. (2011)
		• The maximum number (4.83 per explant) of shoots was generated in MS + 15 μM BAP • Shoot multiplication was highest (4.03-fold) in MS +10 μM BAP + 75 μM adenine sulfate with 27.67 shoots • Optimal rooting (10 roots per propagule) was achieved in shoots cultured on ½ MS + 5 μM IBA + 5 μM NAA	Singh et al. (2012c)
		• Highest bud break (95.83%) was found in MS + 0.5 mgL^{-1} TDZ • High proliferation of shoots (3.12) and leaves (3.29) per plant were witnessed in MS + 4.5 mgL^{-1} BAP • A significant increase in shoot multiplication (6.5-fold in single-shoot plants and 4.45-fold in two shoot plants) was observed when the shoots from MS + TDZ containing media were transferred to MS (liquid) + 3.0 mgL^{-1} BAP • Growth and development of roots were superior in ½ MS + 1.0 mgL^{-1} IBA with 10.75 roots/propagule and average root length of 5.67 cm	Ray et al. (2018)

(continued)

Table 7.1 (continued)

Species	Explant used	Optimal growth response	References
D. giganteus	Seeds	• Seed germination was superior at lower concentrations (0.5 mgL^{-1}) of GA_3 in light conditions • The highest mean number of shoots was noticed in MS + 10.0 mgL^{-1} BAP • The rate of shoot multiplication was highest in MS medium appended with either 3.0 mgL^{-1} BAP or 4.0 mgL^{-1} BAP • Callus induction was maximum (90.2 ± 0.9%) in MS + vitamin B5 + 3.0 mgL^{-1} 2,4 – D + 0.5 mgL^{-1} BAP • Best rooting response (86.0 ± 0.4%) and average root number (13.0 ± 0.6) was demonstrated on ½ MS + 5.0 mgL^{-1} IBA	Waikhom et al. (2012)
D. hamiltonii	Seeds	• MS + 35 μM BAP recorded the highest seed germination response (37.50 ± 2.40%) and shoot number (9.66 ± 0.10) • With (93.93 ± 1.52%) rooting response and (9.77 ± 0.08) root number, MS + 100 μM IBA was the best medium for root induction	Arya et al. (2012)
	Nodal	• Callus induction was highest (80%) in MS + 5.0 μM BAP +5.0 μM 2,4-D • Maximum embryogenesis (93.3%) was attained with the highest number of somatic embryos (38.7 per callus lump) and regenerated plantlets (1 1.9 per callus lump) on MS + 5.0 μM BAP + 7.5 μM 2,4-D • Best rooting (100%) was observed on ½ MS + 5.0 μM IBA	Bag et al. (2012)
		• The best bud break (98.66 ± 0.78%) response was observed from the	Singh et al. (2012a)

(continued)

Table 7.1 (continued)

Species	Explant used	Optimal growth response	References
		explants collected in may • Increased in shoot proliferation (by about 3.8-fold) was witnessed in MS + 1.5 μM TDZ • MS supplemented with 25.0 μM IBA and choline chloride (at 36 μM or 504 μM) resulted in the highest rooting response (89%) with about 12 roots per propagule	
	Somatic Embryo-derived In vitro shoots	• MS + 2% sucrose +0.5 mg L^{-1} BAP + 0.25 mg L^{-1} IBA induced in vitro flowering in 14–35 days from 27 to 80 of shoots • The leaf size/area was reduced during the flowering	Kaur et al. (2014)
D. membranaceus	Seeds	• MS + 8.8 μM BAP + 2.3 μM KN recorded the optimum germination rate (70 ± 13.9%) • Seeds soaked at 50 ppm of GA_3 solution (30 °C, dark condition) for overnight induced a high rate of seed germination (73.3 ± 5.7%) with a corresponding increase in the number of the sprouted shoots (2.1 ± 07)	Brar et al. (2013)
D. stocksii	Leaf, leaf sheath, internode, nodal segments	• Callus induction (>80%) was achieved best in culture inoculated with the nodal segment • MS + 0.55 μML^{-1} 2,3-D + 10% CW achieved the maximum frequency (80%) of calli multiplication • MS + 0.25 μML^{-1} NAA + 0.49 μML^{-1} BAP + 49.9 μML^{-1} ascorbic acid +22.8 μML^{-1} citric acid +24.9 μML^{-1} cysteine +100 μML^{-1} glutamine showed the highest (85%) somatic embryo germination	Somashekar et al. (2018)

(continued)

Table 7.1 (continued)

Species	Explant used	Optimal growth response	References
D. strictus	Nodal	• The highest (69.23%) bud break was witnessed when the explants were treated with 0.2% $HgCl_2$ for 15 minutes • Shoot multiplication was highest (3.33-fold) in MS + 4.0 mg L^{-1} BAP + 15 mg L^{-1} adenine sulfate • Root induction was low (20%) and observed only in MS + 5 mg L^{-1} IBA	Pandey and Singh (2012)
		• MS + 4.0 mg L^{-1} produced the highest bud break (49%) and the most effective shoot regeneration (3.68 ± 0.37) • Superior root multiplication were observed in MS + 3.0 mg L^{-1} NAA (1.36 ± 0.04 per propagule) and MS + 1.0 mg L^{-1} IBA + 3.0 mg L^{-1} NAA (1.32 ± 0.03 per propagule)	Goyal et al. (2015)
		• The highest shooting response (92%) and the maximum number of shoots (4.6 ± 0.18) were induced on MS + 4.0 mg L^{-1} BAP • Shoot proliferation (26.0 ± 0.13) was highest in MS + 3.0 mg L^{-1} BAP + 0.5 mg L^{-1} NAA • The best rooting response (93%) and a high number of roots per shoot (10.2 ± 0.17) were witnessed in ½ MS + 3.0 mg L^{-1} NAA	Rajput et al. (2019a)
Drepanostachyum falcatum	Nodal	• The maximum bud break (90.70 ± 0.40%) was achieved on MS + 4.5 mg L^{-1} BAP with the highest mean shoot number (11.08 ± 0.17) • Shoot multiplication (41.49 ± 0.60 shoots) was highest (7–nine-fold increased) on MS + 3.5 mg L^{-1} BAP • Highest root number (11.34 ± 0.22) with high rooting response	Saini et al. (2016)

(continued)

Table 7.1 (continued)

Species	Explant used	Optimal growth response	References
		(99.0 ± 0.52%) was evidenced on MS + 6.5 mg L^{-1} IBA	
D. Luodianense	Nodal	• Best bud initiation (59.26 ± 27.90%) was noticed in MS + 3.0 mg L^{-1} BAP • Highest frequency of callus formation (65.6%) was shown on MS + 4.0 mg L^{-1} 2, 4-D + 0.5 mg L^{-1} NAA + 0.1 mg L^{-1} TDZ • Callus proliferation was highest in MS + 4.0 mg L^{-1} 2,4-D + 0.5 mg L^{-1} NAA + 0.5 mg L^{-1} TDZ • Maximum rate of shoot multiplication (up to 77.8%) was evident on the MS + 5.0 mg L^{-1} BAP + 1.0 mg L^{-1} NAA • 100% rooting was achieved on MS + 2.0 mg L^{-1} NAA + 0.5 mg L^{-1} IBA	Lin et al. (2019)
Guadua angustifolia	Nodal	• The maximum number of shoots (3.7 ± 0.4) and shoot height (2.5 ± 0.1 cm) were observed in MS (liquid) + 3.0 mg L^{-1} BAP • ½ MS + 3.0 mg L^{-1} IBA witnessed the highest rooting percentage (62.0 ± 0.1), while the maximum root number (2.8 ± 1.2) and the survival during acclimatization (100%) were observed in MS (semisolid) + 3.0 mg L^{-1} IBA	Nogueira et al. (2019)
G. magna	Nodal	• The maximum number of shoots (3.3 ± 0.3) and shoot height (4.3 ± 0.3 cm) were shown in MS (liquid) + 3.0 mg L^{-1} BAP • ½ MS (liquid) + 3.0 mg L^{-1} IBA produced the highest rooting percentage (45.0 ± 0.1) and root number (5.3 ± 2.1), while the survival in acclimatization is	Nogueira et al. (2019)

(continued)

Table 7.1 (continued)

Species	Explant used	Optimal growth response	References
		100% in MS (semisolid) + 3.0 g L^{-1} IBA	
Gigantochloa atroviolacea Widjaja	Nodal	• Best bud break (57.53%) was attained in June–august • The highest shoot number (7.17 ± 0.21) was generated in MS + 20 μM BAP + 3.0 μM NAA • Maximum rooting response (47.67%) with an average of 4.33 ± 0.22 roots per propagule and average root length of 2.90 ± 0.22 cm was obtained in MS + 35 μM IBA	Bisht et al. (2010)
Ochlandra wightii	Embryo, in vitro derived nodal segments	• Seed germination (85%) was high on ½ MS + 0.5 mg L^{-1} BAP • Shoot multiplication (9.8 ± 0.5 shoots) was high on ½ MS + 0.5 mg L^{-1} BAP + 0.5 mg L^{-1} TDZ • Simultaneous rooting with the highest root number (3.3 ± 0.3) was recorded on ½ MS + 0.5 mg L^{-1} TDZ • Multiple shoots from nodal segments was maximum (3.6 ± 0.4) on ½ MS + + 2.0 mg L^{-1} BAP + 0.5 mg L^{-1} KN • 70% rooting was achieved from the in vitro derived shoots (2–4 clump) transplanted on ½ MS + 0.5 mg L^{-1} IBA	Bejoy et al. (2012)
Melocanna baccifera	Nodal	• The highest bud break frequency was witnessed in MS + 3.0 mg L^{-1} BAP • Maximum number of shoots (18.17 ± 0.31) was recorded in MS + 2.0 mg L^{-1} KN + 3.0 mg L^{-1} BAP • Highest rooting (81.67 ± 6.54%) was observed in MS + 3.0 mg L^{-1} IBA + 0.05 mg L^{-1} BAP + 10 mg L^{-1} coumarin	Waikhom and Louis (2014)
Oxytenanthera abyssinica	Seeds	• Seeds treated with 4.0% NaOCl for 25 minutes	Kahsay et al. (2017)

(continued)

Table 7.1 (continued)

Species	Explant used	Optimal growth response	References
		recorded the highest germination rate (23.45%) • MS + 4.0 mg dm^{-3} BAP yielded the best response for shoot initiation (86.67%), shoot number (11.33), and multiplication rate (3.77) per propagule • ½ MS + 8.0 mg dm-^{3} IBA showed the maximum rooting percent (93.33) and root number (9.42) per clump	
Phyllostachys bambusoides	Nodal segments derived in vitro leaf sheaths	• MS + 8.0 mg L^{-1} picloram induced callogenesis • Morphogenesis occurred in yellow-globular callus with red-purple pigmentation • The induction of meristematic tissues was superior in MS + 4% glucose • Rhizogenesis was best in MS + 2% glucose (5.5 roots per callus)	Komatsu et al. (2011)
P. heterocycla var. *pubescens* (mazel ex J. Houz) Ohwi	Zygotic embryos	• Among the various medium tested, callus induction (50.34 ± 2.13%) and embryogenic callus (5.97 ± 1.82%) were best on MS + 4.0 mg L^{-1} 2,4-D • Zygotic embryo germination was highest (6.53 ± 0.79) on N6 + 4.0 mg L^{-1} 2,4-D • Callus induction was optimal on MS +4.0 mg L^{-1} 2,4-D + 0.1 mg L^{-1} ZT that induced embryogenic calli from nearly 15% of explants	Yuan et al. (2013)
P. pubescens	Seeds derived in vitro nodal segments	• Basal MS medium produced 95% seed germination • Nodal explants from in vitro derived seedlings produced highest shoot number (8.60 ± 0.24) on MS + 0.2 mg L^{-1} TDZ + 1.0 mg L^{-1} KN • Best rooting response (up to 75%) was evident on MS + 2.0 mg L^{-1} IBA	Sood et al. (2014)

(continued)

Table 7.1 (continued)

Species	Explant used	Optimal growth response	References
Pseudoxytenanthera stocksii (Munro) T.Q. Nguyen	Nodal	• Shoot induction was maximum (9.0 ± 0.25) in MS + 6.0 mg L^{-1} BAP • Rate of shoot multiplication was highest (41.9 ± 1.00) in MS + 4.0 mg L^{-1} BAP + 0.25 mg L^{-1} NAA • Maximum number of roots (24.3 ± 0.27) was produced in ½ MS + 1.0 mg L^{-1} IBA + 50% sucrose	Rajput et al. (2019b)
Abbreviation: *2,4-D* 2,4-dichlorophenoxy acetic acid, *BAP* 6-benzylamino purine, *CW* coconut water, *IAA* indole-3-acetic acid, *IBA* indole-3-butyric acid, *KN* kinetin, *MS* Murashige and Skoog, *NAA* naphthaleneacetic acid, *PBZ* paclobutrazol, *PPM* plant preservative mixture, *TDZ* thidiazuron *Thyrsostachys siamensis*	Nodal segment derived in vitro shoots	• Shoot formation (13.50 ± 1.08) was optimum in MS + 11.10 μM BAP • Callus induction (0.76 ± 0.06 cm in diameter) from the in vitro raised shoots were best in MS + 11.3 μM 2,4-D+ 4.65 μM KN + 1.96 μM IBA • Highest shoots (27.0 ± 4.6) regenerated from the callus was produced in MS + 11.1 μM BAP + 3.43 μM IBA • Best rooting efficiency (80.0 ± 10%) and root multiplication (2.80 ± 0.1 roots per shoot) were obtained in MS + 26.85 μM NAA	Obsuwan et al. (2019)

watered with 1/fourth solution of MS salts for 25 to 30 days. The hardened plantlets were then acclimatized in the net house for another 4 weeks. The plantlets were then transferred to polybags containing garden soil and vermicompost (1:1) and maintained under the greenhouse. Lin et al. (2019) transplanted *Drepanostachyum luodianense* with a survival rate of 100%. In the acclimatization process, healthy rooted shoots in the open culture bottle were first placed in the greenhouse at room temperature for 7 days. Any residuals of the medium stuck to the plantlets were then removed by first washing in tap water and then soaked in 0.3% carbendazim for 3 min before potting in a mixture of humus soil and perlite (1:1). Acclimatization under the greenhouse at slightly higher irradiance, wavering light, and temperature lower than the field environment has been described to reduce excessive evapotranspiration by steady improvement of physiological adaptation.

7.3 Clonal Fidelity Assessment of Micropropagated Bamboos

7.3.1 Somaclonal Variation

Larkin and Scowcroft (1981) coined the term "somaclonal variation" which was defined as the genetic variation detected in plants derived from any in vitro cultured cells or tissues. While tissue culture generates plants that manifest novel variation beneficial to crop improvement, certain somaclones having undesirable traits like low fertility, stunted growth, and less productivity may also appear (Currais et al. 2013). Appropriate precautions have to be taken during tissue culture to avoid unwanted genetic changes in the regenerants as the technique is extensively exploited for commercial purposes. The presence of any genetic variation in the regenerated plants is undesirable if one wishes to conserve the elite genotypes, and the primary regenerants are the desired end products for commercial applications (Mirani et al. 2020). So, maintenance of genetic stability in the micropropagated plants is highly essential to optimize tissue culture protocols and generate true-to-type clones for field planting. The choice of explant for culture initiation is vital for attaining genetic uniformity between the in vitro regenerated plants. Propagation through seed explants is expected to induce genetic variation as the seed-derived zygotes are formed by fertilizing gametes from two genetically different parents (Krishna et al. 2016). It is assumed that clonal propagation via direct organogenesis using shoot tips, axillary, and stem nodes gives genetically uniform plants, while callus-mediated plantlet generation encompasses higher mutation and chromosome variability (Saravanan et al. 2011). The incidence of variation increases as callus formation involves dedifferentiation phase, followed by abnormal cell proliferation due to uncontrolled cell division (Vazquez 2001). Matured and highly differentiated root, leaf, and stem explants produced higher genetic variation than axillary buds and shoot tips with preexisting meristems (Leva et al. 2012). Plant regeneration via somatic embryogenesis may produce higher genetic uniformity than organogenic differentiation, as less DNA methylation occurs in early embryogenesis (Azizi et al. 2020). Undesirable genetic variation in the regenerants can be either genetic or epigenetic in origin, limiting the broader utility of micropropagation techniques (Sato et al. 2011).

Certain factors that may contribute to somaclonal variation are explant treatment in high disinfectant concentration for a long duration, culture exposure to high hormone concentration, long culture period/subculture cycle, occurrence of a callus transition phase, cell/tissue heterogeneity of explant, and other spontaneous mutations (Rawat et al. 2013; Krishna et al. 2016). The longer exposure of explant tissues to sterilant may trigger mutations leading to the emergence of somaclonal variants. Lengthy culture duration also increases the risk of producing somaclonal variation between the clones. This is because the accumulation of genetic variation may be higher with the longer age of the culture generating variants during successive subcultures (Zayova et al. 2010; Rival et al. 2013). The frequency of subculture may also heighten the rate of somaclonal variation. However, the genetic instability was produced by nucleotide sequence alternation rather than quantitative changes of the genome as depicted by unchanged C-value even after the seventh subculture of olive genotypes (Farahani et al. 2011). Gao et al. (2010) reported the influence of growth hormones on adventitious shoot induction and multiplication, which resulted in the change of somaclonal variation rate in the culture. Several plant growth regulators at particular concentrations/in different combinations may induce mutation leading to genetic variation among the regenerants (Sun et al. 2013; Weckx et al. 2019). The production of somaclonal variants in the prolonged culture increased with 2,4-D in the medium (Silva and Carvalho 2014). The presence of 2,4-D elevated the DNA methylation rate resulting in a change of DNA ploidy level and production of variant genotypes. IAA, when present with inositol in the medium, prompted changes in chromosome arrangement and DNA methylation in the callus culture of carrot (Arnhold-Schmitt 1993). The ratio of different growth hormones also affected the in vitro genetic changes as perceived from the study conducted by Eeuwens et al. (2002) in oil palm in which they observed a low and high incidence of the variant "mantled" flowering in high auxin/cytokinin ratio and high cytokinin/auxin ratio, respectively.

The plant tissues are exposed to various stress conditions during in vitro culture, and most of the variant expression in the regenerants may be related to tissue damage inflicted by oxidative stress. The plants developed a mechanism to eliminate lethal reactive oxygen species (ROS) accumulation due to oxidative stress conditions, through a cascade of phytohormone pathways (Mittler et al. 2011; Tognetti et al. 2012). The crosstalk between plant growth regulators and ROS that occurs through co-regulated and shared signaling components redirect the plant to various morphological, physiological, and biochemical changes, reducing their stress and aligning their growth and development under such unfavorable circumstances to survive (O'Brien and Benková 2013; Xia et al. 2015; Verma et al. 2016). The somaclonal variation may be controlled by refraining from harsh chemical treatment of explants; avoiding an inappropriate concentration of salt, sugar, and growth regulators; evading the use of 2,4-D unless necessary; reducing culture period; and using axillary shoots and juvenile nodal segment as explants rather than differentiated plant parts which might boost genetic variability among the regenerants.

7.3.2 Molecular Markers in Clonal Fidelity Assessment

DNA markers have been successfully employed to assess clonal genetic fidelity of many micropropagated bamboos. RAPD and ISSR are fast and straightforward markers widely applied to detect genetic stability of in vitro regenerated bamboos. Both are dominant markers which do not require prior sequence information of the template DNA (Amom and Nongdam 2017; Giachino 2019). ISSR is more effective than RAPD markers as their primers (15–35 mers) are longer than RAPD, resulting in higher stringency due to high annealing temperature (Nilkanta et al. 2017; Amom

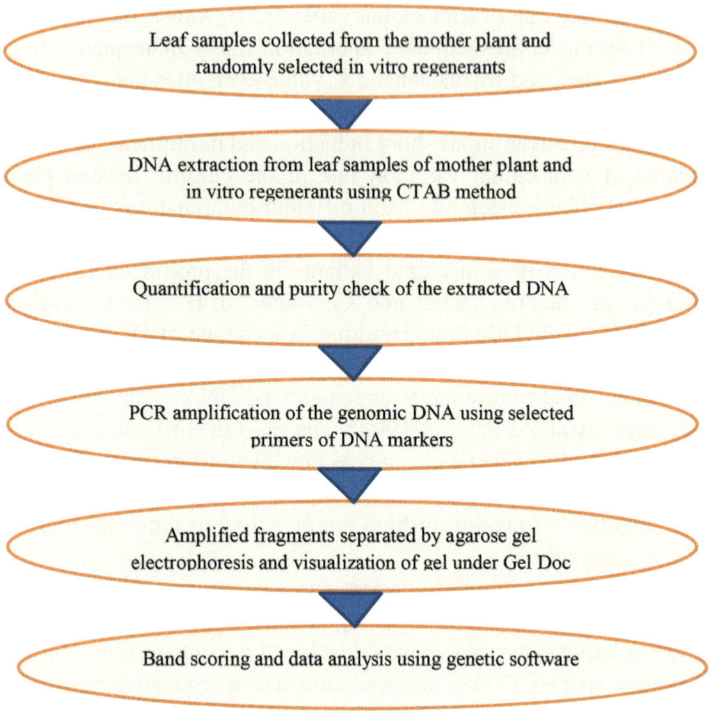

Fig. 7.4 Experimental steps involved in clonal fidelity assessment of the micropropagated bamboos using DNA markers

et al. 2018). However, these markers have limitations in detecting variation as their genetic information is based on DNA noncoding regions not linked to functional traits (Gogoi et al. 2020; Amom et al. 2020). The weaknesses of the arbitrary markers are overcome by using more effective SCoT and retrotransposon markers, which are associated with functional genes (Saboori et al. 2020; Ghonaim et al. 2020). It is pertinent to use a multi-marker system to ascertain the clonal uniformity of the regenerants as a single marker may not provide an accurate assessment (Dey et al. 2019). The combined use of diverse marker types generates reliable and accurate results by validating the outcomes of different marker analysis (Rathore et al. 2014; Tikendra et al. 2019b; Dey et al. 2020b). The various experimental steps which are involved in clonal fidelity assessment are depicted in Fig. 7.4. The genomic DNA is extracted from leaves of randomly selected regenerants and mother plant using a modified CTAB (cetyltrimethylammonium bromide) method. The quality and quantities of extracted DNA are checked using a spectrophotometer at 260 and 280 nm, respectively. The purity and integrity of the DNA are later determined by running 0.8% agarose gel electrophoresis and comparing the intensity of resultant bands with 1 kb DNA ladder. The DNA samples are finally diluted to 50 ng/μl and stored at -20 °C for genetic homogeneity testing. PCR amplification of the template DNA is performed using selected primers of a particular marker. The amplified DNA fragments are separated on a 2.0% (w/v) agarose gel, which is stained with 0.5 μg L^{-1} ethidium bromide. The sizes of unknown DNA fragments are determined by using 1 kb DNA ladder. The banding profile of the molecular marker is obtained by taking a picture of the gel under gel documentation system. The consistent, unambiguous, and reproducible bands generated by the primers were scored, and the data are pooled into a binary matrix based on the presence (1) or absence (0) of bands. The prepared input file is used to generate genetic similarity matrices between the micropropagated plants and construct UPGMA dendrograms and perform PCoA analysis using different genetic software.

Goyal et al. (2015) investigated the clonal stability of in vitro propagated *Dendrocalamus strictus* using RAPD and ISSR markers. The application of two marker systems to assess clonal fidelity was appropriate as the result of RAPD analysis was validated by more efficient ISSR markers. 24 RAPD and 15 ISSR primers were screened, but only 10 RAPD and 9 ISSR primers were selected for genomic DNA amplification as they produced clear and scorable bands. Ten RAPD primers generated 58 scorable bands, while ISSR primers yielded 66 distinct amplified fragments. The bands produced by RAPD primers ranged from 3 to 11, with an average of 5.8 bands per primer. ISSR, on the other hand, gave bands varying from 5 to 11 with an average of 7.33 bands per primer. The size of amplified bands varied from 240 to 1455 bp for RAPD and 183 to 1544 bp for ISSR markers. No polymorphism was detected as monomorphic banding patterns were observed for every primer tested revealing the clonal genetic stability of the micropropagated *D. strictus*. Singh et al. (2012b) used four marker systems (RAPD, ISSR, SSR, and AFLP) to assess the genetic stability of *Dendrocalamus asper* propagated using nodal segments from mature clumps. The application of multiple markers ensured proper screening of the whole genome for the existence of any genetic variation. 25 RAPD primers were tested, but only 22 primers yielded 146 reproducible and scorable bands, which were monomorphic in banding patterns. 25 ISSR primers were tried, but 24 primers gave 170 amplified fragments with sizes ranging from 200 to 3000 bp. The band number produced by the primers varied from 3 to 13, with an average of 7.1 bands per primer. ISSR primers amplified more fragments compared to RAPD, indicating higher efficiency in detecting polymorphism. 21 out of 25 cross-species SSR markers produced 164 scorable bands, which were all monomorphic with band size varying from 100 to 2000 bp. Genetic fidelity was also evaluated using 15 AFLP primer combinations obtained from 5 *Eco*RI and 6 *Mse*I primers. AFLP analysis produced 536 monomorphic and reproducible bands with an average of 48.7 bands per primer combination. The similarity matrix was determined based on Jaccard's similarity coefficient using the scoring data of well-resolved bands of RAPD, ISSR, SSR, and AFLP. There was a 100% similarity between in vitro cultures, regenerated plantlets, and the mother plant as the pair-wise value between them was 1. The study affirmed the retention of genetic identity among the regenerants even after prolonged culture of more than 2 years. It also further acknowledged the possibility of applying DNA markers, which were cost-effective, simple to use, and highly polymorphic for genetic fidelity assessment of micropropagated bamboos.

Genetic uniformity assessment of in vitro clones of *Dendrocalamus hamiltonii* regenerated from single-node cuttings of lateral branches was conducted by Agnihotri et al. (2009) using RAPD markers. Six RAPD primers were employed, which produced 33 scorable bands with fragment size ranging from 0.3 to 2.6 kb. OPA5 and OPA11 generated a maximum of 11 bands, while the least band number was produced by OPC 15. The banding profiles for all the RAPD primers were monomorphic across the in vitro regenerants and the corresponding mother plant. The monomorphic banding profiles revealed the absence of genetic variation and preservation of genetic fidelity between the in vitro cultures at different stages, hardened plantlets, and mother plants. The present finding advocated the use of only RAPD markers for clonal fidelity assessment of in vitro propagated bamboos. However, Singh et al. (2013b) ascertained the genetic uniformity of the micropropagated *D. hamiltonii* using four marker types, viz., RAPD, ISSR, AFLP, and SSR markers. 90 primers which included 25 each of RAPD, ISSR, and SSR and 15 AFLP primers were tested, but only 76 of them produced scorable amplified fragments. 23 RAPD primers produced 162 scorable bands in size range of 200 to 3000 bp. The band number for each RAPD primers ranged from 3 (OPO-6) to 11 (OPJ-04 and OPE-16) with an average of 7.0 fragments per primer. The bands recorded for the RAPD primers were monomorphic, indicating no polymorphism in the regenerants. Of the 25 ISSR primers used, 24 primers generated scorable bands with size varying from 200 to 3000 bp. The band number produced by ISSR primers ranged from 3 (ISSR-23) to 14 (ISSR-16 and ISSR-18), averaging at 7.5 bands per primer. The banding patterns exhibited by ISSR primers were also monomorphic without any genetic variation between the in vitro clones and mother plants. 21 RM microsatellite markers generated 141 amplified fragments ranging from 3 (RM-7 and RM-240) to 14 (RM-44). The size of 141 bands varied from 100 to 2000 bp, with an average of 7.8 bands per primer. The bands produced by SSR markers were monomorphic. Evaluation of genetic fidelity was done with 15 AFLP primer combinations generated from five *Eco*RI and six *Mse*I primers. AFLP markers gave 369 amplified bands with an average of 46 fragments per primer from 8 AFLP primer combinations. The banding profiles for AFLP markers were monomorphic, like the other three markers. The similarity matrix determined through Jaccard's similarity coefficient based on the scoring of the well-resolved bands of 4 marker types disclosed the pairwise value of in vitro shoot cultures, regenerated plants, and mother clump as 1, showing 100% similarity. The study also proved the equal effectiveness of RAPD, ISSR, SSR, and AFLP markers in assessing the clonal fidelity of in vitro clones of *D. hamiltonii*. SSR and AFLP methods, though effective, are limited in use because of higher-cost factors and trained manpower requirements. The simple and less expensive RAPD and ISSR markers can be used as an alternative to more complex and costly SSR and AFLP in testing clonal uniformity.

Negi and Saxena (2011) tested the genetic uniformity of in vitro clones of *B. nutans* in various subculture cycles of shoot multiplication and during the hardening stage before transplantation to the field. 15 ISSR primers that produced reproducible amplified bands were selected from a total of 24 primers screened. 15 ISSR primers generated 93 distinct bands with band numbers ranging from 2 to 13, giving an average band of 6.2 per primer. It was observed that no polymorphic bands formed in the mother plant and the tissue culture raised progenies, while a polymorphic banding pattern was obtained with *Melocanna baccifera* acting as an outlier. The outlier was used for getting surety about the suitability and accountability of ISSR primers employed to detect polymorphism in the species. Genetic variations observed in the outlier due to the polymorphic bands proved that the primers used were competent enough to differentiate the plantlets based on genetic variations. With the micropropagated *B. nutans* exhibiting no morphological differences, the study confirmed the retention of genetic fidelity of bamboos regenerated through axillary branching under prolonged culture duration. Beena and Rathore (2012) examined the clonal fidelity by extracting genomic DNA from the micropropagated plants of *Bambusa pallida* and mother plants. 25 RAPD primers were screened for polymorphism detection, but only 12 oligonucleotides produced clear and visible amplification products, which were monomorphic across all the micropropagated plants. A total of 181 bands were produced ranging from 3 (OPA-02) to 9 (OPB-12), producing an average of 6 bands per primer. The investigation revealed maintenance of genetic stability and absence of morphological variation among the micropropagated plants of *B. pallida*.

Negi and Saxena (2010) ascertained the genetic homogeneity of *Bambusa balcooa* propagated through axillary branching using ISSR markers. The screening of 23 ISSR primers was performed, but only 15 primers gave clear and reproducible bands. 15 primers generated a total of 99 bands with band size extending from 100 to 1600 bp. The bands varied from 3 to 10, with an average bands of 6.6 bands per primer. The primer banding profiles for micropropagated *B. balcooa* analyzed showed monomorphic patterns indicating the absence of somaclonal variation. Only the outlier *B. nutans* exhibited polymorphism demonstrating the competence of ISSR primers in detecting variation among the clones. Molecular analysis revealed the maintenance of genetic identity among the regenerants under prolonged culture duration even after 33 passages. This finding had commercial implications as *B. balcooa* initiation was tough owing to season specificity, persistent contamination, and phenolic exudation (Das and Pal 2005). It is not commercially viable to initiate culture after every 12–15 subculture cycles. If genetic uniformity can be retained for a long culture duration, the plant production cost can be lowered as in vitro shoot multiplication can be continuously maintained. Brar et al. (2014) employed two marker systems (RAPD and ISSR markers) to determine the clonal fidelity of *B. balcooa* plantlets regenerated through axillary bud proliferation. Of the 25 RAPD primers screened, 21 of them produced 61 distinct scorable bands with band sizes ranging from 100 to 1500 kb. In ISSR markers, 15 primers were screened, but only 10 ISSR primers generated 28 scorable bands with size varying from 100 and 1500 kb. The OPO series of RAPD markers provided better amplification products compared to OPT and OPA series. UBC 810, 811, and 888 primers gave maximum amplicons with band size in the range of 250–1500 bp. The banding profiles generated by the two different markers were monomorphic, indicating the nonexistence of genetic variation between the micropropagated *B. balcooa* and the mother plant. The maintenance of genetic stability of the in vitro regenerants suggested axillary bud proliferation as a method of choice for micropropagation of clonally uniformed *B. balcooa*.

Anand et al. (2013) also assessed the genetic homogeneity of plantlets of edible *B. bamboos* regenerated through in vitro axillary branching approach using RAPD and ISSR markers. Genetic testing was performed between 15 randomly selected hardened bamboos and the mother plant. 10 RAPD primers were chosen after scanning 15 primers based on the production of distinct and clear amplified fragments. Only five ISSR primers were selected out of ten markers screened as they could produce scorable and reproducible bands. The combined 20 RAPD and ISSR primers generated 37 bands with band sizes in the range of 100 to 1500 bp. OP series of RAPD and UBC 818 of ISSR markers gave the best amplification producing maximum amplicons of size extending from 200 to 1400 bp. The 37 bands obtained from both the marker systems were all monomorphic suggesting a complete absence of any somaclonal variation between micropropagated *B. bamboos* and mother plant. The present molecular marker analysis affirmed the establishment of clonal fidelity among the regenerants propagated through axillary branching. Desai et al. (2019) examined genetic homogeneity of the plantlets of *Bambusa vulgaris* "wamin" and *B. bamboos* propagated through axillary shoot proliferation using RAPD markers. The ten RAPD primers each were selected for both *Bambusa* species based on the production of reproducible and scorable bands. The 10 RAPD primers generated 97 and 113 amplified fragments in *B. vulgaris* and *B. bamboos*, respectively. The number of bands ranged from 7 to 13 for *B. vulgaris*, while it was extended from 8 to 16 for *B. bamboos*. The exhibition of monomorphic banding patterns established the genetic purity of the two micropropagated *Bambusa* species by ten RAPD primers.

Nadha et al. (2011) assessed the clonal fidelity of in vitro regenerated *Guadua angustifolia* using RAPD and ISSR markers. Out of 30 RAPD primers screened, only 15 primers gave 84 clear and scorable bands having sizes in the range of 200–2500 bp. The number of amplified fragments produced by the RAPD primers extended from 2 (OPO-14) to 11 (OPT-17) with an average of 5.6 bands per primer. No polymorphism was noticed as monomorphic banding patterns were obtained for RAPD primers across in vitro regenerants and mother plants. ISSR primers were tried for amplification, but only 17 primers generated 61 distinct and reproducible bands with band sizes varying from 300 to 2500 bp, producing an average of 5.01 bands per primer. The highest number of bands (8) was given by UBC 808, while the least of only one band was generated by UBC 815, UBC 844, and UBC 850. The banding profiles obtained were monomorphic for ISSR primers tested, indicating the absence of polymorphism between the in vitro clones and mother plants. Nogueira et al. (2019) used only ISSR markers to evaluate clonal uniformity of *Guadua magna* and *G. angustifolia*, which were propagated in liquid MS (LM) and semisolid (SSM) media using nodal segments containing one lateral bud. 20 ISSR primers amplified 223 and 230 bands for *G. magna* and *G. angustifolia*, respectively. The monomorphic banding profiles exhibited by the ISSR primers for both the species suggested the existence of no genetic polymorphism between the in vitro regenerants propagated in LM and SSM culture systems. Genetic uniformity was maintained even after five subcultures of 30 days each, suggesting the possibility of preserving clonal fidelity even after prolonged culture.

7.4 Conclusions

Bamboos are extremely beneficial plants whose applications are found in diverse spheres of life in household item making, construction, handicraft, textile, paper, and food industry. The conventional propagation methods using seeds, culm cutting, and rhizome division have limitations in producing sufficient bamboos for human needs. Mass and rapid production of different bamboo species has been attempted successfully by micropropagating them using mostly seed, nodal, and culm segments as reliable explants. The clonal fidelity of micropropagated bamboos is examined using different DNA markers to ensure production of genetically identical superior plants. ISSR, SSR, and SCoT are more effective for clonal assessment of bamboo regenerants than the less reliable and inconsistent RAPD markers. The two marker system should be used instead of a single-marker type as the result of variability analysis of a marker can be corroborated by the other. The large-scale propagation of genetically stable quality plants will not only help in controlling the falling populations but also fulfill the global bamboo demand.

Acknowledgments LT and PN are thankful to SERB (Science and Engineering Research Board) New Delhi, India, for their financial support.

Conflict of Interest There is no conflict of interest among the authors.

References

Abdulminam AH, Nirmala C, Sharma ML (2009) Control of in vitro contamination in bamboos. Plant Cell Biotechnol Mol Biol 10:119–124

Agnihotri RK, Mishra J, Nandi SK (2009) Improved in vitro shoot multiplication and rooting of *Dendrocalamus hamiltonii* Nees et Arn. Ex Munro: production of genetically uniform plants and field evaluation. Acta Physiol Plant 31:961–967

Ahmadi E, Nasr SMH, Jalilvand H, Savadkoohi SK (2012) Contamination control of microbe *Ziziphus spina* [christi] seed in vitro culture. Trees 26:1299–1304

Alexander MP, Rao TC (1968) In vitro culture of bamboo embryos. Curr Sci 37:415

Ali H, Nirmala C, Sharma ML (2009) Control of in vitro contamination in bamboos. Plant Cell Biotechnol Mol Biol 10:119–124

Amom T, Nongdam P (2017) The use of molecular marker methods in plants: a review. Int J Cur Res Rev 9:1–7

Amom T, Tikendra L, Apana N, Goutam M, Sonia P, Koijam AS, Potshangbam AM, Rahaman H, Nongdam P (2020) Efficiency of RAPD, ISSR, iPBS, SCoT and phytochemical markers in the genetic relationship study of five native and economical important bamboos of north-East India. Phytochemistry 174:112330

Amom T, Tikendra L, Rahaman H, Potshangbam A, Nongdam P (2018) Evaluation of genetic relationship between 15 bamboo species of north-East India based on ISSR marker analysis. Mol Biol Res Commun 7:7–15

Anand M, Brar J, Sood A (2013) In vitro propagation of an edible bamboo *Bambusa bambos* and assessment of clonal fidelity through molecular markers. J Med Bioeng 2:257–261

Ankur V, Meena B, Harsh NSK (2014) Identification and bioassay of fungal contaminants observed during in vitro propagation of *Saraca asoca* (Roxb.) De Wilde. Biotechnol Int 7:35–42

Arnhold-Schmitt B (1993) Rapid changes in amplification and methylation pattern of genomic DNA in cultured carrot root explants (*Daucus carota* L.). Theor Appl Genet 85:793–800

Arshad SM, Kumar A, Bhatnagar SK (2005) Micropropagation of *Bambusa wamin* through proliferation of mature nodal explants. J Biol Res 3:59–66

Arya ID, Kaur B, Arya S (2012) Rapid and mass propagation of economically important bamboo *Dendrocalamus hamiltonii*. Indian J Energy 1:11–16

Arya S, Rana PK, Sharma R, Arya ID (2006) Tissue culture technology for rapid multiplication of *Dendrocalamus giganteus* Munro. Indian For 3:345–357

Arya S, Satsangi R, Arya ID (2008) Direct regeneration of shoots from immature inflorescences in *Dendrocalamus asper* (edible bamboo) leading to mass propagation. Bamboo Sci Cult 21:14–20

Azizi P, Hanafi MM, Sahebi M, Harikrishna JA, Taheri S, Yassoralipour A, Nasehi A (2020) Epigenetic changes and their relationship to somaclonal variation: a need to monitor the micropropagation of plantation crops. Funct Plant Biol 47:508–523

Babaolu M, Yorgancılar M, Akbudak MA (2001) Dokukültürü: Temellaboratuarteknikleri (Plant tissue culture: Basic laboratory techniques).- In: Babaolu M, Gürel E, Özcan S (eds), Bitki Biyoteknolojisi: Doku Kültürüve Uygulamaları (Biotechnology of Plant: Plant Tissue Culture and Application). Selçuk Üniversitesi Vakfı Yayınları Konya pp 1–35

Bag N, Palni LMS, Chandra S, Nandi SK (2012) Somatic embryogenesis in 'maggar' bamboo (*Dendrocalamus hamiltonii*) and field performance of regenerated plants. Curr Sci 102:1279–1287

Banerjee M, Gantait S, Pramanik BR (2011) A two step method for accelerated mass propagation of *Dendrocalamus asper* and their evaluation in field. Physiol Mol Biol Plants 17:387–393

Banik RL (2015) Bamboo silviculture. In: Liese W, Köhl M (eds) Bamboo. Tropical forestry, vol 10. Springer, Cham, pp 113–174

Beena DB, Rathore TS (2012) In vitro cloning of *Bambusa pallida* Munro through axillary shoot proliferation and evaluation of genetic fidelity by random amplified polymorphic DNA markers. Int J Plant Biol 3:6

Bejoy M, Anish NP, Radhika BJ, Nair GM (2012) In vitro propagation of *Ochlandra wightii* (Munro) Fisch.: an endemic reed of southern Western Ghats India. Biotechnol 11:67–73

Bisht MS, Nirmala C, Nongdam P (2012) Bamboo shoot as a source of health food and socio economic development in North –East India. Proceedings of 9th World Bamboo Congress, Antwerp, Belgium, 393–402

Bisht P, Pant M, Kant A (2010) In vitro propagation of *Gigantochloa atroviolacea* Widjaja through nodal explants. J Am Sci 6:1019–1025

Bornman CH, Vogelmann TC (1984) Effect of rigidity of gel medium on benzyladenine-induced adventitious bud formation and vitrification in vitro in *Picea abies*. Physiol Plant 61:505–512

Brand MH (1993) Agar and ammonium nitrate influence hyperhydricity, tissue nitrate and total nitrogen content of serviceberry (*Amelanchier arborea*) shoots in vitro. Plant Cell Tissue Organ Cult 35:203–209

Brar J, Anand M, Sood N (2013) In vitro seed germination of economically important edible bamboo *Dendrocalamus membranaceus* Munro. Indian J Exp Biol 51:88–96

Brar J, Shafi A, Sood P, Anand M, Sood A (2014) In-vitro propagation, biochemical studies and assessment of clonal fidelity through molecular markers in *Bambusa balcooa*. J Trop For Sci 26:115–124

Butenko RG, Popov AS, Volkova LA, Chernyak ND, Nosov AM (1984) Recovery of cell cultures and their biosynthetic capacity after storage of *Dioscorea deltoidea* and *Panax ginseng* in liquid nitrogen. Plant Sci Lett 33:285–292

Chambers SM, Heuch JHR, Pirrle A (1991) Micropropagation and in vitro flowering of the bamboo *Dendrocalamus hamiltonii* Munro. Plant Cell Tissue Organ Cult 27:45–48

Chaturvedi HC (1984) Some aspects of morphogenesis in rapid multiplication and germplasm preservation of economic plants through tissue culture. In: Dutta PC (ed) Proceedings of National Symposium on applied biotechnology of medicinal, aromatic and timber yielding plants. Calcutta University, Calcutta, pp 179–188

Choudhary AK, Priyanka K, Ashish R (2017) Refinement of protocol for rapid clonal regeneration of economical bamboo, *Bambusa balcooa* in the agroclimatic conditions of Bihar, India. Afr J Biotechnol 16:450–462

Choudhury D, Sahu JK, Sharma GD (2012) Value addition to bamboo shoots: a review. J Food Sci Technol 49:407–414

Clark LG, Londoño X, Ruiz-Sanchez E (2015) Bamboo taxonomy and habitat. In: Liese W, Köhl M (eds) Bamboo. Tropical forestry, vol 10. Springer, Cham, pp 1–30

Currais L, Loureiro J, Santos C, Canhoto JM (2013) Ploidy stability in embryogenic cultures and regenerated plantlets of tamarillo. Plant Cell Tissue Organ Cult 114:149–159

Das M, Pal A (2005) In vitro regeneration of *Bambusa balcooa* Roxb., factors affecting changes of morphogenetic competence in the axillary buds. Plant Cell Tissue Organ Cult 81:109–112

Desai P, Desai S, Patel A, Mankad M, Gajera B, Patil G, Narayanan S (2019) Development of efficient micropropagation protocol through axillary shoot proliferation for *Bambusa vulgaris* 'wamin' and *Bambusa bambos* and assessment of clonal fidelity of the micropropagated plants through RAPD markers. ANRES 53:26–32

Devi WS, Sharma GJ (2009) In vitro propagation of *Arundinaria callosa* Munro- an edible bamboo from nodal explants of mature plants. TOPSJ 3:35–39

Dey A, Hazra AK, Nongdam P, Nandy S, Tikendra L, Mukherjee A, Banerjee S, Mukherjee S, Pandey DK (2019) Enhanced bacoside content in polyamine treated in-vitro raised *Bacopa monnieri* (L.) Wettst. S Afr J Bot 123:259–269

Dey A, Nandy S, Nongdam P, Tikendra L, Mukherjee A, Mukherjee S, Pandey DK (2020a) Methyl jasmonate and salicylic acid elicit indole alkaloid production and modulate antioxidant defence and biocidal properties in *Rauvolfia serpentina* Benth. Ex Kurz. In vitro cultures. S Afr J Bot 135:1–17

Dey A, Nongdam P, Nandy S, Mukherjee S, Mukherjee A, Tikendra L, Hazra AK, Pandey DK (2020b) Polyamine elicited aristolochic acid production in in vitro clonally fidel *Aristolochia indica* L.: an ISSR and RAPD markers and HPTLC based study. S Afr J Bot. https://doi.org/10.1016/j.sajb.2020.06.018

Eeuwens CJ, Lord S, Donough CR, Rao V, Vallejo G, Nelson S (2002) Effects of tissue culture conditions during embryoid multiplication on the incidence of "mantled" flowering in clonally propagated oil palm. Plant Cell Tissue Organ Cult 70:311–323

Evans DA, Elgi JE, Kut SA, Sharp WR, Flick CK (1981) In vitro regeneration of the ornamental tobacco *Nicotiana alata*. Hortic Sci 16:425

Farahani F, Yari R, Masoud S (2011) Somaclonal variation in Dezful cultivar of olive (*Olea europaea* subsp. europaea). Gene Conserve 10:216–221

Fisse J, Batalle A, Pera J (1987) Endogenous bacteria elimination in ornamental plants. Acta Hortic 212:87–90

Franclet A (1991) Biotechnology in 'rejuvenation': hope for the micropropagation of difficult woody plants. Acta Hortic 289:273–282

Funada R, Kubo T, Tabuchi M, Sugiyama T, Fushitani M (2001) Seasonal variations in endogenous indole-3-acetic acid and abscisic acid in the cambial region of *Pinus densiflora* Sieb. Et Zucc.

Stems in relation to earlywood/latewood transition and cessation of tracheid production. Holzforschung 55:128–134
Gantait S, Pramanik BR, Banerjee M (2018) Optimization of planting materials for large scale plantation of *Bambusa balcooa* Roxb.: influence of propagation methods. J Saudi Soc Agric Sci 17:79–87
Gao X, Yang D, Cao D, Ao M, Sui X, Wang Q, Kimatu JN, Wang L (2010) In vitro micropropagation of *Freesia hybrid* and the assessment of genetic and epigenetic stability in regenerated plantlets. J Plant Growth Regul 29:257–267
Gaspar T, Kevers C, Bisbis B, Franck T, Crèvecoeur M, Greppin H, Dommes J (2000) Loss of plant organogenic totipotency in the course of in vitro neoplastic progression. In Vitro Cell Dev Biol Plant 36:171–181
George EF (1993) Plant propagation by tissue culture, part 1, technology. Exegetics Ltd., England, pp 121–145
George EF, Debergh PC (2008) Micropropagation: uses and methods. In: George EF, Hall MA, De Klerk GJ (eds) Plant propagation by tissue culture. Springer, New York, pp 29–64
Ghonaim MM, Mohamed HI, Omran AAA (2020) Evaluation of wheat (*Triticum aestivum* L.) salt stress tolerance using physiological parameters and retrotransposon-based markers. Genet Resour Crop Evol. https://doi.org/10.1007/s10722-020-00981-w
Giachino RRA (2019) Investigation of the genetic variation of anise (*Pimpinella anisum* L.) using RAPD and ISSR markers. Genet Resour Crop Evol 67:763–780
Gogoi B, Wann SB, Saikia SP (2020) Comparative assessment of ISSR, RAPD, and SCoT markers for genetic diversity in *Clerodendrum* species of north East India. Mol Biol Rep 47:7365–7377
Goyal AK, Brahma BK (2014) Antioxidant and nutraceutical potential of bamboo: an overview. Int J Fund Appl Sci 3:2–10
Goyal AK, Pradhan S, Basistha BC, Sen A (2015) Micropropagation and assessment of genetic fidelity of *Dendrocalamus strictus* (Roxb.) nees using RAPD and ISSR markers. 3 Biotech 5:473–482
Goyal AK, Sen A (2016) In vitro regeneration of bamboos, the "green gold": an overview. Indian J Biotechnol 15:9–16
Gu D, He W, Huang K, Otieno D, Zhou C, He C, Huang Y (2019) Transpiration of Moso bamboo in southern China is influenced by ramet age, phenology, and drought. For Ecol Manage 450:117526
Hassan AAEL, Debergh P (1987) Embryogenesis and plantlet development in the bamboo *Phyllostachys viridis* (Young) McClure. Plant Cell Tissue Organ Cult 10:73–77
Herman EB (1996) Recent advances in plant tissue culture, IV: Microbial contamination of plant tissue cultures. N.Y. Agritech Consultants, Mohegan Lake
Hu CY, Wang PJ (1983) Meristem shoot tip and bud culture. In: Evans DA, Shard WR, Ammirato PV, Yamada Y (eds) Handbook of plant cell culture. MacMillan Publishing Company, New York, pp 177–227
Huang LC, Huang BL, Chen WL (1989) Tissue culture investigations of bamboo—IV. Organogenesis leading to adventitious shoots and plants in excised shoot apices. Environ Exp Bot 29:307–315
Huang LC, Lee YL, Huang BL, Kuo CI, Shaw JF (2002) High polyphenol oxidase activity and low titratable acidity in browning bamboo tissue culture. In Vitro Cell Dev Biol Plant 38:358–365
Jha A, Das S, Kumar B (2013) Micropropagation of *Dendrocalamus hamiltonii* through nodal explants. Glob J Bio-Sci Biotechnol 2:580–582
Jiménez VM, Castillo J, Tavares E, Guevara E, Montiel M (2006) In vitro propagation of the neotropical giant bamboo, *Guadua angustifolia* Kunth, through axillary shoot proliferation. Plant Cell Tissue Organ Cult 86:389–395
Jiménez VM, Guevara E (2007) Micropropagation of bamboo species through axillary shoot proliferation. In: Jain SM, Häggman H (eds) Protocols for micropropagation of woody trees and fruits. Springer, Dordrecht, pp 465–476

Jullien F, Van KTT (1994) Micropropagation and embryoid formation from young leaves of *Bambusa glaucescens* golden goddess. Plant Sci 98:199–207
Kahsay B, Mekibib F, Teklewold A (2017) In vitro propagation of *Oxytenanthera abyssinica* (a. rich. Munro) from seed culture. BJI 18:1–13
Kane M (2003) Bacterial and fungal indexing of tissue cultures. http://plant-c.cfansumn.edu/listerv/1996/log9612/indeing.htm
Kant A, Arya S, Arya ID (2009) Micropropagation protocol for *Melocanna baccifera* using nodal explants from mature clump. In: 8th World Bamboo Congress, Thailand, pp 2-12
Kapoor P, Rao IU (2006) In vitro rhizome induction and plantlet formation from multiple shoots in *Bambusa bambos* var. gigantea bennet and gaur by using growth regulators and sucrose. Plant Cell Tissue Organ Cult 85:211–217
Kaur D, Thapa P, Sharma M, Bhattacharya A, Sood A (2014) In vitro flowering- a system for tracking floral organ development in *Dendrocalamus hamiltonii* Nees et Arn. Ex Munro. Indian J Exp Biol 52:825–834
Kevers C, Franck T, Strasser RJ, Dommes J, Gaspar T (2004) Hyperhydricity of micropropagated shoots: a typically stress-induced change of physiological state. Plant Cell Tissue Organ Cult 77:181–191
Komatsu YH, Batagin-Piotto KD, Brondani GE, Gonçalves AN, de Almeida M (2011) In vitro morphogenic response of leaf sheath of *Phyllostachys bambusoides*. J For Res 22:209–215
Krishna H, Alizadeh M, Singh D, Singh U, Chauhan N, Eftekhari M, Sadh RK (2016) Somaclonal variations and their applications in horticultural crops improvement. 3. Biotech 6:54
Larkin P, Scowcroft W (1981) Somaclonal variation—a novel source of variability from cell cultures for plant improvement. Theor Appl Genet 60:197–214
Leifert C, Camotta H, Waites WM (1992) Effect of combinations of antibiotics on micropropagated *Clematis, Delphinium, Hosta, Iris* and *Photinia*. Plant Cell Tissue Organ Cult 29:153–160
Leva AR, Petruccelli R, Rinaldi LMR (2012) Somaclonal variation in tissue culture: a case study with olive. In: Leva AR, Rinaldi LMR (eds) Recent advances in plant in vitro culture. INTECH Open Access Publisher, Croatia, pp 123–150
Lin CS, Chang WC (1998) Micropropagation of *Bambusa edulis* through nodal explants of field-grown culms and flowering of regenerated plantlets. Plant Cell Rep 17:617–620
Lin CS, Kalpana K, Chang WC, Lin NS (2007) Improving multiple shoot proliferation in bamboo mosaic virus-free *Bambusa oldhamii* Munro propagation by liquid culture. Hortscience 42:1243–1246
Lin CS, Lin CC, Chang WC (2004) Effect of thidiazuron on vegetative tissue-derived somatic embryogenesis and flowering of bamboo *Bambusa edulis*. Plant Cell Tissue Organ Cult 76:75–82
Lin S, Liu G, Guo T, Zhang L, Wang S, Ding Y (2019) Shoot proliferation and callus regeneration from nodular buds of *Drepanostachyum luodianense*. J For Res 30:1997–2005
Luna C, Collavino M, Mroginski L, Sansberro P (2008) Identification and control of bacterial contaminants from *Ilex dumosa* nodal segments culture in a temporal immersion bioreactor system using 16S rDNA analysis. Plant Cell Tissue Organ Cult 95:13–19
Mayer AM, Harel E (1979) Polyphenol oxidases in plants. Phytochemistry 18:193–215
McClure FA (1966) The bamboos: a fresh perspective. Harvard University Press, Cambridge, MA
Mehta R, Sharma V, Sood A, Sharma M, Sharma RK (2011) Induction of somatic embryogenesis and analysis of genetic fidelity of in vitro-derived plantlets of *Bambusa nutans* Wall., using AFLP markers. Eur J For Res 130:729–736
Mirani Z, Urooj S, Khan M, Khan A, Shaikh I, Siddiqui A (2020) An effective weapon against biofilm consortia and small colony variants of MRSA. Iran J Basic Med Sci 23:1494–1498
Mishra Y, Patel P, Ansari SA (2011) Acclimatization and macroproliferation of micropropagated plants of *Bambusa tulda* Roxb. Asian J Exp Biol Sci 2:498–501
Mittler R, Vanderauwera S, Suzuki N, Miller G, Tognetti VB, Vandepoele K, Gollery M, Shulaev V, Van Breusegem F (2011) ROS signaling: the new wave? Trends Plant Sci 16:300–309

Mudoi KD, Borthakur M (2012) Factors affecting the frequency of in vitro flowering of *Bambusa balcooa* Roxb. Indian J Plant Physiol 17:37–43

Mudoi KD, Saikia SP, Borthakur M (2014) Effect of nodal positions, seasonal variations, shoot clump and growth regulators on micropropagation of commercially important bamboo, *Bambusa nutans* wall. Ex. Munro. Afr J Biotechnol 13:1961–1972

Mudoi KD, Saikia SP, Goswami A, Gogoi A, Bora D, Borthakur M (2013) Micropropagation of important bamboos: a review. Afr J Biotechnol 12:2770–2785

Murashige T, Skoog F (1962) A revised medium for rapid growth and bio assays with tobacco tissue cultures. Physiol Plant 15:473–497

Murata M, Tsurutani M, Hagiwara S, Homma S (1997) Subcellular location of polyphenol oxidase in apples. Biosci Biotechnol Biochem 61:1495–1499

Nadha HK, Kumar R, Sharma RK, Anand M, Sood A (2011) Evaluation of clonal fidelity of in vitro raised plants of *Guadua angustifolia* Kunth using DNA-based markers. J Med Plants Res 5:563–5641

Nadha HK, Kumar R, Sharma RK, Anand M, Sood A (2013) In vitro propagation of *Dendrocalamus asper* and testing the clonal fidelity using RAPD and ISSR markers. Int J Curr Res 5:2060–2067

Negi D, Saxena S (2010) Ascertaining clonal fidelity of tissue culture raised plants of *Bambusa balcooa* Roxb. Using inter simple sequence repeat markers. New For 40:1–8

Negi D, Saxena S (2011) In vitro propagation of *Bambusa nutans* Wall. Ex Munro through axillary shoot proliferation. Plant Biotechnol Rep 5:35–43

Niedz R (1998) Using isothiazolone biocides to control microbial and fungal contaminants in plant tissue cultures. Hortic Technol 8:598–601

Nilkanta H, Amom T, Tikendra L, Rahaman H, Nongdam P (2017) ISSR marker based population genetic study of *Melocanna baccifera* (Roxb.) Kurz: a commercially important bamboo of Manipur, North-East India. Scientifica 2017. Article ID 3757238

Nirmala C, Ali AH, Badal T (2011) De novo organogenesis in the form of rhizome in *Dendrocalamus asper* and *D. membranaceus*. Curr Sci 100:468–470

Nogueira JS, Gomes HT, Scherwinski-Pereira JE (2019) Micropropagation, plantlets production estimation and ISSR marker-based genetic fidelity analysis of *Guadua magna* and *G. angustifólia*. Pesqui Agropecu Trop 49:e53743

Nongdam P (2015) Traditional fermented bamboo shoot foods of North-East India and their characteristic natural microbial flora. 10th World Bamboo Congress, Korea

Nongdam P, Tikendra L (2014) The nutritional facts of bamboo shoots and their usage as important traditional foods of Northeast India. Int Sch Res Notices 2014:1–17

O'Brien JA, Benková E (2013) Cytokinin cross-talking during biotic and abiotic stress responses. Front Plant Sci 4:451

Obsuwan K, Duangmanee A, Thepsithar C (2019) In vitro propagation of a useful tropical bamboo, *Thyrsostachys siamensis* gamble, through shoot-derived callus. Hortic Environ Biotechnol 60:261–267

Ogita S (2005) Callus and cell suspension culture of bamboo plant, *Phyllostachys nigra*. Plant Biotechnol 22:119–125

Ojha A, Verma N, Kumar A (2009) In vitro micropropagation of economically important edible bamboo (*Dendrocalamus asper*) through somatic embryos from root, leaves and nodal segments explants. Res Crops 10:430–436

Omamor IB, Asemota AO, Eke CR, Eziashi EI (2007) Fungal contaminants of the oil palm tissue culture in Nigerian Institute for Oil Palm Research (NIFOR). Afr J Biotechnol 2:534–537

Oprins J, Grunewald W, Gillis K, Delaere P, Peeters H, Gielis J (2004) Micropropagation: a general method for commercial bamboo production. In: 7th world bamboo congress, New Delhi

Pandey BN, Singh NB (2012) Micropropagation of *Dendrocalamus strictus* nees from mature nodal explants. JANS 4:5–9

Pasqualini APA, dos Santos MC, SantAnna-Santos BF, Fraga HPF, Quoirin M (2019) In vitro culture and diversity of endophytic fungi in *Bambusa oldhamii*. Pesqui Agropecu Trop 49:1–9

Patel B, Gami B, Patel N, Bariya V (2015) One step pre-hardening micro propagation of *Bambusa balcooa* Roxb. J Phytology 7:1–9

Rajput BS, Jani MD, Gujjar MR, Shekhawat MS (2019a) Effective and large scale in vitro propagation of *Dendrocalamus strictus* (Roxb.) Nees using nodal segments as explants. World Sci News 130:238–249

Rajput BS, Jani MD, Sasikumar K, Manokari M, Shekhawat MS (2019b) An improved micropropagation protocol for manga bamboo-*Pseudoxytenanthera stocksii* (Munro) T.Q. Nguyen. WNOFNS 25:141–154

Raju RI, Roy SK (2017) Mass propagation of *Bambusa bambos* (L.) Voss through in vitro culture. Jahangirnagar Univ J Biol Sci 5:15–26

Ramanayake SMSD, Yakandawala K (1997) Micropropagation of the giant bamboo (*Dendrocalamus giganteus* Munro) from nodal explants of field grown culms. Plant Sci 129:213–223

Rathore NS, Rai MK, Phulwaria M, Rathore N, Shekhawat NS (2014) Genetic stability in micropropagated *Cleome gynandra* revealed by SCoT analysis. Acta Physiol Plant 36:555–559

Rawat JM, Rawat B, Mehrotra S, Chandra A, Nautiyal S (2013) ISSR and RAPD based evaluation of genetic fidelity and active ingredient analysis of regenerated plants of *Picrorhiza kurroa*. Acta Physiol Plant 35:1797–1805

Ray SS, Ali MN, Banerjee M, Yeasmin L (2018) In vitro and in vivo assessment of thidiazuron mediated micro-clones of *Dendrocalamus asper*, an ornamental bamboo species. J Crop Weed 14:150–157

Ray SS, Ali Md N, Mukherjee S, Chatterjee G, Banerjee M (2017) Elimination and molecular identification of endophytic bacterial contaminants during in vitro propagation of *Bambusa balcooa*. World J Microbiol Biotechnol 33:31

Reed BM, Mentzer J, Tanprasert P, Yu X (1998) Internal bacterial contamination of micropropagated hazelnut: identification and antibiotic treatment. Plant Cell Tissue Organ Cult 52:67–70

Rival A, Ilbert P, Labeyrie A, Torres E, Doulbeau S, Personne A, Dussert S, Beule T, Durand-Gasselin T, Tregear JW, Jaligot E (2013) Variations in genomic DNA methylation during the long-term in vitro proliferation of oil palm embryogenic suspension cultures. Plant Cell Rep 32:359–368

Saboori S, Noormohammadi Z, Sheidai M, Marashi S (2020) SCoT molecular markers and genetic fingerprinting of date palm (*Phoenix dactylifera* L.) cultivars. Genet Resour Crop Evol 67:73–82

Saini H, Arya ID, Arya S, Sharma R (2016) In vitro micropropagation of Himalayan weeping bamboo, *Drepanostachyum falcatum*. AJPS 7:1317–1324

Sandhu M, Wani SH, Jiménez VM (2018) In vitro propagation of bamboo species through axillary shoot proliferation: a review. Plant Cell Tissue Organ Cult 132:27–53

Saravanan S, Sarvesan R, Vinod MS (2011) Identification of DNA elements involved in somaclonal variants of *Rauvolfia serpentina* (L.) arising from indirect organogenesis as evaluated by ISSR analysis. Indian J Sci Technol 4:1241–1245

Sato M, Hosokawa M, Doi M (2011) Somaclonal variation is induced de novo via the tissue culture process: a study quantifying mutated cells in Saintpaulia. PLoS One 6:e23541

Sawarkar AD, Shrimankar DD, Kumar A, Kumar A, Singh E, Singh L, Kumar S, Kumar R (2020) Commercial clustering of sustainable bamboo species in India. Ind Crop Prod 154:112693

Saxena S (1990) In vitro propagation of bamboo (*Bambusa tulda* Roxb.) through shoot propagation. Plant Cell Rep 9:431–434

Saxena S, Bhojwani SS (1993) In vitro clonal multiplication of four year old plants of the *Dendrocalamus longispathus* Kurz. In Vitro Cell Dev Biol Plant 290:135–142

Saxena S, Dhawan V (1999) Regeneration and large-scale propagation of bamboo (*Dendrocalamus strictus* Nees) through somatic embryogenesis. Plant Cell Rep 18:438–443

Scheidt GN, Silva ALL, Oliveira Y, Costa JL, Biasi LA, Soccol CR (2011) In vitro growth of *Melaleuca alternifolia* Cheel in bioreactor of immersion by bubbles. Pak J Bot 43:2937–2939

Sharma P, Sarma KP (2011) In vitro propagation of *Bambusa balcooa* for a better environment. In: International conferences on advances in biotechnology and pharmaceutical sciences (ICABPS'11). Bangkok, pp 248–252

Sharma P, Sarma KP (2013) In vitro propagation of *Bambusa tulda*: an important plant for better environment. J Environ Res Dev 7:1216–1223

Shields R, Robinson SJ, Anslow PA (1984) Use of fungicides in plant tissue culture. Plant Cell Rep 3:33–36

Silva TCR, Carvalho CR (2014) Vertical heterogeneity of DNA ploidy level assessed by flow cytometry in calli of *Passiflora cincinnata*. In Vitro Cell Dev Biol Plant 50:158–165

Silveira AAC, Lopes FJF, Sibov ST (2020) Micropropagation of *Bambusa oldhamii* Munro in heterotrophic, mixotrophic and photomixotrophic systems. Plant Cell Tissue Organ Cult 141:315–326

Singh SR, Dalal S, Singh R, Dhawan AK, Kalia RK (2012a) Seasonal influences on in vitro bud break in *Dendrocalamus hamiltonii* Arn. Ex Munro nodal explants and effect of culture microenvironment on large scale shoot multiplication and plantlet regeneration. Indian J Plant Physiol 17:9–21

Singh SR, Dalal S, Singh R, Dhawan AK, Kalia RK (2012b) Evaluation of genetic fidelity of in vitro raised plants of *Dendrocalamus asper* (Schult. & Schult. F.) Backer ex K. Heyne using DNA-based markers. Acta Physiol Plant 35:419–430

Singh SR, Dalal S, Singh R, Dhawan AK, Kalia RK (2012c) Micropropagation of *Dendrocalamus asper* (Schult. &Schult. F. Backer ex K Heyne): an exotic edible bamboo. J Plant Biochem Biotechnol 21:220–228

Singh SR, Dalal S, Singh R, Dhawan AK, Kalia RK (2013a) Ascertaining clonal fidelity of micropropagated plants of *Dendrocalamus hamiltonii* Nees et Arn. Ex Munro using molecular markers. In Vitro Cell Dev Biol Plant 49:572–583

Singh SR, Singh R, Kalia S, Dalal S, Dhawan AK, Kalia RK (2013b) Limitations, progress and prospects of application of biotechnological tools in improvement of bamboo—a plant with extraordinary qualities. Physiol Mol Biol Plants 19:21–41

Somashekar PV, Rathore TS, Fatima T (2018) In vitro plant regeneration of *Dendrocalamus stocksii* (Munro) M. Kumar, Remesh & Unnikrisnan, through somatic embryogenesis. AJPS 9:2429–2445

Sood A, Nadha HK, Sood S, Walia A, Parkash O (2014) Large scale propagation of an exotic edible bamboo, *Phyllostachys pubescens* mazel ex H. De Lehale (Moso bamboo) using seeds. Indian J Exp Biol 52:755–758

Soreng RJ, Peterson PM, Romaschenko K, Davidse G, Zuloaga FO, Judziewicz EJ, Filgueiras TS, Davis JI, Morrone O (2015) A worldwide phylogenetic classification of the Poaceae (Gramineae). J Syst Evol 53:117–137

Sun S, Zhong J, Li S, Wang X (2013) Tissue culture-induced somaclonal variation of decreased pollen viability in torenia (*Torenia fournieri* Lind.). Bot Stud 54:36

Thakur R, Sood A (2006) An efficient method for explant sterilization for reduced contamination. Plant Cell Tissue Organ Cult 84:369–371

Thapa N, Gauchan DP, Suwal MM, Bhuju S, Upreti A, Byanju B, Lamichhane J (2018) In vitro assessment of *Bambusa balcooa* Roxb. For micropropagation. J Emerg Technol Innov Res 5:464–469

Tikendra L, Amom T, Nongdam P (2019b) Molecular genetic homogeneity assessment of micropropagated Dendrobium moschatum Sw.—a rare medicinal orchid, using RAPD and ISSR markers. Plant Gene 19:100196

Tikendra L, Koijam AS, Nongdam P (2019a) Molecular markers based genetic fidelity assessment of micropropagated *Dendrobium chrysotoxum* Lindl. Meta Gene 20:100562

Tognetti VB, Mühlenbock P, van Breusegem F (2012) Stress homeostasis—the redox and auxin perspective. Plant Cell Environ 35:321–333

Torres GRC, De Lemos EEP (2017) Physical and chemical methods for contaminant control during the in vitro introduction and establishment of *Bambusa vulgaris* Schrad. Ex J. C. Wendl. Científica 45:368

Torres GRC, Houllou LM, De Souza RA (2016) Control of contaminants during introduction and establishment of *Bambusa vulgaris* in vitro. Res Biotechnol 7:58–67

Vazquez AM (2001) Insight into somaclonal variation. Plant Biosyst 135:57–62

Venkatachalam P, Kalaiarasi K, Sreeramanan S (2015) Influence of plant growth regulators (PGRs) and various additives on in vitro plant propagation of *Bambusa arundinacea* (Retz.) Wild: a recalcitrant bamboo species. J Genet Eng Biotechnol 13:193–200

Verma V, Ravindran P, Kumar PP (2016) Plant hormone-mediated regulation of stress responses. BMC Plant Biol 16:86

Waikhom SD, Bengyella L, Sharma GJ (2012) In vitro seed germination and micropropagation of edible bamboo *Dendrocalamus giganteus* Munro using seeds. Biotechnology 11:74–80

Waikhom SD, Louis B (2014) An effective protocol for micropropagation of edible bamboo species (*Bambusa tulda* and *Melocanna baccifera*) through nodal culture. Sci World J 2014:1–8

Wang H, Jin J, Yu P, Fu W, Morrison L, Lin H, Meng M, Zhou X, Lv Y, Wu J (2020) Converting evergreen broad-leaved forests into tea and Moso bamboo plantations affects labile carbon pools and the chemical composition of soil organic carbon. Sci Total Environ 711:135225

Weckx S, Inzé D, Maene L (2019) Tissue culture of oil palm: finding the balance between mass propagation and somaclonal variation. Front Plant Sci 10:722

Wei Q, Cao J, Qian W, Xu M, Li Z, Ding Y (2015) Establishment of an efficient micropropagation and callus regeneration system from the axillary buds of *Bambusa ventricosa*. Plant Cell Tissue Organ Cult 122:1–8

Xia XJ, Zhou YH, Shi K, Zhou J, Foyer CH, Yu JQ (2015) Interplay between reactive oxygen species and hormones in the control of plant development and stress tolerance. J Exp Bot 66:2839–2856

Young PM, Hutchins AS, Canfield ML (1984) Use of antibiotics to control the bacteria in shoot cultures of woody plants. Plant Sci Lett Repo 34:203–209

Yuan JL, Yue JJ, Wu XL, Gu XP (2013) Protocol for callus induction and somatic embryogenesis in Moso bamboo. PLoS One 8:e81954

Zayova E, Vassilevska IR, Kraptchev B, Stoeva D (2010) Somaclonal variations through indirect organogenesis in eggplant (*Solanum melongena* L.). Biol Divers Conserv 3:1–5

Chapter 8
Standardization of Laboratory to Land Transfer Strategies of Micropropagated Plantlets of Bamboo

Kalpataru Dutta Mudoi, Himangshu Lekhak, Nilakshi Borah, Debashish Bora, and Siddhartha Proteem Saikia

Abstract Micropropagation has been widely exploited for the rapid production of different species of bamboo. The plantlets grown in vitro are continuously exposed to a controlled microenvironment with sterile conditions, high humidity, and a medium containing sufficient sugar and nutrients. But its wider application is restricted due to high plant loss prior to field transfer. An efficient hardening, as well as acclimatization approach, leads to shifting rooted in vitro raised bamboo plantlets out of culture on a large scale, at low cost, and with high survival rates under ex vitro condition. After acclimatization, 3–5 months after the micro-plants are transferred to potting soil; the large proliferation can be very suitable for mature bamboo seedlings. Therefore, the transplantation stage is still the main obstacle in field transfer of tissue culture-raised bamboo. To overcome these obstacles, the state-of-the-art information adopted here in this chapter for biotechnological advancement of bamboo will stimulate further research leading to the lab-to-land transfer of micropropagated bamboo.

Keywords Acclimatization · Field establishment · Hardening · In vitro plantlets · Macroproliferation

8.1 Introduction

Bamboo is one of the most wonderful treasures of nature to man. Its versatility has persuaded to the coinage of terms as "green gold" or "poor man's timber." Micropropagation has been broadly utilized for the fast and mass production of several important woody plant species including bamboos. Plants produced under in vitro controlled environment under high humidity, diffused light, and constant

K. D. Mudoi · H. Lekhak · N. Borah · D. Bora · S. P. Saikia (✉)
Agrotechnology and Rural Development Division, CSIR- North East Institute of Science and Technology, Jorhat, Assam, India

Z. Ahmad et al. (eds.), *Biotechnological Advances in Bamboo*,
https://doi.org/10.1007/978-981-16-1310-4_8

temperature need to be acclimatized. For successful field transfer, hardening or acclimatization is the last phase of in vitro propagation where meticulous optimization is important to initiate successful establishment of tissue culture-raised plantlets. The transition from artificially controlled in vitro conditions to a potent ex vitro environment leads to enormous stress on micropropagated plantlets, and it is an obstacle for many tissue-cultured plantlets (Soukup et al. 2004). During lab-to-land transfer, a high mortality rate is recorded mainly due to the utmost variations within in vitro and ex vitro conditions, which are a vital constraint in the mass-scale implementation of this technology (Chandra et al. 2010). For successful field transfer, various procedures have been undertaken for hardening of tissue culture-raised plants (Bhojwani and Dhawan 1989). But the procedures acquire for acclimatization have not been very suitable in imparting quality transplants for the field (Ziv 1995). The micropropagation units had to incur losses due to the death of the plants which are generated with costly inputs. Conventional methods of acclimatization cannot ensure a high percentage of survivability.

Hence, alternative approaches to acclimatization have been evolved with varying successes and popularity. Different techniques have been used by different researchers to establish different micropropagated bamboo plantlets in soil with extreme possible efficacy (Mudoi et al. 2013). The intensity of hardening and acclimatization of plant and photosynthetic apparatus was found to be affected by the hormonal concentration, sucrose level, gelling agent, temperature, and pH of the medium (El-Sherif 2019).

In the present chapter, attempt has been made to discuss the role of basal medium, microenvironment, rooting, hardening, acclimatization, macroproliferation, field performance, etc. for successful in vitro to ex vitro transfer of bamboo plantlets. The state-of-the-art information highlighted in this chapter will minimize the major abnormalities faced during lab-to-land transfer of micropropagated plantlets of bamboo.

8.2 Effect of Different Factors toward Successful Establishment of Bamboo Plantlets

8.2.1 Basal Medium

Basal media are different in their chemical composition. All the plant species usually respond differently to various basal media due to differences in their nutritional requirements. The sole medium could not be optimum for every plant tissue and organ. To begin a new process, it is important to emerge a medium that can fulfill the distinct demands of particular plant tissue (Bhojwani and Razdan 1996). Based on the above concept, sterilized explants of bamboos are inoculated into different types of basal media for a selection of the best basal medium composition.

The majority of work on tissue culture of bamboos has been carried out on Murashige and Skoog 1962 (MS) basal medium (Rout and Das 1994; Arya and Sharma 1998; Arya et al. 2002a, 2002b; Kalia et al. 2004; Das and Pal 2005; Arya et al. 2006; Ramanayake et al. 2008; Yasodha et al. 2008; Mishra et al. 2008; Mudoi and Borthakur 2009; Mudoi et al. 2014; Patel et al. 2015; Kaladhar et al. 2017; Rajput et al. 2019; etc.). However, various researchers reported about positive response received from ½ strength of MS basal medium (Ogita 2005; Shirgurkar et al. 1996; Saxena and Dhawan 1999). But Mukunthakumar et al. (1999), Ndiaye et al. (2006), and Ogita et al. (2008) observed results on modified MS (MMS) medium. Moreover, Arya and Sharma (1998) and Ravikumar et al. (1998) described the effectiveness of WPM medium (Lloyd and McCown 1980) and White (1963) medium, respectively.

8.2.2 Microenvironment in Plant Tissue Culture of Bamboo (Light Intensity, Temperature, and Relative Humidity)

Microenvironment is essential for in vitro culture of bamboos for obtaining healthy plantlets by providing optimum physical environment like light, temperature, and relative humidity. Physical environmental factors are fixed and could be sustained or altered in the course of the growth cycle. Likewise, the chemical environment is settled and adjustable like the pH and the composition of the medium in such a way that the ideal conditions are always provided for the nutriment of young propagules. Physical parameters in culture room conditions can be altered or improved by changing the room temperature, photoperiodic condition, light intensity, and relative humidity. However, various researchers modified the photoperiodic condition and room temperature of culture room (Arya and Sharma 1998; Rajput et al. 2019; Saxena and Dhawan 1999; Sood et al. 2002; Sanjaya et al. 2005; El Hassan and Debergh 1987; Nurhayani et al. 2018). Moreover, varied photoperiodic conditions and relative humidity (50%–90%) were reported by various researchers for in vitro culture of bamboos (Table 8.1).

8.2.3 In Vitro Rooting of Bamboo Shoots

Root induction of in vitro regenerated multifold shoots is a key step of in vitro micropropagation. Generally, it is easy to raise rooting from the shootlets of seedling origin, and quite often, the frequency of rooting reduces with an increase in age of the source plants, particularly in tree species and woody plants. For bamboo root induction, it needs an adjustment in the strengths of both auxins and cytokinins or with auxin alone. Rooting frequency also depends on the level of rejuvenation during the multiplication stage and the quality of the shoots. The frequency and

Table 8.1 Effect of different factors toward successful establishment of in vitro raised bamboo plantlets

Species	Basal medium used	Growth regulators for rooting (as indicated)	Temp	Photoperiod/light intensity	Humidity	Potting mixture	Survival rate	References
Arundinaria callosa	½ MS	25 µM IBA + 0.05µM BAP	–	–	80 ± 5%	Sand/soil/ FYM (farm-yard manure) in 1:1:1	70%	Devi and Sharma (2009)
Bambusa arundinacea	MS	2.0 mg/l AgNO3 + 3.0 mg/l IBA	25 ± 2 °C	16 h/60 µ Emol m^{-2} s^{-1}	65%	Soil/sand (3:1)	92%	Venkatachalam et al. (2015)
B. balcooa	MS	NAA (3.0 mg/l) + BAP (1.0 mg/l)	25 ± 2 °C.	16 h/10 µmol m^{-2} s^{-1}	60–70%	Soil/sand/cow dung mixture (1:1:2)	90%	Mudoi and Borthakur (2009)
B. balcooa	MS	4 mg/L NAA	25 ± 2 °C	16 h/ 10000–12,000 lux	70–75%	Cocopeatt/ vermicompost (3:1)	74.66%	Patel et al. (2015)
B. balcooa	½ MS	NAA (2.5 mg L^{-1})	–	16 hr./1200 lux	80 ± 5%	Cocopeatt/ vermicompost (2:1)	–	Choudhary et al. (2017)
B. balcooa	MS	10 mg L^{-1} IBA+ 5 mg L^{-1} NAA	22 °C	16 h/1000–4000 lux	–	Soil + husks rice + vermicompost (1:1:1)	–	Nurhayani et al. (2018)
B. bambos	MS/ Lloyd and McCown (1980) (WPM)	NAA (3.0 mg L^{-1})	27 ± 1 °C	14 h/3000 lux	80 ± 5%	Soil/organic manure (1:1)	80–90%	Arya and Sharma (1998)
B. bambos	½ MS	2.5 mg/l IBA 2.5 mg/l NAA	24 ± 2 °C	16 h/illumination with florescent light	–	Soil/sand/ compost (1:1:1)	100%	Raju and Roy (2016)

B. nutans	MS	NAA (2 mg/l)	25 ± 2 °C	16 h/10 μmol m^{-2} s^{-1}	60–70%	1:1:2 soil/ sand/cow dung	80%	Mudoi et al. (2014)
B. nutans	MS	IAA (5.0 mgl^{-1}) + IBA (8.0 mgl^{-1}) + NAA (5.0 mgl^{-1})	25 ± 2 °C	16 h/45 μmol m^{-2} s^{-1}	80–85%	Cocopeat/ vermicompost (3:1)	100%	Sharma and Sarma (2014)
B. ventricosa	MS	2.7 μM NAA 4.9 μM IBA 4.4 μM 6-BA	25 °C	16 h cycle	–	Peat soil/ver-miculite/pearl-ite 1:1:1 (v/v/v)	–	Wei et al. (2015)
B. vulgaris	½ MS	NAA (1.0–3.0 mg L^{-1}) IBA (1.0–5.0 mg L^{-1})	–	–	–	Peat soil/ver-miculite/pearl-ite (v/v/v)1:1:1	100%	Islam and Rahman (2005)
B. vulgaris var. *striata*	½ MS	NAA (1.0–3.0 mg L^{-1}) IBA (1.0–5.0 mg L^{-1})	–	–	–	Peat soil/ver-miculite/pearl-ite (v/v/v) 1:1:1	100%	Islam and Rahman (2005)
B. vulgaris	MMS	IBA (20.0 mg L^{-1})	26 °C	16 h/fluorescent light	–	Sterile perlite/ peat (1:2)	100%	Ndiaye et al. (2006)
B. vulgaris 'striata'	MS	IAA+ BAP	25 ± 26 °C	16 h/fluorescent light	50–55%	–	75%	Kaladhar et al. (2017)
Dendrocalamus asper	MS	NAA (3.0 mg L^{-1}) IBA (10.0 mg L^{-1})	26 °C	16 h/3000 μEm^{-2} s^{-1}	85–90%	Sand/farmyard manure/soil (1:1:1)	95%	Arya et al. (2002a, 2002b)
D. asper	MS	IBA (14.76 μM) NAA (3.67 μM)	–	–	80–85%	Sand/soil/ FYM (1:1:1)	95%	Nadha et al. (2013)
D. brandisii	MMS	IBA (1.0 mg L^{-1})	28 ± 2 °C	16 h/2000 lux	–	Soil/farmyard manure (1:1)	–	Mukunthakumar et al. (1999)
D. hamiltonii	jnjnjnjnM½ ½ MS	Chlorine chloride (9 mg L^{-1}) + IBA (1 mg L^{-1}) + coumarin (9 mg L^{-1}) + NAA (0.5 mg $L^{-1)}$	25 ± 2 °C	12 h/20 μmol m^{-2} s^{-1}	–	Garden soil/ river bed Sand/farmyard manure (1:1:1)	80–85%	Sood et al. (2002)

(continued)

Table 8.1 (continued)

Species	Basal medium used	Growth regulators for rooting (as indicated)	Temp	Photoperiod/light intensity	Humidity	Potting mixture	Survival rate	References
D. hamiltonii	½ MS	IBA (25.0 μM) + choline chloride (36.0 μM)	25 ± 2 °C	16 h	–	Dune sand/ vermicompost (3:1)	85%	Singh et al. (2012)
D. latiflorus	½ MS	1 mg/L IAA	26 ± 2 °C	16 h/60–70 μmol m^{-2} s^{-1}.	–	–	98%	Ye et al. (2017)
D. membranaceus	MS	NAA (3.0 mg L^{-1}) IBA (10.0 mg L^{-1})	25 ± 1 ° C	–	–	Vermiculite	–	Arya et al. (2002b)
D. strictus	MS and White	½ MS basal	25 ± 1 °C	16 h/23.34 μmolm^{-2} s^{-1}	90%	Sand/soil (1:1)	90–95%	Shirgurkar et al. (1996)
D. strictus	MS and White	IBA (0.25 mg l^{-1})	25 ± 2 °C	16 h/40 μmol m^{-2} s^{-2}	85–90%	Soil and ver-miculite (1:1)	–	Ravikumar et al. (1998)
D. strictus	½ MS	NAA (5 × 10^{-6} M) IBA (2.5 × 10^{-6} M)	26 ± 2 °C	12 h/37 μmol m^{-2} s^{-1}	60–65%	Soil/farmyard (2:1, v/v)	80%	Saxena and Dhawan (1999)
D. strictus	MS	2.5 mg L^{-1} BAP + 5 mg L^{-1} IAA	25 ± 2 °C	40 μmolm^{-2} s^{-1}	60%	Sand/FYM:/ soil (1:1:1)	–	Kapruwan et al. (2014)
D. strictus	MS	3 mg L^{-1} NAA 1 mg L^{-1} IBA and 3 mg L^{-1} NAA	25 °C	16 h/2000–3000 lux.	–	Perlite/soil/ FYM (1:1:1)	70%	Goyal et al. (2015)
Drepanostachyum falcatum (Nees) Keng f	MS	6.5 mg L^{-1} IBA	25 ± 2 °C	16 h/2500 lux	60%–80%	Sand/soil/ FYM	90–95%	Saini et al. (2016)
Gigantochloa atroviolacea	MS	IBA (35.0Mm)	25 ± 2 °C	16 h/illuminated by cool white fluorescent tubes (Philips, India)	80%	Sand/soil/ FYM in 1:1:1	80%	Bisht et al. (2010)
Phyllostachys viridis	MS	Basal	23 ± 2 °C	24 h/ 30μmolm^{-2} s^{-1}	90%	–	–	El Hassan and Debergh (1987)

P. meyeri	MMS	½ MMS	25 °C	16 h/60 μE		–	–	Ogita et al. (2008)
Pseudoxytenanthera stocksii	HMS	IBA (4.90 μM) + BAP (0.44 μM) + additives	28 ± 2 °C	12 h/60 μmol m^{-2} s^{-1}	70–80%	–	–	Sanjaya et al. (2005)
P. stocksii	MS	1.0 mg L^{-1} IBA	28 ± 2 °C	16 h/40–50 μmol m^{-2} s^{-1}	60–70%	Sterile Soilrite®	96%	Rajput et al. 2019

– indicates that data is not available

quality/quantity of rooting depend on factors like species/genotypes, source of auxins and their concentrations, nutrient media, sucrose concentrations, and incubation conditions. Along with the ideal concentration of growth hormones, the choice of the optimum size of shoot propagule was also a more significant factor for successful root induction. Propagule of 2–5 shoots, having 1.0–2.0 cm in length, is the optimum size of the bamboo shoots used for root induction. Roots were initiated within 20–45 days of culture period either on singular auxin or both auxin and cytokinin fortified composition (Arya et al. 2002a; Mudoi and Borthakur 2009; Mudoi et al. 2014).

The root was induced on zygotic embryo-derived in vitro shoots of different bamboos with indole-3-butyric acid (IBA) augmented medium (Nadgauda et al. 1990; Yasodha et al. 1997; Arya et al. 1999; Saxena 1990; Mascarenhas et al. 1988; Rout and Das 1994; Malini and Anandakumar 2013). [However, IBA was also responsible for in vitro rooting of mature tissue-derived shoots of bamboos (Ramanayake and Yakandawala 1997; Mukunthakumar et al. 1999; Arya et al. 2002b; Das and Pal 2005; Islam and Rahman 2005; Ndiaye et al. 2006; Lin et al. 2006; Shirin and Rana 2007; Arya et al. 2008; Agnihotri et al. 2009). Moreover, singular supplementation of α-naphthaleneacetic acid (NAA) was suitable for inducing roots on mature tissue-derived in vitro bamboo shoots (Suwannamek 1992; Arya et al. 2008; Huang and Huang 1995; Arya and Sharma 1998; Arya et al. 2002b; Lin et al. 2003, 2005; Islam and Rahman 2005; Lin et al. 2006; Lin et al. 2007; Mudoi et al. 2013; Choudhary et al. 2017). The combined effect of plant growth regulators has promoted rooting on various in vitro bamboo shoots, likely 6-benzylaminopurine (BAP) and NAA (Huang and Huang 1995; Mudoi and Borthakur 2009); BAP and IBA (Arya et al. 2006); IBA and coumarin (Ramanayake and Yakandawala 1997); thidiazuron (TDZ) and 2,4-dichlorophenoxyacetic acid (2,4-D) combined medium (Lin et al. 2004); NAA and 2,4-D (Lin et al. 2005); IBA and TDZ (Ramanayake et al. 2006); IBA and glucose (Yasodha et al. 2008); indole-3-acetic acid (IAA), IBA, and coumarin (Saxena and Bhojwani 1993); IBA, TDZ, and coumarin (Ramanayake et al. 2008); BAP and IAA (Kapruwan et al. 2014); IBA and $AgNO_3$ (Venkatachalam et al. 2015); and IBA and NAA (Lin et al. 2005; Nadha et al. 2013; Goyal et al. 2015; Nurhayani et al. 2018). The singular effect of growth regulators has induced rooting in many bamboos, viz., in gibberellic acid (GA_3), coumarin, 2,4-D, TDZ, and BAP-induced medium (Rout and Das 1994; Mishra et al. 2008; Lin et al. 2003; Lin et al. 2004 & Jiménez et al. 2006).

Hormone-free MS basal medium is also responsible for in vitro rooting of some bamboos (Shirgurkar et al. 1996; Godbole et al. 2002; Ramanayake et al. 2008; and Ogita et al. 2008).

8.2.3.1 Effect of Different Phytohormones on Ex Vitro Rooting

Ex vitro rooting reduces one step of in vitro rooting, cost of production, and improvement in the survival rate of plantlets. Auxin type, its concentration and duration of treatment, rooting medium, and incubation conditions are the factors which influence ex vitro rooting. 85–90% of ex vitro rooting was recorded in seedling-derived shoots of *Dendrocalamus strictus*, treated with 200 parts per million (ppm) pulse treatment of IBA within 20–25-day period (Ravikumar et al. 1998). 99% rooting was induced in shoot clumps of *P. stocksii* pulse treated with NAA (1000 ppm) for 10 min (Somashekar et al. 2008). The effect of auxin treatment was significant and increased rooting frequency. Minimum rooting frequency was observed in β-naphthoxyacetic acid (NOA). Amid different tested auxins, i.e., IAA, IBA, NAA, and NOA, at 2000 ppm, pulse treated with two to three shoots/clump of *Bambusa bambos* with NAA in sand medium induced 89.64% rooting under ex vitro (greenhouse) conditions (Kabade 2009). This was followed by IBA and IAA. No root induction was observed in shoots, pulse treated with NOA.

8.2.3.2 Rooting Percentage

Plant growth hormones induced rooting in bamboos; but several differences were noticed amid different species in terms of percent rooting of in vitro microshoots. Totally rooting was not recorded in many bamboos. Different bamboo species showed variations in rooting percentage as they acted differently to the different hormonal combinations. In *Dendrocalamus* species, 20–90% rooting was observed by many researchers (Nadgir et al. 1984; Mascarenhas et al. 1988; Shirgurkar et al. 1996; Ravikumar et al. 1998; Saxena and Bhojwani 1993; Ramanayake and Yakandawala 1997 and Arya et al. 2002b). In *Bambusa* species, different rooting percentages were recorded (Bhojwani and Razdan 1996; Sood et al. 2002; Arya et al. 2002a, b; Lin et al. 2003; Ndiaye et al. 2006; Shirin and Rana 2007; Yasodha et al. 2008; Mudoi and Borthakur 2009; Mudoi et al. 2013 and Kapruwan et al. 2014). Moreover, 100% rooting was reported in a few bamboo species (Jiménez et al. 2006; Kapoor and Rao 2006; Negi and Saxena 2011). But 27.8% rooting was recorded in *P. meyeri* (Ogita et al. 2008).

8.2.4 Hardening

Tissue culture-raised plantlets are fragile, delicate, and vulnerable to the transplanting shocks. Therefore, hardening is one of the most important bottlenecking steps toward field shifting of in vitro bamboo plantlets. It is the adaptation period throughout in vitro to ex vitro environment leading to successive field transfer. During hardening, when the in vitro raised plantlets were shifted to the soil, then subsequently observed decline growth and plant loss are due to environmental changes. In this stage, varied mortality rates of bamboo plantlets were noticed along with abnormal hyperhydric growth. Imperfect growth of cuticular waxes, water loss due to poor uncontrolled transpiration, nonfunctional stomata, and sensitivity to pathogen attack lead to an increase in mortality rate during hardening (Ziv

1995). Later on, in vitro raised plantlets slowly conquered such difficulties and started surviving in ex vitro environments.

8.2.4.1 Survivability during Hardening

Different survivability rates were recorded in different bamboo species after successful hardening (Mascarenhas et al. 1988; Shirgurkar et al. 1996; Sood et al. 2002; Godbole et al. 2002; Arya et al. 2002a, 2002b; Islam and Rahman 2005; Jiménez et al. 2006; Kapoor and Rao 2006; Agnihotri et al. 2009; Mudoi and Borthakur 2009; Brar et al. 2013; Mudoi et al. 2013 & Venkatachalam et al. 2015). In rare case, 100 percent survivability was found in few bamboo species (Lin et al. 2003; Ndiaye et al. 2006; Ramanayake et al. 2006; Gillis et al. 2007 & Negi and Saxena 2011). A two-step hardening process was used by Mudoi and Borthakur 2009 and Mudoi et al. 2014, where the rooted plants were transferred to MS liquid medium for 15–20 days for root elongation, and then the rooted shoots were transferred to semi-strength MS liquid medium for 15 days (Fig. 8.1a). After that, the seedlings were soaked in unsterilized filtered water in the culture room for 15 days and then soaked at ambient temperature (28 ± 2 °C) for another 15 days. In the hardening stage, 30–40% of

Fig. 8.1 In vitro raised plantlets of bamboo during lab-to-land transfer, (**a**) Rooted shoots during hardening, (**b**) mortality during hardening, (**c**) emergence of white-colored new secondary roots, (**d**) acclimatized plantlets, (**e**) 3–5-month-old acclimatized plant ready for macroproliferation, (**f**) rooted healthy tillers, (**g**) splitting of rooted tillers during macroproliferation, (**h**) 5-year-old plant established in field

Table 8.2 General description in hardening stage of *B. balcooa* and *B. nutans*[a]

Sl no.	Factors	Standard	
		B. balcooa	*B. nutans*
1	General appearance	Healthy, greenish	Healthy, greenish
2	Height of plantlets	2.5–6.0 cm	3.0–6.0 cm
3	Tillers/ plantlets	1–4	2–4
4	Leaves/ plantlets	3–7	5–10
5	Visible nodes/ plantlets	3–5	2–5
6	Nature of roots	Fibrous type	Fibrous type
7	Number of roots	1–4 with initiation of white-colored secondary roots	2–6 with initiation of white-colored secondary roots
8	Length of roots	3–10 cm	5–10 cm
9	Survivability	65%	70%

[a]The data is adapted from Mudoi and Borthakur (2009) and Mudoi et al. (2014)

plantlets were lost in *B. balcooa* either for contamination problem or for unhealthy root growth (Fig. 8.1b). To reduce the mortality rate during hardening, the selection of an ideal shoot (i.e., 20–25 days old, 1.5–2.5 cm in length) for rooting is the most crucial point (Mudoi and Borthakur 2009; Mudoi et al. 2013). In this way, plantlet death has been reduced during hardening. When white-colored, new secondary roots were developed during the hardening period, then a higher survival frequency was recorded, and at that stage, plantlets are ready for transferring to the poly-sleeves for acclimatization (Fig. 8.1c; Table 8.2). Similarly, Patel et al. 2015 reported 72% acclimatization rate at the primary and secondary hardening of the same bamboo species.

8.2.5 Acclimatization

After successful hardening, the implanting of plantlets into soils has been a major bottleneck in micropropagation system (Hazarika 2003). Plantlets should be gradually acclimatized to the polyhouse or greenhouse condition providing an unsterile environment by putting them into ex vitro condition. Shifting of plantlets from tissue-cultured vessels to the ex vitro condition can detect the consequence of any micropropagation system. While the plants are acclimatized, these should get physical support and required habitat soil for survival of the plants. Acclimatization process is required in order to establish the hardening plantlets for the survival and growth, prior to transfer to potted mixtures.

8.2.5.1 Effect of Different Soil Mixtures on Acclimatization of Bamboo

The quality of potting mixture used during acclimatization is one of the prime factors that regulate the survival percentage of the plants under ex vitro conditions. In order to overcome the obstacle of hardening, many workers have adopted various strategies. Usually, the healthy rooting seedlings are washed to remove any traces of the rooting medium and then transferred to the potting mixture, viz., soil, sand, Soilrite, perlite, cocopeat, agro-peat, vermiculite, compost, farmyard manure, etc., either alone or in various ratios. Normally, multiple researchers used the soil mixture composition either in 1:1:1 ratio or in a modified ratio (Table 8.1).

Hardened plantlets of bamboos were planted in different soil mixtures in polythene sleeves and maintained in greenhouse/polyhouse before shifting to the field and then recorded their survivability rate (Fig. 8.1d). Hardening plantlets of *D. asper* transferred into potting mixture containing sand/soil/farmyard manure in different ratios (v/v;1:1:1; 1:1:2; 1:0:0; 0:0:1) and kept in greenhouse. The percentage survival was recorded as 95% in the 1:1:1 sand/soil/manure mixture, 90% in 1:1:2 sand/soil/manure mixture, 80% in sand, and 60% in manure, respectively (Nadha et al. 2013). Similarly, Bisht et al. 2010 used 1:1:1 ratio of sand/soil/FYM in *Gigantochloa atroviolacea*. However, the mixture of pea/sand/vermiculate significantly recorded the longest plantlet length (12.20 cm), longest roots (16.30 cm), and largest leaf number (16.0 leaves/explant) in *B. ventricosa* (Wei et al. 2015). Ho et al. 1987 and Sawsan et al. 2010 also experienced the same findings. Rajput et al. 2019 reported 1:1 garden soil and vermicompost composition for transfer of *D. strictus* plantlets, whereas a combination of perlite/soil/farmyard manure (1:1:1) and sterile soil/sand (1:1) was provided by in *D. strictus*. Banik 1987 used moist sterile soil in acclimatization of *B. glaucescens*. Hardening plantlets of *B. balcooa* and *B. nutans* were transferred to polythene sleeves containing 1:1:2 soil/sand/cow dung mixture and kept in polyhouse/net house and recorded different growth parameters, viz., increase the number of height, tillers, nodes, and leaves (Fig. 8.1d). For the preparation of beds, forest top soil, river sand, and cow dung were used in the ratio of 3:2:1 for *B. nutans* (Sharma and Sarma 2014). The adequate results obtained during acclimatization of bamboos could be linked to the simple transformation of the bamboo plantlets in minimal environmental conditions (Crouzet 1981).

8.2.5.2 Survivability during Acclimatization

The rooted plantlets of bamboo showed different survival percentages during acclimatization because transferring of these plantlets from in vitro to ex vitro conditions is the most chilling experience for them (Table 8.1). Nadgir et al. 1984, Pandey and Singh 2012, and Chaturvedi et al. 1993 recorded varied survival rates in the case of *D. strictus*. In *B. bambos*, 80–90% (Arya and Sharma 1998) and 100% survival rate

Table 8.3 General description of tissue culture-raised *B. balcooa* and *B. nutans* plantlets during acclimatization stage[a]

Sl no.	Factors	Standard	
		B. balcooa	*B. nutans*
1	General appearance	Healthy, greenish	Healthy, greenish
2	Height of plantlets	5–12 cm	4–10 cm
3	Tillers/plantlets	1–3	1–3
4	Leaves/plantlets	5–12	3–10
5	Visible nodes/plantlets	1–3	1–3
6	Insect/disease pest infestation	Less than 5%	Less than 10%
7	Survivability	98%	80%

[a]The data is adapted from Mudoi and Borthakur (2009) and Mudoi et al. (2014)

(Raju and Roy 2016) and in *B. tulda* 80–90% (Saxena 1990) rate were recorded. The survival percentage of *D. giganteus* plantlets in greenhouse conditions was found to be 92.5% after 30 days of acclimatization (Hossain et al. 2018). In this stage, *B. balcooa* and *B. nutans* recorded 98% and 80% survivability, respectively (Mudoi and Borthakur 2009; Mudoi et al. 2014) 8] which was described in Table 8.3. However, Patel et al. (2015) reported 74.66% survivability in *B. balcooa*. *B. tulda* and *Melocanna baccifera* resulted 81.8% and 70.31% survival rate, respectively (Waikhom and Louis 2014). In *Drepanostachyum falcatum*, 95% survivability rate was observed (Saini et al. 2016). In *D. hamiltonii*, 79.76% survival rate was achieved (Singh et al. 2012). Similarly, 75% survival rate was achieved in *B. vulgaris* (Table 8.1) by Kaladhar et al. (2017). The rooted plants of *D. strictus* were transferred to eco-friendly paper cups containing Soilrite® irrigated with ¼ MS solution salts and incubated in greenhouse for 4 weeks to acclimatize the plants under ex vitro conditions (Rajput et al. 2019).

8.2.6 *Macroproliferation*

The technique of macroproliferation is a simple solution to the problem that can meet moderate scales of propagation (Kumar 1991). The traditional means of vegetative multiplication through simple techniques like rhizome transplantation and rooting of culm or branch cuttings (Banik 1994), which is the mainstay of bamboo propagation in the absence of seeds, could be further improved by the use of rooting hormones and mist propagators but yet does not meet the increased demand for planting material. In vitro techniques appeared to have great potential to overcome the constraints in the propagation of bamboo on a large scale. Not only is micropropagation a prolific cloning method, but it also has unique advantages like the potential for year-round production of disease- and pest-free planting material.

This method is generally applicable to bamboo seedlings. Bamboo seedlings possess the capacity to proliferate. Macroshoots that regenerated after 3–5 months of transfer of microshoots in polybags were split individually, each containing intact roots having root hairs and a well-developed shoot system. These macroshoots were replanted in the soil in polythene sleeves and were kept in the shade and well-watered conditions. 3–5 months after the transfer of the micro-plants, macroproliferation can be very suitably adapted for mature bamboo plantlets. Rooted tillers of bamboos were split individually to enhance the plant numbers up to three times, and this technique can be further resumed for several years (Fig. 8.1d–g). If micropropagation of bamboos was followed by macroproliferation, then plant production rate could be further increased. As a result, continuous plantlet proliferation was achieved, which were small and easy to handle and transport. Mudoi and Borthakur (2009), Banik (1987), Kumar (1991), and Mudoi et al. (2013) described it as one of the most reliable, fast techniques to proliferate enormous quantities of bamboo plantlets up to three times.

8.2.7 Response of in Vitro Raised Plantlets after Field Transfer

Micropropagation system imparts an unconventional method for the fast and mass production of plantlets, but its optimum success relies on the successful transfer of these plants either in the net house or in the field conditions. Morphological and physiological performance of in vitro raised bamboo plantlets under field conditions is essential for long-term assessment and commercial applications. Plantlets are evaluated periodically by recording different physiological parameters regarding the height of the plant, girth, number of leaves and culms produced/plant, etc.

Comparisons were made between in vitro raised plants and seed–/rhizome-derived check plants of the same age. In bamboo, vigorous growth of in vitro raised plantlets in terms of the girth, height, a number of culms and nodes/plant, etc. was recorded as nearly double with those of seedling-derived plants (Mascarenhas et al. 1988). In *B. balcooa* and *B. nutans*, 98% and 85% survivability were observed after successful field transfer of micropropagated plantlets (Mudoi and Borthakur 2009; Mudoi et al. 2014). Likewise, many researchers (Gupta et al. 1991; Wei and Tien 1995; Sood et al. 2002) also mentioned the superiority of micropropagated plantlets over field raised bamboo plantlets within a study period of 2 years (Fig. 8.1h-i; Table 8.4).

Table 8.4 Field performance of tissue culture-raised *B. balcooa* and *B. nutans* plantlets after field transfer

Sl no.	Factors	Standard							
		6 months		12 months		18 months		24 months	
		B. balcooa	*B. nutans*	*B. balcooa*	*B. nutans*	*B. balcooa*	*B. nutans*	*B. balcooa*	*B. nutans*
1.	General appearance	Healthy, greenish	Healthy, greenish	Healthy, greenish	Healthy, greenish	Healthy, greenish	Healthy, greenish	Healthy, greenish	Healthy, greenish
2.	Height of plantlets	15–22 cm	15–20 cm	20–35 cm	22–25 cm	35–40 cm	28–38 cm	50–65 cm	40–60 cm
3.	Tillers/plantlets	1–6	1–5	1–8	1–7	1–8	1–7	1–10	1–10
4.	Leaves/plantlets	12–18	10–15	20–25	20–30	35–48	35–45	50–65	50–70
5.	Visible nodes/plantlets	5–6	2–6	10–12	6–8	15–18	10–15	20–25	18–23
6.	Insect/ disease Pest infestation	Less than 5%	Less than 10%	Less than 5%	Less than 5%	Less than 5%	Less than 5%	Less than 5%	Less than 5%

8.3 Conclusions

For ages, bamboo or "green gold" has been defined as a matter of incalculable interest, and its demand has been ever increasing. Hence, basic research on micropropagation of bamboo should be encouraged to meet the requirement for the mass propagation of quality planting materials. The benefit of any micropropagation system can be judged by the successful transfer of in vitro plantlets to the ambient ex vitro condition. In such case, hardening and acclimatization are two important aspects of micropropagation where careful optimization is essential for survival and successful establishment of in vitro raised bamboo plantlets. The survivability rate can be increased by choosing appropriate soil mixtures. Thereafter, macroproliferation could be adapted to increase the rate of production of bamboo plantlets. This chapter concluded that for the successful establishment of tissue-cultured plantlets in soil/field conditions, one must carefully adhere to the standardization of lab-to-land strategies. With the increased demand for bamboo as a substitute of timber for sustainable product design, structural application necessitates judicious implementation and preservation of bamboo resources. Therefore, it is necessary to adopt biotechnology tools to wisely use and protect bamboo resources, which can bring more profits to humans all over the world.

Acknowledgments Authors are thankful to Dr. G. N. Sastry, Director, CSIR-North East Institute of Science & Technology, Jorhat, Assam, India, for providing facilities during the study.

Conflicts of Interest The authors declare no conflict of interests.

Author Contributions KDM analyzed data and wrote the manuscript. HL analyzed data and technical support. NB and DB analyzed data. SPS helped in the conceptualization and design of the work and wrote the manuscript.

References

Agnihotri RK, Mishra J, Nandi SK (2009) Improved *in vitro* shoot multiplication and rooting of *Dendrocalamus hamiltonii* Nees et Arn. Ex Munro: production of genetically uniform plants and field evaluation. Acta Physiol Plant 31:961–967

Arya ID, Rana PK, Satsangi R, Muzaffar FS, Sharma S, Arya S (2002b) Rapid and mass multiplication of bamboos through tissue culture techniques. In: Nandi SK, Palni LMS, Kumar A (eds) Role of plant tissue culture in biodiversity conservation and economic development. Gyanodaya Prakashan, Nainital, India, pp 29–39

Arya S, Rana PK, Sharma R, Arya ID (2006) Tissue culture technology for rapid multiplication of *Dendrocalamus giganteus* Munro. Indian For 132:345–357

Arya S, Satsangi R, Arya ID (2002a) Rapid mass multiplication of edible bamboo *Dendrocalamus asper*. J Sustain For 14:103–109

Arya S, Satsangi R, Arya ID (2008) Direct regeneration of shoots from immature inflorescences in *Dendrocalamus asper* (edible bamboo) leading to mass propagation. Bamboo Sci Cult Am Bamboo Soc 21:14–20

Arya S, Sharma S (1998) Micropropagation technology of *Bambusa bambos* through shoot proliferation. Indian For 124:725–731

Arya S, Sharma S, Kaur R, Arya ID (1999) Micropropagation of *Dendrocalamus asper* by shoot proliferation using seeds. Plant Cell Rep 18:879–882

Banik RL (1987) Techniques of bamboo propagation with special reference to prerooted and pre-rhizomed branch cuttings and tissue culture. In: Rao AN, Dhanarajan G, Sastry CB (eds) Proc. Int. Natl. Bamboo Workshop (1985), 'recent research on bamboos' Hongzhou. Peoples of China and IRDC, Canada, pp 160–169

Banik RL (1994) Review of conventional propagation research in bamboos and future strategy. INBAR Technical Report No.5 Constraints to production of bamboo and rattan. INBAR, New Delhi, pp115–142

Bhojwani SS, Dhawan V (1989) Acclimatization of tissue culture raised plants for transplantation to the field. In: Dhawan V (ed) Application of biotechnology in forestry and horticulture. Plenum Pree, New York, pp 249–256

Bhojwani SS, Razdan MK (1996) Plant tissue culture: theory and practice. Elsevier Science Publishers B.V, Amsterdam, p 51

Bisht P, Pant M, Kant A (2010) *In vitro* propagation of *Gigantochloa atroviolacea* Widjaja through nodal explants. J Am Sci 6:1019–1026

Brar J, Ananda M, Sood A (2013) *In vitro* seed germination of economically important edible bamboo *Dendrocalamus membranaceus* Munro. Indian J Exp Biol 51:88–96

Chandra S, Bandopadhyay R, Kumar V, Chandra R (2010) Acclimatisation of tissue cultured plantlets: from laboratory to land. Biotechnol Lett 32:1199–1205

Chaturvedi HC, Sharma M, Sharma AK (1993) *In vitro* regeneration of *Dendrocalamus strictus* Nees through nodal segment taken from field-grown culms. Plant Sci 91:97–101

Choudhary AK, Priyanka K, Ashish R (2017) Refinement of protocol for rapid clonal regeneration of economical bamboo, *Bambusa balcooa* in the agroclimatic conditions of Bihar, India. Afr J Biotechnol 16:450–462

Crouzet Y (1981) Les bambous. INRA, Paris, p 95

Das M, Pal A (2005) *In vitro* regeneration of *Bambusa balcooa* Roxb.: factors affecting changes of morphogenetic competence in the axillary buds. Plant Cell Tissue Organ Cult 81:109–112

Devi WS, Sharma GJ (2009) *In vitro* propagation of *Arundinaria callosa* Munro- an edible bamboo from nodal explants of mature plants. Open Plant Sci J 3:35–39

El Hassan AA, Debergh P (1987) Embryogenesis and plantlet development in the bamboo *Phyllostachys viridis* (Young) McClure. Plant Cell Tissue Organ Cult 10:73–77

El-Sherif NA (2019) Impact of plant tissue culture on agricultural sustainability. In: Negm AM, Abu-hashim M (eds) Sustainability of agricultural environment in Egypt: part II - soil-water-plant nexus, Hdb Env Chem, vol 77. Springer International Publishing AG, pp 93–108

Gillis K, Gielis J, Peeters H, Dhooghe E, Oprins J (2007) Somatic embryogenesis from mature *Bambusa balcooa* Roxburgh as basis for mass production of elite forestry bamboos. Plant Cell Tissue Organ Cult 91:115–123

Godbole S, Sood A, Thakur R, Sharma M, Ahuja PS (2002) Somatic embryogenesis and its conservation into plantlets in a multipurpose bamboo, *Dendrocalamus hamiltonii* Nees et Arm Ex. Munro. Curr Sci 83:885–889

Goyal AK, Pradhan S, Basistha BC, Sen A (2015) Micropropagation and assessment of genetic fidelity of *Dendrocalamus strictus* (Roxb.) nees using RAPD and ISSR markers. 3 Biotech 5:473–482

Gupta PK, Timmis R, Mascarenhas AF (1991) Field performance of micropropagated forestry species. In Vitro Cell Dev Biol Plant 27:159–164

Hazarika BN (2003) Acclimatization of tissue-cultured plants. Curr Sci 85:1704–1712

Ho CK, Change SH, Chen ZZ (1987) Tissue culture and acclimatization in *Anoectochilus formosanus* Hay. Bull Taiwan For Res Inst 2:83–105

Hossain MA, Khan BM, Azhar Uddin M, Rahman MM (2018) *In vitro* propagation of the giant bamboo *Dendrocalamus giganteus* Munro. World Bamboo Net:1–13. http://worldbamboo.net/

wbcxi/papers/Hossain,%20M.A.,%20B.M.%20Khan,%20M.A.%20Uddin,%20M.M.%20Rahman.pdf
Huang LC, Huang BL (1995) Loss of the species distinguishing trait among regenerated *Bambusa ventricosa* McClure plants. Plant Cell Tissue Organ Cult 42:109–111
Islam SAMN, Rahman MM (2005) Micro-cloning in commercially important six bamboo species for mass propagation and at a large scale. Plant Tissue Cult Biotech 15:103–111
Jiménez VM, Castillo J, Tavares E, Guevara E, Montiel M (2006) *In vitro* propagation of the neotropical giant bamboo, *Guadua angustifolia* Kunth, through axillary shoot proliferation. Plant Cell Tissue Organ Cult 86:389–395
Kabade AU (2009) Studies on refinement of protocols for rapid and mass *in vitro* clonal propagation, evaluation of genetic fidelity and growth performance of bamboo species *Bambusa bambos* (L.) Voss and *Dendrocalamus strictus* (Roxb.) Nees. Ph. D. thesis. FRI University, Dehra Dun, pp 198
Kaladhar DSVGK, Tiwari P, Duppala SK (2017) A rapid *in vitro* micropropagation of *Bambusa vulgaris* using inter-node explant. Int J Life Sci Sci Res 3:1052–1054
Kalia S, Kalia RK, Sharma SK (2004) *In vitro* regeneration of an indigenous bamboo (*Bambusa nutans*) from internode and leaf explant. J Bamboo Rattan 3:217–228
Kapoor P, Rao IU (2006) *In vitro* rhizome induction and plantlet formation from multiple shoots in *Bambusa bambos* var. *gigantea* Bennet and Gaur by using growth regulators and sucrose. Plant Cell Tissue Organ Cult 85:211–217
Kapruwan S, Bakshi M, Kaur M (2014) Rapid *in vitro* propagation of the solid bamboo, *Dendrocalamus strictus* nees, through axillary shoot proliferation. Biotechnol Int 7:58–68
Kumar A (1991) Mass production of field planting stock of *Dendrocalamus strictus* through macroproliferation- a technology. Indian For 117:1046–4052
Lin CS, Cheng MJ, Hsiao HW, Hong PI, Jheng FY, Lin CC, Chang WC (2005) Stamen-less inflorescence proliferation of *Bambusa edulis*. Sci Hortic 507:76–80
Lin CS, Liang CJ, Hsaio HW, Lin MJ, Chang WC (2007) *In vitro* flowering of green and albino *Dendrocalamus latiflorus*. New For 34:177–186
Lin CS, Lin CC, Chang WC (2003) *In vitro* flowering of *Bambusa edulis* and subsequent plantlet survival. Plant Cell Tissue Organ Cult 72:71–78
Lin CS, Lin CC, Chang WC (2004) Effect of thidiazuron on vegetative tissue-derived somatic embryogenesis and flowering of *Bambusa edulis*. Plant Cell Tissue Organ Cult 76:75–82
Lin CS, Tseng MC, Hong PI, Chang WC (2006) Albino inflorescence proliferation of *Dendrocalamus latiflorus*. In Vitro Cell Dev Biol Plant 42:331–335
Lloyd G, McCown BH (1980) Commercially feasible micropropagation of mountain laurel, (*Kalmia latifolia*) by use of shoot tip culture. Int Plant Prop Soc Comb Proc 30:421–427
Malini N, Anandakumar CR (2013) Micropropagation of bamboo (*Bambusa vulgaris*) through nodal segment. Int J For Crop Improve 4:36–39
Mascarenhas AF, Nadgir AC, Thengane SR, Phadke CH, Khuspe SS, Shirgurkar MV, Parasharami VA, Nadgauda RS (1988) Potential application of tissue culture for propagation of *Dendrocalamus strictus*. In: Rao IVR, Gnanaharan R, Shastry CB (eds) Bamboos, current research. Proc. Intl. Bamboo workshop, Cochin, India, pp 159–166
Mishra Y, Patel PK, Yadav S, Shirin F, Ansari SA (2008) A micropropagation system for cloning of *Bambusa tulda* Roxb. Sci Hortic 115:315–318
Mudoi KD, Borthakur M (2009) *In vitro* micropropagation of *Bambusa balcooa* Roxb. Through nodal explants from field-grown culms and scope for up-scaling. Curr Sci 96:962–966
Mudoi KD, Saikia SP, Borthakur M (2014) Effect of nodal positions, seasonal variations, shoot clump and growth regulators on micropropagation of commercially important bamboo, *Bambusa nutans* Wall. Ex. Munro. Afr J Biotechnol 13:1961–1972
Mudoi KD, Saikia SP, Goswami A, Gogoi A, Bora D, Borthakur M (2013) Micropropagation of important bamboos: a review. Afr J Biotechnol 12:2770–2785
Mukunthakumar S, Mathur J, Nair PKK, Mathur SN (1999) Micropropagation of *Dendrocalamus brandisii* Kurz. Using *in vivo* nodal explants. Indian For 125:1239–1243

Murashige T, Skoog F (1962) A revised medium for rapid growth and bioassays with tobacco tissue cultures. Physiol Plant 15:473–497

Nadgauda RS, Parasharami VA, Mascarenhas AF (1990) Precocious flowering and seeding behaviour in tissue-cultured bamboos. Nature 344:335–336

Nadgir AL, Phadke CH, Gupta PK, Parsharami VA, Nair S, Mascarenhas AF (1984) Rapid multiplication of bamboo by tissue culture. Silvae Genet 33:219–223

Nadha HK, Kumar R, Sharma RK, Anand M, Sood A (2013) *In vitro* propagation of *Dendrocalamus asper* and testing the clonal fidelity using RAPD and ISSR marker. Int J Curr Res 5:2060–2067

Ndiaye A, Diallo MS, Niang D, Gassama-dia YK (2006) *In vitro* regeneration of adult trees of *Bambusa vulgaris*. Afr J Biotech 5:1245–1248

Negi D, Saxena S (2011) *In vitro* propagation of *Bambusa nutans* wall ex Munro through axillary shoot proliferation. Plant Biotechnol Rep 5:35–43

Nurhayani S, Megia R, Purnamaningsih R (2018) *In vitro* propagation of *Bambusa balcooa* as alternative material of wood. Biosaintifika 10:198–204

Ogita S (2005) Callus and cell suspension culture of bamboo plant, *Phyllostachys nigra*. Plant Biotechnol 22:119–125

Ogita S, Ohki S, Kato Y (2008) Uptake of carbohydrates by suspension cultured cells of bamboo plants. In: da Silva JAT (ed) Floriculture, ornamental and plant biotechnology: advances and tropical issues, vol 5. Global Science Books, Isleworth, pp 240–244

Pandey BN, Singh NB (2012) Micropropagation of *Dendrocalamus strictus* nees from mature nodal explants. J Appl Nat Sci 4:5–9

Patel B, Gami B, Patel N, Baria V (2015) One step pre-hardening micropropagation of *Bambusa balcooa* Roxb. J Phytology 7:1–9

Rajput BS, Jani MD, Sasikumar K, Manokari KM, Shekhawat MS (2019) An improved micropropagation protocol for mango bamboo - *Pseudoxytenanthera stocksii* (Munro) T.Q. Nguyen. World News Natural Sci 25:141–154

Raju R, Roy SK (2016) Mass propagation of *Bambusa bambos* (L.) Voss through *in vitro* culture. Jahangirnagar Univ J Biol Sci 5:15–26

Ramanayake SMSD, Maddegoda KMMN, Vitharana MC, Chaturani GDG (2008) Root induction in three species of bamboo with different rooting abilities. Sci Hortic 118:270–273

Ramanayake SMSD, Meemaduma VN, Weerawardene TE (2006) *In vitro* shoot proliferation and enhancement of rooting for the large-scale propagation of yellow bamboo (*Bambusa vul*garis Striata). Sci Hortic 110:109–113

Ramanayake SMSD, Yakandawala K (1997) Micropropagation of the giant bamboo (*Dendrocalamus giganteus* Munro) from nodal explants of field grown culms. Plant Sci 129:213–223

Ravikumar R, Ananthakrishnan G, Kathiravan K, Ganapathi A (1998) *In vitro* shoot propagation of *Dendrocalamus strictus* Nees. Plant Cell Tissue Organ Cult 52:189–192

Rout GR, Das P (1994) Somatic embryogenesis and *in vitro* flowering of 3 species of bamboo. Plant Cell Rep 13:683–686

Saini H, Arya ID, Arya S, Sharma R (2016) *In vitro* micropropagation of Himalayan weeping bamboo, *Drepanostachyum falcatum*. Am J Plant Sci 7:1317–1324

Sanjaya, Rathore TS, Ravishankar Rai V (2005) Micropropagation of *Pseudoxytenanthera stocksii* Munro. In Vitro Cell Dev Biol Plant 41:333–337

Sawsan SS, Hanan ST, Taie AA, Lobna (2010) Micropropagation antioxidant activity, total phenolic and flavonoids content of *Gardenia jasminoides* Ellis as affected by growth regulators. Int J Acad Res 2:184–191

Saxena S (1990) *In vitro* propagation of bamboo (*Bambusa tulda* Roxb.) through shoot proliferation. Plant Cell Rep 9:431–434

Saxena S, Bhojwani SS (1993) *In vitro* clonal multiplication of four year old plants of the *Dendrocalamus longispathus* Kurz. In Vitro Cell Dev Biol Plant 290:135–142

Saxena S, Dhawan V (1999) Regeneration and large-scale propagation of *bamboo* (*Dendrocalamus strictus* Nees) through somatic embryogenesis. Plant Cell Rep 18:438–443

Sharma P, Sarma KP (2014) *In vitro* propagation of *Bambusa nutans* in commercial scale in Assam, India. J Environ Res Dev 9:348–355

Shirgurkar MV, Thengane SR, Poonawala IS, Jana MM, Nadgauda RS, Mascarenhas AF (1996) A simple *in vitro* method of propagation and rhizome formation in *Dendrocalamus strictus* Nees. Curr Sci 70:940–943

Shirin F, Rana PK (2007) *In vitro* plantlet regeneration from nodal explants of field grown culms in *Bambusa glaucescens* wild. Plant Biotech Rep 1:141–147

Singh SR, Dalal S, Singh R, Dhawan AK, Kalia RK (2012) Seasonal influences on *in vitro* bud break in *Dendrocalamus hamiltonii* Arn. Ex Munro nodal explants and effect of culture microenvironment on large scale shoot multiplication and plantlet regeneration. Indian J Plant Physiol 17:9–21

Somashekar PV, Rathore TS, Shashidhar KS (2008) Rapid and simplified method of micropropagation of *Pseudoxytenanthera stocksii*. In: Ansari SA, Narayanan C, Mandal AK (eds) Forest biotechnology in India. Satish Serial Publishing House, Delhi, pp 165–182

Sood A, Ahuja PS, Sharma OP, Godbole S (2002) *In vitro* protocols and field performance of elites of an important bamboo *Dendrocalamus hamiltonii* Nees et Arn. Ex. Munro. Plant Cell Tissue Organ Cult 71:55–63

Soukup A, Mala J, Hrubcova M, Kalal J, Votrubova O, Cvikrova M (2004) Differences in anatomical structure and lignin content of roots of pedunculate oak and wild cherry-tree plantlets during acclimation. Biol Plant 48:481–489

Suwannamek A (1992) *In vitro* culture of *Dendrocalamus asper* backer. Kasetsart University, Bangkok (Thailand). Faculty of Agriculture

Venkatachalam P, Kalaiarasi K, Sreeramanan S (2015) Influence of plant growth regulators (PGRs) and various additives on *in vitro* plant propagation of *Bambusa arundinacea* (Retz.) Wild: a recalcitrant bamboo species. J Genet Eng Biotech 13:193–200

Waikhom SD, Louis B (2014) An effective protocol for micropropagation of edible bamboo species (*Bambusa tulda* and *Melocanna baccifera*) through nodal culture. Hindawi publishing corporation. Sci World J 2014:345794. https://doi.org/10.1155/2014/345794

Wei CC, Tien HL (1995) Somatic embryogenesis and plant regeneration from roots of bamboo (*Bambusa beecheyana* Munro var *beecheyana*). J Plant Physiol 145:535–538

Wei Q, Cao J, Qian W, Xu M, Li Z, Ding Y (2015) Establishment of an efficient micropropagation and callus regeneration system from the axillary buds of *Bambusa ventricosa*. Plant Cell Tissue Organ Cult 122:1–8

White PR (1963) The cultivation of animal and plant cells. The Ronald Press, New York, p 228

Yasodha R, Kamala S, Ananda Kumar SP, Durai Kumar P, Kalaiarasi K (2008) Effect of glucose on *in vitro* rooting of mature plants of *Bambusa nutans*. Sci Hortic 116:113–116

Yasodha R, Sumathi R, Malliga P, Gurumurthi K (1997) Genetic enhancement and mass production of quality propagules of *Bambusa nutans* and *Dendrocalamus membranaceus*. Indian For 123:303–306

Ye S, Cai C, Ren H, Wang W, Xiang M, Tang X, Zhu C, Yin T, Zhang L, Zhu Q (2017) An efficient plant regeneration and transformation system of Ma bamboo (*Dendrocalamus latiflorus* Munro) started from young shoot as explant. Front Plant Sci 8:1298

Ziv M (1995) The control of bioreactor environment for plant propagation in liquid culture. Acta Hortic 393:25–38

Chapter 9
Management of Bamboo Genetic Resources and Clonal Production Systems

Enéas Ricardo Konzen, Denys Matheus Santana Costa Souza, Sérgio Bruno Fernandes, Gilvano Ebling Brondani, Dulcinéia de Carvalho, and Wellington Ferreira Campos

Abstract Bamboos comprise more than 1600 species distributed throughout the world, according to recent updates. They are used for a wide range of applications such as utensils, charcoal, cellulose extraction, and hybrid microfibers. Bamboos also have an important ecological role in their environment of origin. Therefore, it is critical to recognize, catalogue, and investigate the genetic resources available for the distinct species. Establishing traits of economic value and their heritability is an important step toward breeding programs. Due to the limited sexual reproduction and their natural ability for vegetative propagation, several techniques have been developed for the propagation of genotypes of interest. We provide a brief panorama of methods used for clonal propagation of bamboo species, such as culm-cuttings and culm-layering. In vitro strategies that have been optimized for bamboo propagation is also described. Finally, we outline the significance and application of molecular markers into the management of bamboos. As components of natural communities and as sustainable alternatives for large-scale economic applications, with the advantage of fast growth and high biomass production, conservation and breeding of bamboos certainly must be given attention for future endeavors.

Keywords Genetic resources · Clonal propagation · In vitro culture · Molecular markers

E. R. Konzen (✉)
Center for Coastal, Lymnological and Marine Studies, North Coastal Campus, Federal University of Rio Grande do Sul, Imbé, RS, Brazil

D. M. S. C. Souza · S. B. Fernandes · G. E. Brondani (✉) · D. de Carvalho
Department of Forestry Sciences, Federal University of Lavras, Lavras, MG, Brazil

W. F. Campos
Federal University of Jequitinhonha and Mucuri Valleys, Institute of Agricultural Sciences, Unaí, MG, Brazil

Z. Ahmad et al. (eds.), *Biotechnological Advances in Bamboo*,
https://doi.org/10.1007/978-981-16-1310-4_9

9.1 Introduction

Bamboos comprise a large subfamily of grasses belonging to the Poaceae family, Bambusoideae. Estimates of the number of species vary in literature and have been the object of considerable debate. According to the Bamboo Phylogeny Group (2012), the diversity of bamboos encompasses 1439 species distributed in 115 genera. A further compilation, though, indicated 1662 species distributed in 121 genera (Canavan et al. 2017). Their distribution is worldwide, with species occurring from approximately 46° N latitude to 47° S latitude and from sea level to 4300 meters in elevation (Bamboo Biodiversity, https://www.eeob.iastate.edu/research/bamboo/maps.html). Naturally, they occur in numerous biogeographic regions in the world, encompassing neotropical woody bamboos, north temperate woody bamboos, paleotropic woody bamboos, and herbaceous bamboos (Yeasmin et al. 2015, http://www.eeob.iastate.edu/bamboo/maps.html). Bamboos have also been introduced to various areas, outside of their native range, mainly due to their economic prospects and uses (Canavan et al. 2017). For example, in Brazil, species such as *Bambusa vulgaris* and *Dendrocalamus asper* have been introduced from southeastern Asia.

Several species of bamboos have been used by human populations for a wide range of applications. Two distinct directions of uses could be considered when we refer to the purposes of this chapter in describing genetic resources and their management. First is their use either by traditional communities in small-farming systems or by using resources directly from bamboo forests. Traditional communities use bamboos in accessories and instruments for daily use. A survey conducted in the Province of Yunnan, China, showed that the communities use their shoots as food, wear bamboo as hats and shoes, use their internodes as cooking utensils, and farm with bamboo tools, among other applications (Yuming et al. 2004). Some tribes in India use bamboos to make split mats, threshing trays, and baskets for their livelihood (Patel 2005). Second, numerous industrial uses are documented. Bamboos are also used for cellulose and paper (Li et al. 2015a). Research has been conducted on their use in the production of bioethanol (Kuttiraja et al. 2013). Their hard culm makes bamboos suitable for construction applications, such as sustaining structures (Van der Lugt et al. 2006) and reinforcement of concrete (Ghavami 2005). Medicinal applications, such as antioxidants, present opportunities to be explored by pharmaceutical companies (Nirmala et al. 2018).

It is also important to highlight the ecological significance of bamboos in their environments of origin. Bamboos are the source of food and habitat for local fauna, such as pandas (Li et al. 2015b) and ants (Arruda et al. 2016). Bamboos are also alternatives for containing soil erosion, land rehabilitation, protection of riverbanks, and carbon dioxide sequestration (Ben-Zhi et al. 2005; Ramakrishnan et al. 2018). Bamboos have the potential to provide key ecosystem services with local and global benefits, as a report by Paudyal et al. (2019) presents in detail. The document divides the ecosystem services of bamboos into several categories, which include examples of their regulating functions in endeavors involved in landscape restoration, sediment retention, carbon sequestration, carbon stock, air quality and local climate regulation, water purification, flood control, groundwater recharge, and moderation of extreme events (Paudyal et al. 2019).

In this chapter, we ought to highlight the importance of the germplasm diversity of bamboos, briefly accounting for the main strategies for the ex situ management of bamboo genetic resources and efforts in their breeding. It is not our goal to provide an exhaustive search of the literature, but to describe a panorama of methods that have been used to propagate and cultivate bamboos by distinct approaches. First, we document greenhouse and outdoor vegetative propagation systems for rapidly obtaining propagules and new plantlets. Furthermore, we highlight the main development in tissue culture and their application to large-scale propagation systems. We also provide a briefing on the application of molecular markers in comprehending the diversity of bamboo taxa and assisting in their conservation, management, and breeding.

9.2 Genetic Resources of Bamboos and Essential Traits for Breeding Purposes

Bamboos, in general, have a natural ability for clonal propagation in their habitats. Serving as a food source or refugee for local fauna, bamboo species take part in ecological processes within their communities. This raises the need for in situ practices upon their conservation. One critical case is the giant pandas from China that rely upon bamboos as their primary sources of food. Conservation of bamboo genetic resources in the natural environment assists the conservation of pandas as well (Li et al. 2015b). In neotropical ecosystems, they are an important source of nesting cavities for ant species, such as shown in the Brazilian Cerrado (Arruda et al. 2016). Therefore, conservation practices must consider integrative approaches that come to the level of ecological communities (Shresta et al. 2013).

The social and economic value of bamboos also drives the attention for ex situ conservation and management practices of bamboos. Germplasm banks are aimed at collecting and maintaining the genetic variability and, therefore, the genotypes to be tested in further breeding strategies. Important traits for selection are culm height, diameter, internodal length, number of culms per clump, solidness, and hollowness of culms (Annapurna et al. 2015). Genetic improvement of such traits must consider flowering and breeding behavior, hybridization potential, ploidy level and genetic variation, and selection of desirable populations and individuals (Thakur et al. 2015). It is important to have genetically divergent genotypes to detect variation (Talha et al. 2017) and estimate the heritability of those traits (Singh et al. 2004; Yang et al. 2009). Correlation among traits (Selvan et al. 2014; Bhandari et al. 2015) is also a valuable approach for indirect selection.

Thakur et al. (2015) discuss that classic breeding in bamboos is generally more limited than with other species, especially due to their rare and long flowering cycles.

In general, most studies rely upon the selection of desirable culms and their posterior clonal propagation through an adequate method. The selection is frequently based on a scoring approach aimed at defining which culms are better suited for a specific purpose, the so-called plus clumps. But the various applications of bamboos require specific raw material for their use. The use of bamboo for architectural purposes requires an adequate ratio between wall thickness and collar diameter; for energy, high calorific powers and low ash content are needed (Thakur et al. 2015), as well the knowledge of the contents of cellulose, hemicellulose, and lignin (Li et al. 2015a, c); for pharmaceutical applications, levels of antioxidants are important (Nirmala et al. 2018).

One suitable approach for the selection of bamboos based on traits of interest is performing provenance tests in specific areas. Eight provenances of *Bambusa chungii* were evaluated for fiber dimensions and chemical composition in China. With the support of the analysis of variance and the extraction of genetic variance components, the variation on these traits showed a correlation with their geographic location. Fiber length showed high heritability (0.84), therefore with potential for genetic gains upon selection (Yang et al. 2009). In an experiment with 18 accessions of *Dendrocalamus strictus* from distinct locations, temperature, and precipitation regimes, high heritability was measured for diameter, height, and number of internodes (97.8%, 88.5%, and 84.9%, respectively) (Singh et al. 2004). High heritability values were also calculated for plant height (97.2%), culm basal diameter (98.8%), and the number of culms (89.2%) of seven distinct species of bamboos in India (Selvan et al. 2014). After the evaluation of an experiment with 20 accessions of *D. strictus*, the heritability coefficients were low for variables such as the number of culms (17.6%) and the number of young shoots (25.2%). Relatively high values were encountered for leaf width (68.4%), leaf length (74.6%), and culm length (77.9%) (Bakshi and Rasool 2015). Heritability values varied from 11.4 to 65.6% after evaluating growth attributes of 100 clumps of *D. stocksii* from Western Ghats of India (Rane et al. 2016). The selection process will provide genotypes for further massive cultivation. Large-scale propagation of desirable genotypes has been achieved by diverse vegetative propagation strategies. The next sections of this chapter succinctly describe methods that have been employed to such a goal.

9.3 Major Strategies for Clonal Propagation of Bamboos: Methods Applied to Greenhouse and Outdoor Environments

Bamboo plantations are aimed at obtaining higher productivity for their specific purposes, e.g., fibers for industrial applications. A combination of several strategies for their proper propagation and cultivation are then required. Implementation of breeding programs, selection and production of elite clones, and improvement of vegetative propagation systems are paramount to that goal (Hossain et al. 2018; Lin et al. 2018; Ribeiro et al. 2020). The development of propagation techniques that comply with industrial-level requirements is necessary to achieve the objectives pursued in the multiplication and preservation of selected plants. These advances can consolidate the bamboo productive chain for the composition of forest plantations, aiming at the production of biomass (Bag et al. 2019).

Bamboos can be propagated by means of seeds (sexually, Fig. 9.1a–d) or by vegetative propagation (asexually, Fig. 9.2a–e). However, there is a deficiency in seed production for most species, in addition to the high genetic variability that may be present in plants originated from seeds (Singh et al. 2013; Ray and Ali 2017a), which require selection before clonal production. To fully exploit the value of superior genotypes, the best option is the vegetative propagation, which can provide numerous advantages, such as uniformity of plantations, resistance to pests and diseases, better control of developmental stages, site-specific adaptations of clones, and increased productivity by area, compared to seedlings (Ye et al. 2017; Trueman et al. 2018; Bag et al. 2019). However, difficulties of survival and rooting of propagules with advanced ontogenetic age are some major disadvantages (Singh et al. 2013; Mishra et al. 2019).

Obtaining plants by means of vegetative propagation has been fundamental in the implementation of most bamboo groves (Gulabrao et al. 2012; Yeasmin et al. 2015). Among the main alternatives for the ex situ propagation and rescue of adult material

Fig. 9.1 Propagation of *Dendrocalamus asper* by seeds. (**a**) Reproductive structure. (**b**) Spontaneous germination of seeds right below the mother tree. (**c**–**d**) Detail of seedlings 60 days after transplantation. Credits of images: Gilvano Ebling Brondani

Fig. 9.2 Propagation of *Bambusa vulgaris* through culm sections. (**a**) General view of vertically disposed culms in pots. (**b**) One culm after approximately 30 days, with sprouting. (**c**) One culm after 45 days, with adventitious rooting. (**d**) Single-node culm vertically oriented after approximately 30 days. (**e**) Single-node culm vertically oriented after approximately 60 days, showing adventitious rooting. Credits of images: Enéas Ricardo Konzen

of bamboo species, clump detachment, cuttings (Fig. 9.2), and layering (Fig. 9.3a–d) stand out (Ntirugulirwa et al. 2013; Ray and Ali 2017a; Mishra et al. 2019). These techniques are efficient for obtaining bamboo plants that have difficulties in sexual propagation (Elbasheer and Raddad 2013; Riberio et al. 2016); however, the yield in terms of plant production is lower than desired at the industrial level. Clump detachment can be done by direct transplant or through rhizomes. In the direct transplanting approach, plants are constituted by a complete stem, with rhizomes, roots, branches, and leaves (Deb et al. 2016; Ray and Ali 2017a). This method can be used on sprawling and deafening bamboo and has high survival and rooting. Rhizome propagation can be carried out with or without the presence of part of the stem (Singh et al. 2013; Hossain et al. 2018). This has greater advantages than the previous method, with regard to material savings, transportation, preparation, and obtainment (Mudoi et al. 2014; Ray and Ali 2017a). Propagation by cuttings (Fig. 9.2) can occur by stem or branch cuttings, being the easiest way to multiply the selected genotypes (Ntirugulirwa et al. 2013; Mishra et al. 2019). In stem-cuttings, plants are formed by a portion of the stem, having several nodes with buds or branches. By branch cuttings, these must be removed from as close as to the main stem. Layering (Fig. 9.3) allows branch rooting without separation of the mother plant. The latter is seldom used but of great importance for vegetative rescuing of adult genotypes, in order to avoid losses of high-value materials, based on limited regrowth and rooting of some bamboo species or clones (Riberio et al. 2016; Aier and Khare 2016). Overall, the most used cloning techniques for tropical bamboos are stem and branch cuttings, rhizome splitting, and layering (Hossain et al. 2018; Mishra et al. 2019). Thus, several studies are reported in the literature with satisfactory results, which may favor the ex situ vegetative propagation of bamboo species (Table 9.1).

Fig. 9.3 Propagation of *Dendrocalamus asper* with layering. (**a**, **b**) Preparation of layers. (**c**) Adventitious rooting after 120 days. (**d**) Plantlet originated from layering, transplanted to a pot. Credits of images: Gilvano Ebling Brondani

Due to the difficulty of producing plants on a large scale, the productive potential of several bamboo species has not been properly explored. Therefore, traditional propagation methods are often ineffective for application on a commercial scale (Setiawati et al. 2018). Among the new systems developed for ex situ clonal propagation of bamboo, the use of mini-cliffs and micro-cliffs can be presented as a method of producing plants on a large scale, providing better conditions of environmental control and with lower production cost (Setiawati et al. 2018; Vale et al. 2019). In addition, the use of this type of system provides considerable gains in relation to the production of plants, especially with regard to rooting rates, improvement of the root system, and reduction of time to plant formation, directly influencing its quality and, consequently, in their field performance (Wendling et al. 2014; Bag et al. 2019).

From the aforementioned techniques, it can be inferred that ex situ vegetative propagation techniques are tools of great potential for the maintenance and continuity of bamboo conservation and breeding programs, as well as for the development of new technologies aimed at increasing productivity. The choice of the production method relies upon the final goal, pondering technical and economic advantages as well as limitations. Such techniques are complementary, being applicable in a certain phase of the cloning program and depending on the financial availability, qualified and trained labor, and structures for its execution, which will influence the decision-making process on the course of action to be undertaken.

Table 9.1 Compilation of the main ex situ propagation strategies for bamboo species

Species	Ex situ cultivation method	Major findings	Reference
Arundinaria alpina	Culm-cuttings	Soaking in water is not effective for sprouting. The middle part of culm is the most suitable for propagation	Ntirugulirwa et al. (2013)
A. alpina	One-node culm-cuttings	In comparison to the polythene-shaded nursery, the greenhouse environment provided better growth	Senyanzobe et al. (2013)
Bambusa balcooa	Culm-cuttings	Two-year-old culm-cuttings with intermittent misting provided better responses	Joshi et al. (2012)
B. balcooa	Culm-cuttings	Binodal culm-cuttings with 12–14 cm in diameter were used for producing clonal plants	Mishra et al. (2019)
B. Nagalandiana	Culm-cuttings and branch-cuttings	Higher rooting in culm-cuttings (70.6%) than in branch-cuttings (60.3%) during summer	Deb et al. (2016)
B. vulgaris	Culm-cuttings, two to three nodes, horizontal setting	Cuttings collected in summer showed the highest sprouting and rooting	Gulabrao et al. (2012)
B. nutans	Single-node culm and culm-cuttings and branch-cuttings, horizontal setting	Better performance of rooting was recorded in culm-cuttings (88.3%) compared to branch-cuttings (46.6%)	Singh et al. (2011)
B. balcooa	Culm-cuttings, two to three nodal, horizontal planting	Better performance of rooting in spring (46.7%) than summer (43.3%) and rainy season (40%)	Gulabrao et al. (2012)
B. tulda	Culm-cuttings, two to three nodal, horizontal planting	Culm-cuttings of *B. tulda* presented adventitious rooting only in may (23.3%)	Singh et al. (2011)
B. vulgaris	Culm-cuttings (one, two, and three nodal segments)	Planting of one- or two-noded cuttings horizontally displayed were superior to other alternatives	Bhol and Nayak (2012)
B. vulgaris	Culm-cuttings	The middle part of the culm is more suitable for propagation, and soaking in water is not effective for sprouting	Ntirugulirwa et al. (2013)
B. vulgaris	One nodal culm-cuttings	Propagation of this species is effective in both greenhouse and shaded house	Senyanzobe et al. (2013)
B. vulgaris	Branch-cuttings	The combination of 80% NAA + 20% CW provided better results for propagation	Setiawati et al. (2018)
D. asper	Branch-cuttings	Almost all rooted cuttings formed rhizome at the base from where new culms developed	Hossain et al. (2018)

(continued)

Table 9.1 (continued)

Species	Ex situ cultivation method	Major findings	Reference
Dendrocalamus hamiltonii	Culm-cuttings, two to three nodal, horizontal setting	Cuttings collected in summer showed the highest sprouting (66.7%) and rooting (56.7%)	Gulabrao et al. (2012)
D. giganteus	Culm-cuttings, two to three nodal, horizontal planting	Maximum rooting found in spring (73.3%) compared with summer and rainy seasons (13.3%)	Gulabrao et al. (2012)
D. strictus	Culm-cuttings, two to three nodal, horizontal planting	Cuttings collected in summer showed the highest sprouting (70%) and rooting (53.3%)	Gulabrao et al. (2012)

9.4 In Vitro Strategies for Bamboo Conservation and Large-Scale Propagation

The technology for producing bamboo plants is in constant development, with new advances expected each year. The challenge to produce raw materials will not only be that of investments in the industrial area but also increase productivity through the selection of new varieties and species. Biotechnology, through clonal propagation, is contributing to the development of bamboo conservation and large-scale production strategies.

Vegetative propagation techniques are the key to clonal silviculture, especially for their effectiveness in capturing the genetic gains obtained from breeding programs. Currently, such strategies are best suited for the production of bamboo plants (Ray and Ali 2017a; Ye et al. 2017), especially for species with limitations in sexual propagation. Vegetative propagation is largely used as it allows the multiplication of plants with superior characteristics, germplasm conservation, and research in general, such as stem cuttings and in vitro cultivation (Sandhu et al. 2017; Hossain et al. 2018; Trueman et al. 2018).

As a strategy for the production of clonal bamboo plants, in vitro cultivation through micropropagation has numerous advantages. The possibility of propagating plants in a short time stands out in the list, along with greater nutritional, environmental, and phytosanitary control; possibility of forming and maintaining clonal gardens; transport of clonal materials over long distances without damage; long-term storage; retention of hybrid vigor; acceleration of clonal propagation programs; possibility of cloning hybrids with high heterosis; as well as rejuvenation and reinvigoration of plant tissues with advanced ontogenetic age (Ray and Ali 2017b; Nogueira et al. 2019; Ornellas et al. 2019). Micropropagation can be efficiently used to obtain plants of economically important species that have difficulties in clonal propagation, as is the case of most bamboo species (Table 9.2).

Micropropagation for bamboo species is an excellent alternative, considering the possibility of solving most of the problems observed in traditional propagation

Table 9.2 Summary of the main in vitro propagation strategies in bamboos up to date

Bamboo species	Details of in vitro cultivation method	Major strategies and findings	Reference
Bambusa balcooa	Morphogenetic competence of buds	In vitro regeneration from mature field-grown axillary buds	Das and Pal (2005)
B. nutans	Proliferation of axillary shoots	Plantlets grown in vitro were acclimatized and subsequently transferred to the field	Negi and Saxena (2011)
B. nutans	Proliferation of nodal explants	Plantlets cultivated in the field achieved 95% of survival	Mudoi et al. (2014)
B. oldhamii	In vitro culture establishment	Asepsis of explants inhibited bacterial and fungal growth, allowing the development of shoots	Pasqualini et al. (2019)
B. oldhamii	Proliferation of axillary shoots	The blue LED light had an important effect on the development of shoots from the explants, as shoot proliferation increased	Silveira et al. (2020)
B. oldhamii	In vitro culture establishment	Bacterial isolate was sensitive to 4 mL L^{-1} of PPMTM by minimum inhibitory concentration test	Thiruvengadam et al. (2011)
B. pallida	Proliferation of axillary shoots	Micropropagated plants achieved a height of 25 cm with 3–4 tillers in a 4-month period	Beena and Rathore (2012)
B. pallida	Proliferation of axillary shoots	Rooted plantlets were well established in the greenhouse with survival superior to 95%	Beena et al. (2012)
B. tulda	Proliferation of axillary shoot	Proliferated shoots were successfully rooted on MS medium (liquid) supplemented with coumarin	Bhadrawale et al. (2017)
B. tulda	Proliferation of nodal explants	MS medium supplemented with IBA and sucrose resulted in abundant rooting (81%)	Waikhom and Louis (2014)
B. vulgaris	Proliferation of axillary shoot	Micropropagated plants achieved a height of 30 cm with 3–4 tillers (shoots) with miniatured rhizomes within 4 months	Baskaran et al. (2014)
B. vulgaris	In vitro culture establishment	Active chlorine is recommended for proper sanitization of explants and micropropagation	Furlan et al. (2018)
B. vulgaris	Proliferation of nodal explants	In vitro regenerated plantlets, after acclimatization, showed 80% of survival when transferred to the field	Malini and Anandakumar (2013)

(continued)

Table 9.2 (continued)

Bamboo species	Details of in vitro cultivation method	Major strategies and findings	Reference
B. vulgaris	Proliferation of axillary shoots	Both rooting induction culture medium and mini-incubator use were effective in enabling adventitious rooting	Ribeiro et al. (2020)
B. Wamin	Proliferation of mature nodal explants	In vitro regenerated plantlets, after acclimatization, showed 80–90% of survival when transferred to the field	Arshad et al. (2005)
Dendrocalamus asper	Shoots and roots from calluses	Simultaneous development of shoot and root from calluses	Ali et al. (2009)
D. asper	In vitro culture establishment	More than 25,000 in vitro grown plants were successfully transferred to the field	Arya and Arya (1997)
D. asper	Regeneration of shoots from immature	After hardening and acclimatization, plantlets were transferred to field and reached 80–90% of survival	Arya et al. (2008)
D. asper	Regeneration via calluses	Rooting plantlets showed a survival of 95% when transferred to the greenhouse	Zang et al. (2019)
D. giganteus	Seed germination and micropropagation	In vitro rooting percentage of shoots reached 86% with MS medium containing 5 mg L^{-1} indole-3-butyric acid (IBA)	Waikhom et al. (2012)
D. hamiltonii	Multiplication and rooting approach	Hardened plants, established in the field, exhibited normal growth even after 2 years	Agnihotri and Nandi (2009)
D. hamiltonii	Multiplication and rooting method	Genetic fidelity was confirmed by RAPD markers advocating clonal propagation of this species through nodal segments	Agnihotri et al. (2009)
D. hamiltonii	Proliferation of mature nodal explants	Regenerated plants showed well developed roots and shoots in the field	Jha et al. (2013)
D. hamiltonii	Proliferation of bud breaks	A multiplication rate of about 5.6-fold with healthy cultures was achieved	Singh et al. (2012)
D. latiflorus	Regeneration from young shoots	Important method for propagation and breeding efforts	Ye et al. (2017)
D. membranaceus	Proliferation of axillary shoots	Micropropagation results with potential to ensure the regeneration of large number of plants in a relatively short time	Brar et al. (2013)
D. strictus	Proliferation of mature nodal explants	Only 20% of rooting was found in one treatment, the culture medium with 5 mg L^{-1} indole butyric acid (IBA)	Pandey and Singh (2012)

(continued)

Table 9.2 (continued)

Bamboo species	Details of in vitro cultivation method	Major strategies and findings	Reference
Drepanostachyum luodianense	Shoot proliferation and callus regeneration	Callus induction and regeneration systems for bamboo will be useful for genetic engineering and multiplication	Lin et al. (2018)
Guadua angustifolia	Proliferation of axillary shoots	The temporary immersion system had better performance than the semisolid medium in shoot multiplication and rhizome growth	Gutiérrez et al. (2016)
G. angustifolia	Proliferation of axillary shoots	Spontaneous rooting of 100% of the explants that produced lateral shoots in micropropagation system	Jiménez et al. (2006)
G. chacoensis	Proliferation of nodal explants	Shading improved survival rates, if compared to those under non-shaded conditions	Ornellas et al. (2019)
G. chaparensis	Proliferation of axillary shoots	In plantlets from micropropagation, the plant height does not influence survival rates	Vale et al. (2019)
G. magna	Proliferation of axillary shoots	Rooted plants exhibited a survival rate of up to 100% in acclimatization	Nogueira et al. (2019)
Melocanna baccifera	Proliferation of nodal explants	BAP and TDZ growth hormones are widely used for shoot multiplication, and IBA, NAA, and IAA are used for rooting	Kant et al. (2009)
Mniochloa abersend	Callus induction and plant regeneration	The established regeneration system provides a promising platform for bamboo gene functional characterization	Zang et al. (2016)
Ochlandra wightii	Proliferation of axillary shoots	Around 880 plantlets could be generated in 9 months from 10 embryos isolated from mature seeds using this protocol	Bejoy et al. (2012)
O. abyssinica	In vitro morphogenesis and plant regeneration	The survival rate of plantlets in greenhouse condition was of 91.67% after 30 days of acclimatization	Diab and Mohamed (2008)
Phyllostachys meyeri	Node culture of seedlings	Shoot elongation with rooting was observed within 8 weeks with successful ex vitro multiplication	Ogita et al. (2008)

(continued)

Table 9.2 (continued)

Bamboo species	Details of in vitro cultivation method	Major strategies and findings	Reference
P. pubescens	Proliferation of axillary shoots	In vitro raised plants were successfully transferred to field conditions	Sood et al. (2014)
Thamnocalamus spathiflorus	Comparison of plants originated from seeds and micropropagation	In vitro propagated plants are also morphologically similar to plants grown from seeds at the same stage	Bag et al. (2000)

Fig. 9.4 Micropropagation steps of *Bambusa vulgaris*. (**a**) Harvest of explants from plantlets cultivated in pots. (**b**) In vitro established explants in liquid culture medium. (**c**) Detail of multiplication and elongation. (**d**) In vitro rooting. (**e**) Detail of acclimatization stage of rooted propagules. (**f**) Details of micropagation through a semi-hydroponic system of the *canaletã* (gutters) type, after 60 days. Credits: (**a**) Anatálya dos Santos Ribeiro, (**b**) Gilvano Ebling Brondani, (**c**) Gilvano Ebling Brondani, (**d**) Douglas Santos Gonçalves, (**e**) Anatálya dos Santos Ribeiro, (**f**) Enéas Ricardo Konzen

(Fig. 9.4a–f). However, obtaining bamboo explants free of contaminants has been challenging, making it difficult to obtain in vitro established propagules (Souza et al. 2018; Furlan et al. 2018; Pasqualini et al. 2019; Ribeiro et al. 2020). Therefore, it is paramount to optimize all steps of the in vitro cultivation of microcuttings (e.g., preparation of explants, in vitro introduction, multiplication, elongation, and rooting). Several technologies have been proposed to optimize the micropropagation process, including innovations in the cultivation environment with new lighting

Fig. 9.5 Temporary immersion bioreactor (TIB) for bamboo propagation. Credits: Alzerina Kene Benmuyal Vieira and Palloma Sales de Assunção

sources, with adjusted spectral quality. In addition, novel solutions have also been proposed for the automation of operations of the culture system and routine procedures, through temporary immersion bioreactors (TIB) (Fig. 9.5) (Riberio et al. 2016; Batista et al. 2018; Carvalho et al. 2019; Miranda et al. 2020).

In general, the objectives of the regeneration protocols for bamboo species based on tissue culture focus on achieving large-scale production of plants for commercial cultivation (e.g., biofactories and nursery grids), providing material for breeding programs and future germplasm banks (Gutiérrez et al. 2016; Nogueira et al. 2019; Silveira et al. 2020). Therefore, the technology of bioreactors (Fig. 9.5) consists of an automated system used in the culture of plant tissues and can be applied to improve the multiplication of plants in laboratories and biofactories (Máximo et al. 2018; Ribeiro et al. 2020). This type of system can optimize in vitro cultivation by automated processes (Mendonça et al. 2016), allowing to obtain gains in biomass and reducing time required for propagation (Máximo et al. 2018), in addition to an increase in plant production per unit of area (Gutiérrez et al. 2016; Nogueira et al. 2019).

Despite all the potential for the composition of commercial bamboo plantations in different regions of the world, there are few protocols established for micropropagation on an industrial scale, considering that the reduced knowledge about clonal propagation methods limits advances in breeding. However, different strategies for the propagation and conservation of germplasm have been used in recent years, aiming to provide a more favorable environment for plant development. Thus, several studies are found with satisfactory results, which can favor the vegetative propagation of bamboo species (Table 9.2).

9.5 Application of Molecular Markers to Bamboo Conservation and Breeding

Molecular markers are important tools in bamboo conservation and breeding, as they provide information on genetic variation within populations (Yang et al. 2012; Tian et al. 2012; Konzen 2014), the degree of genetic differentiation among individuals within populations and among populations (Yang et al. 2012; Tian et al. 2012; Konzen 2014), as well as the differentiation and phylogenetic relationships among distinct taxa (Triplett and Clark 2010; Konzen et al. 2017; Ramakrishnan et al. 2020). Due to the high number of species of bamboos, molecular markers are valorous tools for identification at the species and subspecies levels. Several categories of molecular markers have been explored for genetic and phylogenetic analyses with bamboos, such as isozymes, RAPD (random amplified polymorphic DNA), ISSR (inter-simple sequence repeats), microsatellites (SSR (simple sequence repeats)), and SNP (single-nucleotide polymorphisms), among others (Ramakrishnan et al. 2020).

The use of molecular markers has also important application in the design of in situ and ex situ conservation strategies. Populations of *Dendrocalamus membranaceus*, an important woody bamboo in China, showed significant genetic differentiation based on ISSR markers. The low intensity of sexual reproduction in this species, as well as many other bamboos, requires in situ conservation strategies in such populations. The authors also recommended sampling sufficient genotypes for ex situ conservation (Yang et al. 2012). Similar recommendations were given for *D. giganteus*, one of the largest woody bamboos in the world, after ISSR markers revealed low genetic diversity and high differentiation (84.7%) among populations (Tian et al. 2012).

Higher degrees of differentiation require further attention of conservation biologists and breeders as specific genotypes may be exclusive to some regions. Genotyping with molecular markers is important for designing a plan to sustaining the effective population size in in situ conditions. At the same time, it is important to collect enough germplasm to represent the allelic diversity of the regions of focus. One example applicable to that is populations of lowland bamboo (*Oxytenanthera abyssinica*) in Ethiopia, which showed 38.9% of differentiation from the screening of ISSR markers. Additional population structure analysis provided evidence of the association of genetic diversity with the geographic regions that were sampled. One particular region was considerably different from others in the country, based on the molecular markers (Oumer et al. 2020).

As clonal propagation strategies are aimed at maintaining genetic fidelity to selected clones, molecular markers are frequently used to ascertain the clonal fidelity. One example was a study devoted to the micropropagation of *Guadua magna* and *G. angustifolia*. After hundreds of microcuttings were produced, a screening with ISSR markers identified no polymorphisms among all propagules (Nogueira et al. 2019).

9.6 Conclusion and Future Prospects

The demand for bamboos for multiple purposes requires continued improvements in clonal propagation strategies. Micropropagation and bioreactors figure among the main biotechnological approaches with potential for large-scale clonal production. One of the frequent challenges for potential species that have no established protocols is the initial contamination and oxidation of explants, which must be circumvented for proceeding on the next steps. Several protocols have already been adapted to various bamboo species, but much more is to be done to achieve the massive propagation of desirable genotypes. At the same time, micropropagation and other strategies must be forwarded on the premise of germplasm availability, which justifies proper conservation for the representation of the genetic variation of the species. Moreover, conservation aligns well with the ecosystem services provided by bamboos. This integrative approach is needed in a scenario of climate change, and that requires optimization of sustainable management practices.

Improvements in micropropagation also require a lot of work on root emergence in bamboos. Deeper research is needed to better acquaint scientists on the precise mechanisms of rooting initiation, which come to a systems biology approach. Studies on embryogenesis and cryopreservation are scarcely available for bamboos, which also puts into perspective further studies to achieve such goals. Incentives for researching those and other objectives will provide much more efficient approaches for bamboo propagation.

Acknowledgments We thank the National Council for Scientific and Technological Development (CNPq), the Coordination for Improvement of Higher Education Personnel (CAPES), and the Research Support Foundation of the State of Minas Gerais (FAPEMIG) for their financial support and scholarship.

Author Contribution ERK and GEB conceived the general idea of the chapter. ERK, DC, and WFC contributed with the sections of genetic resources and molecular markers. DMSCS, SBF, GEB, and ERK described the main strategies for bamboo propagation. All the authors read, edited, and approved the final version of this chapter.

Conflicts of Interest The authors declare that they have no conflicts of interest.

References

Agnihotri RK, Mishra J, Nandi SK (2009) Improved in vitro shoot multiplication and rooting of *Dendrocalamus hamiltonii* Nees et Arn. Ex Munro: production of genetically uniform plants and field evaluation. Acta Physiol Plant 31(5):961–967. https://doi.org/10.1007/s11738-009-0311-6

Agnihotri RK, Nandi SK (2009) In vitro shoot cut: a high frequency multiplication and rooting method in the bamboo *Dendrocalamus hamiltonii*. Biotechnology 8(2):259–263. https://doi.org/10.3923/biotech.2009.259.263

Aier S, Khare N (2016) Effect on the growth of *Bambusa balcooa* and *Bambusa vulgaris* var. *vitatta* due to type of nursery bed. JIARM 4(7):85–91

Ali AH, Nirmala C, Badal T, Sharma ML (2009) In-vitro organogenesis and simultaneous formation of shoots and roots from callus in *Dendrocalamus asper*. In: 8th World Bamboo Congress, Thailand, pp 31–40

Annapurna D, Muyeed AS, Viswanath S (2015) Morphological and genetic diversity analysis in a germplasm bank of *Dendrocalamus stocksii* (Munro.) - implications on conservation. IJMEC 5 (3):1–8. https://doi.org/10.5376/ijmec.2015.05.0003

Arruda FV, Pesquero MA, Marcelino DG, Leite GA, Delabie JHC, Fagundes R (2016) Size and condition of bamboo as structural factors behind the vertical stratification of the bamboo-nesting ant community. Insect Soc 63(1):99–107. https://doi.org/10.1007/s00040-015-0440-4

Arshad SM, Kumar A, Bhatnagar SK (2005) Micropropagation of *Bambusa wamin* through proliferation of mature nodal explants. J Biol Res 3:59–66

Arya ID, Arya S (1997) In vitro culture establishment of exotic bamboo *Dendrocalamus asper*. IJEB 35:1252–1255

Arya S, Satsangi R, Arya ID (2008) Direct regeneration of shoots from immature inflorescences in *Dendrocalamus asper* (edible bamboo) leading to mass propagation. J Bamboo Sci Cult 21 (1):14–20

Bag N, Chandra S, Palni LMS, Nandi SK (2000) Micropropagation of dev-ringal [*Thamnocalamus spathiflorus* (Trin.) Munro]—a temperate bamboo, and comparison between in vitro propagated plants and seedlings. Plant Sci 156(2):125–135. https://doi.org/10.1016/S0168-9452(00)00212-0

Bag N, Nandi SK, Palni LMS (2019) An efficient method for acclimatization: in vitro hardening of tissue culture- raised tea plants (*Camellia sinensis* (l.) O. Kuntze). Curr Sci 117:288–293. https://doi.org/10.18520/cs/v117/i2/288-293

Bakshi M, Rasool S (2015) Genetic variability in different accessions of *Dendrocalamus strictus* (Roxb.) Nees as assessed through culm characteristics. In: 10th World Bamboo Congress, Korea (pp. 1–16).

Bamboo Phylogeny Group (2012) An updated tribal and subtribal classification of the bamboos (Poaceae: Bambusoideae). J Am Bamboo Soc 24:1–10

Baskaran B, Rathore TS, Thara SKJ (2014) In vitro propagation of *Bambusa vulgaris* cv. *wamin* by axillary shoot proliferation. Int J Fundam Appl Sci 3:53–58

Batista DS, Felipe SHS, Silva TD, de Castro KM, Mamedes-Rodrigues TC, Miranda NA, Ríos-Ríos AM, Faria DV, Fortini EA, Chagas K, Torres-Silva G, Xavier A, Arencibia AD, Otoni WC

(2018) Light quality in plant tissue culture: does it matter? In Vitro Cell Dev Biol Plant 54 (3):195–215. https://doi.org/10.1007/s11627-018-9902-5

Beena DB, Rathore TS (2012) In vitro cloning of *Bambusa pallida* Munro through axillary shoot proliferation and evaluation of genetic fidelity by random amplified polymorphic DNA markers. Int J Plant Biol 3(1):27–33. https://doi.org/10.4081/pb.2012.e6

Beena DB, Rathore TS, Rao PS (2012) Effects of carbohydrates on in vitro axillary shoot initiation and multiplication of *Bambusa pallida* Munro. J Phytology 4:55–58

Bejoy M, Anish NP, Radhika BJ, Nair GM (2012) In vitro propagation of *Ochlandra wightii* (Munro) Fisch.: an endemic reed of Southern Western Ghats. Indian J Biotechnol 11:67–73. https://doi.org/10.3923/biotech.2012.67.73

Ben-zhi Z, Mao-yi F, Jin-zhong X, Xiao-sheng Y, Zheng-cai L (2005) Ecological functions of bamboo forest: research and application. J For Res 16(2):143–147. https://doi.org/10.1007/BF02857909

Bhadrawale D, Mishra JP, Mishra Y (2017) An improvised in vitro vegetative propagation technique for *Bambusa tulda*: influence of season, sterilization and hormones. J For Res 29 (4):1069–1074. https://doi.org/10.1007/s11676-017-0569-2

Bhandari MS, Kaushal R, Banik RL, Tewari SK (2015) Genetic evaluation of nutritional and fodder quality of different bamboo species. Indian For 141(3):265–274

Bhol N, Nayak H (2012) Effect of planting alignment and cutting size on propagation of *Bambusa vulgaris*. J Tree Sci 31(1&2):69–75

Brar J, Anand M, Sood A (2013) In vitro seed germination of economically important edible bamboo *Dendrocalamus membranaceus* Munro. Ind J Exp Biol 51:88–96

Canavan S, Richardson DM, Visser V, Roux JJL, Vorontsova MS, Wilson JRU (2017) The global distribution of bamboos: assessing correlates of introduction and invasion. AoB Plants 9: plw078. https://doi.org/10.1093/aobpla/plw078

Carvalho LSO, Ozudogru EA, Lambardi M, Paiva LV (2019) Temporary immersion system for micropropagation of tree species: a bibliographic and systematic review. Not Bot Hortic Agrobo 47(2):269–277. https://doi.org/10.15835/nbha47111305

Das M, Pal A (2005) In vitro regeneration of *Bambusa balcooa* Roxb.: factors affecting changes of morphogenetic competence in the axillary buds. Plant Cell Tissue Organ Cult 81(1):109–112. https://doi.org/10.1007/s11240-004-3017-x

Deb CR, Sangtam TL, Saku NS (2016) Clonal macropropagation of *Bambusa nagalandiana* Naithani through culm segments and branch cutting: An endemic bamboo of Nagaland, India. Int J Conserv Sci 7(1):147–154

Diab EEE, Mohamed SE (2008) In vitro morphogenesis and plant regeneration of bamboos (*Oxytenanthera abyssinica* a. rich. Munro). Int J Crop Prod 3:72–79. https://doi.org/10.9734/BJI/2017/32715

Elbasheer YHA, Raddad EAY (2013) Vegetative propagation of (*Oxytenanthera abyssinica*) by culm cuttings. J Nat Resour Environ Stud 1(3):1–5

Furlan FC, Gavilan NH, Zorz AZ, de Oliveira LS, Konzen ER, Brondani GE (2018) Active chlorine and charcoal affect the in vitro culture of *Bambusa vulgaris*. Bosque 39(1):61–70. https://doi.org/10.4067/S0717-92002018000100061

Ghavami K (2005) Bamboo as reinforcement in structural concrete elements. Cem Concr Compos 27(6):637–649. https://doi.org/10.1016/j.cemconcomp.2004.06.002

Gulabrao AY, Kaushal R, Tewari KS, Tomar SMJ, Chaturvedi PO (2012) Seasonal effect on rooting behaviour of important bamboo species by culm cuttings. J For Res 23:441–445. https://doi.org/10.1007/s11676-012-0282-0

Gutiérrez LG, Lopez-Franco R, Morales-Pinzon T (2016) Micropropagation of *Guadua angustifolia* Kunth (Poaceae) using a temporary immersion system RITA. Afr J Biotechnol 15(28):1503–1510. https://doi.org/10.5897/AJB2016.15390

Hossain MA, Kumar SM, Seca G, Maheran AA, Nor-Aini AS (2018) Mass propagation of *Dendrocalamus asper* by branch cutting. J Trop For Sci 30(1):82–88. https://doi.org/10.26525/jtfs2018.30.1.8288

Jha A, Das S, Kumar B (2013) Micropropagation of *Dendrocalamus hamiltonii* through nodal explants. IJBB 2:580–582

Jiménez VM, Castillo J, Tavares E, Guevara E, Montiel M (2006) In vitro propagation of the neotropical giant bamboo, *Guadua angustifolia* Kunth, through axillary shoot proliferation. Plant Cell Tiss Organ Cult 86(3):389–395. https://doi.org/10.1007/s11240-006-9120-4

Joshi R, Tewari SK, Kaushal R, Tewari L (2012) Rooting behaviour of *Bambusa balcooa* Roxb. In relation to season, age and growing conditions. Indian For 138(1):79–83

Kant A, Arya S, Arya ID (2009) Micropropagation protocol for *Melocanna baccifera* using nodal explants from mature clump. In: 8th World Bamboo Congress, Thailand, pp 2–12

Konzen ER (2014) Towards conservation strategies for forest tree endangered species: the meaning of population genetic statistics. Adv For Sci 1(1):45–51. https://doi.org/10.34062/afs.v1i1.1405

Konzen ER, Peron R, Ito MA, Brondani GE, Tsai SM (2017) Molecular identification of bamboo genera and species based on RAPD-RFLP markers. Silva Fenn 51(4):1691. https://doi.org/10.14214/sf.1691

Kuttiraja M, Sindhu R, Varghese PE, Sandhya SV, Binod P, Vani S, Pandey A, Sukumaran RK (2013) Bioethanol production from bamboo (*Dendrocalamus* sp.) process waste. Biomass Bioenerg 59:142–150. https://doi.org/10.1016/j.biombioe.2013.10.015

Li G, Fu S, Zhou A, Zhan H (2015a) Improved cellulose yield in the production of dissolving pulp from bamboo using acetic acid in Prehydrolysis. Bioresources 10(1):877–886. https://doi.org/10.15376/biores.10.1.877-886

Li R, Xu M, Wong MHG, Qiu S, Sheng Q, Li X, Song Z (2015b) Climate change-induced decline in bamboo habitats and species diversity: implications for giant panda conservation. Divers Distrib 21(4):379–391. https://doi.org/10.1111/ddi.12284

Li X, Sun C, Zhou B, He Y (2015c) Determination of hemicellulose, cellulose and lignin in Moso bamboo by near infrared spectroscopy. Sci Rep 5(1):17210. https://doi.org/10.1038/srep17210

Lin S, Liu G, Guo T, Zhang L, Wang S, Ding Y (2018) Shoot proliferation and callus regeneration from nodular buds of *Drepanostachyum luodianense*. J For Res 30(6):1997–2005. https://doi.org/10.1007/s11676-018-0772-9

Malini N, Anandakumar CR (2013) Micropropagation of bamboo (*Bambusa vulgaris*) through nodal segment. Int J For Crop Improv 4:36–39

Máximo WPF, Santos PAA, Martins GS, Mendonça EG, Paiva LV, Máximo WPF, Santos PAA, Martins GS, Mendonça EG, Paiva LV (2018) In vitro multiplication of *Eucalyptus* hybrid via temporary immersion bioreactor: culture media and cytokinin effects. Crop Breed Appl Biotechnol 18(2):131–138. https://doi.org/10.1590/1984-70332018v18n2a19

Mendonça EG, Stein VC, Carvalho HHD, Santos BR, Beijo LA, Paiva LV (2016) Uso de biorreatores de imersão continua, temporária e meio de cultura semissólido na produção de clones de *Eucalyptus camaldulensis*. Cienc Florest 26(4):1211–1224. https://doi.org/10.5902/1980509825112

Miranda NA, Xavier A, Otoni WC, Gallo R, Gatti KC, de Moura LC, Souza DMSC, Maggioni JH, Santos SSO (2020) Quality and intensity of light in the in vitro development of microstumps of *Eucalyptus urophylla* in a photoautotrophic system. For Sci 66:1–7. https://doi.org/10.1093/forsci/fxaa027

Mishra JP, Bhadrawale D, Rana PK, Mishra Y (2019) Evaluation of cutting diameter and hormones for clonal propagation of *Bambusa balcooa* Roxb. J For Res 24(5):320–324. https://doi.org/10.1080/13416979.2019.1658699

Mudoi KD, Saikia SP, Borthakur M (2014) Effect of nodal positions, seasonal variations, shoot clump and growth regulators on micropropagation of commercially important bamboo, *Bambusa nutans* Wall. ex. Munro. Afr J Biotechnol 13(19):1961–1972. https://doi.org/10.5897/AJB2014.13659

Negi D, Saxena S (2011) In vitro propagation of *Bambusa nutans* Wall. ex Munro through axillary shoot proliferation. Plant Biotechnol Rep 5(1):35–43. https://doi.org/10.1007/s11816-010-0154-z

Nirmala C, Bisht MS, Bajwa HK, Santosh O (2018) Bamboo: a rich source of natural antioxidants and its applications in the food and pharmaceutical industry. Trends Food Sci Technol 77:91–99. https://doi.org/10.1016/j.tifs.2018.05.003

Nogueira JS, Gomes HT, Scherwinski-Pereira JE, Nogueira JS, Gomes HT, Scherwinski-Pereira JE (2019) Micropropagation, plantlets production estimation and ISSR marker-based genetic fidelity analysis of *Guadua magna* and *G. angustifolia*. Pesqui Agropecu Trop 49:e53743. https://doi.org/10.1590/1983-40632019v4953743

Ntirugulirwa B, Asiimwe T, Gapusi J, Mutaganda A, Nkuba G, Ruzindana A, Ntabana D, Barnabé B, Kahia J, Gahakwa D (2013) Influence of bud position on mother stem and soaking duration on sprouting of bamboo cuttings. Rwanda J 28(1):3–10. https://doi.org/10.4314/rj.v28i1.1

Ogita S, Kashiwagi H, Kato Y (2008) In vitro node culture of seedlings in bamboo plant, *Phyllostachys meyeri* McClure. Plant Biotechnol J 25(4):381–385. https://doi.org/10.5511/plantbiotechnology.25.381

Ornellas TS, Marchetti CK, Oliveira GH, Fritsche Y, Guerra MP, Ornellas TS, Marchetti CK, Oliveira GH, Fritsche Y, Guerra MP (2019) Micropropagation of *Guadua chacoensis* (Rojas) Londoño & P. M. Peterson1. Pesqui Agropecu Trop 49:e55450. https://doi.org/10.1590/1983-40632019v4955450

Oumer OA, Dagne K, Feyissa T, Tesfaye K, Durai J, Hyder MZ (2020) Genetic diversity, population structure, and gene flow analysis of lowland bamboo [*Oxytenanthera abyssinica* (A. Rich.) Munro] in Ethiopia. Ecol Evol 10(20):11217–11236. https://doi.org/10.1002/ece3.6762

Pandey BN, Singh NB (2012) Micropropagation of *Dendrocalamus strictus* nees from mature nodal explants. J Appl Nat Sci 4(1):5–9. https://doi.org/10.31018/jans.v4i1.213

Pasqualini APA, Schneider GX, Fraga HPF, Biasi LA, Quoirin M, Pasqualini APA, Schneider GX, Fraga HPF, Biasi LA, Quoirin M (2019) In vitro establishment of *Bambusa oldhamii* Munro from field-grown matrices and molecular identification of endophytic bactéria. Pesqui Agropecu Trop 49:e53673. https://doi.org/10.1590/1983-40632019v4953673

Patel AB (2005) Traditional bamboo uses by the tribes of Gujarat. Indian J Trad Knowl 4 (2):179–184

Paudyal K, Adhikari S, Sharma S, Samsudin YB, Paudyal BR, Bhandari A, Birhane E, Darche G, Trinh TL, Baral H (2019) Framework for assessing ecosystem services from bamboo forests: Lessons from Asia and Africa. Working Paper 255. Bogor, Indonesia: CIFOR. doi:https://doi.org/10.17528/cifor/007433

Ramakrishnan M, Yrjälä K, Vinod KK, Sharma A, Cho J, Satheesh V, Zhou M (2020) Genetics and genomics of moso bamboo (*Phyllostachys edulis*): current status, future challenges, and biotechnological opportunities toward a sustainable bamboo industry. Food Energy Secur:e229. https://doi.org/10.1002/fes3.229

Ramakrishnan M, Zhou M, Baskar K, Packiam SM (2018) Role of bamboo in ecosystem. Austin J Environ Toxicol 4(1):1023

Rane AD, Chandramouli S, Viswanath S (2016) Can *Dendrocalamus stocksii* (Munro.) be the ideal multipurpose bamboo species for domestication in Peninsular India. J Bamboo Rattan 15 (1/4):23–32

Ray SS, Ali MN (2017a) Factors affecting macropropagation of bamboo with special reference to culm cuttings: a review update. NZ J For Sci 47(1):17. https://doi.org/10.1186/s40490-017-0097-z

Ray SS, Ali N (2017b) Biotic contamination and possible ways of sterilization: a review with reference to bamboo micropropagation. Braz Arch Biol Technol 60:e16160485. https://doi.org/10.1590/1678-4324-2016160485

Ribeiro AS, Figueiredo AJR, Tormen GCR, Silva ALL, Campos WF, Brondani GE (2020) Clonal bamboo production based on in vitro culture. Biosci J 36:4. https://doi.org/10.14393/BJ-v36n4a2020-48169

Riberio AS, Brondani GE, Tormen GCR, Figueiredo AJR (2016) Cultivo in vitro de bambu em diferentes sistemas de propagação. Nativa 4(1):15–18. https://doi.org/10.31413/nativa.v4i1.3241

Sandhu M, Wani SH, Jiménez VM (2017) In vitro propagation of bamboo species through axillary shoot proliferation: a review. Plant Cell Tiss Organ Cult 132(1):27–53. https://doi.org/10.1007/s11240-017-1325-1

Selvan RT, Parthiban KT, Kanna SU, Saravanan V, Palanikumaran B (2014) Biometric and genetic studies in bamboo genetic resources. J Agric Nat Resour 1(4):8

Senyanzobe J, Jennifer R, Grace M, Faustin N, Gaudence M (2013) Growth of *Bambusa vulgaris* and *Arundinaria alpina* under different nursery site conditions at the higher institute of agriculture and animal husbandry, Northern Rwanda. J Biodivers Environ Sci 3(9):9–14

Setiawati T, Keliat APR, Budiono R, Partasasmita R, Iskandar J (2018) Influence of NAA and coconut water with variation of soaking duration on the growth of yellow bamboo branch cutting. Nusantara Biosci 10(3):178–182. https://doi.org/10.13057/nusbiosci/n100308

Shresta PK, Shresta P, Subedi A, Peroni N, de Boef WS (2013) Community biodiversity management: defined and contextualized. In: de Boef WS, Subedi A, Peroni N, Thijssen M, O'Keeffe E (eds) Community biodiversity management, Routledge, pp 53–59

Silveira AAC, Lopes FJF, Sibov ST (2020) Micropropagation of *Bambusa oldhamii* Munro in heterotrophic, mixotrophic and photomixotrophic systems. Plant Cell Tiss Organ Cult 141(2):315–326. https://doi.org/10.1007/s11240-020-01788-4

Singh P, Srivastava SK, Kumari N, Srivastava RJ, Dubey P (2004) Genetic variability and selection parameters in *Dendrocalamus strictus*. J Trop For Sci 16(4):413–418

Singh S, Yadav S, Patel PK, Ansari SA (2011) Adventitious rhizogenesis in *Bambusa nutans* and *Bambusa tulda*: influence of seasonal variation, IBA and cutting type. J For Res 22(4):693–696. https://doi.org/10.1007/s11676-011-0170-z

Singh SR, Dalal S, Singh R, Dhawan AK, Kalia RK (2012) Seasonal influences on in vitro bud break in *Dendrocalamus hamiltonii* arn. Ex munro nodal explants and effect of culture microenvironment on large scale shoot multiplication and plantlet regeneration. Indian J Plant Physiol 17(1):13

Singh SR, Singh R, Kalia S, Dalal S, Dhawan AK, Kalia RK (2013) Limitations, progress and prospects of application of biotechnological tools in improvement of bamboo—a plant with extraordinary qualities. Physiol Mol Biol Plants 19(1):21–41. https://doi.org/10.1007/s12298-012-0147-1

Sood A, Nadha HK, Sood S, Walia S, Parkash O (2014) Large scale propagation of an exotic edible bamboo, *Phyllostachys pubescens* Mazel ex H. De Lehale (Moso Bamboo) using seeds. Indian J Exp Biol 52(7):755–758

Souza DMSC, Xavier A, Otoni WC, Miranda NA, Maggioni JH, Souza DMSC, Xavier A, Otoni WC, Miranda NA, Maggioni JH (2018) Light quality in the in vitro introduction of *Corymbia* hybrid clones. Rev Arvore 42:6. https://doi.org/10.1590/1806-90882018000600004

Talha M, Tewari SK, Harsha ASN, Gahtyari NC (2017) Genetic divergence analysis of Indian bamboo species based on morphological characters. J Pharmacogn Phytochem 6(6):2433–2436

Thakur A, Barthwal S, Ginwal HS (2015) Genetic diversity in bamboos: conservation and improvement for productivity. In: Bamboos in India. ENVIS Centre For, Dehradun, pp 131–146

Thiruvengadam M, Rekha KT, Chung IM (2011) Rapid in vitro micropropagation of *Bambusa oldhamii* Munro. Philipp Agric Sci 94:7–13

Tian B, Yang H-Q, Wong K-M, Liu A-Z, Ruan Z-Y (2012) ISSR analysis shows low genetic diversity versus high genetic differentiation for giant bamboo, *Dendrocalamus giganteus* (Poaceae: Bambusoideae), in China populations. Genet Resour Crop Evol 59(5):901–908. https://doi.org/10.1007/s10722-011-9732-3

Triplett JK, Clark LG (2010) Phylogeny of the temperate bamboos (Poaceae: Bambusoideae: Bambuseae) with an emphasis on Arundinaria and allies. Syst Bot 35(1):102–120. https://doi.org/10.1600/036364410790862678

Trueman SJ, Hung CD, Wendling I (2018) Tissue culture of *Corymbia* and *Eucalyptus*. Forests 9 (2):84. https://doi.org/10.3390/f9020084

Vale PAA, Oliveira Júnior JB, Costa FHS, Scherwinski-Pereira JE, Vale PAA, Oliveira Júnior JB, Costa FHS, Scherwinski-Pereira JE (2019) Height and number of shoots on the survival and development of micropropagated bamboo plantlets during pre-acclimatization. Pesqui Agropecu Trop 49:e53751. https://doi.org/10.1590/1983-40632019v4953751

Van der Lugt P, Van den Dobbelsteen AAJF, Janssen JJA (2006) An environmental, economic and practical assessment of bamboo as a building material for supporting structures. Constr Build Mater 20(9):648–656. https://doi.org/10.1016/j.conbuildmat.2005.02.023

Waikhom SD, Bengyella L, Sharma GJ (2012) In vitro seed germination and micropropagation of edible bamboo *Dendrocalamus giganteus* Munro using seeds. Biotechnology 11:74–80. https://doi.org/10.3923/biotech.2012.74.80

Waikhom SD, Louis B (2014) An effective protocol for micropropagation of edible bamboo species (*Bambusa tulda* and *Melocanna baccifera*) through nodal culture. Sci World J 2014:8. https://www.hindawi.com/journals/tswj/2014/345794/

Wendling I, Trueman SJ, Xavier A (2014) Maturation and related aspects in clonal forestry—part I: concepts, regulation and consequences of phase change. New For 45(4):449–471. https://doi.org/10.1007/s11056-014-9421-0

Yang HQ, An MY, Gu ZJ, Tian B (2012) Genetic diversity and differentiation of *Dendrocalamus membranaceus* (Poaceae: Bambusoideae), a declining bamboo species in Yunnan, China, as based on inter-simple sequence repeat (ISSR) analysis. IJMS 13(4):4446–4457

Yang X, Fu M, Xie J, Li Z (2009) Geographic variation and provenance selection for bamboo wood properties in *Bambusa chungii*. J For Res 20(3):261–267. https://doi.org/10.1007/s11676-009-0045-8

Ye S, Cai C, Ren H, Wang W, Xiang M, Tang X, Zhu C, Yin T, Zhang L, Zhu Q (2017) An efficient plant regeneration and transformation system of ma bamboo (*Dendrocalamus latiflorus* Munro) started from young shoot as explant. Front Plant Sci:8. https://doi.org/10.3389/fpls.2017.01298

Yeasmin L, Ali MN, Gantait S, Chakraborty S (2015) Bamboo: an overview on its genetic diversity and characterization. 3 Biotech 5(1):1–11. https://doi.org/10.1007/s13205-014-0201-5

Yuming Y, Kanglin W, Shengji P, Jiming H (2004) Bamboo diversity and traditional uses in Yunnan, China. Mt Res Dev 24(2):157–165. https://doi.org/10.1659/0276-4741(2004)024[0157:BDATUI]2.0.CO;2

Zang Q, Jiao Y, Guo X, Zhuge F, Yeh K, Lin X (2016) Callus induction and plant regeneration from lateral shoots of herbaceous bamboo *Mniochloa abersend*. J Hortic Sci Biotechnol 92 (2):168–174. https://doi.org/10.1080/14620316.2016.1232610

Zang Q, Liu Q, Zhuge F, Wang Z, Lin X (2019) In vitro regeneration via callus induction in *Dendrocalamus asper* (schult.) backer. Propag Ornam Plants 19(3):66–71

Chapter 10
Polymorphism and Phylogenetic Relationships in Bamboo

Irfan Bashir Ganie, Alin Liana, Zishan Ahmad, and Anwar Shahzad

Abstract A breakthrough in plant systematics began to develop at the end of the nineteenth century, since the development of molecular systematics. This method is considered to contribute to supporting the phylogenetic framework in the plant world. Molecular studies are expected to strengthen existing systematics, not replace them. Until the late 1980s, the bamboo classification system was still based on morphological data. In the early 1990s, identification was started using molecular markers. This identification can provide important information in overcoming various taxonomic constraints. It can determine the taxon level of a type appropriately and corresponds to taxonomic data based on morphological characters. Scientists use various molecular markers to look for similarities or differences between species. Some of the molecular markers used are amplified fragment length polymorphism (AFLP), random amplified polymorphic DNA (RAPD), sequence characterized amplified regions (SCARs), start codon targeted (SCoT), inter-primer binding site (iPBS), and simple sequence repeats (SSR). In addition to molecular markers, bamboo taxonomy is also carried out using DNA sequence-based methods. This method includes sequences of organelle genes and nucleus genes. Furthermore, several chloroplast genes were also found to form molecular relationships in Poaceae. This chapter is aiming to provide a piece of up-to-date information on molecular markers applied in different bamboo species to evaluate the genetic relationships.

I. B. Ganie · A. Shahzad
Plant Biotechnology Section, Department of Botany, Aligarh Muslim University, Aligarh, Uttar Pradesh, India

A. Liana (✉)
Biology Laboratory, STKIP Pembangunan Indonesia, Makassar, Indonesia

Z. Ahmad
Bamboo Research Institute, Nanjing Forestry University, Nanjing, Jiangsu, People's Republic of China

Co-Innovation Center for Sustainable Forestry in Southern China, Nanjing Forestry University, Nanjing, China

Z. Ahmad et al. (eds.), *Biotechnological Advances in Bamboo*,
https://doi.org/10.1007/978-981-16-1310-4_10

Keywords Bamboo · Molecular marker · Polymorphism · Phylogenetic relationship

10.1 Introduction

Gregor J. Mendel put forward the idea of the marker during his experiment back in the nineteenth century. He used phenotype-based genetic markers. Again, during experiments on *Drosophila*, phenotype-based genetic markers were used that led to the establishment of theory of genetic linkage (Agarwal et al. 2008). Although these phenotypic-based markers contributed very much in the past, it has certain drawbacks which limit their use in present time. The foremost limitation is the changing environmental conditions which acutely influence and more importantly cover a limited portion of plant genome which is their major drawback (Amom and Nongdam 2017). These limitations caused the invention of more beneficial markers based on DNA which are now known as molecular markers. A molecular marker is demarcated as a specific unit of DNA that is representative of the variances at the genome level. Also, it is not necessary that molecular markers correlate with phenotypic expression of a trait (Agarwal et al. 2008). The idea of DNA-dependent markers has improved our capability several times in understanding even a minute section of the chromosome. Molecular markers are used for various purposes such as genetic variability evaluation, genome fingerprinting genetic, physical mapping of genomes, population genetic studies, and marker-assisted breeding for crop enhancement. A list of molecular markers has been given in Table 10.1. In spite of the advancement in genome sequencing technologies, molecular markers remain to continue to be a crucial mechanism for extensive analyses of the genome, not solely by enabling assembly of the genome but through their verified worth in high-throughput genotyping, comparative and evolutionary genomics, trait mapping, and breeding in plants. Molecular markers find even the slight variation like a single-base variation in the genome which makes them beneficial in detecting DNA polymorphisms usually related to desirable traits and also in detecting and analyzing involved alleles (Hayward et al. 2015).

The characteristic of a perfect marker comprises of having a highly polymorphic nature, codominant inheritance, and a regular presence in the genome, remaining the same in changing environmental condition, and being easily accessible and highly reproducible (Ibrahim et al. 2010). The best-suited marker can select according to their physical characteristic and location in genome, the cost required, the comfort during use, and the amount of throughput essential (Hayward et al. 2015). The primary marker technique that was used for the physical mapping of plant genomes was RFLP. The method is costly and relies on previous sequence data. After the discovery of PCR technique-based marker systems such as RAPD, AFLP, AP-PCR, etc., the existing strains were relieved. These markers are quick and cheap and do not need knowledge of previous sequences. Methods such as RAPD and AFLP are helpful in population genetics study and breeding resolutions. They are also helpful in tagging a phenotypic trait to a genetic component. SCAR system was intended for

Table 10.1 List of molecular markers

Markers	Name	Reference
AFLP	Amplified fragment length polymorphism	Vos et al. (1995)
AMP-PCR	Anchored microsatellite primed PCR	Wolf et al. (1995)
CAPS	Cleaved amplified polymorphic sequence	Michaels and Amasino (1998)
DALP	Direct amplification of length polymorphism	Desmarais et al. (1998)
ASSR	Anchored simple sequence repeats	Wu et al. (1994)
DAMDPCR	Direct amplification of microsatellite DNA by PCR	Heath et al. (1993)
ASA	Allele-specific amplification	Wu et al. (1989)
DArT	Diversity array technology	Jaccoud et al. (2001)
IRAP	Inter-retrotransposon amplified polymorphism	Kalendar et al. (1999)
SSR	Simple sequence repeats	Litt and Luty (1989)
SSAP	Sequence-specific amplification polymorphism	Waugh et al. (1997)
VNTR	Variable number of tandem repeat	Jeffreys et al. (1985)
SPAR	Single primer amplification reactions	Gupta et al. (1994)
STAR	Sequence-tagged amplified region	Rafalski and Tingey (1993)
SSCP	Single-strand conformational polymorphism	Hayashi (1992)
SNP	Single-nucleotide polymorphism	Landegren et al. (1988)
IM-PCR	Inter-microsatellite PCR	Zietkiewicz et al. (1994)
AP-PCR	Arbitrarily primed PCR	Welsh and McClelland (1991)
DAF	DNA amplification fingerprinting	Caetano-Anolles et al. (1991)
IFLP	Inter-fragment length polymorphism	Hongtrakul et al. (1998)
MP-PCR	Microsatellite-primed PCR	Meyer et al. (1993)
MAAP	Multiple arbitrary amplicon profiling	Caetano-Anolles and Gresshoff (1994)
RAHM	Random amplified hybridizing microsatellites	Ciffarelli et al. (1995)
REM	Retrotransposon microsatellite amplified polymorphism	Kalendar et al. (1999)
SCAR	Sequence characterized amplified regions	Michelmore et al. (1991), Martin et al. (1991)
SSLP	Simple sequence length polymorphism	Tautz (1989)
RBiP	Retrotransposon-based insertion polymorphism	Flavell et al. (1998)
OLA	Oligonucleotide ligation assay	Landegren et al. (1988)
RAM	Random amplified microsatellites	Ender et al. (1996)
REMAP	Retrotransposon microsatellite amplified polymorphism	Kalendar et al. (1999)
SAMPL	Selective amplification of microsatellite polymorphic loci	Morgante and Vogel (1994)
STMS	Sequence-tagged microsatellite site	Beckmann and Soller (1990)

(continued)

Table 10.1 (continued)

Markers	Name	Reference
STR	Short tandem repeats	Edwards et al. (1991)
ISSR	Inter-simple sequence repeat	Zietkiewicz et al. (1994)
RAMP	Random amplified microsatellite polymorphisms	Wu et al. (1994)
RAPD	Random amplified polymorphic DNA	Williams et al. (1993)
RFLP	Restriction fragment length polymorphism	Friar and Kochert (1994)

converting arbitrarily primed PCR products into genomic physical landmarks. Other markers such as microsatellite marker technology exploit the intraindividual and interindividual difference in microsatellites or SSR region for analyzing the fingerprint. For delineating the parental lineage, chloroplast and mitochondrial microsatellite- dependent markers are used which in turn give the best results during breeding and crop improvement (Agarwal et al. 2008). With quick advancement in technique of molecular biology, much effective and greater markers may develop in coming times which can significantly accelerate research in plant breeding.

Almost all molecular methods have become now a basic necessity in the key finding of biological science. Similarly, molecular approaches such as variable number tandem repeats (VNTRs), restriction fragment length polymorphism (RFLP), simple sequence repeats (SSR), random amplified polymorphic DNA (RAPD), and amplified fragment length polymorphism (AFLP) have become an important part of the genetic diversity assays. However, it is very important to understand that different markers have different properties, so it's inevitable that the markers would show different results as well (Karp and Edwards 1995) (Table 10.2). Moreover, molecular data could help us to deal with the more complex questions like taxonomy of the plant (Das et al. 2008). Molecular approaches help us to study the evolutionary phases of a plant and relative diversity of a species (Nayak et al. 2003). Loh et al. (2000) reported that till 2000, the application of molecular approaches in the field of genetic diversity particularly in bamboo was limited. Two types of molecular markers are mostly used in order to study the genetic diversity: (1) hybridization-based, i.e., RFLP, and (2) PCR-based, i.e., AFLP, RAPD, SSR, inter-simple sequence repeats (ISSR), and single-nucleotide polymorphism (SNP). RFLP markers have been reported to be showing a low level of polymorphism as compared to others. Friar and Kochert (1994) employed RFLP for almost 20 *Phyllostachys* species to study genetic variation and evolution. Moreover, it needs a fine quality of DNA. Similarly, RAPD is very easy to use as it does not need any information of plant genome before the application, and this feature has made its use for a large number of plants in order to study the genetic diversity among the species or within the species, whether on basic of ecological or geographical factors (Belaj et al. 2001; Deshwall et al. 2005).

The characters that are commonly used for the identification and classification of plants are morphological, cytological, phytochemical, anatomical, ecological, physiological, and molecular (Stace 1989; Singh 2010) (Table 10.3). The generative

Table 10.2 Application of different markers to study polymorphism and phylogenetic relationship in bamboo

Markers used	Species name	Remarks	Reference
RAPD, ISSR, iPBS, SCoT RAPD	*Bambusa cacharensis*, *B. mizorameana*, *Dendrocalamus manipureanus*, *D. hamiltonii*, and *D. sikkimensis* *Bambusa* sp. *Dinochloa*, *Bambusa* sp., *Dendrocalamus* *Bambusa* sp. *Gigantochloa*, *Arundinaria* sp., and *Dendrocalamus*	Ten primers of each markers were used to examine the genetic polymorphism and relationship between 50 genotypes of 5 important bamboos Phylogenetic relationships among 28 species of *Bambusa* were examined by using 16 RAPD markers Genetic relationship has been identified between bamboo species belonging to five genera The species of *Bambusa* belonging to southeastern China have been investigated, particularly in order to study their genetic relationship Reported the genetic distance between genera *Bambusa* and *Gigantochloa*	Amom et al. (2020) Rong et al. (2020) Nayak et al. (2003) Friar and Kochert (1991) Sun et al. (2006) Ramanayake et al. (2007)
AFLP	*Dendrocalamus*, *Bambusa* sp., *Bamboo* sp., *Guadua angustifolia* *Phyllostachys* sp. *Phyllostachys pubescens*	Phylogenetic and genetic variability among 12 bamboo species belonging to northeastern region of India Phylogenetic relationship among bamboo spread out across regions of the world Analysis of germplasm of *Guadua* particularly in the coffee region of Colombia Phylogenetic studies in genus *Phyllostachys* Ten cultivars of *P. pubescens* were identified which have a highest degree of similarity	Ghosh et al. (2011) Kobayashi (1997) Marulanda et al. (2002) Hodkinson et al. (2000) Lin et al. (2009)
ISSR	15 different bamboo species including *B. mizorameana*, *B. manipureana*, *D. sikkimensis*, and *D. manipureanus*	ISSR markers were used to examine the genetic relationship of 15 various bamboo species of Northeast India	Amom et al. (2018)
CpDNA	*Asian bamboos* *Bamboo* sp. *Bamboo* sp.	CpDNA restriction site mutations were examined in 16 bamboo species of Asia Chloroplast genome sequencing study was conducted Studied polymorphism and genetic relationship among 22 species of bamboo	Watanable et al. (1994) Zhang et al. (2011a) Zhang et al. (2011b)

(continued)

Table 10.2 (continued)

Markers used	Species name	Remarks	Reference
SSR	*Guadua chacoensis*, *Merostachys* *G. chacoensis*	Phylogenetic inference and SSR characterization of tropical woody bamboos tribes Bambuseae (Poaceae: Bambusoideae) were carried out on the basis of complete plastid genome sequences Identification and characterization of SSR markers for genetic studies along with the evaluation of their transferability with the other bamboo species were performed	Vieira et al. (2016) Rossarolla et al. (2020)
SCAR	*B. balcooa*, *B. tulda*	Generated SCAR fragments (species-specific)	Das et al. (2005)
MITEs	*B. multiplex* *B. vulgaris*, *Sasa veitchii*	Ac-like sequence was found Ac-like transposon element was found	Huttley et al. (1995) Gielis (1998)
ITS sequences	*Eremitis*, *Pariana*, and *Parianella* *Arundinaria* sp.	The herbaceous bamboos (tribe Olyreae) were analyzed based on ITS and plastid DNA (*rpl32-trnL* and *trnD-trnT* spacers) to establish phylogenetic relationship within Parianinae Phylogenetic relationships were studied between *Arundinaria* and some related genera such as *Bashania*, *Pleioblastus*, *Pseudosasa*, *Clavinodum*, etc. using special ITS sequences like nrDNA	Ferreira et al. (2019) Qiang et al. (2005)
cDNA library	*B. oldhamii*	Few DNA clones that involve in sucrose synthesis such as BoSus1, BoSus2, BoSus3, and BoSus4 were analyzed from etiolated bamboo shoots	Chiu et al. (2006)
RT-PCR and micro-array analysis	*Phyllostachys praecox*	Several rhizome genes were studied that involve in differentiation of rhizome into rhizome shoots, rhizome buds, bamboo shoots, leaves, etc.	Wang et al. (2010)

RAPD random amplified polymorphic DNA, *ISSR* inter-simple sequence repeats, *iPBS* inter-primer bonding site, *SCoT* start codon targeted, *AFLP* amplified fragment length polymorphism, *CpDNA* cytoplasmic DNA, *SSR* simple sequence repeats, *SCAR* sequence characterized amplified regions, *MITEs* miniature inverted-repeat transposable elements, *ITS* internal transcribed spacers

Table 10.3 Genetic fidelity testing of in vitro raised bamboos

Type of marker	Species name	Remarks	Reference
Morphological descriptors	*Dendrocalamus asper* *D. hamiltonii*	In vitro raised plants were compared with mother plants but found no significant variation Most of the leaf features were found comparable to the mother plant	Singh et al. (2013) Bag et al. (2012)
Biochemical analysis	*D. hamiltonii*	The chlorophyll pigment and leaf mass of the in vitro raised plants were found to be comparable to the mother plant	Bag et al. (2012)
Physiological studies	*D. hamiltonii*	Similarly, the rate of photosynthesis and the water intake efficiency of the in vitro raised and hardened plant were found to be comparable to the mother plant	Agnihotri and Nandi (2009)
Molecular markers RAPD	*Bambusa balcooa,* *B. tulda.* *D. hamiltonii*	Studied the confirmation of genetic fidelity of in vitro raised plants and further suggested that the axillary meristem is the viable part for clonal propagation Genetic fidelity was reported during various stages of development of in vitro raised plant and confirmed the absence of somaclonal variation	Das and Pal (2005) Agnihotri and Nandi (2009)
ISSR	*B. nutans* *G. angustifolia* *B. Balcooa*	The shoot multiplication up to 24th cycle till to the hardening of the in vitro raised plants grown in polyhouse were found genetically similar to the mother plant Similarly, the genetic fidelity was evaluated till to the hardening phase of the in vitro raised plant and was compared with mother plant The monomorphic banding patter was found to be similar with the mother plant	Negi and Saxena (2011) Nadha et al. (2011) Rajput et al. (2020)
SSR	*D. asper*	Similarly, no somaclonal variation was reported, and the in vitro raised plants were genetically similar to that of mother plant	Singh et al. (2012)
AFLP	*B. balcooa*	The tissue culture grown plants emerging from the axillary buds and somatic embryogenesis were having no any epigenetic changes	Gillis et al. (2007)
SCoT	*B. balcooa*	The monomorphic banding pattern of SCoT marker of in vitro derived plants matched with mother plants confirmed the genetic similarity	Rajput et al. (2020)

RAPD random amplified polymorphic DNA, *ISSR* inter-simple sequence repeats, *SSR* simple sequence repeats, *AFLP* amplified fragment length polymorphism, *SCoT* start codon targeted

organs' nature is more ideal for characterization than the vegetative organs because their structure is constant and provides more properties for the differentiation of taxa. In some taxa, vegetative traits have a low taxonomic value, but for taxa that have a low inflorescence frequency, vegetative traits are essential in classification (Jones and Luchsinger 1986). Over the years, botanists have laid the foundations of systematics and identification of bamboos based on morphological and anatomical characters. However, the systematics is based on vegetative characters only. The resulting data is less accurate, so it still needs to be compared with data from other analyses (Das et al. 2008).

Bamboo identification in many countries has been carried out using DNA fingerprint methods, such as random amplified polymorphic DNA (RAPD) (Nayak et al. 2003), amplified fragment length polymorphism (AFLP) (Loh et al. 2000), sequence characterized amplified regions (SCARs) (Das et al. 2005), inter-simple sequence repeat (ISSR) (Negi and Saxena 2010), simple sequence repeats (SSRs) (Nayak and Rout 2005), expressed sequence tag-simple sequence repeat (EST-SSR) (Sharma et al. 2009), and transposons (Keukeleire et al. 2004). Furthermore, Das et al. (2008) succeeded in making dendrogram comparisons of bamboo relationships in India based on morphological and molecular characters.

A taxonomic method based on DNA sequences was also developed to determine genetic diversity, population structure, and phylogenetic relationship between bamboo species. Sun et al. (2005) used the internal transcribed spacer (ITS) rDNA sequence for phylogenetic analysis of *Bambusa* in China. The results of ITS rDNA regional sequences ranged from 637 bp in *Guadua angustifolia* to 696 bp in *Bambusa flexuosa*. The similarity values obtained ranged from 86 to 100%. The identification results can show phylogenetic patterns between *Bambusa* species and their close relatives. Meanwhile, Goh et al. (2010) reported that the phylogenetic relationship analysis of bamboo was also carried out using chloroplast DNA *rps16-trnQ*, *trnC-rpoB*, *trnH-psbA*, and *trnD-T*, and nuclear DNA, namely, the *GBSSI* gene. In a recent study of Liu et al. (2020), double-digest restriction site-associated DNA (ddRAD) sequencing was performed to reveal the phylogenetic relationship of the four important genera of *Bambusa-Dendrocalamus-Gigantochloa* complex.

The present chapter aims to provide the information on different molecular markers, for example, DNA fingerprint-based method, DNA sequence-based method, etc., applied on bamboo to establish their genetic relationships. Moreover, a description of the role of morphological characters for the identification of bamboo has also been discussed.

10.2 Morphological Traits: Key to Bamboo Identification and Characterization

Gamble was probably the first scientist who identified bamboo plants on the basis of morphological characters particularly vegetative and reproductive characters in 1896. However, scientists later on discovered other morphological characters such as culm sheaths that became tools for the early classification of bamboos. Finally, Chatterjee and Raizada (1963) set a culm sheath character, a parameter to identify 22 bamboo taxa. According to Chatterjee and Raizada (1963), the culm sheath characters such as size, texture, blades, and shape of the blade offer a good line of distinction for the classification of bamboos. Similarly, Bennet and Gaur (1990) suggested that the branching pattern could become an important characteristic for the identification of genus. They even suggested that the sprouting vegetative buds could also serve as an important morphological character for the identification of bamboos. Triplett and Clark (2003) have tried to understand the relationship between ecological and geographical variations with the genetic diversity, so they took 7 vegetative and 14 reproductive characters. The principle of their work was that the variations in the characters are a continuous process, so therefore it couldn't act as a kind of parameter to classify the species on the basis of their morphological characters. Their work emphasized the need to conduct more in-depth analysis in order to determine the classification of *C. culeou*. A study was conducted by Das et al. (2007) by means of 32 qualitative and quantitative morphological characters in order to understand the phylogenetic relationship of 15 species of bamboo which were not in agreement with the classification of Gamble (1896).

10.2.1 Limitation of Morphological Characters

The following are some of the limitations that basically guide us that classification on the basis of morphological or vegetative characters would not set a precedent: (1) According to Janzen (1976), the reproductive cycle of the bamboo is too long and that could stretch up to 120 years. So using floral characters for characterization or identification would serve no purpose. (2) Evolutionary studies dictate that the vegetative characters are subjected to environmental impact. So vegetative characters would not be a reliable key for taxonomic classification.

10.3 DNA Fingerprinting-Based Methods

10.3.1 RFLP

The basis of the polymorphisms in the RFLP is the difference in the sequence of the restriction enzyme recognition sites between genomes. This marker is codominant and useful for selection with the help of specific markers. Friar and Kochert (1994) first used RFLP on *Phyllostachys* to study the genetic variability and evolution of its 20 species. This technique is rarely used in bamboo because it requires high-quality DNA and skilled personnel.

10.3.2 RAPD

RAPD is an inexpensive and fast method and does not require preliminary information from the plant genome. This method has been widely used to study plant genetic variation because it is sensitive and effective in obtaining polymorphism data. Random amplified polymorphic DNA (RAPD) molecular markers have been used to reveal that *Bambusa ventricosa* and *B. vulgaris* var. *striata* are the same species (Nayak et al. 2003). The RAPD technique has also separated the spiny *Bambusa* from the *Dendrocalamus* members (Sun et al. 2006). The RAPD technique has also been successful in demonstrating high levels of polymorphism in nine bamboo species in Sri Lanka (Ramanayake et al. 2007). However, the RAPD technique is not suitable for the identification of polymorphisms within species. Bhattacharya et al. (2006) proved that the identification of 17 *B. tulda* populations that experienced geographic isolation did not show any polymorphisms. The same thing happened in the study of Lai and Hsiao (1997), where out of 176 samples of *P. pubescens*, only 9 genotypes were found. These results indicate the genetic diversity in the population is very low.

10.3.3 SCARs

Sequence characterized amplified regions (SCARs) is the development of RAPD (Paran and Michelmore 1993). In the bamboo study, SCARs were used to identify genotypes and varieties, especially for species that have almost the same morphological characteristics. The SCAR marker was developed by Das et al. (2005) for *Bambusa balcooa* and *Bambusa tulda*, in order to assist the paper industry in differentiating the two types of bamboo. Meanwhile, Bhattacharya et al. (2008) conducted a genetic diversity study on 12 populations of *Bambusa balcooa* and 17 populations of *Bambusa tulda* based on morphological characters and molecular marker SCAR in India. The results of these studies indicate a high morphological diversity between *Bambusa* populations. This is known through the coefficient of diversity 49.49% (F value 10.4326; $P = <0.001$). However, the absence of DNA band polymorphisms in SCAR indicates the low intraspecific genetic diversity of the two bamboo types.

10.3.4 AFLP

Another molecular marker technique used in bamboo identification is AFLP (*amplified fragment length polymorphism*). This technique is an RFLP combined with PCR. AFLP analysis allows precise comparisons between taxa to determine genetic distances and phylogenetic relationships, even between closely related taxa, including infraspecies variation. The cluster pattern formed by AFLP has successfully revealed polyphyletic properties in the genus *Bambusa* (Loh et al. 2000). On the other hand, Marulanda et al. (2002) used AFLP for studying the genetic variation of *Guadua*. AFLP has also identified nine species of bamboo in Manipur State, Northeast India (Ghosh et al. 2012). However, this technique is quite tricky to apply, considering the high price. It must be done by skilled personnel, as it is difficult to be analyzed because it produces a lot of data and requires a long working time.

10.3.5 SSR

Simple sequence repeats (SSR) are molecular markers that are also applied to bamboo. SSR is a repeating sequence of tandem nucleotides with a length ranging from 1 to 6 nucleotides; is polymorphic, codominant, and multiple alleles; and is considered a neutral sequence. Therefore, SSR is widely used in the study of plant genetic diversity. SSR primers are designed from conserved genome regions, which enclose these tandem nucleotide sequences. The detected sequence lengths and polymorphisms reflect the variation in the number of repetitions between the genomes.However, all procedures that include genomic construction and screening before primer design are considered impractical and expensive (Das et al. 2008).

This greatly limits the SSR method's application to nonagricultural plants such as bamboo because sufficient genomic information is not yet available in the database. Nayak and Rout (2005) have successfully applied the use of SSR molecular markers to *Bambusa*. Six SSR sequences were isolated from *B. arundinacea* and tested on 18 other bamboo species. Three polymorphic loci are known to identify and characterize bamboo species. These findings suggest that primers designed from the *B. arundinacea* genome could be used to identify other bamboo taxa. Thus, SSR molecular markers can be used to compare taxa without having to do a specific primer design for each bamboo species. This study also shows that SSR molecular markers can be used in the study of population genetics and genetic diversity in bamboo.

10.3.6 SRAP

Zhu et al. (2014) also conducted a genetic diversity study on 13 bamboo accessions in China. A total of 21 vegetative morphological characters and SRAP (sequence-related amplified polymorphism) molecular markers were used to construct the dendrogram. In this study, the similarity coefficient obtained was used to measure genetic diversity. The similarity coefficient of 0.23–0.96 indicates high genetic diversity based on morphological characters. Likewise, the similarity coefficient of 0.36–0.75 shows high genetic diversity in molecular characters compared with research on genetic diversity using molecular markers conducted by previous researchers.

10.4 DNA Sequence-Based Methods

10.4.1 Organelle Genes

The sequence of organelle genes began to develop since the discovery of the *rbcL* gene, which encodes the large subunit protein *ribulose 1,5 biphosphate carboxylase/ oxygenase* (*rbcL*). Using this *rbcL* gene, Barker et al. (1995) describe the position of Bambusoideae among other subfamilies. However, according to Doebley et al. (1990), the *rbcL* gene's use is only suitable for taxa familia and taxa higher than family, not ideal for grasses in subfamily taxa lower than subfamily. Gaut et al. (1997) added that the woody bamboo (Bambuseae) group generally has a long generation time to slow down the rate of nucleotide substitution. It thus becomes unsuitable for an inferior taxonomic analysis. Furthermore, several chloroplast genes were found which were also used to construct molecular relationships in Poaceae, including *ribosomal protein S4* (*rps4*) (Nadot et al. 1994), *NADH-plastoquinone oxidoreductase subunit 5* (*ndhF*) (Clark et al. 1995), *maturase K* (*matK*) (Hilu et al. 1999), and *RNA polymerase b subunit* (*rpoC2*) (Barker et al. 1999).

10.4.2 Nuclear Genes

Sequencing methods with genes from the nucleus use 18S rDNA (Hamby and Zimmer 1988), *granule-bound starch synthase gene* (*GBSSI*) (Mason-Gamer et al. 1998), internal transcribed spacers (ITS) (Hsiao et al. 1999), and phytochrome B (Mathews et al. 2000). Das et al. (2008) argue that ITS is the most popular method to determine phylogenetic relationships at the genus taxon level down because it has a higher rate of nitrogen base substitution than other genetic materials.The ITS sequence data has been used to trace the phylogenetic relationships of *Thamnocalamus* and its close relative species. This sequence shows that members of the *Thamnocalamus* are monophyletic to one another (Guo et al. 2002). ITS sequences have also been used to study 23 alpine bamboo species' genetic diversity from the genus *Thamnocalamus*, *Fargesia*, and *Yushania*. The results of these studies determined *T. spathiflorus* var. crassinodus and *F. spathacea* as alpine bamboo ancestors, although these data are not supported by a useful bootstrap (Guo et al. 2002). ITS sequences have also been used for phylogenetic analysis of 21 species of *Bambusa* (sensu stricto), *Dendrocalamopsis*, *Dendrocalamus*, *Guadua*, *Leleba*, and *Lingnania*. This study concluded that *Bambusa* is closely related to *Dendrocalamus* (Sun et al. 2005).

ITS sequence, which is biparental, has been widely chosen for phylogenetic analysis at the taxon genus level and below because it has a high rate of nucleotide substitution compared to organelle genes. In addition, the ITS sequence also has many duplications, making it easy to amplify by targeting primary adhesions to conserved areas 18S and 26S, using universal primers (Das et al. 2008). However, the results of phylogenetic analyses using ITS sequences are often confusing (Alvarez and Wendel 2003). This can be due to limited information by short sequences (Baldwin et al. 1995) or difficult alignment due to varying sequence lengths (Hsiao et al. 1999).

One of the important prerequisites for phylogenetic studies using ITS sequences is targeting the correct orthologous sequence. However, in the absence of complete homogenization, paralog sequences may appear accidental and bias the results. In Bambuseae, the potential for paralog sequences is very vulnerable due to polyploidization. Another confounding phenomenon discussed by Alvarez and Wendel (2003) is the presence of a large number of rDNA copies and possible contamination due to the use of universal primers. From a number of these problems, contamination is considered a factor that affects the diversity of ITS sequence results. The genetic material (rDNA) of fungi can be accidentally isolated and amplified with the target DNA (Zhang et al. 1997). Epiphyllous fungi are known to be associated with bamboo leaves. Therefore, before DNA isolation, fresh leaves should always be sterilized first, to avoid possible contamination. Besides, researchers should not rely on the results of a single PCR reaction but attempt to be able to clone and amplify DNA products under various reaction conditions (Alvarez and Wendel 2003) to avoid PCR bias or drift (Wagner et al. 1994).

10.5 Bamboo and Molecular Descriptors

New molecular approaches have become an important aspect of the research involving in area of understanding the phenomena of genetic diversity and phylogenetic relationship. As discussed above, the new molecular techniques such as RAPD, SSR, AFLP, and RAPD are actually a trend in determining the genetic pool or genetic population of a particular plant species. Moreover, it is important to understand that all markers are not having the same characteristics and similar functions but rather they are very dissimilar in both characteristically and functionally (Karp and Edwards 1995). These molecular approaches have helped us to generate a data that has significantly helped us to find the exact roots of taxonomic complexities that could probably allow us to deal with plants that are yet to be placed in different classifications (Das et al. 2008).

Till 2000, the study of genetic diversity in bamboo was limited (Loh et al. 2000). However, the previous work done by scientists acted as a source of an encouragement to lay the hands on a large pool of genetic studies in bamboo species. Friar and Kochert (1994) conducted RFLP-based research in *Phyllostachys*. Heng et al. (1996) conducted isozyme-based studies among five genera of bamboo. Similarly,

Kobayashi (1997) conducted research on bamboos belonging to different regions of the world, and Watanable et al. (1994) conducted research based on chloroplast DNA phylogeny of bamboos belonging to Asia. A specific intron sequence of rpl16 was analyzed within *Chusquea* genus (Loh et al. 2000).

As mentioned above, RFLP technique was employed to conduct the research aimed at understanding the genetic evolution of more than 15 species of *Phyllostachys* (Friar and Kochert 1994). But this technique has shown low polymorphism as compared to others. Similarly, RAPD has provided a good alternative, as it does not need any previous information regarding the plant genome. This feature, coupled with easy accessibility in the market, has made it a good choice in studying the genetic variations among various species (Belaj et al. 2001; Deshwall et al. 2005; Ko et al. 1998). It requires very small amount of genomic DNA and can produce very high level of polymorphism and can be effective for diversity analysis in plants (Williams et al. 1993). RAPD analysis has proved its significance for diverse study of field crops like rice (Qian et al. 2001; Rabbani et al. 2008; Pervaiz et al. 2010) and many horticultural plants such as coffee (Orozco-Castillo et al. 1994), tea (Wachira et al. 1995), almond (Shiran et al. 2007), sesame (Akbar et al. 2011), and turmeric (Singh et al. 2012). Recently, large number of scientists employed molecular markers to conduct the characterization and phylogenetic relationship on bamboos (Nayak et al. 2003; Das et al. 2005; Bhattacharya et al. 2006; Ramanayake et al. 2007; Das et al. 2007; Bhattacharya et al. 2009). Moreover, SSR primers derived from rice, sugarcane, etc. were used for the study of genetic diversity among large species of bamboo (Sharma et al. 2009).

Economically bamboo plants are very important because of its multipurpose usage across the globe. China and India are the largest producers of bamboo in Asia. More importantly, bamboo is a genetically diverse plant, so it could serve as a good case of study for the better improvement and production of highly desired plant. Therefore, identification, characterization, and documentation at molecular level of the bamboo plants are essential demands in order to strategize the conservation methods of the bamboo and to improve our understanding about the taxonomy of the plant (Rao and Hodgkin 2002). Loh et al. (2000) explained that the need has arisen to collect different samples of bamboo in order to conserve the plant from further exploitation. Das et al. (2008) further explained that the molecular data of the plant can really help us to classify the plants taxonomically. Moreover, in order to assess the level of interspecies and intraspecies genetic diversity between bamboo plants, the molecular markers are considered to be a hopeful technical asset (Nayak et al. 2003). Molecular marker such as RAPD has been quite useful in revealing some important information regarding the genetic variation existing among various bamboo species. Sun et al. (2006) reported that RAPD markers were quite useful in revealing the genetic relationship between various bamboo species of the southeastern China.

Recent research is conducted by Rong et al. (2020) in which 28 species/varieties of *Bambusa* were subjected to evaluation based on 16 RAPD primers. Amplification of 216 bands were conducted by using 16 RAPD primers, which yielded about 290–3000 bp DNA fragments. It was reported that the percentage of polymorphism were 96.79% and the number of bands was 211 which indicates that the genetic diversity (interspecific and intraspecific) was high among bamboo species. Therefore, the results suggest that the RAPD molecular markers have a practical role in detecting the variation among various species. The reason behind such a higher percentage of polymorphism could be the factors like climate variations, evolutionary changes, and geographic location that eventually set a larger genetic pool of the species (Lou et al. 2011). Nayak et al. (2003) has achieved similar kind of results; however, the number of primers he used while studying the case was higher than the aforementioned work conducted by Rong et al. (2020). It's very significant to realize that each molecular marker has an exclusive property so it would logically display different aspect of diversity of gene (Karp and Edwards 1995).

It is very tough to identify the genetic relationship between various bamboo species because of the absence of phenotypic variance. However, the confirmation of the genotype is essential for both propagators and consumers in order to protect the IPR. RAPD and ISSR markers were employed in order to evaluate the diversity of 13 genotypes of bamboos. A total of 120 RAPD and 63 ISSR primers were tried, among which only 42 polymorphic primers, 30 RAPD, and 12 ISSR reported to have produced amplification profiles. It was reported that 30 RAPD primers generated a total of 645 amplified fragments, among which about 623 were polymorphic and 20.76 polymorphic bands on each primer were detected across 13 genotypes. 12 ISSR primers generate 246 amplified fragments, of which 241 were polymorphic, and 20.08 polymorphic bands per primer were observed across 13 different genotypes. These results indicate that an extensive genetic diversity occurred among 13 genotypes of bamboo. It's very surprising that some top researchers have deemed that RAPD markers illustrate mostly noncoding regions of DNA (Bachmann 1997; Landergott et al. 2001), while some have a very different opinion and consider that RAPDs disseminate throughout the genome and link with only functional loci (Penner 1996).

Another molecular marker that we haven't discussed yet is microsatellite or sometimes also called simple sequence repeats (SSR) which have been proved to be very efficient in revealing the knowledge of genetic polymorphism among various case studies. They have very significant role in genome mapping, population genetic analysis, and genetic diversity and obviously evolutionary study (Brondani et al. 2002; Sharopova et al. 2002; Deutech et al. 2002; Kikuchi and Isagi 2002). Among bamboos, microsatellites have been characterized and identified in bamboo (*Bambusa arundinacea*) (Nayak and Rout 2005), and reportedly three polymorphic sequences have been identified in this plant which could serve as a parameter to study the population genetics among the clones of this plant and other relevant species as well.

10.6 Conclusion and Future Prospects

The use of various taxonomic evidence that includes molecular, morphological, and anatomical will provide answers to population genetics and the taxonomic status of bamboo. Research using various taxonomic evidence can produce information on genetic diversity and population structure, clarity of taxonomic identity, and bamboo relationship. It is hoped that this series of data will become the basis for tracing the evolutionary history of bamboo. Hence, the results of taxonomic identification can become a reference in a bamboo conservation strategy.

Conflict of Interest

No.

Author Contribution Author AL conceived the idea and wrote the manuscript. Author IBG, ZA, and AS reviewed the manuscript and revised the MS as per the requirements. All authors read and approved the manuscript.

References

Agarwal M, Shrivastava N, Padh H (2008) Advances in molecular marker techniques and their applications in plant sciences. Plant Cell Rep 27:617–631

Agnihotri RK, Nandi SK (2009) In vitro shoot cut: a high frequency multiplication and rooting method in bamboo Dendrocalamus hamiltonii. Biotechnology 8:259–263

Akbar F, Rabbani MA, Masood MS, Shinwari ZK (2011) Genetic diversity of sesame (Sesamum indicum L.) germplasm from Pakistan using RAPD markers. Pak J Bot 43:2153–2160

Alvarez I, Wendel JF (2003) Ribosomal ITS sequences and plant phylogenetic inference. Mol Phylogenet Evol 29:417–434

Amom T, Nongdam P (2017) The use of molecular marker methods in plants: a review. Int J Cur Res Rev 9:2–7

Amom T, Tikendra L, Apana N, Goutam M, Sonia P et al (2020) Efficiency of RAPD, ISSR, iPBS and phytochemical markers in the genetic relationship study of five native and economical important bamboos of North-East India. Phytochemistry 174:112330

Amom T, Tikendra L, Rahaman H, Angamba P, Nongdam P (2018) Evaluation of genetic relationship between 15 bamboo species of North-East India based on ISSR markers analysis. MBRC 7:7–15

Bachmann K (1997) Nuclear DNA markers in plant biosystematics research. Opera Bot 132:137–148

Bag N, Palni LMS, Chandra S, Nandi SK (2012) Somatic embryogenesis in 'Maggar' bamboo (Dendrocalamus hamiltonii) and field performance of regenerated plants. Curr Sci 102:1279–1287

Baldwin BG, Sanderson MJ, Porter JM, Wojciechowski MF, Campbell CS, Donoghue MJ (1995) The ITS region of nuclear ribosomal DNA: a valuable source of evidence on angiosperm phylogeny. Ann Mo Bot Gard 82:247–277

Barker NP, Linder HP, Harley EH (1995) Polyphyly of Arundinoideae (Poaceae): evidence from rbcL sequence data. Syst Bot 20:423–435

Barker NP, Linder HP, Harley EH (1999) Sequences of the grass-specific insert in the chloroplast *Rpoc2* gene elucidate generic relationships of the *Arundinoideae* (Poaceae). Syst Bot 23:327–350

Beckmann JS, Soller M (1990) Toward a unified approach to genetic mapping of eukaryotes based on sequence tagged microsatellite sites. Biotechnology 8:930–932

Belaj A, Trujilo I, Rosa R, Rallo L, Gimenez MJ (2001) Polymorphism and discrimination Sci capacity of randomly amplified polymorphic markers in an olive germplasm bank. J Am Soc Hort 126:64–71

Bennet SSR, Gaur RC (1990) Thirty seven bamboos growing in India. Forest Research Institute, Dehradun

Bhattacharya S, Das M, Bar R, Pal A (2006) Morphological and molecular characterization of *Bambusa tulda* with a note of flowering. Ann Bot 98:529–535

Bhattacharya S, Ghosh J S, Mitra A, Pal A (2008) Genetic diversity assessment and search for disease tolerant genotypes across the natural gene pool of *Bambusa balcooa* and *B. tulda* using molecular tools and technique. Proceeding of the national conference on bamboos: management, conservation, Tropical Forest Research Institute

Bhattacharya S, Ghosh JS, Das M, Pal A (2009) Morphological and molecular characterization of Thamnocalamus spathiflorus subsp. spathiflorus at population level. Plant Syst Evol 282:13–20

Brondani RP, Brondani C, Grattapaglia D (2002) Towards a genuswide reference linkage map for Eucalyptus based exclusively on highly informative microsatellite markers. Mol Gen Genomics 267:338–347

Caetano-Anolles G, Bassam BJ, Gresshoff PM (1991) DNA amplification fingerprinting using very short arbitrary oligonucleotide primers. Bio/Technology 9:553–557

Caetano-Anolles G, Gresshoff PM (1994) DNA amplification fingerprinting using arbitrary minihairpin oligonucleotide primers. Biotechnology 12:619–623

Chatterjee RN, Raizada MB (1963) Culmsheaths as an aid to identification of Bamboos. Ind For 89:744–756

Chiu WB, Lin CH, Chang CJ, Hsieh MH, Wang AY (2006) Molecular characterization and expression of four cDNAs encoding sucrose synthase from green bamboo Bambusa oldhamii. New Phytol 170:53–63

Ciffarelli RA, Gallitelli M, Cellini F (1995) Random amplified hybridization microsatellites (RAHM): isolation of a new class of microsatellite-containing DNA clones. Nucleic Acids Res 23:3802–3803

Clark LG, Zhang W, Wendel JF (1995) A phylogeny of the grass family (Poaceae) based on *ndhF* sequence data. Syst Bot 20:436–446

Das M, Bhattacharya S, Basak J, Pal A (2007) Phylogenetic relationship among the bamboo species as revealed by morphological characters and polymorphism analysis. Biol Plant 51:667–672

Das M, Bhattacharya S, Pal A (2005) Generation and characterization of SCARs by cloning and sequencing of RAPD products: a strategy for species-specific marker development in bamboo. Ann Bot 95:835–841

Das M, Bhattacharya S, Singh P, Filgueiras TS, Pal A (2008) Bamboo taxonomy and diversity in the era of molecular markers. Adv Bot Res 47:225–268

Das M, Pal A (2005) In vitro regeneration of Bambusa balcooa Roxb.: factors affecting changes of morphogenetic competence in the axillary buds. Plant Cell Tissue Organ Cult 81:109–112

Deshwall RPS, Singh R, Malik K, Randhawa GJ (2005) Assessment of genetic diversity and genetic relationships among 29 populations of Azadirachta indica A. Juss. using RAPD markers. Genet Resour Crop Evol 52:285–292

Desmarais E, Lanneluc I, Lagnel J (1998) Direct amplification of length polymorphisms (DALP), or how to get and characterize new genetic markers in many species. Nucleic Acids Res 26 (6):1458–1465

Deutech C, Seiter J, Petronelli P, Joly HI, Jarne P (2002) Evidence of gene flow in a geotropically clustered tree species in two rainforest stands of French Guiana. Mol Ecol 11:725–738

Doebley J, Durbin M, Golenberg EM, Clegg MT, Ma DP (1990) Evolutionary analysis of the large subunit of carboxylase (Rbcl) nucleotide sequence data among the Grasses (Gramineae). Evolution 44:1097–1108

Edwards K, Johnstone C, Thompson C (1991) A simple and rapid method for the preparation of genomic plant DNA for PCR analysis. Nucleic Acids Res 19:1349

Ender A, Schwenk K, Stadler T, Streit B, Schierwater B (1996) RAPD identification of microsatellites in Daphnia. Mol Ecol 5:437–441

Ferreira FM, Oliveira RP, Welker CAD, da Costa DM et al (2019) Phylogenetic relationship within Parianinae (Poaceae: Bambusoideae: Olyreae) with emphasis on *Eremitis*: evidence from nuclear and plastid DNA sequences, macromorphology, pollen ectexine patterns. Mol Phylogenet Evol 139:106541

Flavell AJ, Knox MR, Pearce SR, Ellis TH (1998) Retrotransposon-based insertion polymorphisms (RBIP) for high throughput marker analysis. Plant J 16:643–650

Friar E, Kochert G (1991) Bamboo germplasm screening with nuclear restriction fragment length polymorphisms. Theor Appl Genet 82:697–703

Friar E, Kochert G (1994) A study of genetic variation and evolution of Phyllostachys (Bambusoideae: Poaceae) using nuclear restriction fragment length polymorphisms. Theor Appl Genet 89:265–270

Gamble JS (1896) The Bambuseae of British India. Ann R Bot Gard Calcutta 7:1–133

Gaut BS, Clark LG, Wendel JF, Muse SV (1997) Comparisons of the molecular evolutionary process at rbcL and ndhF in the grass family (Poaceae). Mol Biol Evol 14:769–777

Ghosh S, Devi SW, Mandi S, Talukdar NC (2011) Amplified fragment length polymorphism based study of phylogenetic relationship and genetic variability among some edible bamboo species of North-East India. J Plant Mol Biol Biotechnol 2:8–15

Ghosh S, Somkuvar B, Mandi SS, Talukdar NC (2012) Genetic variability and phylogenetic relationship among some bamboo species of north-East India by AFLP analysis. Asean J Plant Sci Res 2:478–485

Gielis J (1998) Up-stream fundamental research in bamboo-possibilities and directions. Keynote lecture at Vth International Bamboo Congress, San Jose, Costa Rica. http://www.bamboonetwork.org/ downloads/gielis03.pdf

Gillis K, Gielis J, Peeters H, Dhooghe E, Oprins J (2007) Somatic embryogenesis from mature Bambusa balcooa Roxb as basis for mass production of elite forestry bamboos. Plant Cell Tissue Organ Cult 91:115–123

Goh WL, Chandran S, Lin RS, Xia NH, Wong KM (2010) Phylogenetic relationship among Southeast Asian climbing bamboos (Poaceae: Bambusoideae) and the *Bambusa* complex. Biochem Syst Ecol 38:764–773

Guo ZH, Chen YY, Li DZ (2002) Phylogenetic studies on Thamnocalamus group and ITS allies (Bambusoideae: Poaceae) based on ITS sequence data. Mol Phylogenet Evol 22:20–30

Gupta VS, Ramakrishna W, Rawat SR, Ranjekar PK (1994) (CAC) detects DNA fingerprints and sequences homologous to gene transcripts in rice. Biochem Genet 32:1–8

Hamby RK, Zimmer EA (1988) Ribosomal RNA sequences for inferring phylogeny within the grass family (Poaceae). Plant Syst Evol 160:29–37

Hayashi K (1992) PCR-SSCP: a method for detection of mutation. Genet Anal Tech Appl 9:73–79

Hayward AC, Tollenaere R, Dalton-Morgan J, Batley J (2015) Molecular marker applications in plants. In Plant Genotyping Humana Press, New York, pp 13–27

Heath DD, Iwama GK, Devlin RH (1993) PCR primed with VNTR core sequence yields species specific patterns and hypervariable probes. Nucleic Acids Res 21:5782–5785

Heng HP, Yeoh HH, Tan CKC, Rao AN (1996) Leaf isozyme polymorphisms in bamboo species. J Singapore Nat Acad Sci 22:10–14

Hilu KW, Alice LA, Liang H (1999) Phylogeny of Poaceae inferred from matK sequences. Ann Mo Bot Gard 86:835–851

Hodkinson TR, Renvoize SA, Chonghaile GN (2000) A comparison of ITS nuclear rDNA sequence data and AFLP markers for phylogenetic studies in Phyllostachys (Bambusoideae, Poaceae). J Plant Res 113:259–269

Hongtrakul K, Goodband RD, Behnke KC, Nelssen JL, Tokach MD, Bergström JR, Nessmith WB Jr, Kim IH (1998) The effects of extrusion processing of carbohydrate sources on weanling pig performances. J Anim Sci 76:3034–3042

Hsiao C, Jacobs SWL, Chatterton NJ, Asay KH (1999) A molecular phylogeny of the grass family (Poaceae) based on the sequences of nuclear ribosomal Dna (ITS). Aust Syst Bot 11:667–688

Huttley GA, McRae AF, Clegg MT (1995) Molecular evolution of the Ac/Ds transposable element family in pearl millet and other grasses. Genetics 139:1411–1419

Ibrahim AA, ABr M, Haseeb AK, Ahmad HA, Ali AA, Ali HB, Mohammad AM (2010) A brief review of molecular techniques to assess plant diversity. Int J Mol Sci 11:2079–2096

Jaccoud D, Peng K, Feinstein D, Kilian A (2001) Diversity arrays: a solid state technology for sequence information independent genotyping. Nucleic Acids Res 29:e25

Janzen DH (1976) Why bamboos wait so long to flower. Annu Rev Ecol Syst 7:347–391

Jeffreys AJ, Wilson V, Thein SL (1985) Hypervariable 'minisatellite' regions in human DNA. Nature 314:67–73

Jones SB, Luchsinger AE (1986) Plant systematics. McGraw-Hill, New York

Kalendar R, Grob T, Regina M, Suoniemi A, Schulman A (1999) IRAP and REMAP: two new retrotransposon-based fingerprinting techniques. Theor Appl Genet 98:704–711

Karp A, Edwards KJ (1995) Molecular techniques in the analysis of the extent and distribution of genetic diversity. In: IPGRI workshop on molecular genetic tools in plant genetic resources 9–11

Keukeleire PD, De S, Schepper J, Gielis GT (2004) A PCR-based assay to detect hAT-like transposon sequences in plants. Chromosom Res 12:117–123

Kikuchi S, Isagi Y (2002) Microsatellite genetic variation in small and isolated populations of Magnolia sieboldii ssp. japonica. Heredity 88:313–321

Ko MK, Yang J, Jin YH, Lee CH, Oh BJ (1998) Genetic relationships of Viola species evaluated by random amplified polymorphic DNA analysis. J Hort Sci Biotechnol 74:601–605

Kobayashi M (1997) Phylogeny of world bamboos analyzed by restriction fragment length polymorphisms of chloroplast DNA. In: Chapman GP (ed) The bamboos. Linean Society Symposium series. Linean Society of London, UK, pp 61–81

Lai CC, Hsiao JY (1997) Genetic variation of *Phyllostachys pubescens* (Bambusoideae, Poaceae) in Taiwan based on DNA polymorphisms. Bot Bull Acad Sin 38:145–152

Landegren U, Kaiser R, Sanders J, Hood L (1988) A ligase-mediated gene detection technique dinucleotide repeat within the cardiac muscle actin gene. Am J Hum Genet 44:397–340

Landergott U, Holderegger R, Kozlowski G, Schneller JJ (2001) Historical bottlenecks decrease genetic diversity in natural populations of *Dryopteris cristata*. Heredity 87:344–355

Lin XC, Ruan XS, Lou YF, Guo XQ, Fang W (2009) Genetic similarity among cultivars of Phyllostachys pubescens. Plant Syst Evol 277:67–73

Litt M, Luty JA (1989) A hypervariable microsatellite revealed by in vitro amplification of a science. 241:1077–1080

Liu J-X, Zhou M-Y, Yang G-Q, Zhang Y-X et al (2020) ddRAD analyses reveal a credible phylogenetic relationship of the four main genera of *Bambusa-Dendrocalamus-Gigantochloa* complex (Poaceae: Bambusoideae). Mol Phylogenet Evol 146:106758

Loh JP, Kiew R, Set O, Gan LH, Gan YY (2000) A study of genetic variation and relationships within the bamboo subtribe Bambusinae using amplified fragment length polymorphism. Ann Bot 85:607–612

Lou Y, Yang H, Zhang Y, Li X, Lin X, Fang W (2011) Analysis of genetic variation of some bamboo species by AFLP, ISSR and SRAP. J Fujian Coll For 31:38–43

Martin GB, Williams JGK, Tanksley SD (1991) Rapid identification of markers linked to a Natt Pseudomonas resistance gene in tomato by using random primers and near-isogenic lines. Proc Natl Acad Sci U S A 88:2336–2340

Marulanda ML, Márques P, Londono X (2002) AFLP analysis of *Guadua angustifolia* (Poaceae: Bambusoideae) in Colombia with emphasis on the coffee region. Bamboo science and culture. J Am Bamboo Soc 16:32–42

Mason-Gamer RJ, Weil CF, Kellogg EA (1998) Granule-bound starch synthase: structure, function, and phylogenetic utility. Mol Biol Evol 15:1658–1673

Mathews S, Tsai RC, Kellogg E (2000) Phylogenetic structure in the grass family (Poaceae): evidence from the nuclear gene Phytochrome B. Am J Bot 87:96–107

Meyer JP, Allen NJ, Smith C (1993) Commitment to organizations and occupations: extension and test of a three-component conceptualization. J Appl Psychol 78:538–551

Michaels SD, Amasino RM (1998) A robust method for detecting single-nucleotide changes as polymorphic markers by PCR. Plant J 14:381–385

Michelmore RW, Paran I, Kesseli RV (1991) Identification of markers linked to disease resistant genes by bulked sergeant analysis: a rapid method to detect markers in specific genomic regions using segregating populations. Proc Natl Acad Sci U S A 88:9828–9832

Morgante M, Vogel J (1994) Compound microsatellite primer for the detection of genetic polymorphism. US Patent APP No 08/326456

Nadha HK, Kumar R, Sharma RK, Anand M, Sood A (2011) Evaluation of clonal fidelity of in vitro raised plants of Guadua Physiol Mol Biol Plants (January–March 2013) 19(1):21–41 39 angustifolia Kunth using DNA-based markers. J Med Plants Res 5:5636–5641

Nadot S, Bajon R, Lejeune B (1994) The chloroplast gene *Rps4* as a tool for the study of Poaceae phylogeny. Plant Syst Evol 191:27–38

Nayak S, Rout GR (2005) Isolation and characterization of micro satellites in *Bambusa Arundinacea* and cross species amplification in other bamboos. Afr J Biotechnol 4:151–156

Nayak S, Rout GR, Das P (2003) Evaluation of genetic variability in bamboo using RAPD markers. Plant Soil Environ 49:24–28

Negi D, Saxena S (2010) Ascertaining clonal fidelity of tissue culture raised plants of *Bambusa Balcooa* Roxb. Using inter simple sequence repeat markers. New For 40:1–8

Negi D, Saxena S (2011) In vitro propagation of Bambusa nutans wall. Ex Munro through axillary shoot proliferation. Plant Biotechnol Rep 5:35–43

Orozco-Castillo C, Chalmers KJ, Wauh R, Powell W (1994) Detection of genetic diversity and selective gene introgression in coffee using RAPD markers. Theor Appl Genet 8:934–940

Paran I, Michelmore RW (1993) Development of reliable PCR-based markers linked to downy mildew resistance genes in lettuce. Theor Appl Genet 85(8):985–993

Penner GA (1996) RAPD analysis of plant genomes. In: Jauhar PP (ed) Methods of genome analysis in plants. CRC, Boca Raton, pp 251–268

Pervaiz ZH, Rabbani MA, Shinwari ZK, Masood MS, Malik SA (2010) Assessment of genetic variability in rice (Oryza sativa L.) germplasm from Pakistan using RAPD markers. Pak J Bot 42:3369–3376

Qian W, Ge S, Hong DY (2001) Genetic variation within and among populations of a wild rice Oryza granulata from China detected by RAPD and ISSR markers. Theor Appl Genet 102:440–449

Qiang Z, Yu-long D, Chen X, Hui-yu Z, Min-ren H, Ming-xiu W (2005) A preliminary analysis of phylogenetic relationships of Arundinaria and related genera based on nucleotide sequences of nrDNA (ITS region) and cpDNA (trnL-F intergenic spacer). J For Res 16:5–8

Rabbani MA, Pervaiz ZH, Masood MS (2008) Genetic diversity analysis of traditional and improved cultivars of Pakistani rice (Oryza sativa L.) using RAPD markers. Electron J Biotechnol 11:1–10

Rafalski JA, Tingey SV (1993) Genetic diagnostics in plant breeding: RAPDs, microsatellites and machines. Trends Genet 9:275–280

Rajput SB, Jani M, Ramesh K, Manokari M, Phanikanth J et al (2020) Large-scale clonal propagation of Bambusa balcooa Roxb.: An industrially important bamboo species. Ind Crop Prod 157:112905

Ramanayake SMSD, Meemaduma VN, Weerawardene TE (2007) Genetic diversity and relationship between nine species of bamboo in Sri Lanka, using random amplified polymorphic DNA. Plant Syst Evol 269:55–61

Rao RV, Hodgkin T (2002) Genetic diversity and conservation and utilization of plant genetic resources. Plant Cell Tissue Organ Cult 68:1–19

Rong JD, Zhang YH, Fan LL, Yu ZJ et al (2020) Genetic diversity and phylogenetic relationships in the *Bambusa* genus as revealed by RAPD markers. Appl Ecol Env Res 18:5011–5021

Rossarolla MD, Tomazetti TC, Vieira LN et al (2020) Identification and characterization of SSR markers of *Guadua chacoensis* (Rojas) Londoño & P.M. Peterson and transferability to other bamboo species. 3 Biotech 10:273

Sharma V, Bhardwaj P, Kumar Sharma RK, Sood A, Ahuja PS (2009) Identification and cross-species amplification of EST derived SSR markers in different bamboo species. Conserv Genet 10:721–724

Sharopova N, McMullen MD, Schultz L, Schroeder S, Sanchez-Villeda H, Gardiner J, Bergstorm D, Houchins K, Melia Hancock S, Musket T, Duru N, Polacco M, Edwards K, Ruff T, Register JC, Brouwer C, Thompson R, Velasco R, Chin E, Lee M, Woodman-Clikeman W, Long MJ, Liscum E, Cone K, Davis G, Coe EH Jr (2002) Development and mapping of SSR markers for maize. Plant Mol Biol 48:463–481

Shiran B, Amirbakhtiar N, Kiani S, Mohammadi S, Sayed-Tabatabaei BE, Moradi H (2007) Molecular characterization and genetic relationship among almond cultivars assessed by RAPD and SSR markers. Sci Hortic 111:280–292

Singh G (2010) Plant systematics. An integrated approach, 3rd edn. Science Publishers, Enfield, NH, pp 1–358

Singh SR, Dalal S, Singh R, Dhawan AK, Kalia RK (2012) Evaluation of genetic fidelity of in vitro raised plants of Dendrocalamus asper (Schult. & Schult. F.) Backer ex K. Heyne using DNA-based markers. Acta Physiol Plant. https://doi.org/10.1007/s11738-012-1084-x

Singh SR, Dalal S, Singh R, Dhawan A, Kalia RK (2013) Evaluation of genetic fidelity of in vitro raised plants of Dendrocalamus asper (Schult. & Schult. F.) Backer ex K. Heyne using DNA-based markers. Acta Physiol Plant 35:419–430

Stace CA (1989) Plant taxonomy and biosystematics, 2nd edn. Chapman and Hall Inc USA, Routledge

Sun Y, Xia N, Lin R (2005) Phylogenetic analysis of Bambusa (Poaceae: Bambusoideae) based on internal transcribed spacer sequences of nuclear ribosomal DNA. Biochem Genet 43:603–612

Sun Y, Xia N, Stapleton CMA (2006) Relationships between Bambusa species (Poaceae, ecology Bambusoideae) revealed by random amplified polymorphic DNA. Biochem Syst Ecol 34:417–423

Tautz D (1989) Hypervariability of simple sequences as a general source for polymorphic DNA markers. Nucleic Acids Res 17:6463–6471

Triplett J, Clark LG (2003) Ambiguity and an American Bamboo: the Chusquea culeou species complex. Bamboo Sci Cult J Am Bamboo Soc 17:21–27

Vieira LN, Dos Anjos KG, Faoro H, Fraga HP, Greco TM, Pedrosa Fde O, de Souza EM, Rogalski M, de Souza RF, Guerra MP (2016) Phylogenetic inference and SSR characterization of tropical woody bamboos tribe Bambuseae (Poaceae: Bambusoideae) based on complete plastid genome sequences. Curr Genet 62(2):443–453

Vos P, Hogers R, Bleeker M, Reijans M, van de Lee T, Hornes M, Frijters A, Pot J, Peleman J, Kuiper M, Zabeau M (1995) AFLP: a new technique for DNA fingerprinting. Nucleic Acids Res 23:4407–4414

Wachira FN, Waugh R, Powell W, Hackett CA (1995) Detection of genetic diversity in tea (Camellia sinensis) using RAPD markers. Genome 38:201–210

Wagner A, Blackstone N, Cartwright P, Dick M, Misof B, Snow P, Wagner GP, Bartels J, Murtha M, Pendleton J (1994) Surveys of gene families using polymerase chain reaction: PCR selection and PCR drift. Syst Biol 43:250–261

Wang K, Peng H, Lin E, Jin Q, Hua X, Yao S, Bian H, Han N, Pan J, Wang J, Deng M, Zhu M (2010) Identification of genes related to the development of bamboo rhizome bud. J Exp Bot 61:551–561

Watanable M, Ito M, Kurita S (1994) Chloroplast DNA phylogeny of Asian bamboos (Bambusoideae, Poaceae) and its systematic implication. J Plant Res 107:253–261

Waugh R, Bonar N, Baird E, Thomas B, Graner A, Hayes P, Powell W (1997) Homology of AFLP products in three mapping populations of barley. Mol Gen Genet 255:311–321

Welsh J, McClelland M (1991) Genomic fingerprints produced by PCR with consensus tRNA gene primers. Nucleic Acids Res 19:861–866

Williams JGK, Hanafey MK, Rafalski JA, Tingey SV (1993) Genetic analysis using random amplified polymorphic DNA markers. Methods Enzymol 218:705–740

Wolf ME, Dahlin SL, Hu XT, Xue CJ, White K (1995) Effects of lesions of prefrontal cortex, amygdala, or fornix on behavioral sensitization to amphetamine: comparison with N-methylD-aspartate antagonists. Neuroscience 69:417–439

Wu D, Meydani SN, Sastre J, Hayek M, Meydani M (1994) In vitro glutathione supplementation enhances interleukin-2 production and mitogenic response of peripheral blood mononuclear cells from young and old subjects. J Nutr 124:655–663

Wu MM, Kuo TL, Hwang YH, Chen CJ (1989) Dose–response relation between arsenic concentration in well water and mortality from cancers and vascular diseases. Am J Epidemiol 130:1123–1132

Zhang HY, Yang YM, Liu XZ (2011b) Bamboo species relations revealed by random amplified polymorphism of chloroplast DNA. Afr J Agric Res 6:1241–1245

Zhang W, Wendel JF, Clark LG (1997) Bamboozled again inadvertent isolation of fungal rDNA sequence from bamboos (Poaceae: Bambusoideae). Mol Phylogenet Evol 8:205–217

Zhang YJ, Ma PF, Li DZ (2011a) High-throughput sequencing of six bamboo chloroplast genomes: phylogenetic implications for temperate woody bamboos (Poaceae: Bambusoideae). PLoS One 6(5):e20596

Zhu S, Liu T, Tang Q, Fu L, Tang SH (2014) Evaluation of bamboo genetic diversity using morphological and SRAP analyses. Russ J Genet 50:267–273

Zietkiewicz E, Rafalski A, Labuda D (1994) Genome fingerprinting by simple sequence repeat (SSR)- anchored polymerase chain reaction amplification. Genomics 20:176–183

Chapter 11
Transgenic Approaches in Bamboo

Irfan Bashir Ganie, Anwar Shahzad, Zishan Ahmad, Najat A. Bukhari, and Kahkashan Parveen

Abstract The abiotic stress (salt, drought, and light) dangers coupled with the virus infection (Bamboo mosaic virus) have been a challenge in the growth and development of the bamboo. Transgenic studies of bamboo have revealed that the abiotic stress often caused by salt, drought, and the high light intensity challenge could be avoided by introducing abiotic stress-tolerant genes such as Aquaporin gene PeTIP4;1–1 which regulates the salt stress response in the genome of the bamboo, PeVDE genes in case of the light-sensitive plants. Moreover, CDKs such as AtCPK4 and AtCPK11 have also shown positive regulation against salt and drought stress. Most of the transgenic studies have revealed that the plant activates the antioxidant enzyme system in order to contain ROS generated by various abiotic stresses. Similarly, enzymes that are involved in xanthophyll cycle, which regulate the response to high light intensity, have been studied to lay the emphasis on the need to control the rising trend of light stress. However, bamboo has yet to be explored scientifically; otherwise, it could break many challenges that a man faces in the day-to-day spheres of life.

Keywords Aquaporin · ASR genes · Valine-glutamine (VQ) motif · Bamboo mosaic virus · CDKs · PheDi19-8 gene · Xanthophyll cycle

I. B. Ganie · A. Shahzad (✉)
Plant Biotechnology Section, Department of Botany, Aligarh Muslim University, Aligarh, Uttar Pradesh, India
e-mail: ashahzad.bt@amu.ac.in

Z. Ahmad
Bamboo Research Institute, Nanjing Forestry University, Nanjing, Jiangsu, China

Co-Innovation Center for Sustainable Forestry in Southern China, Nanjing Forestry University, Nanjing, China

N. A. Bukhari · K. Parveen
Department of Botany and Microbiology, College of Science, King Saud University, Riyadh, Saudi Arabia

Z. Ahmad et al. (eds.), *Biotechnological Advances in Bamboo*,
https://doi.org/10.1007/978-981-16-1310-4_11

11.1 Introduction

Bamboos (family, Poaceae; subfamily, Bambusoideae) are early maturing and fast-growing plants that mostly grow in the tropical areas of the world. It has been reported that the members of the Bambusoideae grow in various geographical regions ranging from subarctic to tropical regions (Gratani et al. 2008). Moreover, bamboo has been found to be highly flexible, so much that it can bend in strong winds, but would rarely break into pieces; thus, the nature of flexibility gives the plant enough strength to withstand against the fast wind. Further reports have suggested that Brazil alone represents about 89% genera and 65% species of bamboos in the world. Tabocais and pacales are the two natural bamboo forests in Brazil, which cover around 600,000 ha across Brazil.

At the economic point of view, India's capital is growing from the export of bamboo shoots to the countries which are having rising domestic demands of bamboo, and this business has successfully rescued many lives from the dangers of neglect and marginalization. India's domestic size of the economy coming from bamboo exportation has been reported to be growing from 450 to 26,000 million Indian rupees, and moreover, it has been suggested that such a business could uplift five million families from poverty in India. In Asian countries, particularly in China and Japan, bamboo plants have so much wide applications that it can be better comprehended from the famous lines of William Edgar Geil's book *A Yankee on the Yangtze* that "a man can sit in a bamboo house under a bamboo roof, on a bamboo chair at a bamboo table, with a bamboo hat on his head and bamboo sandals on his feet. He can at the same time hold in one hand a bamboo bowl, in the other hand bamboo chopsticks and eat bamboo sprouts." It is used in furniture, utensils, and fiber and construction purposes.

Having low fat content and high dietary fibers, vitamins, potassium, and other active materials, bamboo shoots have taken a space on our dining table as well and, therefore, have been consumed by people in various forms—raw, boiled, fermented, liquid, and medicinal forms. However, all bamboo shoots are not edible. It has been reported that B. pallida, B. tulda, B. polymorpha, B. balcooa, *D. hamiltonii*, D. giganteus, and M. bambusoides are the species of bamboo that have edible shoots. This is quite surprising that beside all the good attributes bamboo plants have there is nowhere bamboo processing units or bamboo industries which can serve best for the people in terms of both social and economic aspect.

Since time immemorial, bamboo parts have been in use for medicinal purposes particularly in tribal areas. Bamboo contains flavones, glycosides and antioxidants, and antiaging potential elements. Shoot sap of *Bambusa vulgaris* has found to be effective against jaundice disease. Bamboo manna is a silicious concretion of bamboo shoots, particularly found in abundance in the shoots of *Bambusa arundinacea* species which was reported to be an excellent tonic for respiratory disorders. Leaf juice has often been given to patients suffering from osteoporosis and has found to be effective in strengthening the cartilage (Vanithakumari et al. 1989). Similarly, *Bambusa arundinacea* extract has played an important role in Indian folk medicine while treating a large number of inflammatory conditions. Because of its antimicrobial activity, it has served well as a bioactive agent for a number of applications, including bamboo juice, bamboo salt, and bamboo charcoal. A tall bamboo, *Pleioblastus amarus*, growing in the southern part in China has a slightly pungent taste but has proven to be effective against fever and lung inflammation (Kiruba et al. 2007). Moreover, because of its surprising potential for the world market and certainly for its commercial value, bamboo has given livelihood to around 2.5 billion people who are earning around US$ 7 billion per annum. It has been an inseparable part of various traditions and cultures particularly that it has proven to be sources of musical instruments which has hugely contributed in strengthening and influencing various cultures of the world. Due to technological advancement, bamboo has become more important in the global markets, particularly in the form ply bamboo, parquet, and canned vegetable.

11.2 Brief Account of Its Potential for Micro-Propagation

Traditionally, bamboos have been propagated through culm cuttings and seeds. However, the method of seed propagation provides very less productive results because of the several problems that are natural to the plant like poor seed set, too long flowering cycle, the monocarpic nature of plants, highly heterogeneous seedling population, and consumption of seeds by wild animals. Moreover, the seasonal dependence of the plant, limited propagules, low survival rate, and transportation are the major constraints that we face in propagating the bamboo through vegetative means. The current pace of bamboo supply has hardly met the growing demands of the global market and the bamboo areas that have been rendered desolate. So the vegetative reproduction of the bamboo alone could not meet up the rising demands of the bamboo. In addition, bamboo has to be researched both wide and broadly on the lines of its potential to cure many diseases.

Besides the abiotic and biotic factors, the success or failure of any species of bamboo nursery also depends on how plant evolves a mechanism to defend itself from pests and diseases, and, however, nothing has remained aloof from diseases; therefore, the existing study has revealed that the three species of bamboo, namely, *B. blumeana*, *Bambusa* sp., and *D. latiflorus*, are highly susceptible to tar spot disease-causing pathogens and leaf rust disease-causing *Phakopsora louditiae* (Dayan 1988). Similarly, Bamboo mosaic virus (BaMV), caused by *Potexvirus*, has infected about 13 species of bamboo alone and has highly reduced the quantity, quality, and yield of the plant. It has severely impacted the production of useful bamboos and therefore has become the challenging factor for bamboo growers. Surprisingly, so far no effective chemical has successfully been able to tackle BaMV (Hsu et al. 2000).

So in such circumstances, the sole solution to protect the plant from disease-causing organisms is to develop in vitro resistant varieties of bamboo, and as said earlier research must be broadened for the species that are having utmost importance in environment, capital, and medicine. Moreover, traditional breeding methods can offer a quick and easy solution to disease-susceptible bamboo plants (John and Nadgauda 1999). However, the flowering habits of bamboo species are so peculiar that it has become difficult to conduct the breeding experiments for desired traits among bamboo plants. In addition, the clumps of the plant die just after flowering stage, which has added to the pain. Therefore, tissue culture has opened some doors to rescue the hybrid seeds that have produced through traditional methods of breeding. A number of attempts have been made in propagating the bamboo through in vitro techniques in order to save the precious lines of having high economic value. In fact, some attempts have been laid to induce flowering in bamboo through in vitro methods. Moreover, as we shall discuss, attempts have also been made to establish transgenic bamboo plants that are particularly resistant to abiotic stresses. The choice of its seed propagation is rarely being practiced because of its natural long flowering season, which can extend up to 120 years. Moreover, seed sterility and short seed viability have also made its propagation an uphill task.

11.3 Stress-Tolerant Genes in Bamboo

Sucrose is a major form of carbohydrate for transportation in higher plants and is an essential substrate for the proper growth of the plant. The principal use of sucrose in the sink organs is to provide carbon and energy source. Enzyme invertase breaks sucrose into two sugar molecules: glucose and fructose. Sucrose through various products enters into a range of metabolic pathways. Moreover, glucose, sucrose, and fructose can act as signaling molecules and could help in the regulation of various important genes. Based on their site of location, plant invertases have been categorized into cell wall, vacuolar, and cytoplasmic invertases. The vacuolar invertases play functions like cell expansion, osmoregulation, and sugar storage. Similarly, cell wall invertase entails in assimilating partitioning, plant development, and sink strength regulation. However, the most important role of cell wall invertases which has been supported by many molecular genetic studies is the vegetative and reproductive (Roitsch et al. 2003; Ruan et al. 2010).

More importantly, the above-studied enzymes have a crucial role in generating the stress response. The gene expression of these enzymes is regulated by stresses like phytohormone, temperature, light, and pathogen and wounding. Genes of such enzymes exhibit differential pattern in various tissues of the bamboo. Further insights have been drawn from the study conducted on Boβf ruct1 and Boβf ruct2 cell wall invertase genes (Chiu et al. 2006; Hyun et al. 2009). In order to gain more knowledge about the role of invertases in the bamboo shoot, promoter regions of cell wall invertase genes, Boβf ruct1 and Boβf ruct2, and a vacuolar invertase gene, Boβf ruct3, from *Bambusa oldhamii* have been cloned and analyzed in silico. Moreover, studies related to the effect of abiotic factors and different phytohormones in an in vitro culture have also been conducted.

It was reported that like cell wall invertase, gene of various plant species is induced by sucrose and glucose molecules; similarly the gene Boβf ruct1 is induced by sucrose and glucose molecules. However the Boβruct2 was found to be enhanced once the mannitol was added in the media replacing the sucrose as a source of carbon. More surprisingly, the level of Boβruct2 mRNA was found to be getting decreased when the bamboo shoots were transferred from a medium devoid of sucrose to the sucrose-rich medium. It is a well-known fact that regulation of transcription involves cis and trans elements, and it was observed that no any responsive element within the promoter region of Boβf ruct1 is involved in the induction of gene regulation. However, still more research needs to be done.

Similarly, several responsive elements within the promoter region of Boβf ruct2 which were found to be involved in gene expression are SURE-1, SP8b, GARE, and a pyrimidine box. In addition, phytohormones like cytokinins, GA, and auxin enhance the expression of cell wall invertases in the growing tissues of bamboo, and Boβf ruct2 genes were observed to exhibit different responses to ethylene, abscisic acid (ABA), IA, and GAs. GAs were observed to be showing quite wide differential gene expression patterns in bamboo plant. Gene Boβf ruct1 was found to be hardly impacted by GA and was observed to be predominantly expressing in the base culms of bamboo shoots, whereas Boβf ruct 2, which has been observed to be highly expressing in different regions of the bamboo shoot, was upregulated by GA (Hsieh et al. 2006). Moreover, it was further revealed that the Boβf ruct2 gene was upregulated by cold conditions, which clearly suggest that the bamboo plant can resist the cold conditions of the habitat.

The presence of a CGCG element within the promoter region of the 0.8 kb Boβf ruct1 is most striking feature of the gene. The CGCG element has shown diverse functions in a number of plants, particularly its active involvement in hormone signaling, like ethylene and ABA signaling (Yang and Poovaiah 2002). Moreover ERE ethylene-responsive element (ERE), GCC box, and CE1 were also found to be associated with the Boβf ruct1. The gene expression studies of cell wall invertase (Boβf ruct genes) reveal that Boβfruct1 and Boβfruct2 are involved in unloading and translocation of sucrose from medium to organs in order to maintain the sink.

11.4 Salt- and Drought-Tolerant Genes in Bamboo

It has been studied that an aquaporin gene -PeTIP4;1–1 is involved in the regulation of various abiotic stresses. Keeping this potential in mind, Aquaporin gene PeTIP4;1–1 has been overexpressed in *Arabidopsis* and has successfully shown positive drought and salt tolerance results. The Aquaporin gene of Moso bamboo (*Phyllostachys edulis*) comprises of 756 bp ORF that encodes a protein unit of 251 amino acids. Moreover, PeTIP4;1–1 gene expresses constitutively in the shoot culms of the bamboo, and the expression level increases with the increase in the height of the bamboo shoot. More importantly the gene was upregulated in leaves and roots under saline and drought stress conditions. Transgenic *Arabidopsis* has shown considerable enhancement of tolerance against the drought and salinity stress, particularly when the aquaporin genes were allowed to overexpress under the control of CaMV 35S promoter. This tolerance was both morphologically and physiologically quite visible, particularly at the physiological point of view which clearly suggested that the level and the activities of SOD, POD, and CAT were observed to be higher along with the higher absorption of water content.

Because of the well-established rhizome system and rapid growth, bamboo has shown immense potential of water transport, which is quite visible during the days of spring, when the bamboo plant shows a rapid elongation in its shoots and its validation has been further supported by the fact that the water content of bamboo shoot was found to vary from environment to environment. Therefore, it could be argued that the water content and atmospheric conditions have a key part to play in the morphological study of the plant. Aquaporins are the most essential proteins in the bamboo and were found to be regulating about 70–80% water transportation. Recently, numerous aquaporin genes were identified in barley (Utsugi et al. 2015) *Jatropha*, and alone in Moso bamboo around 26 aquaporins were identified (Sun et al. 2016). More interestingly, aquaporin PeTIP4;1–1 is the first aquaporin gene that has been cloned and characterized so far. Its structure comprises of six transmembrane α-helices and two conserved "NPA" motifs, which has a quite semblance to TIPs of other plant species, which mostly perform the function of water transportation and osmotic regulation (Utsugi et al. 2015).

Nonstructural carbohydrates (NSCs) from the mature portions of the bamboo support the plant in achieving the height and diameter (Song et al. 2016b). It has been studied that NCPs transport along with the sap flow, thereby indicating the importance of water availability. Moreover, some recent studies have further suggested that aquaporins play an important role in the elongation of cells that rapidly promotes the shoot growth, via the co-regulation of gibberellin and auxin phytohormones. Furthermore, few meristem expression elements like CAT box were identified in the promoter regions of PeTIP4;1–1 gene which support the bamboo growth and development. The accumulation of higher transcript levels of PeTIP4;1–1 in the leaf and root parts of the plant was reported to have higher tolerance to salinity and drought stress.

To further investigate and evaluate the biological capability of PeTIP4;1–1 gene in leaves and roots under stress conditions, in vitro transgenic *Arabidopsis* plants were generated in which the expression of the gene has been upregulated. The transgenic plants were exhibiting long taproots and more green leaves. Antioxidant system is a vital system that defends the plant from the damage often caused by stress conditions. Antioxidant system is composed of a number of enzymes such as SOD, POD, and CAT which destroy reactive oxygen species. More importantly, the activities of such antioxidant enzymes were enhanced in transgenic *Arabidopsis* under salt and drought conditions. This clearly implies that the gene PeTIP4;1–1 effectively improves the activities of SOD-, POD-, and CAT-like enzymes (Chang et al. 2016).

Moreover, lipid peroxidation generates MDA which is an indicator of damage caused by salt stress via ROS system (Niu et al. 2012). Transgenic *Arabidopsis* was reported to have a lower concentration of MDA which indicates that the transgenic plants experience less lipid peroxidation reaction and cellular damage under salinity and drought stress (Moore and Roberts 1998). Mostly a network of proteins works together to bring out a change that one specifically desires to bring about, and such a network needs a trigger which usually comes through the expression of a desired gene. Same is the case with PeTIP4;1–1 gene, which is believed to have an indirect effect on a number of proteins that are involved in the regulations of protein transporters. Gene expression study of PeTIP4;1–1 has revealed that the stress-responsive genes such as NHX, P5CS, and LEA were reported to have shown undergoing upregulation in transgenic plants particularly under abiotic stress conditions.

Rauf et al. (2014) have reported that the overexpression of LfNHX1 in tobacco has led to tolerance against salt and drought stress, and similar results were observed in groundnut when AtNHX1 genes were overexpressed (Asif et al. 2011). Therefore, the higher transcript levels of AtNHX1 suggest that PeTIP4;1-1-overexpressing plants may regulate ion transporters, which contribute in the stress responses against stress responses. Similarly, another gene AtP5CS leads to the accumulation of proline under the conditions of salt and drought stress (Zhang et al. 1995; Szabados and Savouré 2010); thus, overexpression of PeTIP4;1–1 may have enhanced the proline content in transgenic *Arabidopsis*.

Moreover, desiccation-tolerant proteins LEAs were studied to have correlation with PeTIP4;1-1-overexpressing *Arabidopsis* plants, as expression of AtLEA and PeTIP4;1–1- genes resulted in desiccation tolerance (Pedrosa et al. 2015). These studies suggest that there might be an unknown mechanism that interlinks the functionality between PeTIP4;1–1 and LEA. To elucidate the overarch effect of PeTIP4;1–1 genes, further research is needed to conduct in bamboo plants.

11.5 Drought Tolerance Shown by Bamboo

During the growth cycle of a plant, exposure to abiotic stress conditions affects the total yield and production of the plant. Statistical analysis has shown that the abiotic stresses cause a total loss of about 50% bamboo per year. Among various stress conditions, drought stress has become a very problematic condition as it deeply affects the overall growth and development of a plant. Moreover, it has been further studied that the drought has become an acute constrain in the growth and development of the bamboo, as it affects the very important physiological processes such as photosynthesis and fluorescence characteristics of the plant. With the increase in the intensity of drought, the rate of photosynthesis and transpiration rapidly decreases, which eventually affects the growth of the plant. To understand the stress responses in bamboo plant at the standpoint of its response under drought stress, it is important to study the genes that are involved in abiotic stress responses in a bamboo plant. The drought-related genes are categorized as (a) functional genes, e.g., LEA, and (b) osmotic protectant synthetase genes, whose products are reported to having a vital importance for the plant under the conditions of drought stress. However, TFs and CDKs also play a part in positively regulating the stress responses under drought stress.

11.6 The ASR (Abscisic Acid-, Stress-, and Ripening-Induced) Gene Family

The ASR (abscisic acid-, stress-, and ripening-induced) gene family is a type of transcription factor. The first ASR was isolated and identified from tomato, since then many ASR TFs have been identified in large numbers among plant species, including woody and herbaceous plants; no arthologues of ASR so far have successfully been identified in *Arabidopsis* (González and Iusem 2014). It has been reported that ASR proteins are involved in fruit ripening, senescence, and plant growth (Chen et al. 2011; González and Iusem 2014; Wimmer 2003). Moreover, several research works have supported the view that several ASR proteins respond to abiotic stresses, including ABA (Philippe et al. 2010; Saumonneau et al. 2008). For example, transgenic *Arabidopsis* expressing ASR gene from banana (MaASR), *Musa paradisiaca* (MpASR), and *Lilium longiflorum* has shown improved results of salt and drought tolerance (Table 11.1).

It has been reported that although ASR genes responded to various abiotic stresses, however, their exact role in abiotic stress responses is not well known in Moso bamboo. Therefore, further studies were conducted in which ASR gene was isolated from bamboo, and its role in abiotic stress responses was characterized. It has been reported that ASR (PheASR2) genes involve ABA signaling while conferring drought tolerance. ASR gene PheASR2 was observed to have quite well homology with other ASR stress-responsive genes, particularly in two conserved domains. The phylogenetic relationship of PheASR2 gene has shown quite similar with another cluster of stress-responsive genes of ASR genes (Birsen et al. 2003; Hu et al. 2013; Li et al. 2016; Vivekanand et al. 2015). Structurally PheASR2 comprises

Table 11.1 ASR genes showing tolerance to different abiotic stresses

Plant source	ASR type gene	Transgenic plant	Tolerance	Reference
Musa paradisiaca	MpASR	Arabidopsis	Salt and drought tolerance	Dai et al. (2011)
Banana	MaASR	Arabidopsis	Salt and drought tolerance	Hsu et al. (2000)
Lilium longiflorum	LLA23	Arabidopsis	Salt and drought tolerance	Zhang et al. (2014)
Solanum lycopersicum	SlASR1	Tobacco	Osmotic stress	Kalifa et al. (2010); Jha et al. (2012)
Wheat	TaASR1	Tobacco	Water stress	Hu et al. (2013)
Foxtail millet	SiASR1	Arabidopsis	Drought and oxidative stress	Hu et al. (2013)

of a zinc binding domain and putative nuclear targeting domain on N and C terminal, respectively. It was further reported that the expression of transient PheASR2:GF in a transformed tobacco leaf has revealed that PheASR2 is localized in the nucleus, and previous studies had revealed that such pattern was also observed in other plants such as lily, grape, and wheat (Birsen et al. 2003; Hu et al. 2013; Yang et al. 2005a, b). In transgenic rice, the PheASR was observed to having displayed strong tolerance to drought stress. The PheASR was reported to be getting overexpressed, particularly at the germination stage of the transgenic plant, therefore helping the plant in exhibiting the drought tolerance at the most crucial stage of the plant growth. So the germination rate was observed quite large in number as well as the plant height, compared to normal plants. Under water deficit conditions, the drought tolerance of the 4-week-old transgenic plants, particularly in its vegetative growth phase, was studied to be unpredictably quite high compared to non-transgenic plant with the vegetative stage. It has been well studied that under stress conditions the content of MDA and H2O2 rises to the level that it severely damages the membrane system of the cell. Therefore, MDA is a crucial parameter to assess whether a particular drought-tolerant gene functions in transgenic plant or not (Hu et al. 2012; Reilly and Aust 1998). Under the conditions of drought stress, the MDA and H2O2 content was observed to be reduced in PheASR2-overexpvressing plants, therefore implying that the transgenic plant experiences less membrane damage and lipid peroxidation as compared to non-transgenic plants, as studied earlier that the content of ROS species increases to the extent that it poses a serious threat to the membrane system of the plant under a stressful environment. However, the transgenic plants in such circumstances switch on their antioxidant system by increasing the activities of CAT and SOD to scavenge the ROS during the drought stress. Similar results were reported in tobacco line expressing another type of ASR gene—TaASR1 (Hu et al. 2013). Similarly, ROS levels were regulated by ROS-producing and ROS-scavenging genes (Abbasi et al. 2007; Mittler 2002).

It was reported that the gene PheASR2 in case of transgenic plants increases the activity of antioxidant enzymes like CAT, SOD, and APX while reducing the expression of ROS genes like RbohA and RbohB. ASR drought resistance genes such as OsASR5, OsASR1, SiASR4, and VvMSA, PheASR2 genes were stimulated by ABA signaling (Joo et al. 2013; Li et al. 2017, 2016; Hu et al. 2013; Birsen et al. 2003). Other important genes that are involved in drought tolerance such as OsAREB, P5CS1, OsLEA, and OsNCED3 are categorized as ABA-induced marker genes (Yasunari et al. 2005; Zhu et al. 2009). Existing knowledge guides us that ABA-responsive element binding functions as major ABA-responsive element which can activate ABRE binding genes. Similarly, OsNCED2 enzymes also take active part in ABA-mediated drought tolerance response (Zhu et al. 2009). Moreover, under drought stress, the expression of OsAREB and OsNCED2 was increased by PheASR2-overexpressing lines, while no any marked change in non-transgenic plants was observed. Consistent with this, the expression of other genes such as OsDREB1A and OsERD that are involved in drought tolerance through ABA-pathway was also reported to be increasing in transgenic plants expressing PheASR2 genes.

11.7 Salt Tolerance in Bamboo

The environment is an important factor in determining the overall development of the plant. However, negative factors of an environment cannot be ignored such as insects, diseases, and abiotic stresses (salinity, drought, and temperature) which could have a serious impact on the growth of the medicinally and economically important bamboo plants. Fortunately, plants have evolved various mechanisms to tackle both abiotic and biotic stresses in order to adapt in a changing environment. During the period of its adaptation, plants regulate various process such as osmotic balance and synthesis of stress-related proteins and antioxidants to support the growth and development of the plant (Shinozaki and Yamaguchi-Shinozaki 1997; Zhu 2002). It has been reported that valine-glutamine (VQ) motif-containing proteins are involved in stress responses (Song et al. 2016a). Plant-specific VQ proteins were reported to have a close association with WRKY transcription factors for an effective transcription (Lai et al. 2011). VQ proteins have also been identified in *Arabidopsis*, maize, grape, and soybean (Song et al. 2016a). VQ proteins regulate many biochemical processes in plants by interacting with a wide variety of transcription factors (Wu et al. 2017). This has been validated by a TF AtVQ15 which interacts with WRKY25 and WRKY51 specifically in order to regulate drought and salt stress (Perruc et al. 2004). Moreover, protoplasmic studies from the leaves of *Arabidopsis* reveal that VQ proteins can interact with each other also (Wang et al. 2015).

Further studies have revealed that the accumulation of salt, low N2, light stress, and ABA can either inhibit or induce the VQ gene expression, indicating that VQ proteins play a vital role in regulating the stress responses (Hu et al. 2013b; Wang et al. 2014b). It was further reported that the AtVQ9 mutants of *Arabidopsis* have shown a higher growth of germination and a superior growth of seedling under the treatment of NaCl indicating its involvement in negative regulation (Perruc et al. 2004). Similarly, the low N2 stress triggers GmVQ6 and GmVQ53 gene expression significantly in root and stem and resulted in the overall growth of a soybean plant (Wang et al. 2014b). Collectively, the aforementioned studies indicate that the plant with changing environment evolves a system, particularly at the molecular level, which significantly alters in best way both the physiological and biochemical processes to support the growth of the plant.

Similarly, PeVQ28 cDNA – a type of VQ protein isolated from Moso bamboo – was transformed into *Arabidopsis* through *Agrobacterium* bacteria in order to obtain a transgenic with a trait of overexpressing PeVQ28 gene. This transformation method has helped us a lot in elucidating the intricacies of molecular mechanisms that involve in stress responses induced by PeVQ28 gene expression in Moso bamboo. Under high salt stress conditions, gene PeVQ28 was so highly regulated that the expression level was observed to about 80-fold increase, therefore indicating that the gene might be induced by salt stress adversity. Second, the VQ protein family can interact with WRKY transcription factors in ***Arabidopsis*** (Hu et al. 2013a). To study the interaction of verify VQ protein with WRKY transcription factor in bamboo, yeast two-hybrid and BiFC assays were performed. It has been reported that PeVQ28 protein, both in vivo and in vitro, interacts with WRKY transcription factors that results in the formation of a complex in the nucleus. The VQ-assisted WRKY transcription factors are reported to be involved in the regulation of plant development and stress response (Perruc et al. 2004; Wang et al. 2015; Wu et al. 2017).

As mentioned before, plants evolve many mechanisms to tackle abiotic stresses. It is a known fact now that plants suffering from cellular damage during abiotic stresses often accumulate ROS which causes detrimental reactions such as peroxidation of lipids and oxidative stress (Xiong et al. 2002; Mittler et al. 2004). Moreover, MDA content has also become an indicator to determine the level of stress impact, as it provides information about the level of ROS in a cell. Under salinity stress conditions, the MDA content of non-transgenic plants was higher as compared to transgenic plants which were expressing PeVQ28 gene. Further studies have suggested that the accumulation of the PRO is an important physiological activity to stabilize the subcellular structures in times of stress (Xiang et al. 2007). In case of PeVQ28 transgenic plants, the data has revealed that the accumulation of PRO was much higher compared to the non-transgenic plant which indicates that the PeVQ28 has a role in activating the PRO system (Song et al. 2011).

Besides, an antioxidant enzyme system CAT, SOD, and POD functions during stress conditions to protect the cell from any cellular damage. Under salt stress conditions, the activities of antioxidant enzymes like POD, CAT, and SOD in transgenic plants were reported to be high which eventually helps in reducing the MDA content. The activation of ABA-dependent signaling under salt stress leads to the accumulation of ABA (Nambara and Marion-Poll 2005). Overexpression of some genes induces the activation of stress-responsive genes in plants, which in turn leads to the resistance of the plant to various abiotic stresses of diverse stresses. Moreover, under normal conditions, AtRD29A, AtRD29B, AtABF1, AtAAO3, and AtNCED2 have shown low expression both in PeVQ28-overexpressing plant and non-transgenic plant. However, the expression level of these genes was observed to be quite high in transgenic plants under salt stress conditions, thereby indicating a vital role that the gene PeVQ28 does play in salt tolerance (Fig. 11.1).

Furthermore, the expression levels of various ABA signaling genes such as AtP5CS1, ABI1, and AtPP2CA were separately studied in transgenic and non-transgenic plants under both stress and normal conditions (Hirayama and Shinozaki 2007). Gene AtABI1 and AtPP2CA codes for an important protein called phosphatase 2C, which is a key enzyme in stress responses conducted through the ABA-mediated signaling. Moreover, the expression level of both ABI1 and AtPP2CA was reported to be very higher in PeVQ28-overexpressing plants than the non-transgenic plants. In addition, gene AtP5CS1, once induced, leads to the accumulation of PRO in PeVQ28 transgenic plants, thus contributing to the enhancement of salt tolerance. Therefore, the studies conducted suggest that the resistance of PeVQ28-overexpressing plants to abiotic stress may also be due to the expression of other genes. Fundamentally the interaction between proteins is the basic for a number of key metabolic activities. Therefore, so is the case with VQ proteins, as

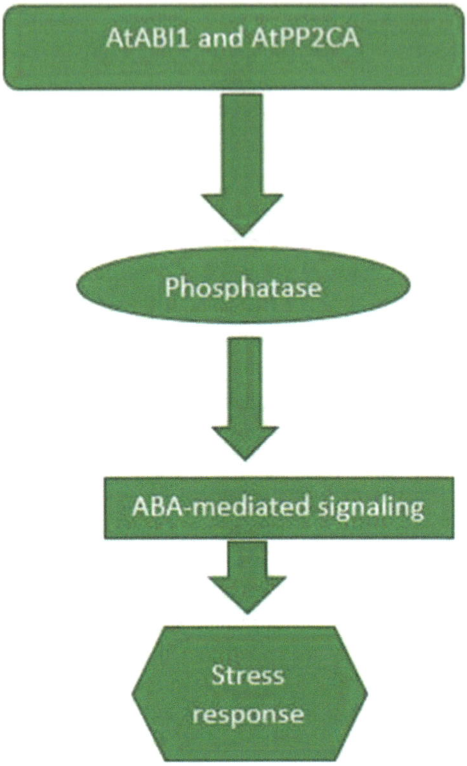

Fig. 11.1 Stress response mediated by ABA signaling

it interacts with other proteins too (Hu et al. 2013a, b; Wu et al. 2017). For example, interaction between AtVQ14 and AtWRKY10 resulted into a formation of a complex that significantly affected the size of the seed (Wang et al. 2010). AtVQ14 also interacts with AtWRKY28 to affect the jasmonic acid-mediated signaling pathways (Hu et al. 2013a). Keeping this principle of protein interaction into consideration, research on interaction PeVQ28 and PeWRKY83 was conducted using a yeast two-hybrid assay method, which confirmed that protein PeVQ28 does interact with PeWRKY83. Thus, we can conclude that VQ family proteins in Moso bamboo interact with each other in order to respond abiotic stresses.

11.8 Role of PheDi19-8, a Type of Di19 Gene Family

As we have studied earlier, plant growth is strongly impacted by salt, drought, and light stresses. Moreover, different TFs have also been the part of stress regulation in plants, and such transcription factors include Cys2/His2-type zinc finger proteins, NAC, MYB, bZIP, and WRKY family members (Chung et al. 2002; Mao et al. 2012; Oh et al. 2011; Uno et al. 2000; Marè et al. 2004).

PheDi19-8-is a type of Di19 gene family studied from Moso bamboo which has shown salt and drought tolerance when the plant is subjected to such stresses. Phylogenic studies have revealed that the gene PheDi19-8 is closely associated with OsDi19-4 and TaDi19A genes, and the existing knowledge has revealed that both these genes are involved in abiotic stress responses (Wang et al. 2014a). Phylogenetic studies and expressional analysis suggest that PheDi19-8 may be involved in stress responses in Moso bamboo. To find the answer of this question, the stress regulation role of PheDi19-8 was examined in plants that are expressing a foreign gene, particularly for drought tolerance. PheDi19-8 in rice and *Arabidopsis* were showing tolerance to drought. It was observed that at vegetative stage, the PheDi19-8 gene-overexpressing plants were showing higher survival rate under drought stress. Moreover, the tolerance was equally good, even after the stage of germination of PheDi19-8 gene-overexpressing plant, particularly when the transgenic plant was subjected to mannitol treatment. Gene Atdi19 also played a part in the regulation of drought tolerance. It was observed that PheDi19-8 gene has successfully replaced Atdi19 mutant, when subjected to drought stress (Liu et al. 2013). Further studies reveal that to date so many Di19 proteins were studied which are positively regulating the multiple number of abiotic stress responses, particularly drought tolerance responses observed in transgenic rice plants where the expression of OsDi19-4 has been studied to touch expectations (Wang et al. 2014a). However, it would sheer injustice to not to mention that a majority of the Di19 proteins function as negative regulators for stress responses (Feng et al. 2015).

11.9 Overexpression of PheDi19-8 Increases Drought Tolerance in *Arabidopsis*

Was it really PheDi19-8 gene or any other factor/genes that probably had triggered the good results of drought tolerance? To answer the question, some of the experiments were conducted; firstly, 3-week-old transgenic PheDi19-8-overexpressing plant and Col plants were subjected to water-deficient conditions for 18 days; in this experiment, it was observed that the transgenic plants remained healthy, while the Col plants suffered so much that they died and failed to stand the stress.

Leaf excision test too revealed that the water loss was quite high from mutant and non-transgenic plants as compared to PheDi19-8-overexpressing plant; in fact, all the plants with di19 mutations died. Moreover, the stomatal apertures of leaves were observed to getting reduced, which eventually results in the decrease of transpiration. However, in case of di19 mutant plants, the stomatal apertures were larger and thus failed to withstand the water-deficient conditions for a long time.

The above results demonstrate that the regulation of stomatal size may be due to the expression of PheDi19-8 genes, therefore contributing to drought tolerance systemically. To find out the exact knowledge behind the high level of drought tolerance in case of transgenic PheDi19-8 genes over-expressing plants, the seedlings of gene expressing plant and a non-transgenic plant were further subjected to osmotic stress. When the seedlings of both transgenic and non-transgenic plants were then placed on MS medium plates devoid of any particular stress inducer treatment, unsurprisingly the root growth of both transgenic PheDi19-8-overexpressing plants and non-transgenic plants were same. However, under the conditions of abiotic stress, transgenic seedlings were able to induce rooting more superior than non-transgenic plant. Furthermore, the root growth could be hardly distinguished between non-transgenic and di19 mutant plants.

11.10 Effect of Drought Stress on the Physiology of PheDi19-8-Overexpressing Lines

To study the impact of drought stress on the physiology of the transgenic plant, some important physiological activities were studied. Furthermore, studies conducted have revealed that the plant (transgenic) has remarkably shown tolerance when it was subjected to water-deficient conditions and has comparatively shown higher content of sugar molecules than non-transgenic plants or mutant plants. Moreover, the relative electrolyte leakage (REL) and malondialdehyde (MDA), which cause membrane injury, were reported to be low in number in transgenic *Arabidopsis* seedlings than the mutants and non-transgenic plants.

The gene PheCDPK22 was reported be induced by abiotic stress. However, its function has been reported to be antagonistically working, better to say, negatively regulating responses to abiotic stresses by controlling the most key proteins in interacting with the crucial gene, i.e., PheDi19-8 (Liu et al. 2013). Growing scientific research supports the thought that the CDPK gene family members work as a key enhancer of abiotic stress responses (Harmon et al. 2001). For example, AtCPK4 and AtCPK11 mutants of *Arabidopsis* display a low level of tolerance to abiotic stress and cpk3 mutants display a salt-sensitive phenotype (Zhu et al. 2007). The calcium-dependent protein kinase (CDPK) having conserved serine/kinase domain and two EF domains play an important role in enhancing tolerance to abiotic stresses, and the CDPK family members have been identified in number of plant species, including rice, wheat, and *Arabidopsis thaliana*. CDPK genes have largely contributed in bringing out tolerance to plants against a large number of abiotic stresses, such as cold, salt, and drought stresses (Xu et al. 2010). Similarly, gene OsCDPK7 have shown tolerance against drought, salt, and cold stresses in transgenic rice plants (Saijo et al. 2010). Moreover, genes OsCDPK13 and OsCPK21 have also shown tolerance to cold and salinity stress (Abbasi et al. 2004; Asano et al. 2011). However, the expression of PheCDPK22 has shown sensitivity to drought and salt stress. Xu et al. (2010) report that two genes AtCPK6 and AtCPK23 are of vital importance in responding to abiotic stresses. Similarly, OsCDPK7-overexpressing plants have also shown a tolerance to abiotic stresses (Abbasi et al. 2004; Asano et al. 2011). These findings contribute to our knowledge and implicitly suggest us that the stress responses is significantly regulated by CDPK-mediated stress response (Fig. 11.2).

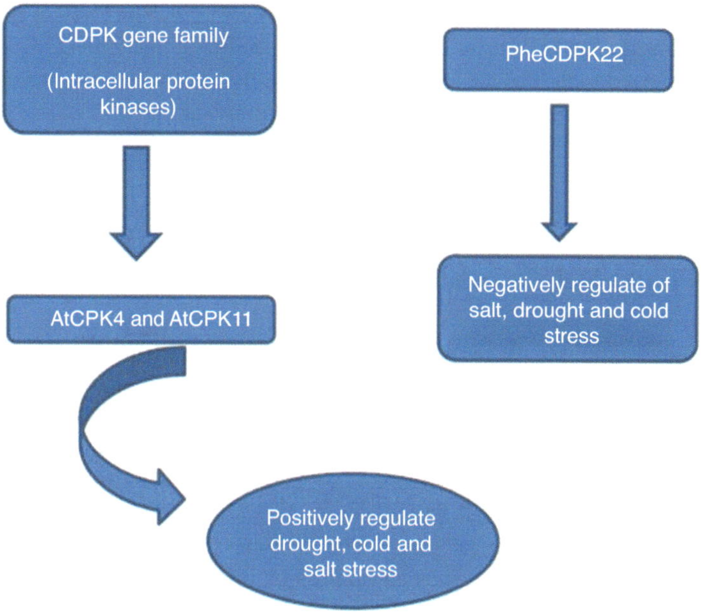

Fig. 11.2 Positive and negative regulation of salt, drought, and cold stress

Wei et al. (2018) studied how the PheDi19-8 and PheCDPK22 genes mediate the stress response against drought stress; the biochemistry and physiology (REL and MDA) of transgenic plants were studied under drought and normal environmental conditions. Plant accumulates soluble sugars to adapt with or withstand the environmental stress (Abraham et al. 2003). Under draught stress conditions, the transgenic PheDi19-8-overexpressing plants were reported to having higher content of soluble sugars compared to non-transgenic plants. However, the PheCDPK22-overexpressing plants have shown a very low content of soluble sugars under drought stress conditions. Furthermore, plants often suffer membrane damage under the conditions of salt stress, which can be determined with REL and MDA levels. It has been studied that REL content is mostly used for the study of the amount of damage the membrane has undergone, and MDA as said earlier is used to study the degree of membrane lipid injury (Smirnoff 1993; Hu et al. 2012). Correspondingly, the PheCDPK22-overexpressing plant was reported to having 1.65-fold increase of REL and 1.35-fold increase of MDA content than Col plants. Moreover, the expression of some stress-responsive genes such as LEA, RD29A, DREB2A, and RD22 was enhanced in transgenic PheDi19-8-overexpressing plants and reduced in PheCDPK22-overexpressing *Arabidopsis*. LEA, RD29A, RD22, and DREB2A are well-studied marker genes that are responsive to abiotic stresses (Finkelstein and Gampala 2002). Further results suggest that both PheDi19-8 and PheCDPK22 work as positive and negative regulators of drought tolerance. Bioinformatics information dictates us that the promoter region of DREB2A has a two

DIBS element, and this promoter is easily recognized by PheDi19-8 in a yeast one-hybrid assays and EMSAs. DREB2A gene specifically responds to water stress and was believed to induce the transcription of various TFs which act in a concerted manner against abiotic stresses (Yamaguchi and Shinozaki 2006).

11.11 Xanthophyll Cycle: The Center Where the Excess Light Regulation Takes Place

Needless to say, light is the most important component for the overall development of plants. However, sometimes light that do not fall in the tolerant range of a plant could cause a serious damage to the structural organization of a plant; nevertheless, the plant possesses an inbuilt capacity to tackle the light stress, probably through the physiological mechanism of dissipation to protect the PSII from any inhibitory damage. It has been reported that bamboo has a more than thousand species distributed across geographical places and ecotypes. Bamboo species over their lifetime have to adapt with various abiotic light stresses, as it is one of the serious environmental stresses that poses a serious challenge to the bamboo for its growth and production.

The physiological epicenter where the light stress is exceedingly well regulated is a xanthophyll cycle (Fig. 11.3). As said earlier, the xanthophyll cycle is a type of photoprotection regulation system in plants and algae, which converts V to A then Z

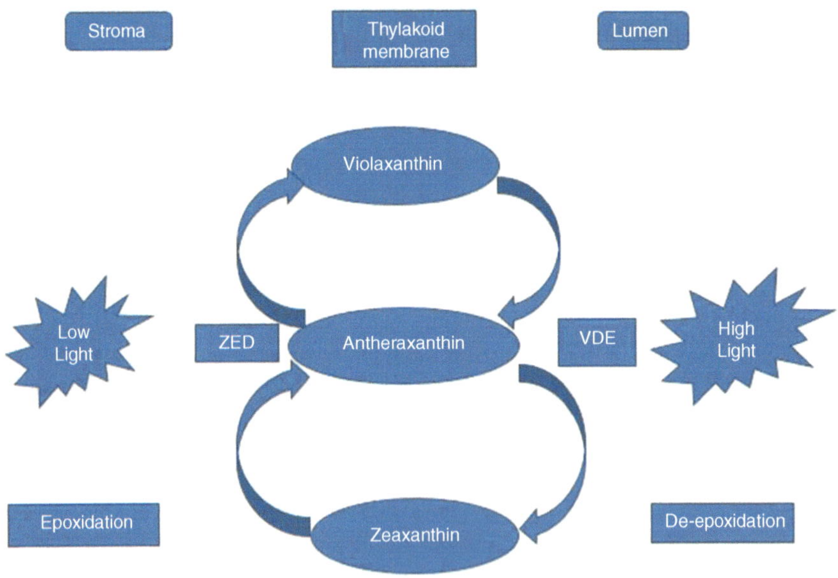

Fig. 11.3 Regulation of light stress by xanthophyll cycle

by VDE during the extreme hours of light stress phase to protect the photosynthetic apparatus of the plant and reverse the reaction when the threat of high light hours fades away (North et al. 2005), more precisely, to say, the physiological reaction system where the dissipation of light energy via de-epoxidation reactions of the xanthophyll cycle takes place. This is one of the key routes through which plants mitigate any chances of having threat of light stress (Demmig-Adams and Adams 1996). In the xanthophyll cycle, VDE is a very important enzyme, as it synthesizes zeaxanthin, which responds to extreme light responses. In light of this extraordinary photoprotection system, the efforts have been made to comprehend the light stress regulation in bamboos more decisively. Therefore, function and characterization of PeVDE was analyzed. It was observed that the length of an isolated cDNA of PeVDE was 1723 bp long, heavily interrupted with cis-acting regulatory elements. Moreover, the genomic sequence was interrupted with various motifs like CBFHV, CCAA TBOXI, GT1GMSCAM4, MYB2AT, and MYBCORE, which suggested that the transcription factors might be having influence over PeVDE. **The full-length cDNA 1723 bp long was interrupted with four introns containing many cis-acting regulatory elements.**

The recombinant protein of VDEs successfully catalyzed the reactions that ultimately lead to the dissipation of excess light. So these results suggest that the expressed protein has shown the enzymatic activities of converting V into Z through A, in the xanthophyll cycle. More importantly, it confirms that the isolated cDNA actually codes for VDE. However, when it was compared with other VDEs from other plant species (Lin et al. 2002), the enzymatic activities were relatively low. The reason for this difference might be the difference in the sequence of amino acids.

PeVDE were observed to have higher similarity with other plant VDEs, and a slight change in the sequencing would certainly affect the overall structure and the activity of an enzyme. The lipocalin domain has revealed that the PH strongly affects its conformity and hence the function of an enzyme (Arnoux et al. 2009). Saga et al. (2010) **have studied that the residues like Asp-177 and Tyr-198 have a catalytic function in the enzyme**. However, in case of PeVDEs, more research is needed to be done particularly on the role of various residues of the protein, which probably could be confirmed through the mechanisms like site-directed mutagenesis.

However, further studies to understand the enzymatic activity of PeVDE, particularly with its residues and other factors, are a need of the hour. Real-time PCR of PeVDE suggests that the transcription of PeVDE was reported to be higher in the leaves, as the accumulation of protein was found to be higher in leaves among the various green parts of the plant (North et al. 2005). Furthermore, the localization of PeVDE in the thylakoid lumen of the chloroplast was confirmed by Western blotting (Hager and Holocher 1994). Under light stress conditions, the level of PeVDE mRNA was reported to be increasing gradually particularly in leaves during the first 2-h treatment of extreme light stress conditions (Huang et al. 2007). However, the transcription level of mRNA was gradually observed to be decreasing under high light stress conditions for a long period of 8 h. After the treatment of 8 h, the transcription level reached to its stability. These results were inconsistent with that of

GVDE (Huang et al. 2007) and VDE in *Arabidopsis* (Woitsch and Romer 2003) which kept increasing their transcription levels during the treatment phase of 8 hours. Similarly, in the case of tomato, the LeVDE overexpression has significantly reduced the impact of light-induced photo-inhibition on PSII and PSI. Similarly in ***Arabidopsis***, the overexpression of VDE leads to de-epoxidation reaction that serves as a precursor to many important reactions which in turn promotes the growth of the plant (Chen and Gallie. 2012).

11.12 Conclusions

Keeping the huge economic and social benefits of bamboo into consideration, well-planned and an organized system of its cultivation should be set as a topmost priority. Moreover, bamboo plants that having immense medicinal and economic potential should be given separate care and treatment important plant. Dense plantation at tribal locations should not be disturbed unless a consensus is not generated between tribal and interest group; otherwise, it would amount to an unwanted invasion. Diseases have become the major threat to the bamboo cultivation, which have cost a huge economic loss. Bamboo has been prone to fungal pathogens such as *Alterneria*, *Colletotrichum*, *Curvularia*, *Dactylaria*, *Drechslera*, *Ganoderma*, and *Fusarium*. Therefore, developing resistant varieties of various bamboo species is a need of the hour. Moreover, the strategies for the rapid production of bamboo species should be also taken into consideration. Before reaching to the germination stage, bamboo seeds take 2–3 months of resting time under natural conditions, which has made its vegetative reproduction at a rapid rate an uphill task. However, the longevity of the seeds could be shortened by artificially controlling the moisture contents of the seeds. In addition, conventional methods of cultivation of bamboo species should be encouraged and supported. Tissue culture can pave us the best way to increase the rate of propagation and develop some resistant varieties of bamboo at a rapid rate, particularly such plants which are prone to pathogens. Modern methods need to be involved in order to identify molecular markers associated with higher biomass. Moreover, the new cases like the coloration and regulation and management of excess sunlight by bamboo species should also be made a case in research study.

Acknowledgments Financial assistance in the form of UGC-Non-Net Fellowship from UGC, India, to author IBG is highly acknowledged.

Conflicts of Interest There are no conflicts of interest among the author.

References

Abbasi F, Onodera H, Toki S, Tanaka H, Komatsu S (2004) OsCDPK13, a calcium-dependent protein kinase gene from rice is induced by cold and gibberellin in rice leaf sheath. Plant Mol Biol 55:541–552

Abbasi AR, Hajirezaei M, Hofius D, Sonnewald U, Voll LM (2007) Specific roles of alpha- and gamma-tocopherol in abiotic stress responses of transgenic tobacco. Plant Physiol 143:1720–1738

Abraham E, Rigo G, Szekely G, Nagy R, Koncz C, Szabados L (2003) Light-dependent induction of proline biosynthesis by abscise acid and salt stress is inhibited by brassinosteroid in Arabidopsis. Plant Mol Biol 51:363–372

Arnoux P, Morosinotto T, Saga G, Bassi R, Pignol D (2009) A structural basis for the pH dependent xanthophyll cycle in *Arabidopsis thaliana*. Plant Cell 21:2036–2044

Asano T, Hakata M, Nakamura H, Aoki N, Komatsu S, Ichikawa, Hirochika H, Ohsugi R (2011) Functional characterization of OsCPK21, a calcium-dependent protein kinase that confers salt tolerance in rice. Plant Mol Biol 75:179–191

Asif MA, Zafar Y, Iqbal J, Iqbal MM, Rashid U, Ali GM, Arif A, Nazir F (2011 Nov) Enhanced expression of AtNHX1, in transgenic groundnut (Arachis hypogaea L.) improves salt and drought tolerence. Mol Biotechnol 49(3):250–256. https://doi.org/10.1007/s12033-011-9399-1

Birsen C, Alice A, Cécile G, Amélie S, Serge D, Rossitza A (2003) A grape ASR protein involved in sugar and abscisic acid signaling. Plant Cell 15:2165–2180

Chang W, Liu X, Zhu J, Fan W, Zhang Z (2016) An aquaporin gene from halophyte *Sesuvium portulacastrum*, SpAQP1, increases salt tolerance in transgenic tobacco. Plant Cell Rep 35:385–395

Chen JY, Liu DJ, Jiang YM, Zhao ML, Shan W, Kuang JF, Lu WJ (2011) Molecular characterization of a strawberry FaASR gene in relation to fruit ripening. PLoS One 6:e24649

Chiu WB, Lin CH, Chang J, Hsieh MH, Wang AY (2006) Molecular characterization and expression of four cDNAs encoding sucrose synthase from green *Bambusa oldhamii*. New Phytol 170:53–63

Chung HR, Schäfer U, Jackle H, Böhm S (2002) Genomic expansion and clustering of ZAD-containing C2H2 zinc-finger genes in Drosophila. EMBO Rep 3:1158–1162

Dai JR, Liu B, Feng DR, Liu HY, He YM, Qi KB, Wang HB, Wang JF (2011) MpAsr encodes an intrinsically unstructured protein and enhances osmotic tolerance in transgenic Arabidopsis. Plant Cell Rep 30:1219–1230

Dayan MP (1988) Survey, identification and pathogenicity of pests and diseases of bamboos in the Philippines. Forest Res J 13:61–77

Demmig-Adams B, Adams WW (1996) The role of xanthophyll cycle carotenoids in the protection of photosynthesis. Trends Plant Sci 1:21–26

Feng ZJ, Cui XY, Cui XY, Chen M, Yang GY, Ma YZ, He GY, Xu ZS (2015) The soybean GmDi19-5 interacts with GmLEA and increases sensitivity of transgenic plants to abiotic stresses. Front Plant Sci 6:179

Finkelstein RR, Gampala SS (2002) Abscisic acid signaling in seeds and seedlings. The Plant Cell S15-S45, supplement 2002

González RM, Iusem ND (2014) Twenty years of research on Asr (ABA-stress-ripening) genes and proteins. Planta 239:941–949

Gratani L, Crescente MF, Varone L, Fabrini G, Digiulio E (2008) Growth pattern and photosynthetic activity of different bamboo species growing in the botanical garden of Rome. Flora 203:77–84

Hager A, Holocher K (1994) Localization of the xanthophyll-cycle enzyme violaxanthin de-epoxidase within the thylakoid lumen and abolition of its mobility by a (light-dependent) pH decrease. Planta 192:581–589

Harmon AC, Gribskov M, Gubrium E, Harper JF (2001) The CDPK superfamily of protein kinases. New Phytol 151:175–183

Hirayama T, Shinozaki K (2007) Perception and transduction of abscisic acid signals: keys to the function of the versatile plant hormone ABA. Trends Plant Sci 12:343–351

Hsieh CW, Liu LK, Yeh SH, Chen CF, Lin HI, Sung HY, Wang AY (2006) Molecular cloning and functional identification of invertase isozymes from green bamboo *Bambusa oldhamii*. J Agric Food Chem 54:3101–3107

Hsu YH, Annamalai P, Lin CS, Chen YY, Chang WC, Lin NS (2000) A sensitive method for detecting bamboo mosaic virus (BaMV) and establishment of BaMV-free meristem tip cultures. Plant Pathol 49:101–107

Hu L, Li H, Pang H, Fu J (2012) Responses of antioxidant gene, protein and enzymes to salinity stress in two genotypes of perennial ryegrass (*Lolium perenne*) differing in salt tolerance. J Plant Physiol 169:146–156

Hu W, Huang C, Deng XM, Zhou S, Chen L, Li Y, Wang C, Ma Z, He G (2013) TaASR1, a transcription factor gene in wheat, confers drought stress tolerance in transgenic tobacco. Plant Cell Environ 36:1449–1464

Hu P, Zhou W, Cheng Z, Fan M, Wang L, Xie D (2013a) JAV1 controls jasmonate-regulated plant defense. Mol Cell 50:504–515

Hu Y, Chen L, Wang H, Zhang L, Wang F, Yu D (2013b) Arabidopsis transcription factor WRKY8 functions antagonistically with its interacting partner VQ9 to modulate salinity stress tolerance. Plant J 74:730–745

Huang JL, Cheng LL, Zhang ZX (2007) Molecular cloning and characterization of violaxanthin de-epoxidase (VDE) in *Zingiber officinale*. Plant Sci 172:228–235

Hyun TK, Hoffmann A, Sinha AK, Roitsch T (2009) Tomato mitogen activated protein kinases regulate the expression of extracellular invertase Lin6 in response to stress related stimuli. Funct Plant Biol 36:1088–1097

Jha B, Tiwari V, Yadav SK, Agarwal PK (2012) The SbASR-1 gene cloned from an extreme halophyte *Salicornia brachiata* enhances salt tolerance in transgenic tobacco. Mar Biotechnol 14:782–792

John CK, Nadgauda RS (1999) Review: in vitro-induced flowering in bamboos. In Vitro Cell Dev Biol Plant 35:309–315

Joo J, Lee HY, Song SI, Kim Y-K, Nahm BH (2013) Abiotic stress responsive rice ASR1 and ASR3 exhibit different; tissue-dependent sugar and hormone-sensitivities. Mol Cells 35:421–435

Kalifa Y, Perlson EA, Konrad Z, Scolnik PA, Bar Zvi D (2010) Over-expression of the water and salt stress-regulated Asr1 gene confers anincreased salt tolerance. Plant Cell Environ 27:1459–1468

Kiruba S, Jeeve S, Manohardas S, Kannan D (2007) Bamboo seed as a mean to sustenance of the indigenous community. Indian J Tradit Knowl 1:199–203

Lai Z, Li Y, Wang F, Cheng Y, Fan B, Yu JB, Chen Z (2011) Arabidopsis sigma factor binding proteins are activators of the WRKY33 transcription factor in plant defense. Plant Cell 23:3824

Li J, Dong Y, Li C, Pan Y, Yu J (2016) SiASR4, the target gene of SiARDP from *Setariaitalica*, improves abiotic stress adaption in plants. Front Plant Sci 7:2053

Li J, Li Y, Yin Z, Zhang J, Guo M, Ye X, Zhao Z, Xiong Y, Zhang H (2017) OsASR5 enhances drought tolerance through a stomatal closure pathway associated with ABA and H2 O2 signaling in rice. Plant Biotechnol J 15:183–196

Lin R, Li L, Kuang T (2002) Molecular cloning, in vitro expression and enzyme activity analysis of violaxanthin de-epoxidase from *Oryza sativa* L. Chin Sci Bull 47:915–917

Liu XY, Zhang FC, Zhang WZ, Song LF, Wu WH, Chen YF (2013) Arabidopsis Di19 functions as a transcription factor and modulates PR1, PR2, and PR5 expression in response to drought stress. Mol Plant 6:1487–1502

Mao X, Zhang H, Qian X, Li A, Zhao G, Jing R (2012) TaNAC2, a NAC-type wheat transcription factor conferring enhanced multiple abiotic stress tolerances in Arabidopsis. J Exp Bot 63:2933–2946

Marè C, Mazzucotelli E, Crosatti C, Francia E, Stanca AM, Cattivelli L (2004) A new transcription factor involved in cold- and drought-response in barley. Plant Mol Biol 55:399–416
Mittler R (2002) Oxidative stress, antioxidants and stress tolerance. Trends Plant Sci 7:405–410
Mittler R, Vanderauwera S, Gollery M, Breusegem FV (2004) Reactive oxygen gene network of plants. Trends Plant Sci 9:490–498
Moore K, Roberts LJ (1998) Measurement of lipid peroxidation. Free Radic Res 6:659–671
Nambara E, Marion-Poll A (2005) Abscisic acid biosynthesis and catabolism. Annu Rev Plant Biol 56:165–185
Niu CF, Wei W, Zhou QY, Tian AG, Hao YJ, Zhang WK, Ma B, Lin Q, Zhang ZB, Zhang JS, Chen SY (2012) Wheat WRKY genes TaWRKY2 and TaWRKY19 regulateabiotic stress tolerance in transgenic Arabidopsis plants. Plant Cell Environ 35:1156–1170
North HM, Frey A, Boutin JP, Sotta B, Marion-Poll A (2005) Analysis of xanthophyll cycle gene expression during the adaptation of Arabidopsis to excess light and drought stress: changes in RNA steady-state levels do not contribute to short term responses. Plant Sci 169:115–124
Oh JE, Kwon Y, Kim JH, Noh NH, Hong SW, Lee H (2011) A dual role for MYB60 in stomatal regulation and root growth of *Arabidopsis thaliana* under drought stress. Plant Mol Biol 77:91–103
Pedrosa AM, Martins Cde P, Gonçalves LP, Costa MG (2015) Late embryogenesis abundant (LEA) constitutes a large and diverse family of proteins involved in development and abiotic stress responses in sweet orange (*Citrus sinensis L.*). PLoS One 10:e0145785
Perruc E, Charpenteau M, Ramirez BC, Jauneau A, Galaud JP, Ranjeva R, Ranty B (2004) A novel calmodulin-binding protein functions as a negative regulator of osmotic stress tolerance in *Arabidopsis thaliana* seedlings. Plant J 38:410–420
Philippe R, Courtois B, Mcnally KL, Mournet P, Malki R, Paslier MCL, Fabre D, Billot C, Brunel D, Glaszmann JC (2010) Structure, allelic diversity and selection of Asr genes, candidate for drought tolerance, in *Oryza sativa* L. and wild relatives. Theor Appl Genet 121:769–787
Rauf M, Shahzad K, Ali R et al (2014) Cloning and characterization of Na+/H+ antiporter (LfNHX1) gene from a halophyte grass Leptochloa fusca for drought and salt tolerance. Mol Biol Rep 41:1669–1682
Reilly CA, Aust SD (1998) Measurement of lipid peroxidation. Free Radic Res 28:659–671
Roitsch T, Balibrea ME, Hofmann M, Proels R, Sinha AK (2003) Extracellular invertase: key metabolic enzyme and PR protein. J Exp Bot 54:513–524
Ruan YL, Jin Y, Yang YJ, Li GJ, Boyer JS (2010) Sugar input, metabolism, and signaling mediated by invertase: role in development, yield potential, and response to drought and heat. Mol Plant 3:942–955
Saga G, Giorgetti A, Fufezan C, Giacometti GM, Bassi R, Morosinotto T (2010) Mutation analysis of violaxanthin de-epoxidase identifies substrate-binding sites and residues involved in catalysis. J Biol Chem 285:23763–23770
Saijo Y, Hata S, Kyozuka J, Shimamoto K, Izui K (2010) Over-expression of a single Ca2+-dependent protein kinase confers both cold and salt/drought tolerance on rice plants. Plant J 23:319–327
Saumonneau A, Agasse A, Bidoyen MT, Lallemand M, Cantereau A, Medici A, Laloi M, Atanassova R (2008) Interaction of grape ASR proteins with a 14 DREB transcription factor in the nucleus. FEBS Lett 582:3281–3287
Shinozaki K, Yamaguchi-Shinozaki K (1997) Gene expression and signal transduction in water-stress response. Plant Physiol 115:327–334
Smirnoff N (1993) The role of active oxygen in the response of plants to water deficit and desiccation. New Phytol 125:27–58
Song S, Chen Y, Zhao M, Zhang WH (2011) A novel *Medicago truncatula* HD-zip gene, MtHB2, is involved in abiotic stress responses. Environ Exp Bot 69:1–9
Song W, Zhao H, Zhang X, Lei L, Lai J (2016a) Genome-wide identifcation of VQ motif-containing proteins and their expression profiles under abiotic stresses in maize. Front Plant Sci 6:1177

Song X, Peng C, Zhou G, Gu H, Li Q, Zhang C (2016b) Dynamic allocation and transfer of non-structural carbohydrates, a possible mechanism for the explosive growth of Moso bamboo (*Phyllostachys heterocycla*). Sci Rep 6:25908
Sun H, Li L, Lou Y, Zhao H, Wang S, Gao Z (2016) The bamboo aquaporin gene PeTIP4;1-1 confers drought and salinity tolerance in transgenic Arabidopsis. Plant Cell Rep 36:597–609
Szabados L, Savouré A (2010) Proline: a multifunctional amino acid. Trends Plant Sci 15:89–97
Uno Y, Furihata T, Abe H, Yoshida R, Shinozaki K (2000) Arabidopsis basic leucine zipper transcription factors involved in an abscisic aciddependent signal transduction pathway under drought and high-salinity conditions. Proc Natl Acad Sci U S A 97:11632–11637
Utsugi S, Shibasaka M, Maekawa M, Katsuhara M (2015) Control of the water transport activity of barley HvTIP3;1 specifically expressed in seeds. Plant Cell Physiol 56:1831–1840
Vanithakumari G, Manonayagi S, Padma S, Malini T (1989) Antifertility effect of Bambusa aruninacea shoot extract in male rat. J Ethanopharmocol 25:173–180
Vivekanand T, Amit Kumar C, Avinash M, Bhavanath J (2015) Introgression of the SbASR-1 gene cloned from a halophyte *Salicornia brachiate* enhances salinity and drought endurance in transgenic groundnut (*Arachis hypogaea*) and acts as a transcription factor. PLoS One 10: e0131567
Wang A, Garcia D, Zhang H, Feng K, Chaudhury A, Berger F, Peacock WJ, Dennis ES, Luo M (2010) The VQ motif protein IKU1 regulates endosperm growth and seed size in Arabidopsis. Plant J 63:670–679
Wang L, Yu C, Chen C, He C, Zhu Y, Huang W (2014a) Identification of rice Di19 family reveals OsDi19-4 involved in drought resistance. Plant Cell Rep 33:2047–2062
Wang X, Zhang H, Sun G, Jin Y, Qiu L (2014b) Identification of active VQ motif-containing genes and the expression patterns under low nitrogen treatment in soybean. Gene 543:237–243
Wang H, Hu Y, Pan J, Yu D (2015) Arabidopsis VQ motif-containing proteins VQ12 and VQ29 negatively modulate basal defense against *Botrytis cinerea*. Sci Rep 5:14185
Wei W, Cui M, Hu Y, Gao K, Xie Y, Jiang Y, Feng J (2018) Ectopic expression of FvWRKY42, a WRKY transcription factor from the diploid woodland strawberry (*Fragaria vesca*), enhances resistance to powdery mildew, improves osmotic stress resistance, and increases abscisic acid sensitivity in Arabidopsis. Plant Sci 275:60–74
Wimmer H (2003) The nuclear localization signal of a pollen-specific, desiccation associated protein of lily is necessary and sufficient for nuclear targeting. Bot Bull Acad Sinica Taipei 44:171–188
Woitsch S, Romer S (2003) Expression of xanthophyll biosynthetic genes during light-dependent chloroplast differentiation. Plant Physiol 132:1508–15017
Wu M, Liu H, Han G, Cai R, Pan F, Xiang Y (2017) A moso bamboo WRKY gene PeWRKY83 confers salinity tolerance in transgenic Arabidopsis plants. Sci Rep 7:11721
Xiang Y, Huang Y, Xiong L (2007) Characterization of stress-responsive CIPK genes in rice for stress tolerance improvement. Plant Physiol 144:1416–1428
Xiong L, Schumaker KS, Zhu JK (2002) Cell signaling during cold, drought, and salt stress. Plant Cell 14:S165–S183
Xu J, Tian YS, Reng RH, Xiong AS, Zhu B, Jin XF, Gao F, Fu XY, Hou XL, Yao QH (2010) AtCPK6, a functionally redundant and positive regulator involved in salt/drought stress tolerance in Arabidopsis. Planta 231:1251–1260
Yamaguchi SK, Shinozaki K (2006) Transcriptional regulatory networks in cellular responses and tolerance to dehydration and cold stresses. Annu Rev Plant Biol 57:781–803
Yang T, Poovaiah BW (2002) A calmodulin-binding/CGCG box DNA-binding protein family involved in multiple signaling pathways in plants. J Biol Chem 277:45049–45058
Yang CY, Chen YC, Guang Yuh J, Wang Co S (2005a) A lily ASR protein involves abscisic acid signaling and confers drought and salt resistance in Arabidopsis. Plant Physiol 139:836–846
Yang X, Feng Y, He Z, Stoffella PJ (2005b) Molecular mechanisms of heavy metal hyperaccumulation and phytoremediation. J Trace Elem Med Biol 18:339–353

Yasunari F, Miki F, Rie S, Kyonoshin M, Parvez MM, Motoaki S, Keiichiro H, Masaru OT, Kazuo S, Kazuko YS (2005) AREB1 is a transcription activator of novel ABRE-dependent ABA signaling that enhances drought stress tolerance in Arabidopsis. Plant Cell 17:3470

Zhang CS, Lu Q, Verma DP (1995) Removal of feedback inhibition of delta 1-pyrroline-5-carboxylate synthetase, a bifunctional enzyme catalyzing the first two steps of proline biosynthesis in plants. J Biol Chem 270(35):20491–20496

Zhang L, Hu W, Wang Y, Feng R, Zhang Y, Liu J, Jia C, Miao H, Zhang J, Xu B (2014) The MaASR gene as a crucial component in multiple drought stress response pathways in Arabidopsis. Funct Integr Genomics 155:247–260

Zhu JK (2002) Salt and drought stress signal transduction in plants. Annu Rev Plant Biol 53:247–273

Zhu SY, Yu XC, Wang XJ, Zhao R, Li Y, Fan RC, Shang Y, Du XF, Wang FQ, Wu (2007) Two calcium-dependent protein kinases, CPK4 and CPK11, regulate abscisic acid signal transduction in Arabidopsis. Plant Cell 19:3019–3036

Zhu G, Ye N, Zhang J (2009) Glucose-induced delay of seed germination in rice is mediated by the suppression of ABA catabolism rather than an enhancement of ABA biosynthesis. Plant Cell Physiol 50:644–651

Chapter 12
Advances in the Conservation of Bamboo Genetic Resources Through Whole Seed Cryopreservation

Jonny Everson Scherwinski-Pereira, Jênifer Silva Nogueira, Janaína Medeiros Vasconcelos, Joane dos Santos Neves, and Talita Aparecida Balzon

Abstract Cryopreservation is the conservation of living organisms in liquid nitrogen at a temperature of −196 °C or slightly lower; this phase may also constitute an intermediate stage for definitive freezing at −196 °C. At this temperature, cellular metabolism and biochemical processes are substantially reduced, and biological deterioration is practically paralyzed. Consequently, the biological deterioration of the material during storage is minimal, and it can be stored for indefinite periods of time. Apparently, most seeds of bamboo species do not have any kind of numbness but lose viability quickly when kept under normal environmental conditions. Thus, the use of cryogenic techniques or cryopreservation is an alternative for the conservation of germplasm in the long term, especially when conventional conservation, based on the storage of seeds at low temperatures, is not efficient. In this chapter, we evaluate techniques capable of providing ex situ conservation of whole bamboo seeds. In it, we indicate an effective and safe alternative of conserving the genetic variability of the species to avoid or at least minimize the process of genetic erosion and loss of genetic resources of bamboo, from the cryopreservation of whole seeds. In the end, we also indicate regeneration alternatives, in addition to describing in detail the protocol for cryopreservation of seeds that, which followed correctly, can be extended to the various bamboo species.

J. E. Scherwinski-Pereira (✉)
Embrapa Recursos Genéticos e Biotecnologia, Brasília, Distrito Federal, Brazil
e-mail: jonny.pereira@embrapa.br

J. S. Nogueira · J. dos Santos Neves
Universidade de Brasília, Brasília, Distrito Federal, Brazil

J. M. Vasconcelos
Universidade Federal do Acre, Rio Branco, Acre, Brazil

T. A. Balzon
BiotecT Biotecnologia, Brasília, Distrito Federal, Brazil

Z. Ahmad et al. (eds.), *Biotechnological Advances in Bamboo*,
https://doi.org/10.1007/978-981-16-1310-4_12

Keywords Poaceae · Bambusoideae · Ex situ conservation · Seed conservation · Cryobiotechnologies

12.1 Introduction

Plant genetic resources can be conserved in situ, which is the conservation of the species in its habitat, and ex situ, which consists of the removal of plants from their natural habitat and their transfer to artificial storage conditions. The preservation of ex situ germplasm of plants, through the use of seeds, has been considered one of the most efficient forms of conservation (Hong and Ellis 1996). However, each species has its own characteristics regarding tolerance to low temperatures and low humidity, which prevents seed storage from being applied to all plant species with the same effectiveness, being determining factors for conservation (Roberts 1973). Therefore, to fill this research gap, it is imperative to find ex situ conservation methods suitable for bamboo (Bahru et al. 2015; Ruiz-Sanchez et al. 2020), an area that has been little studied until now due to its importance, and the intrinsic characteristics of the species. In addition, one cannot forget the difficulties that the various species present in relation to propagation, which greatly hinders the development of efficient and safe conventional protocols for the conservation of bamboo.

Bamboos have reproductive cycles ranging from one to tens of years. Basically, there are two types of flowering on bamboo: one sporadic and the other gregarious. The first occurs in isolated culms, contrasting with the gregarious flowering, where all the culms of the same species existing in a given region bloom, bear fruit, and die, regardless of their ages. This flowering is undesirable, and its constant occurrence in a given species limits its use for any commercial or industrial activity (Janzen 1976; Azzini et al. 1982; Thapliyal et al. 2015). In fact, this gregarious phenomenon can be observed in most species and centers of origin of species, such as those occurring in the Brazilian Amazon and Cerrado biomes, where in species of *Guadua* this can be easily observed (Fig. 12.1). In these cases, the regeneration of the bamboo forest depends on the natural regeneration of the seeds, a fact that can put natural populations at risk, mainly due to the extension of areas destined for agricultural exploitation, as has happened in the Brazilian Cerrado biome. In this sense, it seems of great importance that genetic resources of the species are conserved ex situ, from efficient and adapted protocols and that can quickly be used, especially in those moments when the bamboo populations begin to present flowering in natural conditions.

With the long reproductive cycles and the low seed production, the propagation of bamboos has preferably occurred vegetatively, by dividing the clumps, parts of rhizomes, or sections of culms, varying according to the species (Banik 1995). These techniques require a large amount of material and space, becoming costly and low-yielding. Thus, micropropagation techniques may constitute a viable alternative for the propagation of these plants, in addition to being an important tool for conservation studies, once seeds could be conserved for indeterminate periods and,

Fig. 12.1 Flowering and fruiting of *Guadua magna* Londoño and Filg., a woody bamboo species from Cerrado biome, Santo Antônio de Goiás, GO, Brazil. November 2012

when regenerated, be multiplied in vitro, generating tens, hundreds, or even thousands of plants, using little number of seeds. In fact, in vitro cultivation has already been used to propagate bamboos, especially for Asian species. In the literature, several protocols have already been established (Sandhu et al. 2018).

Thus, for ex situ conservation to be done rationally, micropropagation also seems to be fundamental for conservation protocols to be in fact viable and regenerated plants to be used effectively and optimally, in a practical, safe, and economical manner. In addition to having excellent physical, chemical, and mechanical characteristics, bamboos are considered efficient carbon sequesters and can be used in reforestation, in the restoration of riparian forests, and, also, as environmental protectors and regenerators (Yeasmin et al. 2015; Akwada and Akinlabi 2018). Due to the physical quality of its culms, several species of bamboo are routinely used as raw materials in engineering and construction works (Akwada and Akinlabi 2015), in addition to being important raw materials for the manufacture of paper (Chen et al. 2019), flooring, and laminates (Sharma et al. 2015), besides being an important resource for food (Satya et al. 2010).

In this chapter, we evaluate several techniques capable of providing ex situ conservation and regeneration of bamboo seeds, addressing the advantages, disadvantages, and difficulties of each of them. In the end, we indicate the technique that seems to be the most effective for the conservation of genetic resources of the species and that can be extended to different species.

12.2 The Importance of Plant Diversity and Genetic Resources

Vegetables are the main source of human and animal food. Of the approximately 300,000 species of plants known, 30% are potentially edible; however, only 0.2% of these are part of the basic diet of humans. Currently, about 15 species supply most of the human diet, and due to this importance, the emphasis on plant genetic improvement has been directed, in large part, to increase agricultural productivity, due to the need to satisfy the demand for food, since the population grows constantly in a world of limited area (Conway and Barbier 1990). Only a few regions in the world have new areas to be added to food production areas. Therefore, there is a consensus in the world scientific community that the technology currently used will not allow food production to be increased on a sufficient scale to meet the food need of this growing population.

It is important to emphasize that approximately 80% of the Earth's biodiversity is located in only 12 countries, including Brazil, where it is estimated that there are about 20% of the total number of species on the planet, and considering only plant species, this value increases to 30% (Forzza et al. 2012). Through biodiversity, it has been possible to generate products of high added value and improve the quality of human life, favoring the production of food, drugs, energy generation, and leisure, among others. For this reason, its conservation and rational use become of fundamental importance, since with the emergence of new techniques, it is possible for new products to be created from the access and the exchange of existing genetic material between different species.

Bamboo populations are naturally distributed from the tropics to temperate regions (Fig. 12.2). However, they have a higher occurrence in hot areas and heavy rainfall. It is estimated that bamboo species occupy between 14 and 18 million hectares in global forest ecosystems, including Asia, Africa, and the Americas (Brias

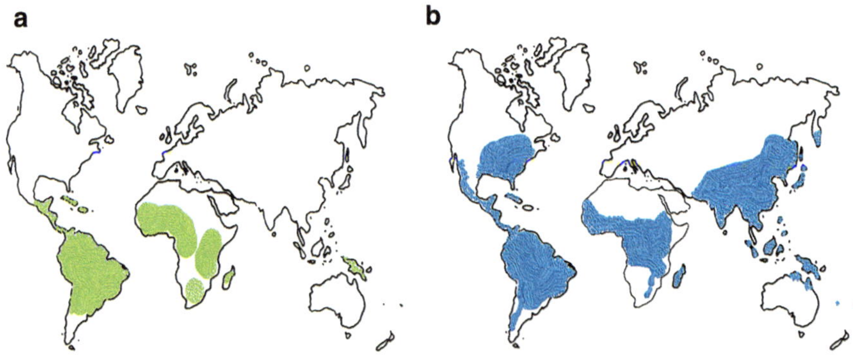

Fig. 12.2 The diversity of herbaceous (**a**) and woody (**b**) bamboos in the world (Source: Adapted of Yeasmin et al. 2015; Canavan et al. 2017; https://www.eeob.iastate.edu/research/bamboo/maps.html)

and Hunde 2009; Guerreiro and Lizarazu 2010; Yeasmin et al. 2015). Of the total of more than 1650 species distributed in 120 known bamboo genera worldwide (Soreng et al. 2015), Latin America has about 39% of species and 31% of genera (BPG 2012). In Brazil alone, 35 genera and 258 native species are found, representing 18% of all bamboo species in the world (Filgueiras and Viana 2017). Of this total, 173 species are endemic, which makes Brazil the country with the greatest diversity of bamboos in the American continent (Filgueiras and Gonçalves 2004; Greco et al. 2015; Filgueiras and Viana 2017; Silva et al. 2020; Leal et al. 2021).

The distribution occurs in the main Brazilian biomes, which include the Atlantic Forest, with 65% of the species, the Amazon, with 26%, and the Cerrado, with 9% (Filgueiras and Gonçalves 2004). Studies conducted by satellite images have demonstrated the existence of a native bamboo forest in the state of Acre (Smith and Nelson 2011), where bamboos of the genus *Guadua* predominate. In addition to covering a large range of the state, natural bamboo populations also cover areas of Bolivia and Peru (Filgueiras and Gonçalves 2004).

12.3 Germplasm Conservation

Currently, genetic resources are conserved in germplasm banks and collections that are germplasm conservative units for immediate use or with potential for future use (Nass et al. 2012). Plant genetic resources can be conserved in situ, which is the conservation of the species in its habitat, allowing the plant to continue its evolutionary process, and ex situ, which consists of the withdrawal of genetic resources from its habitat and their transfer to artificial storage conditions, such as seed banks, field collections, as well as in vitro collections (Kapai et al. 2010). According to Veiga (2006), ex situ conservation unfolds in several modalities, among which in vitro conservation and field collections, in cold chambers and in liquid nitrogen (−196 °C) stand out.

Ex situ conservation is a viable option for species of economic interest or that have propagation difficulties, as what occurs with bamboo species, which have long flowering cycles, thus reducing the availability of seeds for short-, medium-, and long-term conservation. The conservation of ex situ germplasm of plants, through the use of seeds, has been considered one of the most efficient forms of conservation (León-Lobos et al. 2012; Nass et al. 2012). However, each species has its own characteristics regarding tolerance to low temperatures and low humidity, which prevents seed storage from being applied to all plant species with the same effectiveness, as they are determining factors for conservation. Thus, when it is intended to conserve a particular species by seeds, it is essential to know the storage behavior so that the most appropriate storage strategies and conditions can be determined (Hong and Ellis 1996).

Seeds can be classified into orthodox, recalcitrant, and intermediate, in relation to the physiological responses related to tolerance to desiccation and cooling (Ellis et al. 1990). Orthodox seeds can be desiccated at low moisture levels (7% or less), without cell damage, and they tolerate storage at low temperatures for long periods, with little or no loss of viability (Roberts 1973). Recalcitrant seeds have high moisture content and are sensitive to water loss, not surviving when subjected to the same conditions used to store orthodox seeds (Walters et al. 2013). Intermediate seeds can be stored under the same conditions as orthodox seeds, albeit for shorter periods (Bonner 1990). Intermediate seeds are moderately sensitive to desiccation, that is, they tolerate the loss of water enough to prevent the formation of ice crystals; however, they do not tolerate cooling for long periods, making them susceptible to temperature-caused injuries (Ellis et al. 1990).

The use of cryopreservation has been considered as an interesting strategy for long-term conservation of species of medicinal, agroforestry, horticultural, and biotechnological interest. It consists of maintaining the plant material of interest under ultralow temperature, usually in liquid nitrogen (−196 °C). These temperatures provide a drastic reduction of cellular metabolism, keeping intact the conserved biological material and ensuring high genetic and physiological stability in the presence or absence of cryopotential substances (Engelmann 2004). Panis et al. (2001) stated that the key to the success of cryopreservation is not with tolerance to freezing, but with tolerance to dehydration and its induction, since the biggest challenge for cryopreservation protocols is to perform a freeze without the formation of ice crystals inside the cells. According to Yamazaki et al. (2008), ultrafast freezing promotes dehydration of cells before freezing, preventing the formation of large ice crystals inside the cells.

Cryopreservation of plant material has advanced significantly in recent decades, especially for tropical crops. However, although there are several cryogenic procedures for different species, the routine use of cryopreservation in plant biodiversity is still considered limited (Panis and Lambardi 2006), especially for bamboo.

12.4 Characteristic of Seeds and Germination of Bamboo

The first physiological process of establishing a new plant is seed germination, where there is a predominance of catabolic activities and mobilization of reserves, which will culminate in the development of the embryonic axis. From a physiological point of view, germination begins with the soaking or rehydration of the seed tissues, with subsequent reactivation of the energy metabolism (respiration), stimulating the division, cell elongation, and the development of the embryo, which culminates with the appearance of the primary root or other embryonic structure on the surface of the diaspora, this characteristic being the first indication of germination (Bewley 2001). In laboratory conditions, a germinated seed is considered when there is a protrusion of 1–2 mm from the primary root. Under a practical concept, the diaspora is considered germinated when the plant emerges from the soil (Bewley et al. 2013).

As for the morphophysiological aspect, the germination of grasses is of the hypogean type, where the cotyledon remains below the ground, the epicotyl lengthens with the emergence of the aerial part, the epicotyl folds, and, as the seedling grows, it is taken to the surface of the soil. The reserves present in the cotyledons are used by the developing seedling until they are exhausted, and the rest of the diaspora decomposes with time (Bewley 2001). The composition of reserve tissues is governed genetically, and seeds, during their formation, can accumulate carbohydrates, proteins, and lipids. Among the carbohydrates, starch, hemicellulose, and sugars such as sucrose, fructose, galactose, stachiosis, and raffinose stand out. Lipids are a kind of reserve present in Poaceae seeds. Proteins can be soluble or insoluble in water. Albumins are soluble proteins, while globulins and polyamines, which are stored in aleurone grains present in cereals, are insoluble in water (Bewley 2001; Sert et al. 2009). During germination, the reserves are degraded and subsequently mobilized to different parts of the embryo, helping the growth of the seedling until it becomes autotrophic. The metabolic process by which the reserves are degraded depends on their chemical composition (Mazzottini-dos-Santos et al. 2020).

The chemical, physical, and mechanical characteristics of bamboo are defined by its structural constitution, and the properties of the culm are determined mainly by anatomical structure (Londoño 2002; Luis et al. 2017). Thus, the anatomical study can be an important tool to assist in defining the potential and the best way of the use of different species, besides being a determining area for taxonomy studies in bamboo (Leandro et al. 2016). Therefore, the anatomical characterization of germination can help in a better understanding of the processes involved in this step. However, there are still few and fragmented works describing the anatomy of bamboo vegetative organs, and the existing bibliography tends to focus on the culms, due to the greater economic importance of this organ.

Despite the importance of germination for plant production, little is known about the main germination, anatomical features, and mobilization of reserves during this stage. It is known that woody bamboo seeds that have gregarious flowering are produced at long intervals, soon after flowering, culminating in the death of the clump (Azzini et al. 1982; Thapliyal et al. 2015). These seeds do not have any kind of numbness but lose viability quickly, when kept in normal environmental conditions. Although bamboo seeds are excellent material for propagation and conservation of germplasm, the limited availability of seeds, limitations in viability, and little knowledge about suitable methods for storage are practical problems that directly interfere with the propagation on a bamboo scale, from the use of seeds (Bahru et al. 2015).

The percentage of germination can be high (80–100%) if the sowing of the seeds occurs under shades, soon after harvesting (Thapliyal et al. 2015; Singh et al. 2017). The germination period varies from 4 to 20 days in orthodox seeds, while for recalcitrant seeds of *Melocanna* and *Ochlandra*, it may be shorter (Singh et al. 2017). Seed viability studies can be performed using biochemical methods such as the tetrazolium test, while germination can be tested in a variety of conditions, such as between or on sheets of absorbent paper, in sand, or in water-agar medium

Fig. 12.3 Bamboo seeds before and after cleaning (**a–b**). (**c**) Bamboo seed germination in vitro after 7 days. (**d**) Bamboo seed germination in vitro after 15 days. (**e**) Pre-acclimatization of plantlets. (**f**) Germination of seeds on substrate

(Ashmore et al. 2015), as well as different substrates and growing medium (Fig. 12.3).

12.5 Cryopreservation of Bamboo Seeds for Long-Term Conservation

Apparently, most bamboo seeds do not have any kind of numbness but lose viability quickly when kept in normal environmental conditions. Thus, it is strongly recommended that the conservation of bamboo seeds be done as soon as possible after the collection or obtaining the seeds from the field. Once in the laboratory, the seeds receive the first cleaning treatment, where the caryopsis that covers the seeds is removed manually. This process, in addition to allowing the initial cleaning of the seeds, also assists in the disposal of those hatched and/or malformed seeds from the lot, preventing unviable seeds from being added to the conservation treatments.

After removal of the caryopsis, the seeds are then taken for desiccation, in order to eliminate excess moisture that may compromise conservation. For this, the seeds are packed in a desiccator containing silica gel for several hours or when the moisture content reaches desirable values for cryopreservation. The moisture content (*MC*) of

Fig. 12.4 The proportion of germinability of *Dendrocalamus asper* seeds at different moisture contents (Nogueira et al. 2020)

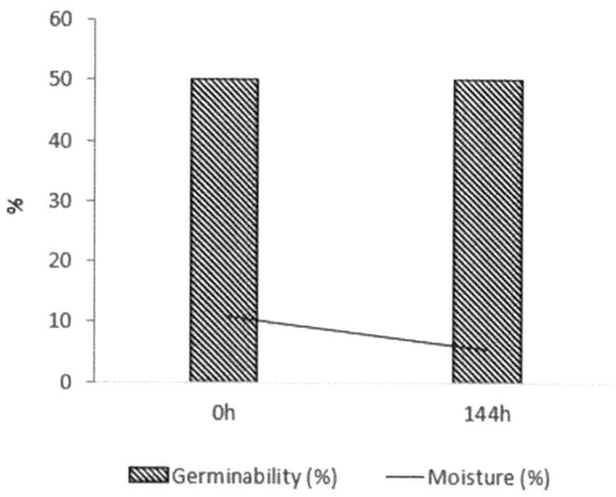

the seeds is determined after each drying period by means of a greenhouse test at 105 °C for 24 h. The percentage of seed moisture is calculated based on the wet weight, applying the following formula:

$$MC(\%) = \frac{100(iW - fW)}{iW - t}$$

where

iW = initial seed weight after desiccation hours

fW = final seed weight after 24 h in a greenhouse at 105 °C

t = weight of the aluminum plate where the seeds are weighed (tare weight of container).

In studies conducted by Nogueira et al. (2020), it was found that the desiccation of bamboo seeds between 0 and 144 h allowed the gradual and constant reduction of moisture content, with a difference of 5.6% between the initial time and the final time of desiccation (Fig. 12.4). The reduction of the moisture of the seeds to just over 5% of moisture did not affect the viability, since it was observed 50% of germinability in the initial time and the same germinability after 144 h of desiccation of the seeds in silica gel, suggesting that bamboo seeds can be preserved with humidity between 5 and 10% of moisture, without losing the characteristics of germinability.

12.6 Cryopreservation of Bamboo Seeds

The protocol described below was established from our laboratory and describes part of the work developed by Nogueira et al. (2020). After determining the lowest seed moisture, but also allowing the seeds to germinate, the seeds are then immersed in liquid nitrogen (+LN) (−196 °C) conditions. For this, the seeds are sampled and

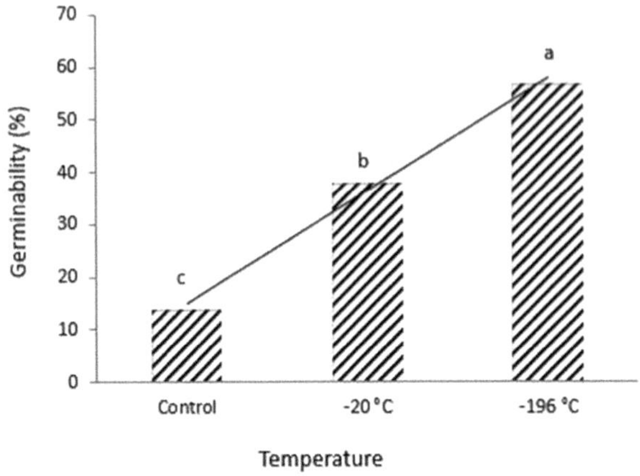

Fig. 12.5 Seed germinability of *Dendrocalamus asper* after different storage conditions (−25 °C—control, −20 °C, and −196 °C) (Nogueira 2018)

packaged in cryotubes of 2 mL capacity. These are then placed in aluminized trifoliate envelopes which are then packaged in cylindrical aluminum tubes (canisters) and gently immersed in liquid nitrogen (+LN) contained in cylinders. In this condition, the materials remain for indeterminate periods of time, without losing the germinability conditions. Nogueira (2018), in a conservation study of *Dendrocalamus asper* seeds in different storage conditions (−25 °C—control, −20 °C and −196 °C), found that after different conservation periods, the germination rates of seeds did not show a decrease when they were cryopreserved, contrary to the other conservation treatments tested that showed a decrease in the germination of seeds with the passage of conservation time (Fig. 12.5). After the required storage period in +LN, the envelopes with the seeds are removed from the bottles. The preserved seeds are then thawed in a water bath at 40 °C for 90 s, when they will then be ready to germinate. Although germination can be performed using substrates and/or germitest paper in the laboratory, it is strongly recommended that the seeds be germinated under in vitro conditions because they find a more favorable environment for germination. In addition, once germinated in vitro, the plants in this condition can be used as a source of propagules for their micropropagation, allowing tens or hundreds of new identical plants to be obtained from a single seed, helping in the regeneration of forests and increasing the availability of plants. Additionally, it makes it possible to understand the physiological behavior during germination. According to Bahru et al. (2015), studying the ecology of seed germination allows us to select favorable conditions for faster germination, producing vigorous plants, which can be used in the production of seedlings. This possibility should be taken into account precisely because most of the bamboo species do not show regularity of flowering, or many of them have only one flowering event in life, as already discussed earlier. Thus, from a few seeds, it is possible to reproduce appreciable amounts of plants from that cryopreserved sample, from micropropagation (Fig. 12.6).

In the Laboratory of Tissue Culture II (Laboratório de Cultura de Tecidos II) of Embrapa Genetic Resources and Biotechnology (Embrapa Recursos Genéticos e

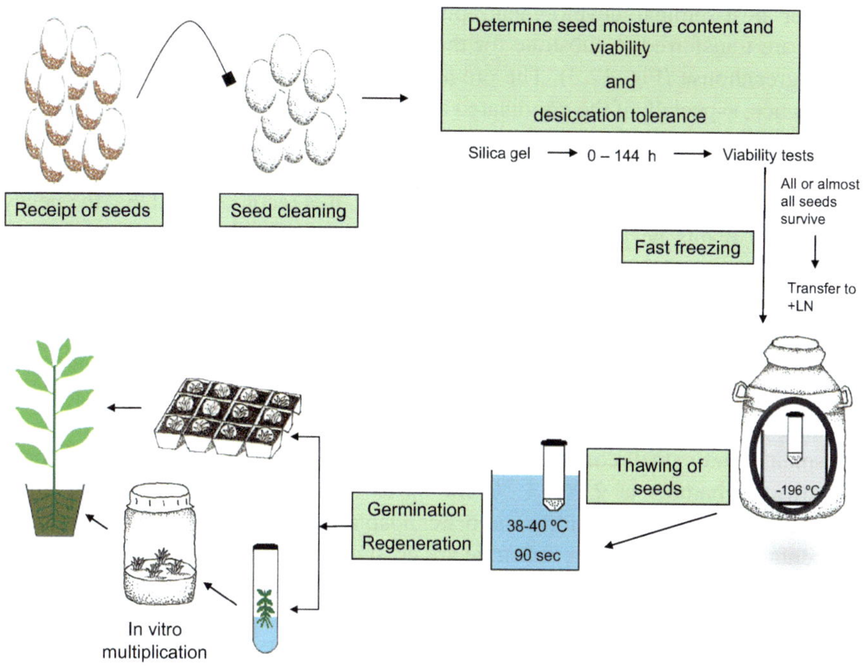

Fig. 12.6 Workflow for the cryopreservation of seeds for the conservation and sustainable use of bamboo genetic resources

Biotecnologia), in Brasilia, Brazil, after the cryopreservation step, bamboo seeds are disinfested by immersion in 70% ethyl alcohol (v/v) for 1 min and immersed for 20 min in mercury bichloride ($HgCl_2$) 0.1% or sodium hypochlorite (2.5% active chlorine) for 15–20 min. Subsequently, the seeds are rinsed three times in distilled and autoclaved water. Soon after, the seeds are inoculated into test tubes (25 × 150 mm), containing MS culture medium (Murashige and Skoog 1962), where the pH of the medium is set to 5.8 ± 0.1 before autoclaving at 121 °C at 1.5 atm for 20 min. The materials are then kept in a growth room with a temperature of 25 ± 2 °C, luminosity of 100μm m^{-2} s^{-1}, and photoperiod of 16 h until the seeds germinate and form complete plants, when then they are either pre-acclimatized or transferred to a new appropriate culture medium to be multiplied by cloning (Nogueira et al. 2019).

The viability of seed conservation, after cryogenic storage, is verified by germination. The germination of bamboo seeds generally does not differ if they are germinated in light or dark, and the germinated seeds can be counted in up to 30 days, in which time the viable seeds have already responded. Under in vitro conditions, the criterion used to verify germination may be the radicle emission, whereas if it is carried out on substrate, it is suggested to mark the germination by the epicotyl emission, since it is not possible to visualize the radicle emission, as in in vitro germination.

After post-seminal development, plants that have well-developed roots and shoots are transferred to substrate for the pre-acclimatization stage, before transfer to the greenhouse (Fig. 12.3). The pre-acclimatization stage in bamboo is of great importance, especially if the germinated and formed plants originate from laboratory conditions, especially from the in vitro condition, since young bamboo plants are fragile and easily stressed by water conditions in adverse conditions, such as in acclimatization (Vale et al. 2019). For this reason, it is recommended that the newly germinated plants be pre-acclimatized under growth chamber conditions, with a temperature around 25 °C and maintaining conditions of high relative humidity of the air, at least in the first 2–3 weeks of transplantation into the soil. Such care guarantees minimal loss of plant death due to stress. Therefore, if the seeds are germinated in laboratory conditions, the change of plants from artificial growing conditions to the substrate begins by removing them from the cultivation jars and washing the roots to remove residues of culture medium. Then the seedlings are transplanted into small containers (300 mL capacity), with substrate, being kept in the growth chamber at 25 ± 1 °C for up to 30 days. For the formation of an environment with high humidity and better adaptation of plants to this environment, the containers containing the substrate are covered with a polyethylene plastic bag. During pre-acclimatization, openings are made in the plastic bags that cover the plants each week so that the plants adapt to the natural conditions. After 3 weeks, the plastic bags that cover the plants are then removed completely, leaving the plants in direct contact with the external environment until they complete the 30 days of pre-acclimatization, and then they can be transferred to the greenhouse to complete the development. Following this methodology, Nogueira et al. (2020) obtained 100% survival of *D. asper* plants after cryopreservation of seeds for 360 days, with subsequent germination in vitro.

12.7 Conclusive Remarks and Challenges

This chapter addressed a topic that we consider to be of great importance to bamboo. In it, we indicate an effective and safe way to conserve the genetic variability of bamboo to avoid, or at least minimize, the process of genetic erosion and loss of genetic resources of bamboo using the cryopreservation of whole seeds. In the literature, there are scarce and practically nonexistent subjects related to this topic. The existing studies show weaknesses in the strategies because they allow samples to be kept for only limited periods of time. The importance of the results obtained by our team during cryopreservation of seeds is also justified by the difficulty of the various research groups in the area in creating safe conditions for ex situ conservation, since conventional techniques, such as the conservation of seeds by traditional methods or those using conservation chambers at −20 °C, do not guarantee that the seeds remain viable for long periods of time. On the other hand, the ex situ conservation from vegetative propagules, except in fields and collections, is extremely limiting, due to the intrinsic characteristics of bamboo species, which easily present physiological stress depending on the conservation treatment and, therefore, have low capacity to be maintained, for example, under conditions of minimal growth in in vitro conservation laboratories. Additionally, because most of the bamboo species cannot be propagated in vitro by isolated lateral buds, much less by meristematic apex, cryopreservation techniques, using these types of propagules, are still a challenge for scientific research. In this sense, the development of somatic embryogenesis protocols, cells in suspension, or calluses with efficient regeneration protocols are short- and medium-term alternatives that, once developed, can provide cryopreservation techniques to be applied. However, due to a large number of existing species, it seems evident that still, in this area, we are far from achieving efficient and broad protocols for conserving bamboo genetic resources. Thus, it seems evident that, at firsthand, conservation guarantees of genetic variability from seeds are the closest technique to maintain germplasm of the species, despite the uncertainties of flowering, seed production, and seed viability. In Brazil, where the agricultural border advances annually, but which has one of the largest areas of natural bamboo in the world, as are the cases of species of the genus *Guadua* in the southwest of the Amazon and Cerrado, species threatened of disappearing are a reality, with the risk that they will never be characterized, monitored or used. Finally, it should be emphasized that the methodology proposed in this chapter can be quickly extended to the various species, with only the need to determine the best moisture condition, without the loss of germinability of the species in question.

Acknowledgments The authors thank the Conselho Nacional de Desenvolvimento Científico e Tecnológico (CNPq Grant 458151/2013-0), Coordenação de Aperfeiçoamento de Pessoal de Nível Superior (Capes/Embrapa 001-2011/Grant 39), and Centro de Pesquisa e Aplicação de Bambu e Fibras Naturais (CPAB), for financial support and fellowships.

Conflicts of Interest The authors declare that they have no conflict of interest.

References

Akwada DR, Akinlabi ET (2015). Bamboo Use in Construction Industry: How Sustainable is it? In: Proceedings of the DII-2015 conference on infrastructure development and Livingstone, Zambia investment strategies for Africa, 16–18 September 2015

Akwada DR, Akinlabi ET (2018) Bamboo an alternative wood to reducing tropical deforestation in Ghana 2018. In: Proceedings of the DII-2018 conference on infrastructure development investment strategies for Africa - readying Africa for sustainable development: Livingstone, Zambia. A capacity building/empowerment and strategy discourse

Ashmore SE, Martyn A, Sommerville K, Errington G (2015) Seed biology. In: Crop wild relatives and climate change. Wiley, New York

Azzini A, Aranha C, Mary R (1982) Florescimento e frutificação em bambu. Bragantia 41:175–180

Bahru T, Mulatu Y, Kidane B (2015) Germination ecology of *Arundinaria alpina* (K. Schum) and *Oxytenanthera abyssinica* (A. Rich.) Munro seeds: Indigeneous bamboo species in Ethiopia. Int J Biodivers 2015:323128. https://doi.org/10.1155/2015/323128

Bamboo Phylogeny Group (2012) An updated tribal and subtribal classification for the Bambusoideae (Poaceae). In: Gielis J, Potters G (eds) Proceedings of the 9th World Bamboo Congress, 10–12 April 2012. Antwerp, Belgium, pp 3–27
Banik RL (1995) A manual for vegetative propagation of bamboos. Intl. Agric. Res. Tech. Rep. 6, Bangladesh Forest Research Institute, Dhaka, 66p
Bewley JD (2001) Seed germination and reserve mobilization. In: Encyclopedia of life sciences. Wiley, Chichester, p 7pp. https://doi.org/10.1038/npg.els.0002047
Bewley JD, Bradford KJ, Hilhorst HWM, Nonogaki H (2013) Seeds: physiology of development, germination and dormancy. Springer, New York, 392p
Bonner F (1990) Storage of seeds: potential and limitations for germplasm conservation. For Ecol Manag 35:35–43
Brias V, Hunde T (2009) Bamboo cultivation manual. In: Guidelines for cultivating Ethiopian highland bamboo. United Nations Industrial Development Organization, Vienna, 58p
Canavan S, Richardson DM, Visser V, Roux JJL, Vorontsova MS, Wilson JRU (2017) The global distribution of bamboos: assessing correlates of introduction and invasion. AoB Plants 9:plw078
Chen Z, Zhang H, He Z, Zhang L, Yue X (2019) Bamboo as an emerging resource for worldwide pulping and papermaking. BioResources 14(1):3–5
Conway G, Barbier E (1990) After the green revolution. Earthscar Press, London, 204p
Ellis RH, Hong TD, Roberts EH (1990) An intermediate category of seed storage behaviour? J Exp Bot 41:1167–1174
Engelmann F (2004) Plant cryopreservation: Progress and prospects. In Vitro Cell Dev Biol Plant 40:427–433. https://doi.org/10.1079/IVP2004541
Filgueiras TS, Gonçalves APS (2004) A checklist of the basal grasses and bamboos in Brazil (Poaceae). J Am Bamboo Soc 18:7–18
Filgueiras TS, Viana PL (2017) Bambus brasileiros: morfologia, taxonomia, distribuição e conservação. In: Drumond PM, Wiedeman G (eds) Bambus no Brasil: da biologia à tecnologia, 1st edn. ICH, Rio de Janeiro, pp 10–27
Forzza RC, Baumgratz JFA, Bicudo CEM, Canhos DAL, Carvalho AA Jr, Costa A, Costa DP, Hopkins M, Leitman PM, Lohmann LG, NicLughadha E, Maia LC, Martinelli G, Menezes M, Morim MP, Coelho MAN, Peixoto AL, Pirani JR, Prado J, Queiroz LP, Souza S, Souza VC, Stehmann JR, Sylvestre LS, Walter BMT, Zappi D (2012) New Brazilian floristic list highlights conservation challenges. Bioscience 62:39–45. https://doi.org/10.1525/bio.2012.62.1.8
Greco TM, Pinto MM, Tombolato AFC, Nianhe X (2015) Diversity of bamboo in Brazil. J Trop Subtrop Bot 23:1–16
Guerreiro CI, Lizarazu MA (2010) Flowering of *Bambusa tuldoides* (Poaceae, Bambusoideae, Bambuseae) in southern South America. Darwin 48(1):25–31
Hong TD, Ellis RH (1996) A protocol to determinate seed storage behaviour. Int Plant Gen Res Tech Bull 1(1):1–55
Janzen DH (1976) Why bamboos wait so long to flower. Ann Rev Ecol Syst 7:347–391
Kapai VY, Kapoor P, Rao IU (2010) *In vitro* propagation for conservation of rare and threatened plants of India – a review. Int J Biol Technol 1(2):1–14
Leal GSA, Leal FA, Gomes HT, Souza AM, Ribeiro SC, Scherwinski-Pereira JE (2021) Structure and genetic diversity of natural populations of *Guadua weberbaueri* in the southwestern Amazon, Brazil. J For Res 32:755–763
Leandro TD, Shirasuna RT, Filgueiras TS, Scatena VL (2016) The utility of Bambusoideae (Poaceae, Poales) leaf blade anatomy for identification and systematics. Braz J Biol 76 (3):708–717
León-Lobos A, Wayc M, Arandad PD, Lima-Junior M (2012) The role of *ex situ* seed banks in the conservation of plant diversity and in ecological restoration in Latin America. Plant Ecol Divers 5:245–258. https://doi.org/10.1080/17550874.2012.713402
Londoño X (2002) Distribuicion, morfologia, taxonomia, anatomia, silvicultura y usos de los bambues del nuevo mundo. Universidad Nacional de Colombia, Bogotá. http://www.maderinsa.com/guadua/taller.html

Luis ZG, Nogueira JS, Ribeiro DG, Scherwinski-Pereira JE (2017) Caracterização anatômica de órgãos vegetativos de bambu (Poaceae, Bambusoideae). In: Drumond PM, Wiedeman G (eds) Bambus no Brasil: da biologia à tecnologia, 1st edn. ICH, Rio de Janeiro, pp 10–27

Mazzottini-dos-Santos HC, Ribeiro LM, Oliveira DMT, Paiva EAS (2020) Ultrastructural aspects of metabolic pathways and translocation routes during mobilization of seed reserves in *Acrocomia aculeata* (Arecaceae). Braz J Bot 43:589–600. https://doi.org/10.1007/s40415-020-00622-7

Murashige T, Skoog F (1962) A revised medium for rapid growth and bio assay with tobacco tissue culture. Physiol Plant 15:473–497

Nass LL, Sigrist MS, Ribeiro CSC, Reifschneider FJB (2012) Genetic resources: the basis for sustainable and competitive plant breeding. Crop Breed Appl Biotechnol 12:75–86

Nogueira JS (2018) Estratégias para a conservação *ex situ* de *Dendrocalamus asper* e micropropagação de espécies do gênero *Guadua* (Bambusoideae, Poaceae). PhD Thesis, Universidade de Brasília

Nogueira JS, Gomes HT, Scherwinski-Pereira JE (2019) Micropropagation, plantlets production estimation and ISSR marker-based genetic fidelity analysis of *Guadua magna* and *G. angustifolia*. Pesq Agrop Trop 49:1–9. https://doi.org/10.1590/1983-40632019v4953743

Nogueira JS, Nogueira GF, Silva-Cardoso IMA, Scherwinski-Pereira JE (2020) Tolerância à dessecação e criopreservação de sementes de bambu (*Dendrocalamus asper*). In: VI Congresso Brasileiro de Recursos Genéticos, vol 1, p. 577

Panis B, Lambardi M (2006) Status of cryopreservation technologies in plants (crops and forest trees). In: Ruane J, Sonnino A (eds) The role of biotechnology in exploring and protecting agricultural genetic resources. FAO, Corporate Document Repositoy, Rome, pp 61–78

Panis B, Swennen R, Engelmann F (2001) Cryopreservation of plant germplasm. Acta Hort 560:79–86

Roberts EH (1973) Predicting the storage life of seeds. Seed Sci Technol 1:499–514

Ruiz-Sanchez E, Munguía-Lino G, Vargas-Amado G, Rodríguez A (2020) Diversity, endemism and conservation status of native Mexican woody bamboos (Poaceae: Bambusoideae: Bambuseae). Bot J Linn Soc 192:281–295. https://doi.org/10.1093/botlinnean/boz062

Sandhu M, Wani SH, Jiménez VM (2018) *In vitro* propagation of bamboo species through axillary shoot proliferation: a review. Plant Cell Tissue Organ Cult 132:27–53. https://doi.org/10.1007/s11240-017-1325-1

Satya S, Bal LM, Singhal P, Naik SN (2010) Bamboo shoot processing: food quality and safety aspect (a review). Trends Food Sci Technol 21(4):181–189. https://doi.org/10.1016/j.tifs.2009.11.002

Sert MA, Bonato CM, Souza LA (2009) Germinação da semente. In: Sementes e plântulas: germinação, estrutura e adaptação. Todapalavra, Ponta Grossa, PR, pp 88–118

Sharma B, Gatoo A, Maximilian B, Mulligan H, Ramage M (2015) Engineered bamboo: state of the art. Proc Inst Civ Eng Constr Mater 168(2):57–67. https://doi.org/10.1680/coma.14.00020

Silva SMM, Martins K, Costa FHS, Campos T, Scherwinski-Pereira JE (2020) Genetic structure and diversity of native *Guadua* species (Poaceae: Bambusoideae) in natural populations of the Brazilian Amazon rainforest. An Acad Bras Cienc 92:e20190083

Singh G, Richa, Sharma ML (2017) A review article – bamboo seed viability. Int J Curr Adv Res 6:1690–1691

Smith M, Nelson BW (2011) Fire favours expansion of bamboo-dominated forests in the south-west Amazon. J Trop Ecol 27:59–64

Soreng RJ, Peterson PM, Romaschenko K, Davidse G, Zuloaga FO, Judziewicz EJ, Filgueiras TS, Davis JI, Morrone O (2015) A worldwide phylogenetic classification of the Poaceae (Gramineae). J Syst Evol 53:117–137

Thapliyal M, Joshi G, Behera F (2015) Bamboo: flowering, seed germination and storage. In: Kaushik S, Singh YP, Kumar D, Thapliyal M, Barthwal S (eds) Bamboos in India. National Forest Library and Information Centre Forest Research Institute, Dehradun, pp 89–108

Vale PAA, Oliveira Júnior JB, Costa FHS, Scherwinski-Pereira JE (2019) Height and number of shoots on the survival and development of micropropagated bamboo plantlets during pre-acclimatization. Pesq Agrop Trop 49:e53751. https://doi.org/10.1590/1983-40632019v4953751

Veiga RFA (2006) Crioconservação de sementes de recursos genéticos hortícolas no Instituto Agronômico (IAC). O Agronômico 58:19–21

Walters C, Berjak P, Pammenter N, Kennedy K, Raven P (2013) Preservation of recalcitrant seeds. Science 339(6122):915–916

Yamazaki T, Kawamura Y, Uemura M (2008) Cryobehavior of the plasma membrane in protoplasts isolated from cold-acclimated *Arabidopsis* leaves is related to surface area regulation. Plant Cell Physiol 49:944–957

Yeasmin L, Ali MN, Gantait S, Chakraborty S (2015) Bamboo: an overview on its genetic diversity and characterization. 3 Biotech 5:1–11. https://doi.org/10.1007/s13205-014-0201-5

Chapter 13
Application of Biotechnological Tool in Bamboo Improvement

Anwar Shahzad, Sabaha Tahseen, Adla Wasi, Zishan Ahmad, and Akil A. Khan

Abstract Bamboos belonging to the family Poaceae are one of the most versatile, natural, and renewable resources among the plant kingdom. Bamboos are generally found in tropical and subtropical parts of the world. They play an important role in the bioenergy and the bioeconomy of many Asian countries. Although bamboos are early maturing and fast-growing plants, due to a lack of suitable regulation, the annual yields are not sufficient to meet the annual demands. Increasing demand for bioenergy; exploitation by industries for paper, pulp, timber, fiber, biofuel, food, and medicine; and lack of sufficient efforts to sustain the cultivation of bamboo are the main causes of rapid reduction in its population. Bamboos are conventionally propagated through seed, culm cuttings, rhizome cuttings, and clump division, but these methods are inadequate due to large demands of propagules of new crop along with limited availability, low rate of multiplication, a low percentage of rooting, and seasonal dependence. This can be accomplished by the employment of advanced biotechnological tools. Through biotechnological tools, bamboos may be improved by applying micropropagation, genomics, proteomics, transgenic technology, and nanotechnology. This chapter is dealing with the available information on the applications of the biotechnological tools applied in bamboo.

Keywords Bamboo · Genetic transformation · Gene editing · Genome map · Nanotechnology

---

A. Shahzad (✉) · S. Tahseen · A. Wasi
Plant Biotechnology Section, Department of Botany, Aligarh Muslim University, Aligarh, Uttar Pradesh, India
e-mail: ashahzad.bt@amu.ac.in

Z. Ahmad
Bamboo Research Institute, Nanjing Forestry University, Nanjing, Jiangsu, China

Co-Innovation Center for Sustainable Forestry in Southern China, Nanjing Forestry University, Nanjing, China

A. A. Khan
Department of Botany, GF (PG) College, Shahjahanpur, Uttar Pradesh, India

Z. Ahmad et al. (eds.), *Biotechnological Advances in Bamboo*,
https://doi.org/10.1007/978-981-16-1310-4_13

13.1 Introduction

Food security and more recently the debate around climate change have gained much attention as the key elements of sustainable development (Karp and Richter 2011; Lal 2010). There are numerous solutions to encounter the various challenges, in which bamboo stays ahead among the useful species of plant that can endure the coming generation with its diverse properties (Ramakrishnan et al. 2018). Bamboos have their place in Poaceae, a family of angiosperm under the Bambusoideae subfamily embracing 123 genera comprising a total of 1642 species worldwide according to the *World Checklist of Bamboos and Rattans* (Vorontsova et al. 2016). Speedy growth, flowering, and superior physical and mechanical properties are the unique characteristics of bamboo (Sawarkar et al. 2020). The root, culm, and leaves are its main parts. Due to its fast growth and long juvenility, woody grass bamboos are chiefly significant. Bamboo touched countless surfaces of rural living, whereas it is also helpful in firming the urban sector that is why it is known as "forest's green gold" (Devi 2013). Traditionally bamboo has been utilized as a building material for village housing, foodstuff, and domestic decorative items. In the present era, bamboo utilization further widens. In addition to giving ecological answers to worldwide climate variation with its carbon dioxide seclusion capacity, around two and half billion of population of the world rely on bamboo (Thapa et al. 2018). Numerous recognized applications of bamboo have been exploiting its rapid growth and high renewable tendency (Nayak and Mishra 2016). However, wide anthropogenic pressures on its local habitats led to the decline of the natural stands and prohibited their regeneration. Having a reproductive cycle of 120 years due to its monocarpic nature and recalcitrant seed production make it difficult to grow on such a big scale. Propagating bamboo through the nodal segment is cumbersome and labor-intensive (Sood et al. 2013). To enhance the production and to fulfill the large industrial demand for bamboo, micropropagation through culturing its various tissues is a useful technique.

Bamboos are affected by pests and diseases which also restrict its successful establishment of the nursery. Schizotetranychus floresi (Mite) was observed to be most alarming species affecting the bamboo plants (Singh et al. 2013a, b). Similarly, 13 species of bamboo are prone to the Bamboo mosaic virus (BaMV), and no chemical is known which efficiently prevents the infection (Alazem et al. 2014). To overcome this problem, disease-resistant bamboo should develop. It is tough to obtain disease-resistant plants through conventional breeding (Singh et al. 2013a, b).

Peculiar flowering makes it difficult to obtain superior traits in bamboo through breeding programs. After flowering, the typical end of bamboo clumps makes the understanding of the flowering phenomenon of bamboo rather problematic, and it also limits the study of their reproductive organs. Here tissue culture technology is well acknowledged for the saving of genetically hybrid seeds obtained by traditional approaches of breeding besides the study of floral detail through in vitro flowering. Also, biotechnology tools added great advancement to the breeding technique in

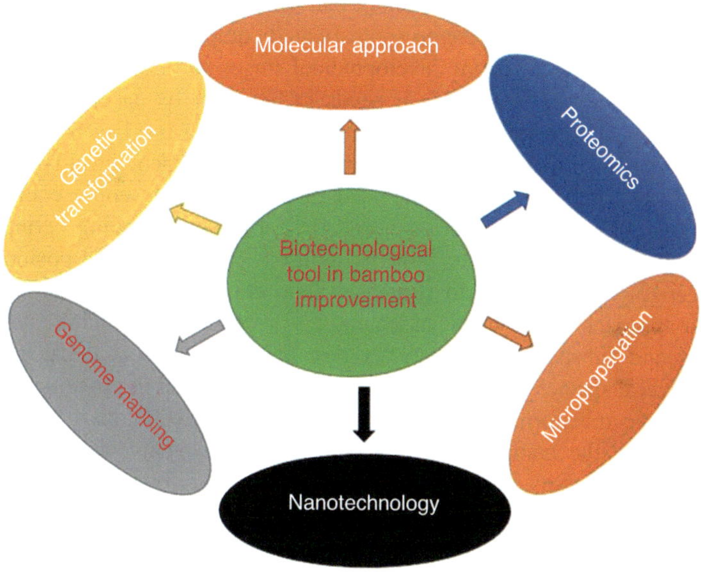

Fig. 13.1 Biotechnological tools applied in bamboo improvement

obtaining better varieties within less time. This can be accomplished by applying the transgenic, molecular breeding and tissue culture approach. Biotechnological approaches are generally concerned with protoplast fusion to obtain somatic hybrids, usage of DNA markers to choose a trait of interest, and gene transfer to get genetically engineered organisms (Adlak et al. 2019). A diverse technological tool that is involved in cellular engineering, bioinformatics, nanobiotechnology, and bioprocess engineering has revolutionized present-day biotechnology (Fig. 13.1). These have unlocked several paths for environmental and socioeconomic advances as well as enabled expanded utilization of plant resource. By means of these technologies, numerous less popular plants having fast growth and ample biomass have appeared as a substitute for wood, food, fabric, and fuel resources (Thapa et al. 2018). Therefore, these biotechnological tools are helpful in balancing the demand and supply of natural resources.

13.2 Improvement Through Molecular Approaches

13.2.1 Polymorphism and Delineating Phylogenetic Relationships

In the majority of plant species to evaluate diversity inside and between populations and for taxonomic studies, vegetative and floral features have been used since ancient times. Morphological characters can be observed visually, and for data documentation of these features, no sophisticated equipment is required. But these characters may vary at different developmental stages and are also affected by the changed environmental factors, so it needs to be taken by an expert taxonomist (Kalia et al. 2013). In bamboo, there are limited characters that have been described yet, and evidence on their mechanism of regulation at a genetic level is still mostly ambiguous. The key obstacle is extended flowering round, ample equipment or facilities, and sufficient time period needs for caring and handling germplasm or populations since their lives are short in addition to its disordered position under diverse geography of the world (Ramakrishnan et al. 2020). In these situations, molecular approaches are beneficial techniques designed for taxonomic delineation of species and subspecies for characterizing the differences at the genetic level between several species to decipher the gene of interest and improvement with the help of genetic transformation studies. To understand evolutionary relations and diversity among the bamboo, the molecular marker system plays a crucial role through genetic diversity studies (Nirmala and Bisht 2012). Various marker systems such as morphological, biochemical, and molecular have been used to decipher the genetic diversity of bamboo (Ramakrishnan et al. 2020). The foremost purpose of studying the diversity of bamboos is their population's clonal structure, as bamboo has evolved through a natural tendency of vegetative propagation. Random Amplified Polymorphic DNA (RAPD), Restriction Fragment Length Polymorphism (RFLP), Polymerase Chain Reaction, Sequence-Tagged Sites (STS), Macro nuclear (MAC) Ribozymes, Gene Tagging and Leucoanthocyanidin Reductase (LCR) and Deoxyribose Nucleic Acid (DNA) fingerprinting, STS are the technologies which meant for the genetic engineering of bamboos. For identification, taxonomy, study of variability, and early assessment AFLP patterns are used in bamboo (Kaushik et al. 2015). Similarly, 12 species of bamboos had been investigated with the help of RAPD marker (Nayak et al. 2003). When cpDNA is used to draw phylogeny in bamboo with the help of 15 restriction enzymes, 2 major lineages were discovered (Stapleton et al. 1995). While the studies by means of organelle genomes are inadequate, cpDNA was helpful in studying the relationships within bamboo species (Zhang et al. 2011). In 176 genotypes of Moso bamboo, second-generation molecular marker was applied in which RAPD comes out more appropriate within 13 RAPD, 3 SSRs, and 1 minisatellite marker. It is best suited to recognize the clones and delineate the genetic relationship between bamboo cultivars (Zhang et al. 2011). Similarly, RAPD markers were used to differentiate genera such as *Bambusa*, *Dendrocalamus*, *Dinochloa*, and *Cephalostachyum* (Nayak et al. 2003). During the study of the genetic relationship of Moso bamboos in 17 states of China with the help of ISSR and AFLP markers, low polymorphism was stated at the species level. Initially, genome-based SSR markers with the help of 127,593 SSR motifs recognized along the genome of moso bamboo (Zhao et al. 2015). By utilizing 1451 primers of which 1098 markers were mapped physically, also 917 markers in this can be authorized in 9 accessions with 39.8% transferability. Later, a group of 24 SSRs from the effective set can separate an assembly of 78 accessions to their taxonomic arrangement. Likewise, Jiang et al. (2017) studied the genetic diversity and differentiation using twenty fluorescently labeled microsatellite markers in *P. edulis* including 34 representative populations (803 individuals). It was reported that in bamboo the gene sequencing has revealed a unique long sequences of DNA of around 4,700,803 and 268,150 called unique single-nucleotide polymorphisms (Uni-SNPs) and unique InDels (Uni-Indels) respectively. Moreover, the important pathways like caffeine metabolism and ribosome biogenesis are believed to be associated with these genes. Recently, in bamboo 16 inter-retrotransposon amplified polymorphism (IRAP) markers are developed (Li et al. 2019). By means of these markers, 58 Asian bamboo accessions (*Phyllostachys*) were separated into 4 sub-groups PhSP1, PhSP2, PhSP3, and PhSP4 depending on genetic diversity and population structure. These markers perceive the use of a retrotransposon-based marker in defining the interspecific differences of bamboos.

13.2.2 Genetic Fidelity

For the large-scale production of plants, micropropagation is one of the best approaches. However, explant source, culture conditions, ploidy level, and time of culture bring somaclonal variation during micropropagation. Somaclonal variation produced numerous genomic changes in in vitro raised plants (Peredo et al. 2006). These variations occur because of externally applied growth regulators, enhanced mutation rate, and accumulation of mutations in due course of time. It also caused a change in patterns of DNA methylation, DNA destruction, and mutation and also changes the capability of cells to restore damage and mutation (Singh et al. 2012). Therefore, it is really significant to determine the clonal homogeneity of the micropropagated plants. Biochemical, physiological, morphological, and molecular markers are applied for the determination of fidelity. In all this determining system, determination through molecular markers is most effective, because it is not affected by the stage of culture, time of culture, and the prevailing environmental conditions of in vitro raised plants (Singh et al. 2013a, b).

This determination also considers the structure of the leaf, leaf mass, chlorophyll, ratio of water, leaf anatomy, and photosynthetic parameters (Singh et al. 2013a, b). In vitro raised *D. asper* and *D. hamiltonii* genetic fidelity has been tested with the help of markers based on DNA such as rapid amplified polymorphic DNA (RAPD), ISSR, AFLP, and simple sequence repeat (Agnihotri et al. 2009; Bag et al. 2012; Singh et al. 2013a, b). Amplified monomorphic bands of in vitro-grown plant and mother plant established the fact the previous were genetically uniformed or true to type to the mother plant. It is also necessary to ascertain the genetic fidelity of plants obtained through the process of somatic embryogenesis. While sampling of somatic embryos raised *B. nutans* and *D. hamiltonii* species of bamboo, the outcome shows genetic uniformity in RAPD as well as AFLP markers. In addition, when the field-established plants were evaluated with AFLP markers by means of six primer combinations, a greater level of genetic stability was recorded. Out of 407 scorable fragments verified, 402 (98.8%) were shown conservation at various morphogenetic stages leading to plantlet regeneration (Sood et al. 2013). Similarly, AFLP identified 22 constituent clones in the population of dwarf bamboo, *Sasa senanensis* (Suyama et al. 2000). In *D. Strictus* species of bamboo, screening of somaclonal variation was done with the help of 24 RAPD and 15 ISSR primers. However, in amplifying genomic DNA, only ten RAPD and nine ISSR primers had given positive results. During amplification, 10 RAPD primers produced 58 scorable bands, while ISSR primers produced 66 scorable bands. In parent as well as in vitro-grown plantlets, the produced bands were found to be monomorphic with help of both RAPD and ISSR primers. Also, in the case of RAPD, the size of the band ranges from 240 to 1455 bp, and with ISSR marker size varied between 183 and 1544 bp. This homogeneity in banding patterns of micropropagated *D. strictus* plantlets established genetic fidelity (Goyal et al. 2015). Several sequence-based PCR and hybridization markers have been used these days each having different benefits and disadvantages (Kalia et al. 2011). For example, to detect polymorphism, both ISSR and RAPD markers are utilized, but ISSR ruled over RAPD markers. ISSR markers displayed more polymorphism and also are more reproducible due to the occurrence of a large number of SSR regions when equated to RAPD (Ray and Roy 2007). For a better analysis of genetic stability, more than one marker system has been recommended to be utilized, as diverse genome regions consist of diversity in markers (Palombi and Damiano 2002; Lakshmanan et al. 2007). It is also important to select a particular marker and technique according to requirement and experimental design (Parveen et al. 2016).

13.3 Genetic Transformation and Gene Editing

Pragmatic application of genetic transformation is very commonly applied in numerous plants these days (Oramas et al. 2000). Particle bombardment, silicon carbide, electroporation, polyethylene glycol (PEG), and *Agrobacterium*-mediated transformation are numerous approaches that can be used to transfer foreign genes in plant cells. Various aspects are carefully observed during transformation in which an efficient tissue regeneration system is most important. Above all, genetic fidelity of the regenerated plants is extremely necessary, for confirming the efficiency of the regeneration and foreign gene expression (Saeed and Shahzad 2016). In bamboo such as *D. hamiltonii*, *D. farinosus*, and *D. Latiflorus* (ma bamboo), several efforts have been done on genetic transformation (Jiang and Zhou 2014; Qiao et al. 2014; Sood et al. 2014). A successful genetic transformation has been achieved in *D. Latiflorus* through the culture of the anther. It was aimed for CodA, bacterial gene transformation that encodes choline oxidase and is useful in establishing cold tolerance into the bamboo genome (Qiao et al. 2014). However, transformation through *Agrobacterium* and particle bombardment show less transformation frequency (Sood et al. 2014). Recently, the *Agrobacterium*-mediated transformation protocol was developed for Ma bamboo. It takes 8 months to obtain vigorous calli and a 7–8 months' time period for the successful transformation. Although genetic engineering improved agronomic traits, it remains restricted to only one bamboo species, i.e., *D. latiflorus*. The major problem associated with bamboo for successful transformation is its low efficiency and slow regeneration.

Clustered regularly interspaced short palindromic repeat (CRISPR) and CRISPR-associated endonuclease (Cas9) are a major breakthrough in gene-editing technologies. This CRISPR/Cas9 editing process is also applied in Ma bamboo species by aiming total homo-alleles or precisely one allele of a gene (Ye et al. 2020). By means of this technology, it is determined that knocking out one gene which is responsible for the expression of gibberellin in Ma bamboo could change plant height. In *D. hamiltonii*, both *Agrobacterium tumefaciens* and microprojectile bombardment-mediated protocols were developed for genetic transformation (Sood et al. 2013). The *gfp*or *gus* reporter genes were used to observe the successful transformation in globular somatic embryos. Somatic embryo subjected to GUS assay and for gfp reporter gene is detected under a confocal microscope. The selection of transformed somatic embryos had been done on kanamycin (200 mg/l) or hygromycin (50 mg/l) selection medium which are allowed to germinate (Sood et al. 2013). This is a significant accomplishment since the transformation of monocots; exclusively bamboos have always been a challenging area of research. In bamboo or other plants, development of transformation protocols greatly depends on the formation of reliable and effective regeneration schemes for somatic tissues or cells. Recently bamboo genome is released which is anticipated to enhance the experiments on genetic transformation and gene editing in Moso bamboo. However, intensive research is essential to know the molecular mechanism of somatic embryogenesis and regeneration for accomplishing an effective genetic transformation scheme to gain profit in editing the genome. Many successful attempts of the transformation have already been achieved in bamboo, though a large amount of work is still needed to be finished for real exploitation of an improved transgenic bamboo (Ramakrishnan et al. 2020).

13.4 Genome Map of Bamboo

These days, understanding of the genome of the plants at its physical, functional, epigenomic, and comparative stages has been significantly increasing because of genomics which pervaded every aspect of plant biology. Several model plants sequenced successfully, and a wide series of their data sets are presented (Bennetzen et al. 2012). Recently many scientists are working on non-model plants which are economically very important. Sequencing of nearly 328 vascular plants (consisting of 323 angiosperms, 5 gymnosperms, 3 lycophytes), 3 nonvascular terrestrial plant (2 mosses and 2 liverworts), and 60 green algae has been done. In all the angiosperms, Poales is the top sampled order with 104 genome assemblies of which 102 assemblies placed in the Poaceae family, 38 were present in the genus *Oryza*, and 24 are assemblies of *Oryza sativa* (Kersey 2019). In all the angiosperm, bamboos are considered as one of the exceptional, timber-less, monocarpic member of the family of grass. Nearly all important grass lineage genomic data have assembled quickly, but in database bamboos which belongs to one of the big subfamilies of Poaceae have only little genomic data available (Zhang et al. 2012). In the previous few years, genomic experiments have been done on various species of bamboo, and efforts have been put to delineate the complication in the regulation of gene and bamboo networking. The initial understanding of the gene and assemblies of the bamboo genome was achieved with the help of cloning and sequencing 10,608 putative complete cDNAs (FL-cDNAs), mainly from the Moso bamboo, *Phyllostachys heterocycla* cv. *pubescens* (Peng et al. 2010). It characterizes the third major FL-cDNA assembly of all plant species. This FL-cDNA sequence established the fact that bamboo deviated from its nearest lineage, i.e., paddy, barley, and wheat, due to adaptive radiation. Rice and sorghum could be possible models for cracking Bambusoideae genomes because both are showing high genomic synteny with bamboo (Gui et al. 2010). The comparative study also supported a hexaploid beginning of ma bamboo. Unpredictable flowering time in bamboo is problematic for its genetic map construction. Knowing the genome of bamboo comparative genome analysis is more useful. In order to decipher bamboo and rattan (BR) genetics in the direction of a bioeconomy, the International Network for Bamboo and Rattan (www.inbar.int) developed a giant project termed "Genome Atlas of Bamboo and Rattan (GABR)" in 2017. Chromosome-based assemblies of Moso bamboo and two rattan species were released under this program (Zhao et al. 2018). This assembly covered a substantial genome and further enhancement of 243 times when compared with the draft genome. The genome size is 1908 Mb with an N50 scaffold and a dimension of 79.90 Mb. The genome reference reads comprise a sum of 51,074 superior protein-coding genes. For *Oropetium thomaeum*, the genome is the smallest known grass genome of only 245 Mb long compared to bamboo; the size of bamboo genome comes 7.8 times larger (Bartels and Mattar 2002; VanBuren et al. 2018). Likewise, it is 4.4 times greater than the genome of rice which consists of 430 Mb and 2.5 times bigger than the genome of sorghum which consists of 772 Mb (Zhou et al. 2007; Paterson 2010). Recently, a genetic map of bamboo has been derived with the help of 3627 ddRAD markers. It is obtained from 190 progenies of an inbred population of ma bamboo (*Dendrocalamus latiflorus* Munro) (Liu et al. 2020). This gene map depicted 36 linkage groups equivalent to the 36 chromosome pairs occupying 93.3% of the genome. The whole map extends up to 3113 centimorgan (cM), with an average marker length of 0.93 cm for each marker. Also, 4.91 and 131.69 cM are the measured dimension range of linkage group. Various noncoding RNAs such as microRNA (miRNA), transfer RNA (tRNA), small nuclear RNA (snRNA), ribosomal RNA (rRNA), and small nucleolar RNA (snoRNA) genes are also reported to be present in bamboo genomes. For the investigation of the methylation of DNA, for analysis of transcriptome and alteration in histone protein of the bamboo genomes, these noncoding genetic resources play crucial roles.

13.5 Micropropagation

Generally, bamboos are conventionally propagated through seed propagation, culm cuttings, rhizome cuttings, clonal propagation, and clump division (Banik 1994; Banik 1995). But these classical methods are inadequate and unproductive for extensive propagation of bamboo. Due to the lack of sufficient amounts of seeds, propagation through seed is very limited. Furthermore, conventional propagation is very costly because it requires large nurseries and setup. So, micropropagation is the best alternative for the improvement and production of bamboo at a large scale (Nadgauda et al. 1990). Micropropagation is a commercially well-known technology for the quick augmentation of hard to proliferate plants for commercial production, germplasm conservation, and production of enormous number of the hereditarily indistinguishable plants. Micropropagation depends on perception of the cell totipotency in which a single cell regenerates into a whole organism, and this technique is advantageous over other inadequate and unproductive techniques that are used for the extensive propagation of bamboo (Singh et al. 2013a, b; Thapa et al. 2018). Bamboo production through micropropagation was firstly reported by Alexander and Rao (1968) by means of explants from zygotic embryos in *Dendrocalamus strictus*. Mudoi et al. (2013) and Singh et al. (2013a, b) reported that numerous efforts have been made that proved different regeneration techniques such as somatic embryogenesis, adventitious shoot formation, and organogenesis, applied to different species and species cultivars and also at explant's age. Kaur et al. (2012) reported a slow growth method by means of the liquid paraffin overlay method in *D. hamiltonii* for medium-term preservation. To understand the effects of factors on the transition of flower and flower development, in vitro flowering system is also used in *D. Hamiltonii* (Kaur et al. 2014). Kaur et al. (2015) also did an effort to detect genetic markers and allocators of flowering in *D. hamiltonii* by studying changes in proteome associated with the floral transition. In bamboo, homozygous diploid plants can also be acquired in one generation by means of anther culture by Qiao et al. (2013). But this technique depends upon different developmental stages of immature microspore or pollen grain for different genotypes (Niazian and Shariatpanahi 2020). Because of these differences, a common protocol is not applicable for enormous production and resourceful regeneration of bamboo through micropropagation (Ye et al. 2017). But, two frequently used micropropagation techniques for bamboo production are somatic embryogenesis and organogenesis (Singh et al. 2013a, b). For organogenesis, inflorescences and nodal segments are commonly used as explants (Lin and Chang 1998; Lin et al. 2005). Generally, it is found that plants are free from somaclonal variations and genetically stable, which are regenerated from nodal segments or shoot tips. Mehta et al. (1982) initiated research associated with somatic embryogenesis in bamboo. They produced plantlets in *Bambusa arundinacea*. In *D. hamiltonii*, maximum embryogenesis (93.3% and 90.0%) was reported on the above Murashige and Skoog medium augmented with 5.0μM BAP and 7.5μM 2,4-D by Bag et al. (2012). They also reported in vitro regenerated plantlets (11.9 and 11.3 per callus lump) and highest somatic embryo number (38.7 and 37.3 per callus lump) from *D. hamiltonii* (10- and 45-year-old), respectively, on above Murashige and Skoog medium augmented with 5.0μM BAP and 7.5μM 2,4-D. For the initiation of embryo development in bamboos, cell proliferation is arrested through auxin and cytokinin removal and provides them medium-free from PGR (Godbole et al. 2002). They also reported in a plant growth regulator-free medium; germination of mature somatic embryos takes place and converts into plantlets. In some studies, it was found that cytokinin is essential for somatic embryo germination. Yeh and Chang (1986a, b, 1987) reported that kinetin promotes the somatic embryo germination of *Sinocalamuslatiflora*, *B. beecheyana*, and *B. oldhamii*. A description of in vitro regeneration of bamboo using different explants via organogenesis and embryogenesis is given in Tables 13.1 and 13.2.

For in vitro flowering, rooting, and shoot regeneration, plant growth regulators such as 2,4-dichlorophenoxyacetic acid (2,4-D), thidiazuron (TDZ), 1-naphthaleneacetic acid (NAA), and 1-aminocyclopropane-1-carboxylic acid (ACC) are testified appropriate. Likewise, for somatic embryogenesis, zygotic seed embryos, in vitro spikelets, and nodal segments are used with PGRs such as zeatin, TDZ, kinetin, 2,4-D, and coconut milk. From the observations, it is clear that TDZ is most suitable for in vitro flowering and somatic embryogenesis (Lin et al. 2004a, b), whereas NAA is the best choice for rooting (Ramakrishnan et al. 2020). In the proliferation medium, some additives like amino acids, activated charcoal, and adenine sulfate have also been used with PGRs. Due to the release of phenolic compounds, the blackening or lethal browning is controlled by using antioxidant or polyphenol adsorbents. Ascorbic acid has been found most effective in *D. hamiltonii*, whereas activated charcoal and PVP were found ineffective (Singh et al. 2012). In other cases, PVP was found effective in *D. strictus* for shoot health improvement (Saxena and Dhawan 1999).

Although plant tissue culture is a widely used technique for the large-scale production of bamboos, however, it is less applicable at the commercial level because of the extensive plant loss when transferred to natural condition. As regards to successful field transfer, only limited reports are available. It was reported by Sood et al. (2002) and Agnihotri et al. (2009) that 70% survival percentage was possible in ***D. hamiltonii*** in the field, while Mishra et al. (2011) reported 91% survival percentage in *B. tulda* in the greenhouse, and Negi and Saxena (2011) reported 95.83% acclimatization rate up to garden center and transferred 12 plants with 100% survival percentage in the field. Singh et al. (2011, 2012) reported 92.34% and 100% survival percentages for *D. asper* and *D. hamiltonii* in the greenhouse while 79.76% and 85% survival percentages in the field.

13.6 Nanotechnology

Nanotechnology is an emerging scientific field that ushers an era of innovation at the very basic level, i.e., atomic level. As we know, the recent improvements in agriculture have come at the cost of our ecological balance which poses a serious

Table 13.1 In vitro propagation in bamboo using different explants

Species	Explant used	Treatments	Remarks	References
Arundinaria callosal	Nodal segments	MS + BAP 13.3 μM + IBA 1.0 μM ½ MS + IBA 25.0 μM+ BAP 0.05 μM	Shoot multiplication Rooting	Devi and Sharma (2009)
Bambusa edulis	Nodal segments	MS + TDZ 0.1 mg/l	Inflorescence proliferation	Lin et al. (2003, 2004a)
Bambusa edulis	Inflorescence	MS + TDZ (0.1 mg/l), MS + NAA (5.0 mg/l)	Inflorescence proliferation	Lin et al. (2004b)
Bambusa balcooa, B. nutans, B. salarkhanii, *B.vulgaris, B. vulgaris var. striata and Thyrsostachys Oliveri*	Nodal segments	MS + BAP 4.4–22.0 μM ½ MS + NAA 5.4–16.2 μM + IBA 4.9–24.5 μM	Shoot multiplication Rooting	Islam and Rahman (2005)
Bambusa edulis	Inflorescence	MS + 2,4-D (1 mg/l), MS + NAA (10 mg/l), MS + NAA (5 mg/l) + ACC (1 mg/l)	Shoot regeneration, re-flowering, and post flowering	Lin et al. (2005)
Bambusa oldhamii	Shoot apices	MS + TDZ 0.45 μM MS + NAA 26.85 Mm	Shoot initiation Rooting	Lin et al. (2007)
Bambusa oldhamii	Node	MS + BAP 4.4 μM MS + IBA 9.84 μM + NAA 2.69 μM	Shoot multiplication Rooting	Thiruvengadam et al. (2011)
Bambusa tulda, *Melocanna baccifera*	Nodal segments	MS + BAP 13.2 μM MS + Kn 9.4 μM + BAP 12.12 μM MS + IBA 14.7 μM + coumarin 68.4 μM	Bud breaking Shoot multiplication Production of rooting and rhizome	Waikhom and Louis (2014)
Bambusa arundinacea	-Do-	MS + BAP 13.2 μM + IBA 2.45 μM + 4% sucrose + 4% CM MS+ IBA 14.7 μM + 2.0 mg/l $AgNO_3$	Initiation of shoot bud Rooting	Venkatachalam et al. (2015)
Bambusa vulgaris	Internodes	MS + NAA 3 mg/l + BAP 0.3 mg/l MS + IAA 3mg/l+ BAP 0.3 mg/l	Shoot multiplication and Rooting	Kaladhar et al. (2017)
Bambusa balcooa	Nodal segment		Shoot multiplication Rooting	Rajput et al. (2020)

(continued)

Table 13.1 (continued)

Species	Explant used	Treatments	Remarks	References
		MS + BAP 4.0 mg/l + ascorbic acid 50 mg/l + L-arginine, citric acid, adenine sulphate (25 mg/l Each) ½ MS + NAA 6.0 mg/l + activated charcoal 100 mg/l		
Dendrocalamu sstrictus	Zygotic embryo	White major and minor elements	Shoot formation	Alexander and Rao (1968)
Dendrocalamus farinosus	Seed embryo, young shoots	MS + 2,4,5-T 2.0 mg/l +Kn0.2 mg/l + IBA 0.4 mg /l MS + Kn2.5 mg/l + IAA 0.5 mg/l	Callus formation organogenesis	Hu et al. (2011)
Dendrocalamus giganteus	Seeds	MS + BAP 8.9–13.3 μM ½MS + BAP 13.3 μM + 1.0 IBA μM 15.0 IBA μM ½ MS with IBA 25.0 μM and BAP 0.05 μM	Bud break Shoot multiplication Rooting	Devi et al. (2012)
Dendrocalamus latiflorus	Anthers	M8+ NAA 5.37 μM+ BA 1.33 μM+ PAA 110.17 μM M8 + KT 2.32 μM+ BA 8.89 μM+ NAA 1.08 μM+ PAA 110.17 μM	Callus induction and plant regeneration	Qiao et al. (2013)
Dendrocalamus. hamiltonii	Shoot tips	MS + 2,4-D 13.5 μM, BAP 4.4 μM, glutamine 500 mg/l, casein hydrolysate 500 mg/l MS + BAP 4.4 μM, Kn 1.41 μM, NAA 1.62 μM ½ MS + IBA 14.7 μM	Callus induction Differentiation of callus Rooting and acclimatization	Zang et al. (2016)
Dendrocalamus asper	Nodal explant	MS + TDZ 0.25 mg/l + BA 3 mg/l MS + TDZ 0.25 mg/l + BA 4.5 mg/l MS + TDZ 0.5 mg/l + BA 3 mg/l MS + TDZ 0.5 mg/l + BA3 mg/l ½ MS + IBA 1 mg/l	Shoot multiplication Rooting	Ray et al. (2018)
Gigantochloa atroviolaceae	Nodal segment	MS (liquid) + BAP 25.0 μM MS + BAP 20.0 μM + NAA 3.0 μM MS + IBA 35.0 μM	Bud break Shoot multiplication Rooting	Bisht et al. (2010)
Pseudoxytenanthera stocksii	Nodal segment	MS + BAP 6.0 mg/l	Shoot multiplication	Rajput et al. (2019)

Table 13.2 In vitro propagation in bamboo through somatic embryogenesis

Species	Explant used	Treatments	Remarks	References
Bambusa nutans	In-vitro shoots	MS + 2,4-D 5.0 μM + BAP 2.5 μM + ABA 1.0 μM MS + 2,4-D 5.0 μM + BAP 2.5 μM MS + BAP 5.0 μM + NAA 1.25 μM	Somatic embryogenesis Maturation Germination	Kalia et al. (2004)
Bambusa balcooa	Pseudo-spikelet	MS + 2,4-D 4.5 μM MS + BAP 22.2 Mm	Somatic embryogenesis Germination	Gillis et al. (2007)
Bambusa nutans	-Do-	MS +2,4-D 4.52 mg/l + BAP 4.44 mg/l MS + TDZ 0.499 mg/l+ NAA 10.74 mg/l MS +2% sucrose + NAA 16.11 mg/l	Somatic embryogenesis Germination Rooting	Mehta et al. (2011)
Bambusa arundinaceae	Mature seed embryo	MS + 2,4-D 1 mg/l + BAP 1 mg/l MS + 2,4-D 1 mg/l + NAA 1 mg/l + BAP or KIN (0.5- 4.0 mg/l)	Indirect somatic embryogenesis (Embryogenic callus) Somatic embryo maturation and formation	Venkatachalam and Kalaiarasi (2016)
Dendrocalamus hamiltonii	Nodal segments	MS + Sucrose 8% + BAP 4.44 μM + 2,4-D 4.53 μM MS + Sucrose 8% + BAP 11.1 μM	Somatic embryogenesis Maturation Germination	Godbole et al. (2002)
Dendrocalamus asper	In vitro shoots	MS + BAP 0.88μM + 2,4-D 9μM + IAA 2.85μM MS + BAP 4.4μM +GA 2.8μM	Somatic embryogenesis Germination	Arya et al. (2008)
Dendrocalamus hasper	Roots, leaves and nodal segments	MS + 2,4-D 30.0 μM MS + BAP 20.0 μM MS + IAA and MS + NAA 5.0–25.0 μM	Embryogenic callus, Germination of somatic embryos Rooting	Ojha et al. (2009)
Dendrocalamus hamiltonii	Seeds	MS + 2,4-D 4.5–13.5 μM MS + BAP 8.88 μM + Kn 4.65 μM + NAA5.37μM MS + IBA 24.5 μM	Embryogenic calli Somatic embryos proliferation, shoot differentiation and subsequent growth rooting	Zhang et al. (2010)

(continued)

Table 13.2 (continued)

Species	Explant used	Treatments	Remarks	References
Dendrocalamus asper *Drepenostachyum falcatum* *Dendrocalamus hamiltonii*	Nodal segment	MS + BAP 10 mg/l MS + NAA 1–5 mg/l MS + IBA 10 mg/l MS + 2,4-D 20-30 μM	Shoot multiplication Rooting Somatic embryogenesis	Arya and Arya (2015)
Dendrocalamus stocksii	Leaf sheath, shoot tip, nodal shoot segments and inter node segments	MS + 2,4-D 0.44 μM +Additives Ascorbic acid 8.8 μM + Citric acid 4.8 μM + Cysteine 3.02 μM + Glutamine 14.6 μM MS + NAA 0.55 μM + BAP 0.22 μM	Callus induction Somatic embryogenesis	Somashekar et al. (2018)
Ochlandra wightii	Seeds	½ MS + BAP 2.22 μM ½ MS + BAP 2.22 μM + TDZ 2.22 μM MS + Kn 4.65 μM MS +BAP 8.8 μM + Kn 2.35 μM	Germination of embryos Shoot multiplication Rhizome induction Shoot multiplication from nodal segments	Bejoy et al. (2012)
Phyllostachys heterocycla var. pubescens	Zygotic seed Embryos	MS + 2,4-D (4.0 mg/l) + Zeatin (0.1 mg/l) MS + 2,4-D (4.0 mg/l) MS + Zeatin (7.0 mg/l) MS + NAA (2.0 mg/l)	Callus initiation Somatic embryo formation Regeneration Rooting	Yuan et al. (2013)

threat to the global climate. However, nanotechnology offers a balance and eco-friendly approach to meet the agricultural demands (Sugunan and Dutta 2008). This is expected as a potential complement to molecular plant breeding and genetic engineering in addition to traditional plant breeding in the near future. The growing insecurity about food safety and security has been our serious concern since the population of our country is yet to take its dip phase; therefore, this challenge could be dealt by employing the nanotechnological tools in the field of agriculture. To understand the matter's properties and functions at the nanoscale, nanotechnology is one of the most important scientific areas dealing with nanometric (>100 nm) dimensions which provides a unique and innovative intuition in many scientific fields, i.e., material sciences, mathematics, engineering, medicine, biology, physics, chemistry, etc. (Roco et al. 1999; Scott and Chan 2002; Kulzer and Orrit 2004). At the present time, this technology is evolving as an energetic new industry associated with multi-sectors like materials, energy, and electronics and biomedical and manufacturing to support the global economy. Since nanotechnology has been employed, more than 800 nanomaterials have been made available in the market (Al-Halafi 2014; Safiuddin et al. 2014; Zhou et al. 2014). The first bamboo derivative, i.e., carbon nanostructure, has shown electrical and structural strength (Erkoc 2006). Based on bamboos' high antifungal and antimicrobial properties, many scientists reported carbon nanosphere and silver nanoparticle formation from bamboo charcoal and bamboo leaf extract, respectively (Yasin et al. 2013; Das and Saha 2012). Moreover, nanoparticles and nanospheres have added both physical and chemical properties to the macroparticles of bamboo, and these particles are reported to having wide applications in biosensing (Mirkin et al. 1996; Lu and Lieber 2007; Mitin et al. 2008); pharmaceuticals, optoelectronics, and photonics (Baruwati et al. 2009; Ahmad et al. 2011); cosmetics and DNA sequencing (Cao et al. 2001); water treatment and textiles (Murphy et al. 2008); prevention of HIV; and wound-healing (Singla et al. 2017; Murphy et al. 2008). Ahmad et al. (2015) and Singla et al. (2017) reported two new applications of bamboo in construction and nanocomposite dressings. They are used for the first-time nanocrystals from plant cellulose as wound dressing material made from leaves of ***D. hamiltonii*** **and** ***B. bambos***. For this purpose, bamboo was found to be the right choice due to its cellulosic properties and faster growth (David 1984). However, bamboos' hydrophilic nature of CNCs lacks antimicrobial activity (Peng et al. 2016). But nanocomposites (NCs) in ointment and film forms have in vitro antimicrobial and in vivo topical wound-healing properties and showed that its hydrophilic nature plays important role in keeping wounded tissue moist, ultimately enhancing tissue repair chances, along with anti-inflammatory and antibacterial properties of AgNPs accelerating tissue repair.

13.7 Conclusions

Bamboo acts as a natural source for various aspects of human life. Humans have an inseparable relationship with bamboo because of its capability to enhance the environment and economy. The higher significance of bamboo causes its overexploitation and destruction. Therefore, it is a crucial job to conserve and search means to improve various methods in order to fulfill the demand of humankind. Crucial features for the sustainable growth of the bamboo industry are the characteristics of bamboo resources and their geographical distribution (Zhaohua and Wei 2018). The main challenge in Moso bamboo improvement occurs due to its slow breeding which is primarily associated with its long flowering cycle and large population size. In addition, asynchronous flowering also produces extra hurdles during hybridization. In these situations, biotechnological tools offer an alternative that could help in escaping some natural barriers and promote flowering, variability, etc. under controlled conditions. It helps in the induction of flowering in controlled conditions, mainly in situ or ex situ gardens; thus, natural recombination opportunities might be exploited in achieving extra variability. Furthermore, micropropagation and cryopreservation techniques help in rescuing some agronomically important cultivars despite the existence of natural calamities which may otherwise risk the germplasm preserved in gene banks. Similarly, cryopreservation has turned into an effective tool to preserve some important explants such as seeds, syn-seeds, ovules, embryos, callus, etc. can be used effectively for breeding purpose in future. Since plants have been classified on the basis of morphological, physiological, and chemical attributes which keeps various phylogenetic studies aside, so by incorporating the phylogeny as a basis of classification, further studies could become very easy to conduct (Thapa et al. 2018). To check whether an in vitro raised plants are true to the type or not, molecular markers are often used to check the genetic fidelity. These are another important aspect of biotechnology that needs further refinement in the future. Genetic transformation occurs in various species of bamboos, but its actual utilization still needs to improve to obtain different varieties of bamboos resistant to biotic and abiotic stresses. All the previous and current approaches in bamboo have culminated into improved understanding, conservation, and advancement of bamboos and their various applications for a sustainable future.

Acknowledgments Financial assistance from the University Grant Commission, India, in the form of fellowship to the author ST and AW is gratefully acknowledged.

Conflict of Interest No

References

Adlak T, Tiwari S, Tripathi MK, Gupta N, Sahu VK, Bhawar P, Kandalkar VS (2019) Biotechnology: An advanced tool for crop improvement. Curr J Appl Sci Technol 33:1–11

Agnihotri RK, Mishra J, Nandi SK (2009) Improved in vitro shoot multiplication and rooting of *Dendrocalamus hamiltonii* Nees et Arn. ex Munro: production of genetically uniform plants and field evaluation. Acta Physiol Plant 31:961–967

Ahmad N, Sharma S, Singh VN, Shasmi SF, Fatma A, Mehta BR (2011) Biosynthesis of silver nanoparticles from *Desmodiumtriflorum*: a novel approach towards weed utilization. Biotechnol Res Int 2011:1–8

Ahmad S, Khushnooda RA, Jagdalec P, Tulliani JM, Ferroa GA (2015) High performance selfconsolidating cementitious composites by using micro carbonized bamboo particles. Mater Des 76:223–229

Alazem M, Lin KY, Lin NS (2014) The abscisic acid pathway has multifaceted effects on the accumulation of bamboo mosaic virus. Mol Plant-Microbe Interact 27:177–189

Alexander MP, Rao TC (1968) In vitro culture of bamboo embryo. Curr Sci 37:415

Al-Halafi AM (2014) Nanocarriers of nanotechnology in retinal diseases. Saudi J Ophthalmol 28:304–309

Arya I D, Arya S (2015) In vitro shoot proliferation and somatic embryogenesis: means of rapid bamboo multiplication. In: 10th World Bamboo Congress, propagation, plantation and management, Korea

Arya S, Satsangi R, Arya ID (2008) Large scale plant production of edible bamboo *Dendrocalamus asper* through somatic embryogenesis. Bamboo Sci Cult 21(1):13–23

Bag N, Palni LMS, Chandra S, Nandi SK (2012) Somatic embryogenesis in 'Maggar' bamboo (*Dendrocalamus hamiltonii*) and field performance of regenerated plants. Curr Sci 102:1279–1287

Banik RL (1994) Review of conventional propagation research in bamboos and future strategy. INBAR Technical Report No. 5 Constraints to production of bamboo and rattan. INBAR, New Delhi, pp 115–142

Banik RL (1995) Selection criteria and population enhancement of priority bamboos. In: Williams JT, Rao IVR, Rao AN (eds) Genetic enhancement of bamboo and rattan. INBAR Technical Report No. 7, New Delhi, pp 99–110

Bartels D, Mattar MZM (2002) *Oropetium thomaeum.* A resurrection grass with a diploid genome. Maydica (Italy)

Baruwati B, Polshettiwar V, Varma RS (2009) Glutathione promoted expeditious green synthesis of silver nanoparticles in water using microwaves. Green Chem 11:926–930

Bejoy M, Anish NP, Radhika BJ, Nai GM (2012) In vitro propagation of *Ochlandrawightii* (Munro) Fisch: an endemic reed of Southern Western Ghats India. Biotechnology 11:67–73

Bennetzen JL, Schmutz J, Wang H, Percifield R, Hawkins J, Pontaroli AC, Jenkins J (2012) Reference genome sequence of the model plant Setaria. Nat Biotechnol 30:555–561

Bisht P, Pant M, Kant A (2010) In vitro propagation of *Gigantochloaatroviolaceae* Widjaja through nodal explants. J Am Sci 6:1019–1025

Cao YW, Jin R, Mirkin CA (2001) DNA-modified core-shell ag/au nanoparticles. J Am Chem Soc 123:7961–7962

Das S, Saha M (2012) Preparation of carbon Nanosphere from bamboo and its use in water purification. Curr Trends Tech Sci 2:2279–0535

David F (1984) The book of bamboo: a comprehensive guide to this remarkable plant, its uses, and its history. Sierra Club Books, San Francisco

Devi YR (2013) Bamboo forest resources of India and its role in food security-a review. Agric Rev 34:236–241

Devi WS, Sharma GJ (2009) In vitro propagation of *Arundinariacallosa* Munroan edible bamboo from nodal explants of mature plants. Open Plant Sci J 3:35–39

Devi WS, Bengyella L, Sharma GJ (2012) In vitro seed germination and micropropagation of edible bamboo *Dendrocalamusgiganteus* Munro using seeds. Biotechnology 11:74–80

Erkoc S (2006) Structural and electronic properties of bamboo-like carbon nanostructure. Phys E 31:62–66

Gillis K, Gielis J, Peeters H, Dhooghe E, Oprins J (2007) Somatic embryogenesis from mature *Bambusabalcooa* Roxburgh as basis for mass production of elite forestry bamboo. Plant Cell Tissue Organ Cult 91:115–123

Godbole S, Sood A, Thakur R, Sharma M, Ahuja PS (2002) Somatic embryogenesis and its conversion into plantlets in a multipurpose bamboo, *Dendrocalamus hamiltonii* Nees et Arn. Ex Munro. Curr Sci 83:885–889

Goyal AK, Pradhan S, Basistha BC, Sen A (2015) Micropropagation and assessment of genetic fidelity of *Dendrocalamusstrictus* (Roxb.) nees using RAPD and ISSR markers. 3 Biotech 5:473–482

Gui YJ, Zhou Y, Wang Y, Wang S, Wang SY, HuY ZTZ (2010) Insights into the bamboo genome: syntenic relationships to rice and sorghum. J Integr Plant Biol 52:1008–1015

Hu S, Zhou J, Cao Y, Lu X, Duan N, Ren P, Chen K (2011) In vitro callus induction and plant from mature seed embryo and young shoots in a giant sympodial bamboo, *Dendrocalamus farinosus* (Keng et Keng f.) Chia et H. L. Fung. Afr J Biotechnol 10:3210–3215

Islam N, Rahman MM (2005) Micro-cloning in commercially important six bamboo species for mass propagation and at a large-scale cultivation. Plant Tissue Cult Biotech 15:103–111

Jiang K, Zhou M (2014) Recent advances in bamboo molecular biology. J Trop Subtrop Bot 22:632–642

Jiang W, Bai T, Dai H, Wei Q, Zhang W, Ding Y (2017) Microsatellite markers revealed moderate genetic diversity and population differentiation of moso bamboo (*Phyllostachys edulis*)—a primarily asexual reproduction species in China. Tree Genet Genomes 13:130

Kaladhar DSVGK, Tiwari P, Duppala SK (2017) A rapid in vitro micro propagation of *Bambusa Vulgaris* using inter-node explant. Int J Life Sci Sci Res 3:1052–1054

Kalia S, Kalia RK, Sharma SK (2004) In vitro regeneration of an indigenous bamboo (*Bambusa nutans*) from internode and leaf explant. J Bamboo Rattan 3:217–228

Kalia RK, Rai MK, Kalia S, Singh R, Dhawan AK (2011) Microsatellite markers: an overview of the recent progress in plants. Euphytica 177:309–334

Kalia RK, Singh SR, Singh R, Kalia S, Dalal S, Dhawan AK (2013) Limitations, progress and prospects of application of biotechnological tools in improvement of bamboo—a plant with extraordinary qualities. Physiol Mol Biol Plants 19:21–41

Karp A, Richter GM (2011) Meeting the challenge of food and energy security. J Exp Bot 62:3263–3271

Kaur D, Ogra RK, Bhattacharya A, Sood A (2012) Changes in sugar levels during slow growth of *Dendrocalamus hamiltonii* somatic embryos due to liquid paraffin overlay. In Vitro Cell Dev Biol Plant 48:120–126

Kaur D, Thapa P, Sharma M, Bhattacharya A, Sood A (2014) In vitro flowering- a system for tracking floral organ development in *Dendrocalamus hamiltonii* Nees et Arn ex Munro. Indian J Exp Biol 52:825–834

Kaur D, Dogra V, Thapa P, Bhattacharya A, Sood A, Sreenivasulu Y (2015) In vitro flowering associated protein changes in *Dendrocalamus hamiltonii*. Proteomics 15:1291–1306

Kaushik S, Singh YP, Kumar D, Thapliyal M, Barthwal S (2015) Bamboos in India. ENVIS Centre on Forestry. National Forest Library and Information Centre, FRI, Dehradun

Kersey PJ (2019) Plant genome sequences: past, present, future. Curr Opin Plant Biol 48:1–8

Kulzer F, Orrit M (2004) Single-molecule optics. Annu Rev Phys Chem 55:585–611

Lakshmanan V, Reddampalli Venkataramareddy S, Neelwarne B (2007) Molecular analysis of genetic stability in long-term micropropagated shoots of banana using RAPD and ISSR markers. Electron J Biotechnol 10:106–113

Lal R (2010) Managing soils and ecosystems for mitigating anthropogenic carbon emissions and advancing global food security. Bioscience 60:708–721

Li S, Ramakrishnan M, Vinod KK, Kalendar R, Yrjälä K, Zhou M (2019) Development and deployment of high throughput inter-retrotransposon amplified polymorphism (IRAP)-based markers reveal genetic diversity and population structure of Phyllostachys bamboo. Forests

Lin S, Chang WC (1998) Micropropagation of *Bambusa edulis* through nodal explants of field-grown culms and flowering of regenerated plantlets. Plant Cell Rep 17:617–620

Lin CS, Chen CT, Lin CC, Chang WC (2003) A method of inflorescence proliferation. Plant Cell Rep 21:838–843

Lin CS, Lin CC, Chang WC (2004a) Effect of thidiazuron on vegetative tissue-derived somatic embryogenesis and flowering of bamboo *Bambusa edulis*. Plant Cell Tissue Organ Cult 76:75–82

Lin CS, Vidmar J, Chang WC (2004b) Effects of growth regulators on inflorescence proliferation of *Bambusa edulis*. Plant Growth Regul 43:221–225

Lin CS, Lin CC, Chang WC (2005) Shoot regeneration, re-flowering and post flowering survival in bamboo inflorescence culture. Plant Cell Tissue Organ cult 82:243–249

Lin CS, Kalpana K, Chang WC, Lin NS (2007) Improving multiple shoot proliferation in bamboo mosaic virus-free *Bambusa oldhamii* Munro propagation by liquid culture. HortScience 42:1243–1246

Liu JX, Zhou MY, Yang GQ, Zhang YX, Ma PF, Guo C, Li DZ (2020) ddRAD analyses reveal a credible phylogenetic relationship of the four main genera of *Bambusa-Dendrocalamus*-Gigantochloa complex (Poaceae: Bambusoideae). Mol Phylogenet Evol 146:106758

Lu W, Lieber CM (2007) Nanoelectronics from the bottom up. Nat Mater 6:841–850

Mehta U, Rao IVR, Ram HYM (1982) Somatic embryogenesis in bamboo. In: Fujiwara A (ed) Jpn. Assoc. Proc 5th Intl Cong Plant Tiss Cell Cult, pp 109–110

Mehta R, Sharma V, Sood A, Sharma M, Sharma RK (2011) Induction of somatic embryogenesis and analysis of genetic fidelity of in vitro derived plantlets of *B. nutans* wall., using AFLP markers. Eur J For Res 130:729–736

Mirkin CA, Letsinger RL, Mucic RC, Storhoff JJ (1996) A DNA based method for rationally assembling nanoparticles into macroscopic materials. Nature 382:607–609

Mishra Y, Patel P, Ansari SA (2011) Acclimatization and macroproliferation of micropropagated plants *of Bambusatulda* Roxb. Asian J Exp Biol Sci 2:498–501

Mitin VV, Kochelap VA, Stroscio MA (eds) (2008) Introduction to nanoelectronics: materials for nanoelectronics. Cambridge University Press, Cambridge, pp 65–108

Mudoi KD, Saikia SP, Goswami A, Gogoi A, Bora D, Borthakur M (2013) Micropropagation of important bamboos: a review. Afr J Biotechnol 12:2770–2785

Murphy CJ, Gole AM, Hunyadi SE, Stone JW, Sisco PN, Alkilany A, Kinard BE, Hankins P (2008) Chemical sensing and imaging with metallic nanorods. Chem Commun 5:544–557

Nadgauda RS, Parasharami VA, Mascarenhas AF (1990) Precocious flowering and seeding behaviour in tissue-cultured bamboos. Nature 344:335

Nayak L, Mishra SP (2016) Prospect of bamboo as a renewable textile fiber, historical overview, labeling, controversies and regulation. Fash Text 3:2

Nayak S, Rout GR, Das P (2003) Evaluation of the genetic variability in bamboo using RAPD markers. Plant Soil Environ 49:24–28

Negi D, Saxena S (2011) In vitro propagation of *Bambusa nutans* Wall. ex Munro through axillary shoot proliferation. Plant Biotechnol Rep 5:35–43

Niazian M, Shariatpanahi ME (2020) In vitro-based doubled haploid production: recent improvements. Euphytica 216:69

Nirmala C, Bisht MS (2012) Molecular markers in bamboo systematics and germplasm screening. Plant Cell Biotechnol Mol Biol 13:73–82

Ojha A, Verma N, Kumar A (2009) In vitro micropropagation of economically important edible bamboo (*Dendrocalamus asper*) through somatic embryos from root, leaves and nodal segments explants. Res Crops 10:430–436

Oramas P, Ramírez N, Ayala M, García R, Herrera L, Falcón V, Gavilondo J (2000) Molecular farming of pharmaceutical and veterinary proteins from transgenic plants: CIGB experience. In: Plant genetic engineering: towards the third millennium: proceedings of the international symposium on plant genetic engineering, Havana, 6–10 December 1999, pp 222–228

Palombi M, Damiano C (2002) Comparison between RAPD and SSR molecular markers in detecting genetic variation in kiwifruit (Actinidiadeliciosa A. Chev). Plant Cell Rep 20:1061–1066

Parveen S, Shahzad A, Yadav V (2016) Molecular markers and their application in plant biotechnology. In: Biotechnological strategies for the conservation of medicinal and ornamental climbers, pp 389–413

Paterson AH (2010) Comparative genomics in crop plants. In: Molecular techniques in crop improvement, pp 23–61

Peng Z, Lu T, Li L, Liu X, Gao Z, HuT FD (2010) Genome-wide characterization of the biggest grass, bamboo, based on 10,608 putative full-length cDNA sequences. BMC Plant Biol 10:116

Peng N, Wang Y, Ye Q, Liang L, An Y, Li Q (2016) Biocompatible cellulose- based superabsorbent hydrogels with antimicrobial activity. Carbohydr Polym 137:59–64

Peredo EL, Revilla MÁ, Arroyo-García R (2006) Assessment of genetic and epigenetic variation in hop plants regenerated from sequential subcultures of organogenic calli. J Plant Physiol 163:1071–1079

Qiao G, Li H, Liu M, Jiang J, Yin Y, Zhang L, Zhuo R (2013) Callus induction and plant regeneration from anthers of *Dendrocalamus latiflorus* Munro. In Vitro Cell Dev Biol Plant 49:375–382

Qiao G, Yang H, Zhang L, Han X, Liu M, Jiang J, Zhuo R (2014) Enhanced cold stress tolerance of transgenic *Dendrocalamus latiflorus* Munro (Ma bamboo) plants expressing a bacterial CodA gene. In Vitro Cell Dev Biol Plant 50:385–391

Rajput BS, Jani MD, Sasikumar K, Manokari M, Shekhawat MS (2019) An improved micropropagation protocol for manga bamboo-*Pseudoxytenanthera stocksii* (Munro) TQ Nguyen. World News Nat Sci 25:141–154

Rajput BS, Jani M, Ramesh K, Manokari M, Jogam P, Allini VR, Shekhawat MS (2020) Large-scale clonal propagation of *Bambusabalcooa* Roxb.: an industrially important bamboo species. Ind Crop Prod 157:112905

Ramakrishnan M, Zhou M, Baskar K, Packiam S (2018) Role of bamboo in ecosystem. Austin J Environ Toxicol 4:1023

Ramakrishnan M, Yrjälä K, Vinod KK, Sharma A, Cho J, Satheesh V, Zhou M (2020) Genetics and genomics of moso bamboo (*Phyllostachys edulis*): current status, future challenges, and biotechnological opportunities toward a sustainable bamboo industry. Food and Energy Secur 9: e229

Ray T, Roy SC (2007) Phylogenetic relationships between members of Amaranthaceae and Chenopodiaceae of lower gangetic plains using RAPD and ISSR markers. Bangladesh J Bot 36:21–28

Ray SS, Ali MDN, Banerjee M, Yeasmin L (2018) In vitro and in vivo assessment of thidiazuron mediated micro-clones of *Dendrocalamus asper*, an ornamental bamboo species. J Crop Weed 14:150–157

Roco MC, Williams RS, Alivasatos P (1999) Nanotechnology research direction: IWGN workshopreport. Kluwer Academic Publishers, Norwell

Saeed T, Shahzad A (2016) Basic principles behind genetic transformation in plants. In: Biotechnological strategies for the conservation of medicinal and ornamental climbers. Springer, Cham, pp 327–350

Safiuddin M, Gonzalez M, Cao JW, Tighe SL (2014) State of-the-art report on use of nanomaterials in concrete. Int J Pavement Eng 15:940–949

Sawarkar AD, Shrimankar DD, Kumar A, Kumar A, Singh E, Singh L, Kumar R (2020) Commercial clustering of sustainable bamboo species in India. Ind Crop Prod 154:112693

Saxena S, Dhawan V (1999) Regeneration and large-scale propagation of bamboo (*Dendrocalamusstrictus* Nees) through somatic embryogenesis. Plant Cell Rep 18:438–443

Scott N, Chan H (2002) Nanoscale science and engineering for agriculture and food system report. National Planning Workshop, Washington, DC

Singh SR, Dalal S, Singh R, Dhawan AK, Kalia RK (2011) Micropropagation of *Dendrocalamus asper* (Schult. & Schult. F. Backer ex K Heyne): an exotic edible bamboo. J Plant Biochem Biotechnol 21:220–228

Singh SR, Dalal S, Singh R, Dhawan AK, Kalia RK (2012) Seasonal influences on in vitro bud break in *Dendrocalamus hamiltonii* Arn. ex Munro nodal explants and effect of culture microenvironment on large scale shoot multiplication and plantlet regeneration. Indian J Plant Physiol 17:9–21

Singh SR, Singh R, Kalia S, Dalal S, Dhawan AK, Kalia RK (2013a) Limitations, progress and prospects of application of biotechnological tools in improvement of bamboo—a plant with extraordinary qualities. Physiol Mol Biol Plants 19:21–41

Singh SR, Dalal S, Singh R, Dhawan AK, Kalia RK (2013b) Evaluation of genetic fidelity of in vitro raised plants of *Dendrocalamus asper* (Schult. & Schult. F.) backer ex K. Heyne using DNA-based markers. Acta Physiol Plant 35:419–430

Singla R, Sonib S, Markand P, Avnesh K, Mahesh K, Patial V, Padwad YS, Yadava SK (2017) In situ functionalized nanobiocomposites dressings of bamboo cellulose nanocrystals and silver nanoparticles for accelerated wound healing. Carbohydr Polym 155:152–162

Somashekar PV, Rathore TS, Fatima T (2018) In vitro plant regeneration of *Dendrocalamusstocksii* (Munro) M. Kumar, Remesh& Unnikrisnan, through somatic embryogenesis. Am J Plant Sci 9:2429–2445

Sood A, Ahuja PS, Sharma M, Sharma OP, Godbole S (2002) In vitro protocols and field performance of elites of an important bamboo *Dendrocalamus hamiltonii* Nees et Arn. ex Munro. Plant Cell Tissue Organ Cult 71:55–63

Sood A, Bhattacharya A, Sharma M, Sharma RK, Nadha HK, Sood P, Ahuja PS (2013) Somatic embryogenesis and *Agrobacterium* mediated genetic transformation in bamboos. Somatic embryogenesis and genetic transformation in plants

Sood P, Bhattacharya A, Joshi R, Gulati A, Chanda S, Sood A (2014) A method to overcome the waxy surface, cell wall thickening and polyphenol induced necrosis at wound sites-the major deterrents to *Agrobacterium*-mediated transformation of bamboo, a woody monocot. J Plant Biochem Biotechnol 23:69–80

Stapleton CMA, Rao VR, Ramanatha V (1995) Progress and prospects in genetic diversity studies on bamboo and its conservation. In: Proceedings of the IVth International Bamboo Congress, Bali, pp 23–44

Sugunan A, Dutta J (2008) Pollution treatment, remediation and sensing. In: Harald K (ed) Nanotechnology, vol 3. Weinheim, Wiley, pp 125–143

Suyama Y, Obayashi K, Hayashi I (2000) Clonal structure in a dwarf bamboo (Sasasenanensis) population inferred from amplified fragment length polymorphism (AFLP) fingerprints. Mol Ecol 9:901–906

Thapa P, Bhattacharya A, Sood P, Devi K, Sood A (2018) Advances in bamboo biotechnology: present status and future perspective. In: Biotechnologies of crop improvement, pp 243–265

Thiruvengadam M, Rekha KT, Chung IM (2011) Rapid in vitro micropropagation of *Bambusa oldhamii* Munro. Philipp Agric Sci 94:7–13

VanBuren R, Wai CM, Keilwagen J, Pardo J (2018) A chromosome-scale assembly of the model desiccation tolerant grass *Oropetium thomaeum*. Plant Direct 2(11):e00096

Venkatachalam P, Kalaiarasi K (2016) Indirect somatic embryogenesis and plantlet development from mature seed embryo explants of *Bambusaarundinacea* (Retz.) wild. In: Plant tissue culture: propagation, conservation and crop improvement, pp 509–519

Venkatachalam P, Kalaiarasi K, Sreeramanan S (2015) Influence of plant growth regulators (PGRs) and various additives on in vitro plant propagation of *Bambusaarundinacea* (Retz.) wild: a recalcitrant bamboo species. J Genet Eng Biotechnol 13:193–200

Vorontsova MS, Clark LG, Dransfield J, Govaerts R, Baker WJ (2016) World checklist of bamboos and rattans: in celebrating of INBAR's 20th anniversary. International Centre for Bamboo and Rattan, Beijing

Waikhom SD, Louis B (2014) An effective protocol for micropropagation of edible bamboo species (*Bambusatulda* and *Melocannabaccifera*) through nodal culture. Sci World J 2014:345794

Yasin S, Liu L, Yao J (2013) Biosynthesis of silver nanoparticles by bamboo leaves extract and their antimicrobial activity. J Fiber Bio Eng Inform 6:77–84

Ye S, Cai C, Ren H, Wang W, Xiang M, Tang X, Zhu Q (2017) An efficient plant regeneration and transformation system of ma bamboo (*Dendrocalamus latiflorus* Munro) started from young shoot as explant. Front Plant Sci 8:1298

Ye S, Chen G, Kohnen MV, Wang W, Cai C, Ding W, Lin C (2020) Robust CRISPR/Cas9 mediated genome editing and its application in manipulating plant height in the first generation of hexaploid Ma bamboo (*Dendrocalamus latiflorus* Munro). Plant Biotechnol J 18:1501–1503

Yeh ML, Chang WC (1986a) Plant regeneration through somatic embryogenesis in callus culture of green bamboo (*Bambusa oldhamii* Munro). Theor Appl Genet 73:161–163

Yeh ML, Chang WC (1986b) Somatic embryogenesis and subsequent plant regeneration from inflorescence callus of *Bambusabeecheyana* Munro var. beecheyana. Plant Cell Rep 5:409–411

Yeh ML, Chang WC (1987) Plant regeneration via somatic embryogenesis in mature embryo derived callus culture of *Sinocalamuslatiflora* (Munro) McClure. Plant Sci 51:93–96

Yuan JL, Yue JJ, Wu XL, Gu XP (2013) Protocol for callus induction and somatic embryogenesis in Moso bamboo. PLoS One 8:81954

Zang QL, Zhou L, Fei ZG, Yang HY, Wang XQ, Lin XC (2016) Callus induction and regeneration via shoot tips of *Dendrocalamus hamiltonii*. Springerplus 5:1799

Zhang N, Fang W, Shi Y, Liu Q, Yang H, Gui R, Lin X (2010) Somatic embryogenesis and organogenesis in *Dendrocalamus hamiltonii*. Plant Cell Tissue Organ Cult 103:325–332

Zhang YJ, Ma PF, Li DZ (2011) High-throughput sequencing of six bamboo chloroplast genomes: phylogenetic implications for temperate woody bamboos (Poaceae: Bambusoideae). PLoS One 6:20596

Zhang XM, Zhao L, Larson-Rabin Z, Li DZ, Guo ZH (2012) De novo sequencing and characterization of the floral transcriptome of *Dendrocalamus latiflorus* (Poaceae: Bambusoideae). PLoS One 7:e42082

Zhao H, Yang L, Peng Z, Sun H, Yue X, Lou Y, Gao Z (2015) Developing genome-wide microsatellite markers of bamboo and their applications on molecular marker assisted taxonomy for accessions in the genus Phyllostachys. Sci Rep 5:1–10

Zhao H, Gao Z, Wang L, Wang J, Wang S, Fei B, Lou Y (2018) Chromosome-level reference genome and alternative splicing atlas of moso bamboo (*Phyllostachys edulis*). GigaScience 7: giy115

Zhaohua Z, Wei J (2018) Sustainable bamboo development. CABI, Wallingford

Zhou S, Bechner MC, Place M, Churas CP, Pape L, Leong SA, Schwartz DC (2007) Validation of rice genome sequence by optical mapping. BMC Genomics 8:278

Zhou X, Torabi MLJ, Shen R, Zhang K (2014) Nanostructured energetic composites: synthesis, ignition/combustion modeling, and applications. ACS Appl Mater Interfaces 6:3058–3074

Chapter 14
Ethnobamboology: Traditional Uses of Bamboos and Opportunities to Exploit Genomic Resources for Better Exploitation

Souradeep Banerjee, Mridushree Basak, Smritikana Dutta, Chhandam Chanda, Abhijit Dey, and Malay Das

Abstract Bamboos, a member of the family Poaceae and subfamily Bambusoideae, represent approximately 1670 species within 125 genera. Bamboos are widely distributed across the world and grow naturally in all the continents except Europe and Antarctica. The plant group had evolved from grasses approximately 30 million years ago, long before human existence. In the eventual development of human society, bamboos had been playing an important role, like always. Bamboo has an enormous potential to contribute to the growth of the rural economy worldwide and hence is known as "green gold of poor." Due to the high rate of clonal propagation, bamboos can be a good, renewable biosource to meet various demands to cater to the human lifestyle. They have been widely used for many important purposes such as food, fodder, construction, paper, pulp, textile, and pharmaceuticals. However, this chapter is particularly focused on the utility of bamboos as ethnomedicines and medicinally important phytochemicals. Also, multiple, ethnobotanical utilities of bamboos suitable for the daily life of indigenous people such as in the making of handicrafts, artifacts, jewelry, baskets, furniture, and constructions have also been discussed. As a future perspective, prospects of exploiting the available genome resources to make better exploitation of bamboo germplasm have also been discussed.

Keywords Bamboo · Ethnic uses · Handicrafts · Medicinal use · Phytochemicals

---

S. Banerjee · M. Basak · S. Dutta · A. Dey · M. Das (✉)
Department of Life Sciences, Presidency University, Kolkata, India
e-mail: malay.dbs@presiuniv.ac.in

C. Chanda
Botanical Survey of India, Arunachal Pradesh Regional Centre, Itanagar, Arunachal Pradesh, India

Z. Ahmad et al. (eds.), *Biotechnological Advances in Bamboo*,
https://doi.org/10.1007/978-981-16-1310-4_14

14.1 Introduction

Bamboos are members of the family Poaceae, having more than a thousand species, and are regarded as the tallest grass on the Earth. It is one of the fastest-growing plants on Earth and can reach its final height of 5–20 m within a span of 2–4 months due to rapid expansion of internodes (Magel et al. 2005; He et al. 2013). For instance, the giant bamboo *Dendrocalamus giganteus* can grow as tall as 25 m from the ground (World Bamboo Resources 2005). Bamboos are distributed from 51° North in Japan to 47° South in South Argentina (Yeasmin et al. 2015). Most of the species are found in Asia followed by South America and the least are present in Africa (Bystriakova et al. 2003). In the Neotropics of Brazil, Paraguay, Mexico, and West Indies, herbaceous bamboos are found. The tribe Bambuseae is further subdivided into three groups (Das et al. 2008b). The Paleotropical woody bamboo is distributed in Africa, Sri Lanka, Madagascar, India, South China and Japan, and Oceania, whereas Neotropical woody bamboos are distributed in Argentina, Chile, Southern Mexico, and West Indies, while north temperate woody bamboo are distributed in Madagascar, India, Sri Lanka, and Africa (http://www.eeob.iastate.edu/bamboo/maps.html, Das et al., 2008b). They can grow from the sea level up to approximately 4000 m height both in the tropics and temperate zone but are more common in the tropics (https://www.inbar.int/scientific-efforts-to-understand-bamboo-better-underway/). Approximately, 24 million hectares of the area spreading across 16 Asian countries and ten million hectares in 10 Latin American countries are covered by bamboo vegetation (http://www.fao.org/3/a1243e/a1243e00.htm). The bamboo-growing area in China is approximately 6.01 million hectares, which is comprised of 43 genera and 861 species (Liu et al., 2018). In India, bamboo covers approximately 1,60,037 sq. km area, with an increase of 3229 sq. km land in the past 2 years (India State of Forest Report 2019, https://fsi.nic.in/forest-report-2019?pgID=forest-report-2019).

Historically bamboo has played a very important role in the growth of old civilizations like in China and India. Prehistoric records suggest that even before the establishment of human civilizations, people were dependent upon these plant resources. Historians claim that bamboos were used even as early as sixteenth- to eleventh-century BC, when the Shang Dynasty reigned over China (Mei et al. 2015). The famous Taj Mahal was camouflaged with green bamboos during the Indo-Pak war of 1971 (Das and Kumar 2019). It would be almost impossible to name a tribe in the Southeast Asia, who, in one way or the other, is not dependent on bamboos. Due to their enormous utility and wide availability, they are quite popular among the tribes for various purposes such as food, fodder, medicine, fishing gears, daily use articles, and many more (Yuming et al. 2004; Patel 2005; Ju et al. 2013; Dhyani and Dhyani 2016). Even the giant pandas survive on a special bamboo diet (Jin et al. 2011). Bamboos are popularly used in construction due to the excessive tensile strength of their fibers (approximately 28,000 pounds per square inch), which is much higher than low-quality steels used in structures (23,000 pounds per square inch, Kaur 2018). It has been proven that bamboo houses can withstand earthquakes much better than houses made of other materials (Earthquake of Costa Rica 1991, Gutiérrez 1993).

Due to their excessive utilities, bamboo trades grew leaps and bounds over the decades. For instance, in China, the production value increased to US$ 19.5 billion from US$ 13.1 billion only in 2 years from 2010 to 2012 (International Trade of Bamboo and Rattan 2012, http://www.aha-kh.com/wp-content/uploads/2017/01/5-inbar-international-trade-of-bamboo-and-rattan-2012.pdf). Similarly, the value of bamboo industry in India was US$ 34 million in 2003, which took a large leap in the last few years (International Trade of Bamboo and Rattan 2012, http://www.aha-kh.com/wp-content/uploads/2017/01/5-inbar-international-trade-of-bamboo-and-rattan-2012.pdf). China is the biggest producer and exporter of different bamboo products. As per the information available from the Association of Bamboo Industry of China, it was worth USD 39.6 billion in 2018 only (https://www.inbar.int/resources/inbar_publications/trade-overview-2018-bamboo-and-rattan-commodities-in-china/). The second highest exporter was the European Union with value ~ USD 153 million, and the third one was Indonesia with ~USD 130 million. The European Union was the highest importer of bamboo products (~37% of the global total) followed by Asia (~32%) and North America (~25%) (https://www.inbar.int/resources/inbar_publications/trade-overview-2017-bamboo-and-rattan-commodities-in-the-international-market/).

There are countless ethnic uses of bamboos such as in making fishing gears, manufacturing transport vehicles, treating and healing wounds and diseases, and making artifacts and handicrafts. Therefore, it is not surprising at all that the plant group has been termed as "poor man's timber," "friend of the people," and "green gold." In this chapter, we particularly focus on various ethnobotanical utilities of bamboos with a major focus on their uses by ethnic people of India. In addition to published reports, we also add data collected during survey work in two Indian states, Assam and Arunachal Pradesh, which are known as hotspots for bamboo biodiversity as well as the diversity of the local ethnic group. We introduce a new term "ethnobamboology," which we define as collective uses of bamboos to cater to diverse needs of the local, ethnic population. We also discuss how genomic advances can be exploited to reshape "ethnobamboology" research in the future.

14.2 Various Ethnobotanical Uses of Bamboos

14.2.1 Various Medicinal Utilities of Bamboos on Human

14.2.1.1 Reports from India

Bamboos are used by different ethnic groups to treat a plethora of diseases. To name a few of them are *Bambusa arundinacea*, *B. tulda*, *B. balcooa*, *B. blumeana*, *D. hamiltonii*, *Dinochloa compactiflora*, *Indosasa pingbianensis*, *M. baccifera*, *Phyllostachys glauca*, and *P. heterocycla* cv. *pubescens* (Table 14.1). In addition,

Table 14.1 Summary of various ethnomedicinal utilities of bamboos on human. The available local names are indicated after species names within third brackets

Bamboo species	Place from where reported	Tribe/ethnic people	Plant parts and their use as medicine	References
B. arundinacea (Retz.) Roxb [Bongu veduru]	Vishakhapatnam, Andhra Pradesh, India	N.A.	Tender stems for diabetes, stem barks for piles	Rao et al. (2011)
B. arundinacea (Retz.) Willd. [Vaans]	Valsad district, Gujarat, India	N.A.	Young bamboo shoot juice in asthma treatment	Shah et al. (2012)
B. arundinacea (Retz.) Willd.	Hingoli district, Maharashtra, India	N.A.	Paste of bamboo nodal stems to treat inflamed finger joints	Patil and Biradar (2011)
B. arundinacea (Retz.) Willd	Sahiwal district, Punjab, Pakistan	N.A.	Leaves of bamboo to treat helminthosis	Hussain et al. (2008)
B. arundinacea Willd.	Adilabad District, Telengana, India	Traditional healers	Leaves to reduce inflammation	Gurrapu and Mamidala (2016)
B. arundinacea Willd. [Moongil]	Thoppampatti, Dindigul district, Tamil Nadu, India	N.A.	Inflammation, ophthalmia treatment	Sivasankari et al. (2014)
B. bambos (L.) Voss.	Ranchi, Jharkhand, India	Traditional herbal practitioners	Bamboo seeds to make contraceptive pills	Chandra et al. (2007)
B. tulda Roxb.	Lawachara National Park, Bangladesh	N.A.	Cooked stem to treat impotency	Uddin et al. (2017)
B. tulda Roxb. [Bah]	Bongaigaon, Assam, India	Koch, Rajbanshi	Leaves as coagulant	Tamuli and Sharma (2010)
B. tulda Roxb. [Bans]	Assam, India	Hajong	Root decoction for piles and constipations	Sharma et al. (2012)
B. tulda Roxb. [Ruo]	Cachar District, Assam, India	Chiru	Leaves as coagulant	Singh et al. (2011)
B. vulgaris [Kawayan]	Cagayan, Philippines	N.A.	Shoots to treat headache	Baddu and Ouano (2018)
B. vulgaris [Okuther]	Rivers State, Nigeria.	Ogba/ Egbema/ Ndoni	Roots to treat gonorrhea and miscarriage	Oladele and Elem (2018)
B. vulgaris Schrad. [yellow bamboo]	North Sumatra	Karo	Treatment of diabetes	Situmorang et al. (2015)
B. vulgaris Schrader. ex Wendland [Bambu kuning]	West Java, Indonesia	Sundanese	Roots and stems in hepatitis and hookworm treatment	Roosita et al. (2008)

(continued)

Table 14.1 (continued)

Bamboo species	Place from where reported	Tribe/ethnic people	Plant parts and their use as medicine	References
B. blumeana, *B. tulda*	Lao PDR	Kry, Brou, Saek	Leaves to treat headache, dizziness, fever, cough	de Boer et al. (2012)
B. blumeana Schult.f. [Kawayan]	Dinalupihan, Bataan, Philippines	Ayta	Root and leaf extracts for cold, cough, kidney stone, dengue	Tantengco et al. (2018)
B. balcooa Roxb. [Bholuka Bah]	Nalbari, Assam, India	N.A.	Stem paste and leaves to cure menstrual problems	Das et al. (2008a)
Bambusa sp.	Northeast Brazil	N.A.	Roots in erysipelas treatment	Albuquerque et al. (2007)
Bambusa sp. [Bamboo]	Philippines	Ayta	Leaf extracts to treat coughs and colds	Tantengco et al. (2018)
D. hamiltonii Gamble.	Darjeeling Hills, India	Tea garden workers	Stems to treat fractures	Chettri and Chowdhury (2018)
Dinochloa compactiflora (Kurz.) McClure [Sairil]	Indo-Burma Hotspot Region	N.A.	Stems used in influenza, cough, and chest complaints	Rai and Lalramnghinglova (2011)
Indosasa pingbianensis	Yunnan, China	N.A.	Shoots in common cold and headache treatment	Yuming et al. (2004)
M. baccifera (Roxb.) Kurz [Mautak]	Indo-Burma Hotspot Region	N.A.	Stems as coagulant	Rai and Lalramnghinglova (2011)
Phyllostachys glauca; *P. heterocycla* cv. *pubescens*	Yunnan, China	N.A.	Leaves and sap of young culms in cough and lung inflammation	Yuming et al. (2004)
N.A.	Thailand	N.A.	Shoots as abortion inducer	Chaveerach et al. (2006)
N.A.	Maharashtra, India	Gond	Bamboo nodes to relief pain of burning urination	Gupta et al. (2010)
N.A.	Lohit, Dibang districts, Arunachal Pradesh, India	N.A.	Internodes in fracture and joint pain	Shankar and Rawat (2008)
N.A.	Kans, Uttar Kannada; "Cumindad" lands, Goa, India	N.A.	Leaves in placental expulsion and diarrhea treatment	Dixit and Goyal (2011)

(continued)

Table 14.1 (continued)

Bamboo species	Place from where reported	Tribe/ethnic people	Plant parts and their use as medicine	References
N.A.	Lao PDR	Kry	Bamboo fiber strings to tie umbilical cord after pregnancy	Laxmay et al. (2011)
N.A.	Gujarat, India	Kotwalia	Leaves in urinary infection	Patel (2005)

Abbreviations used: *N.A.* not available

there are numerous studies which refer to multiple, medicinal properties of bamboos without mentioning the scientific names of those bamboos. The thorny bamboos, *Bambusa arundinacea*, are used to treat diabetes, piles (Rao et al. 2011), asthma (Shah et al. 2012), inflammation (Gurrapu and Mamidala 2016), and ophthalmia (Sivasankari et al. 2014) and also to make contraceptive pills (Chandra et al. 2007). Similarly, another commonly available bamboo *B. tulda* can help prevent piles and constipation (Sharma et al. 2012), whereas *B. vulgaris* is quite effective against headache (Baddu and Ouano 2018), gonorrhea, miscarriage (Oladele and Elem 2018), diabetes (Situmorang et al. 2015), and hepatitis and in preventing hookworm infection (Roosita et al. 2008). Although young shoots and leaves were of utility in maximum instances, almost each and every organ of the plant body demonstrated some medicinal properties (Fig. 14.1a, b, Table 14.1). This shows the enormous promise of the plant group for the pharmaceutical industry. It is also noteworthy that the ethnic groups, which have reported these uses, are quite scattered across different states in India. For instance, one ethnic group in Vishakhapatnam, Andhra Pradesh, treats diabetes and piles with medicines prepared from stem and bark of *B. arundinacea* (Rao et al. 2011). In Gujarat, juices extracted from young bamboo shoots of *B. arundinacea* are used to treat asthma (Shah et al. 2012) and also provide relief to urinary diseases, deafness, and cough (Patel 2005, Table 14.1). The ethnic groups of the Hingoli district in Maharashtra prepare medicines from bamboo stem to treat inflamed finger joints (Patil and Biradar 2011). The HIV/AIDS-related inflammations are treated by the traditional healers in Telangana region (Gurrapu and Mamidala 2016). Inflammation and ophthalmia are treated with bamboo parts in Dindigul district of Tamil Nadu (Sivasankari et al. 2014). In Maharashtra, powdery white material collected from bamboo nodes is used to cure burning feelings during urination (Gupta et al. 2010). Contraceptive pills are made from bamboo seeds by traditional herbal practitioners in Jharkhand (Chandra et al. 2007), whereas in Assam, menstrual problems are treated by bamboo plant parts (Das et al. 2008a). The Koch Rajbongshi and Chirutribe of Assam and some ethnic groups of Arunachal Pradesh and Sikkim prepare coagulants from bamboo (Tamuli and Sharma 2010; Singh et al. 2011; Bam et al. 2015). The fractured body parts are supported by bamboo stems by the tea garden workers of Darjeeling hills (Chettri and Chowdhury 2018), Lohit and Dibang Valleys in Arunachal Pradesh (Shankar and Rawat 2008) and the north-central region of the Western Ghats (Upadhya et al.

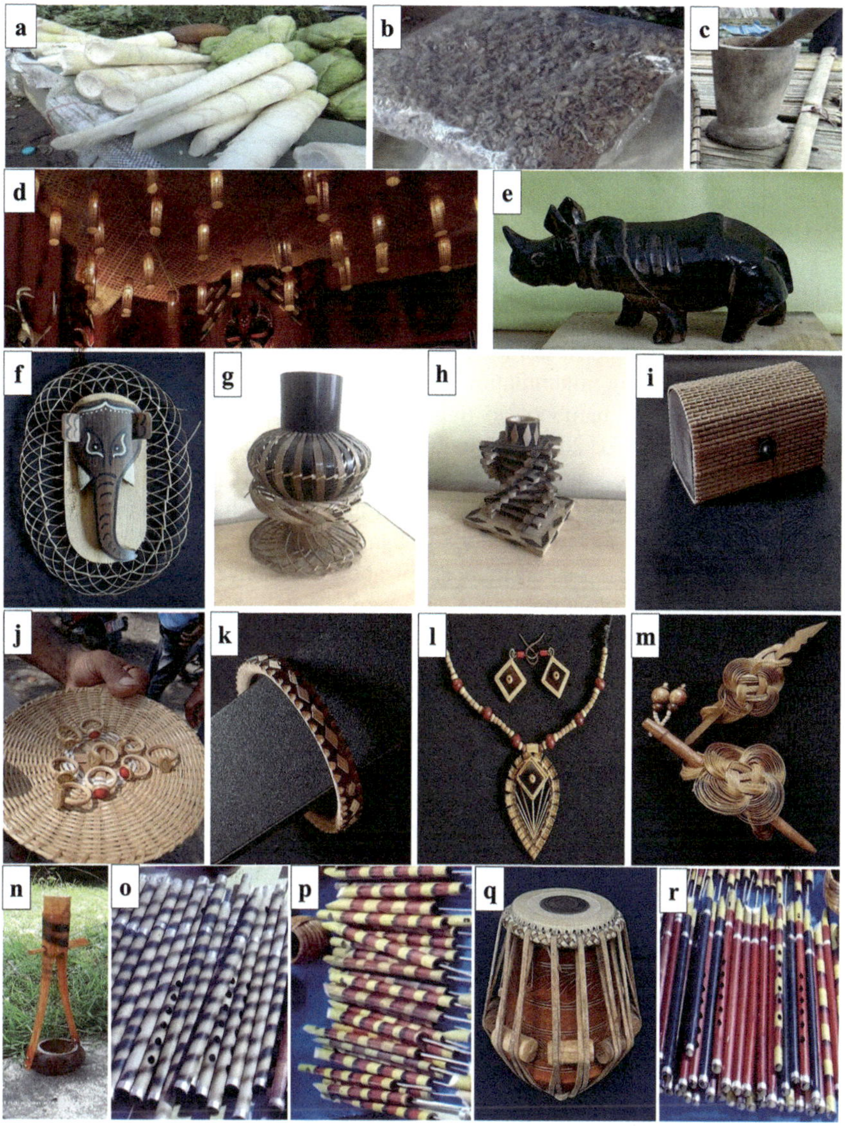

Fig. 14.1 Usage of bamboos in making medicine, handicrafts, jewelry, artifacts, and musical instruments. (**a**) Young shoot, (**b**) dried shoot used for preparation of medicine, (**c**) mortar pestle used for preparation of medicines, (**d**) lamp shed, (**e**) rhino, (**f**) wall hanger, (**g**) flower vase, (**h**) candle stand, (**i**) jewelry box, (**j**) finger ring, (**k**) bangel, (**l**) neck piece and earring, (**m**) hair clips, (**n**) "ektara," musical instrument, (**o**) musical flute, (**p**) whistle, (**q**) ring of "tabla," (**r**) whistle flute. (**a**, **b**, **c**, **d**, **f**, **i**, **k**, **l**, **m**, and **o**) were obtained from the Nyishi tribal community of Senki view, Itanagar, Arunachal Pradesh; (**e**, **g**, **h**, and **j**) from Boro tribal community of Kamrup, Assam; and (**n**, **p**, **q**, and **r**) from Purulia, West Bengal, India

2012). In the Birbhum district of West Bengal and Dumka district of Jharkhand, bamboo sticks are used to apply medicine pastes (Mondal and Rahaman 2012). Some ethnic people use crude extracts obtained from bamboos in the treatment of diarrhea and dysentery (Saran et al. 2015).

14.2.1.2 Reports from Other Countries

Bamboos are used as important ethnomedicinal resources outside India too. In the Yunnan province of China, *Indosasa pingbianensis*, *Phyllostachys glauca*, and *P. heterocycla* cv. *pubescens* are used to treat common cold, headache, skin inflammation, cough, and lung inflammation (Yuming et al. 2004). The Ayta communities of the Philippines use bamboos to treat spasm, cough, colds, kidney stone, and dengue (Tantengco et al. 2018). In Indonesia, the Sundanese community of West Java uses bamboo roots and stems to treat hepatitis and hookworm infestation (Roosita et al. 2008). The tribes living around Lawachara National Park in Bangladesh treat impotence with cooked bamboo stems (Uddin et al. 2017). In Thailand, bamboo is used as an inducer for abortion (Chaveerach et al. 2006). In North Sumatra and Indonesia, the Karo people use bamboos to cure diabetes (Situmorang et al. 2015). Headache is treated with bamboo by tribes of Cagayan in the Philippines (Baddu and Ouano 2018). The ethnic groups of Rivers State in Nigeria treat gonorrhea and miscarriage with bamboo (Oladele and Elem 2018).

14.2.2 Isolation of Phytochemicals having Potential Medicinal Importance

While a lot of knowledge has been gathered on medicinal uses of crude extracts obtained from various bamboo species, very little is known regarding the active compounds, which are actually conferring these properties. A few studies have been conducted, which actually indicate the possible group of compounds that might be systematically studied in the future (Table 14.2). For instance, *B. arundinacea* has flavonoids, glycosides, alkaloids, phytosterols, amino acids, enzymes, carbohydrates, and many other compounds which make it an ethnomedicinally enriched species (Nazreen et al. 2011; Rathod et al. 2011; Muniappan et al. 2014). It has antibacterial wound-healing properties (Muniappan et al. 2014) and antidiabetic properties (Nazreen et al. 2011; Rathod et al. 2011) and also helps in regulating estrogenic activity (Jawaid et al. 2015). Similarly, *B. vulgaris* shows antibacterial and antifungal properties (Owolabi and Lajide 2015) and antidiabetic activity (Senthilkumar et al. 2011) and is frequently used to cause abortions (Yakubu and Bukoye 2009). Therefore, *B. arundinacea* and *B. vulgaris* can be treated as "reference bamboos" to conduct extensive, metabolic profiling to identify the active compounds having potential medicinal properties.

Table 14.2 Phytochemicals isolated from various bamboo species and their potential medicinal properties

Bamboo species	Parts used	Phytochemicals	Extraction/solvents	Medicinal activity	Model organism to study efficacy	References
B. arundinacea	Leaves	Flavonoids, glycosides, alkaloids, phyosterols	Methanol extract, petroleum extract	Antibacterial activities, wound-healing property	Albino Rats.	Muniappan et al. (2014)
B. arundinacea	N.A.	N.A.	Aqueous ethanolic extract	Reduction of blood glucose level	Alloxan induced diabetic rats	Rathod et al. (2011)
	Bamboo shavings		Water-phase extract	Antimicrobial activity against *Staphylococcus aureus*, *Bacillus subtilis*, *Escherichia coli*, *Aspergillus niger*, *Penicillium citrinum*, and *Saccharomyces cerevisiae*		
B. arundinacea	Fresh leaves	N.A.	n-Hexane	Antimicrobial against *E. coli*, *P. multocida*, *B. subtilis*	N.A.	Zubair et al. (2013)
			Chloroform extract	Antimicrobial against S. *aureus* and *B. subtilis*		
			Methanol extract	Antimicrobial against *G. lucidum* and *A. alternata*		
			Acetone extracts	Antimicrobial against *G. lucidum*		
B. arundinacea	Matured leaves	Flavonoids, steroids, carbohydrates, proteins	N.A.	Estrogenic activity	Immature and mature ovariectomized rats	Jawaid et al. (2015)
B. arundinacea (Retz.)	Leaves	Flavonoids, tannins, steroids, and phenolic glycosides	Ethanolic extract, chloroform fraction, ethyl	Antidiabetic	Diabetic Albino wistar rats	Nazreen et al. (2011)

(continued)

Table 14.2 (continued)

Bamboo species	Parts used	Phytochemicals	Extraction/solvents	Medicinal activity	Model organism to study efficacy	References
			acetate fraction, and standard glibenclamide			
B. arundinacea (Retz) Roxb.	Dried seeds	Flavonoids, phenols, steroids, tannin, and quinines	Methanolic extracts	Effects on CNS, hypoglycemic, cardiotonic, lipid-lowering, antiulcer, hepatoprotective, anti-inflammatory, antineoplastic, antimicrobial, antioxidant	N.A.	Thamizharasan et al. (2015)
B. vulgaris	Leaves	Phytosterols and tannins	Petroleum ether extract	Antidiabetic	Streptozotocin induced diabetic rats	Senthilkumar et al. (2011)
B. vulgaris	Leaves	Alkaloids and phenolics	Aqueous extract	Implantation, changes in hormone levels, and partly estrogenicity	N.A.	Yakubu and Bukoye (2009)
B. vulgaris Schrad. Ex J.C. Wendl	Fresh leaves	Alkaloids, tannins, flavonoids, phenols (only effective in ethyl alcohol extracts), and terpenoids	n-Hexane extract	Antifungal against *A. niger*, antibacterial against *B. cereus*	N.A.	Owolabi and Lajide (2015)
			Ethyl acetate extract	Antifungal against *V. albo-atrum*, antibacterial against *S. aureus*, *E. coli*, and *K. pneumoniae*		
D. strictus	Multiple parts	Flavonoids, steroids, glycosides, tannins, morin, flavanone, 6-hydroxy flavones	Aqueous and methanolic extracts	N.A.	N.A.	Daswad et al. (2017)

Abbreviations used: *N.A.* not available

14.2.3 Various Medicinal Utilities of Bamboos on Domestic Animals and Plants

14.2.3.1 Reports from India

Bamboos are being used to treat animals for quite some time. Traditionally, bamboo culms are used to fix fractured bones of cattle (Saha et al. 2014, Table 14.3) and placental retention of goats, as have been reported from Malda, West Bengal, and Uttarakhand, India (Khadda et al. 2018, Table 14.3). In Birbhum district of West Bengal, loose motion and fractured bones of cattle are treated with *B. bambos* (Mandal and Rahaman 2014). Fractured bone and broken horns are treated with bamboo chip bandages by ethnic people of southern Odisha (Rautray et al. 2015). The "Korku" tribe of Maharashtra uses bamboo strips for healing cattle bone fractures (Jagtap et al. 2006). The Bhatra, Bhumia, Gadaba, and Paroja tribes of Koraput, Odisha, feed their cattle with leaves of *Dendrocalamus strictus* to treat diarrhea (Mishra and Chaudhury 2012, Table 14.3). Bamboo-made utensils are used to prepare and feed cattle medicine by some ethnic groups of Uttar Pradesh (Fig. 14.1c, Kumar and Bharati 2012a, b) and Sikkim, India (Bharati and Sharma 2012). In Cumindad lands of Goa, India, some tribes use bamboos for placental expulsion of cattle and treatment of diarrhea (Dixit and Goyal 2011). Leaf and root extracts of *B. bambos* are used to treat foot and mouth diseases of cattle (Kumar and Bharati 2013).

14.2.3.2 Reports from Other Countries

In the Indo-Burma hotspot region, labor pains of backyard pigs and birds have been treated with leaf decoction of bamboo (Rai and Lalramnghinglova 2011). A few tribes in the Sahiwal district of Pakistan use bamboo leaves in the treatment of helminthosis (Hussain et al. 2008). Some ethnic groups of Jimma Zone in Ethiopia use bamboo syringes to diagnose animal diseases (Yigezu et al. 2014). Compared to veterinary use, phytopathological utilities of bamboos are much less. There is only one report from Serampas of Sumatra, Indonesia, on the successful application of bamboo to treat diseased rice plants (Hariyadi and Ticktin 2012, Table 14.3).

14.2.4 Uses of Bamboo as Raw Material for Handicrafts, Artifacts, and Handloom Industries

14.2.4.1 Reports from India

One of the most promising and commercially viable utilities of bamboos is their usage in making handicrafts, artifacts, handloom products, and musical instruments

Table 14.3 Summary of various ethnomedicinal utilities of bamboos on plants and domestic animals. The available local names are indicated after species names within third brackets

Bamboo species	Place from where reported	Tribe/ethnic people	Specific use	References
B. arundinacea (Retz.) Wild.	Tikamgarh district, Bundelkhand, India	N.A.	Leaves and rhizomes used for easier delivery of pregnant buffaloes and diarrhea treatment	Verma (2014)
B. arundinacea Willd. [Bans]	Jalaun district, Uttar Pradesh, India	N.A.	Leaves fed to cattle for placenta and umbilical cord expansion	Kumar and Bharati (2012b)
B. bambos [Bans]	Birbhum district, West Bengal, India	N.A.	To cure loose motion and fractures of cattle	Mandal and Rahaman (2014)
B. bambos (L.) Voss [Baas]	Bareilly district, Uttar Pradesh, India	N.A.	Crushed roots in treatment of foot and mouth disease (Khur pakka or Chapara), crushed leaves in prolonged placental expulsion, bamboo glass (Naar/ Tharka in case of flatulence)	Kumar and Bharati (2013)
B. vulgaris Schrad.	Uttarakhand, India	Goatkeepers	Green leaves to treat retention of placenta; fractured bones are tied with bamboo sticks	Khadda et al. (2018)
B. vulgaris Schrad. ex J.C. Wendl	Indo-Burma Hotspot Region	N.A.	Labor pains of backyard pigs and birds are treated; postpartum cleansing is done by feeding leaves	Rai and Lalramnghinglova (2011)
B. tulda Roxb. [Bans]	Assam, India	Hajong	After parturition, cows are fed leaves	Sharma et al. (2012)
Bambusa sp. [Aur Gajah]	Serampas, Sumatra, Indonesia	N.A.	Used to treat diseased rice	Hariyadi and Ticktin (2012)
D. strictus	Amravati district, Maharashtra, India	Korku	To treat healing of bone fracture in cattle	Jagtap et al. (2006)
D. strictus (Roxb.) Nees [Bamboo]	Koraput, Odisha, India	Bhatra, Bhumia, Gadaba, and Paroja	Diarrhea treatment of cattle	Mishra and Chaudhury (2012)
N.A.	Southern Odisha, India	N.A.	Chips used to bandage the areas of bone fracture and broken horn	Rautray et al. (2015)

(continued)

Table 14.3 (continued)

Bamboo species	Place from where reported	Tribe/ethnic people	Specific use	References
N.A.	Kachchh district, Gujarat, India	N.A.	Sticks used to apply paste in fractured bones	Mistry et al. (2003)
N.A.	Malda, West Bengal, India	N.A.	Medicinal paste is tied using bamboo sticks on broken bones	Saha et al. (2014)
N.A.	Sitapur district, Uttar Pradesh, India	N.A.	Bamboo glass (Naar) is used to the cattle during fever	Kumar and Bharati (2012a)
N.A.	Sikkim Himalayas, India	N.A.	Bamboo cups are used to medicate decoction during indigestion	Bharati and Sharma (2012)

Abbreviations used: *N.A.* not available

(Fig. 14.1d–r). There are numerous reports from India on bamboo-made handicrafts and their uses in a different spectrum of daily life. Bamboo case, "Shabri," is made by ethnic people in the Kandi belt of Jammu (Slathia and Paul 2012). Bamboo is traditionally used in Assam in shed maintenance (Teron and Borthakur 2012), in looms (Borthakur 1979), and in making containers and pipes (Jain and Borthakur 1980). Various kinds of handicraft products are made from bamboos in different parts of India including the Garhwal Himalayas (Fig. 14.1d–m, Dhyani and Dhyani 2016). In Meghalaya, "Biates" tribes use bamboos in traditional dresses and hairbands (Karolia and Ladia 2012), whereas "Kom" tribes use them in handloom weaving (Khatoon et al. 2014). The "Chakhesang Naga" and "Lotha Naga" ethnic groups of Nagaland use bamboos for weaving purposes (Karolia and Prakash 2014). The "Lepcha" tribes residing in Dzongu Tribal Reserved Area (DTRA) of Sikkim make hats and holy crafts from bamboo (Fig. 14.2a–d, Lepcha et al. 2012). In Bhagirathi Valley of Western Himalayas, bamboos are used for igniting the fire and making household articles (Fig. 14.2e–o, Uniyal et al. 2002). The "Kotwalia" tribes of Gujarat make various articles of daily use such as bags, umbrellas (Fig. 14.2k–l), and mat (Fig. 14.2e) from bamboo (Patel 2005). In Orissa, the "Bhatra," "Bhumia," "Gadaba," and "Paroja" tribes of Koraput make distinct rain hats from bamboo (Fig. 14.2k–l, Mishra and Chaudhury 2012). The "Nicobari" tribe of Nicobar group of islands uses bamboos to make "Hodi" (Fig. 14.3a, b, traditional craft for fishing) and some other articles for poultry (Fig. 14.3c–d, Ravikumar et al. 2015).

14.2.4.2 Reports from Other Countries

The "Lhoba" people in Tibet use bamboo fiber to make traditional clothes (Li et al. 2015). The tribes living in Bumdeling Wildlife Sanctuary of Trashiyangtse in

Fig. 14.2 Ethnobotanical usage of bamboos in daily life of the tribal people in Assam and Arunachal Pradesh, India. (**a**) Flower "Jhapi" used for felicitation; (**b**) "Tokdi," used for felicitation; (**c**) "Bonta," used for serving brittle nut for felicitation; (**d**) used for felicitation; (**e**) sitting mat; (**f**) floor mat; (**g**) hand fan; (**h**) handle of brush; (**i**) coffee mug; (**j**) cup and plate; (**k**) "Japi," sun and rain guard; (**l**) rain guard; (**m**) arrow holder with arrows; (**n, o**) bamboo-made long scabbard and set. (**b**, **c**, **d**, **e**, **f**, **g**, and **j**) were obtained from the Boro tribal community of Kamrup, Assam; (**h**, **i**, **k**, **l**, **m**, **n**, and **o**) from Nyishi tribal community of Senki view, Itanagar, Arunachal Pradesh, and a from Shilong, Meghalaya, India

Fig. 14.3 Usage of bamboos in making different baskets used by ethnic people of Assam and Arunachal Pradesh, India. (**a**) Fishing items made up of bamboo; (**b**) fish catcher (jakoi) and keeper (khaloi); (**c**) poultry bird cage; (**d**) used to transport poultry birds; (**e**) different types of baskets and sitting stools made of bamboos; (**f**) strainer used in making beverage; (**g**) used in grain cleaning; (**h**) "Hura," used for grain storage or carrying objects; (**i**) "Tukuri," fruit hanger; (**j**) "Kuki," used for keeping things; (**k**) used in grain drying; (**l**) used as road side garbage bin; (**m**) used for vegetable storage; (**n**) used for carrying wood logs; (**o**) used for carrying leaves and vegetable; (**p**) used in fish catching and vegetable storing. (**a**, **b**, **c**, **e**, **h**, **i**, **j**, **k**, **l**, and **m**) captured from the tribal communities of Assam and (**d**, **f**, **g**, **n**, **o**, and **p**) from Nyishi tribal community of Senki view, Itanagar, Arunachal Pradesh, India

Bhutan make bows and arrows by using bamboo (Fig. 14.2m–o, Wangyal 2012). Few tribes in Bangladesh make festive coin boxes by using bamboo (Partha 2014). Tibetans living in Shangri-la region of Yunnan province of China manufacture bamboo wares (Ju et al. 2013). The ethnic people living in Yunnan province of China use bamboos to make household articles, handicrafts, ornaments, and musical instruments (Fig. 14.1n–r, Yuming et al. 2004, Table 14.4). The "Sasak" tribes in Indonesia make bamboo containers for storing rice (Sukenti et al. 2016). In Southwest Ethiopia, bamboos are used widely in handloom products (Worku 2015).

14.2.5 Uses of Bamboo as Raw Materials for Making Baskets and Furnitures

14.2.5.1 Reports from India

Bamboos are used in making baskets (Fig. 14.3e–p) and furnitures (Fig. 14.4a–h) by various tribes allover India. For instance, in the Kandi belt of Jammu, ethnic people make baskets and chairs from bamboos (Slathia and Paul 2012). The Shaakhsaazi communities of Kashmir use bamboos in wicker handicrafts, i.e., in basket making and furniture weaving (Islam and Sheikh Shah 2017). Baskets are made from different bamboo species in the Upper Kedarnath valley of Garhwal Himalayas (Fig. 14.3h–p, Dhyani and Dhyani 2016, Table 14.5). In Arunachal Pradesh, the traditional basket "Frokpa" is made from bamboo by Yak pastoralists (Fig. 14.3e–g, Bora et al. 2013). The "Kotwalia" tribes of Gujarat use bamboo in making baskets (Patel 2005). Bamboo-made baskets are used for different purposes in different parts of India such as in the preparation of different oils by "Pawra" tribes of Maharashtra (Mukherjee et al. 2013), for mushroom collection by the "Santals" of Eastern India (Manna et al. 2014) and for rice storage in southern Assam (Fig. 14.3h, Das and Das 2014). Bamboo baskets are used in the traditional duckeries of some ethnic people residing in Tamil Nadu (Fig. 14.3c, Gajendran and Karthickeyan 2011, Table 14.5). The Jarawa folks of Andaman islands use bamboos in making planks for sitting and sleeping purposes (Sharief and Panda 2018).

14.2.5.2 Reports from Other Countries

Bamboo-made baskets and furnitures are popular not only in India but also in other countries. The ethnic communities residing in Yunnan province of China and Milin County of Tibet use bamboos to make furnitures (Fig. 14.4a–h), tools, and baskets (Fig. 14.3e–p, Yuming et al. 2004, Li et al. 2015, Table 14.5). In Ethiopia, baskets and furnitures such as chairs (Fig. 14.4d), shelves, (Fig. 14.4a, c, e) and beds are made from bamboos (Worku 2015, Table 14.5).

Table 14.4 Ethnobotanical usages of bamboos for making handicrafts and artifacts and in handloom industries. The available local names are indicated after species names within third brackets

Name of species	Place from where reported	Tribe/ ethnic people	Specific use	Reference
Arundinaria sp.; *Dendrocalamus* sp. Nees.	Bumdeling Wildlife Sanctuary, Trashiyangtse, Bhutan	N.A.	Bows and arrows	Wangyal (2012)
A. falcata	Bhagirathi valley (Western Himalayas), India	N.A.	Mats, broom, winnow	Uniyal et al. (2002)
Arundinaria sp.; *Chimonobambusa falcata*; *Thamnocalamus spathiflorus*	Upper Kedarnath valley, Garhwal, India	N.A.	Various handicrafts	Dhyani and Dhyani (2016)
B. polymorpha Munro. [Bethua-lakla]	Bangladesh	N.A.	Devoted coin boxes	Partha (2014)
B. pallida [Sho]; *B. tulda*; *D. hamiltonii*	Arunachal Pradesh, India	Yak pastoralists	Inner-outer layers of milking cans, storage boxes (Zai), milk churners (Zopu), churpy separators (Churchuk)	Bora et al. (2013)
B. blumeana; *B. lapidea*; *B. sinospinosa*; *B. vulgaris*	Yunnan, China	N.A.	Rafters, frames for lifting, scaffoldings, beanpoles, mine props	Yuming et al. (2004)
B. multiplex; *B. ventricosa*; *B. vulgaris* cv. *vittata*; *B. vulgaris* cv. *wamin*; *Chimonobambusa* spp.; *F. yuanjiangensis*; *Yushania nana*; *P. aurea*; *P. heterocycla*; *P. nigra*; *Qiongzhuea tumidinoda*; *Thyrsostachys siamensis*			Ornaments	
B. textiles; *Cephalostachyum scandens*; *Neosinocalamus*			Mats, fans, slippers, curtains, fences, hats, ribbons, woven	

(continued)

Table 14.4 (continued)

Name of species	Place from where reported	Tribe/ ethnic people	Specific use	Reference
affinis; *Schizostachyum funghomii*			bamboo products, ropes, straps	
D. sinica			Drums	
Fargesia spp.; *Yushania* spp.			Reed pipes	
F. papyrifera; *F. utilis*; *F. yunnanensis*			Umbrella handles, walking sticks, vaulting poles, nails, hoops, smoking pipes, measuring implements	
Phyllostachys decora; *P. bambusoides*; *P. heteroclata*			Bows, arrows, crossbows, brooms, fishing poles, carrying pole	
P. bambusoides			Flutes	
Cephalostachyum capitatum [Po-young]	Dzongu Tribal Reserved Area, India	Lepcha tribes	Sumok-thyaktuk (Lepcha hat)	Lepcha et al. (2012)
D. giganteus Munro. [Unan]	Manipur, India	Kom tribes	Bamboo bow (Patsai), beam for weaving, Taru, Kothai (shuttles), Tako, Trai	Khatoon et al. (2014)
D. strictus (Roxb.) Nees [Bamboo]	Koraput, Odisha, India	Bhatra, Bhumia, Gadaba, and Paroja tribes	Rain hats	Mishra and Chaudhury (2012)
D. strictus (Roxb.) Nees. [Saliabanso]	Odisha, India	N.A.	Mats	Sahu et al. (2013)
F. melanostachys (Hand.-Mazz.) T.P. Yi [Sunzi]	Shangri-la region, Yunnan province, China	Tibetans	Wares	Ju et al. (2013)
N.A.	Nicobar, India	Nicobari	Tools like Kinlah roon ap (bamboo), Laneiny (bamboo scale), Kunan (thin bamboo scale), Harah laneiny (double-pointed bamboo arrow), Kinlahpamo (single-pointed bamboo arrow)	Ravikumar et al. (2015)
N.A.	Dzongu Tribal Reserved Area, India	Lepcha	Bamboo crafts to keep away evil spirits	Lepcha et al. (2012)
N.A.	Kandi belt, Jammu, India	N.A.	Cases (Shabri), mats	Slathia and Paul (2012)

(continued)

Table 14.4 (continued)

Name of species	Place from where reported	Tribe/ ethnic people	Specific use	Reference
N.A.	Manipur, India	N.A.	Spools, shuttles	Pandya and Thoudam (2010)
N.A.	Meghalaya, India	Biates	Rows of bamboo in traditional dress (rua), Ritai (hair band)	Karolia and Ladia (2012)
N.A.	Assam, India	Karbis	Bamboo stick (Barlim) used to maintain asei (so created shed)	Teron and Borthakur (2012)
N.A.	Manipur, India	Nagas and Kukis	Fans, trays, vases, ash-trays, gift boxes, seat bases, etc.	Yumkham and Singh (2013)
N.A.	Southwest Ethiopia	N.A.	Beehive, floor mat, flutes, drinking cups, water container (dollo), cups, traditional tray (gamo), pipe for smoking tobacco, utensils	Worku (2015)
N.A.	Milin County, Nanyi, Tibet	Lhoba	Clothes, mats, cages, bowls, rain gears, bows and arrows	Li et al. (2015)
N.A.	Lombok Island, Indonesia	Sasak	Kitchen, container for storing rice (raru)	Sukenti et al. (2016)
N.A.	Assam, India	Mikir (Karbis)	Bamboo pipes	Jain and Borthakur (1980)
N.A.	Gujarat, India	Kotwalia	Supadas (used in folk dance), bags, umbrellas, mats, measurement utensil (Mapiyu), device to protect eggs (Kolaju), bows and arrows, flute, Pawa, Pihooda, container to store arrows (Bhatho)	Patel (2005)

Abbreviations used: *N.A.* not available

Fig. 14.4 Usage of bamboos in making furniture. (**a**) Book shelf; (**b**) cloth hanger; (**c**) utensil shelf; (**d**) chair; (**e**) shelf; (**f**) plant pot keeper; (**g**) stool; (**h**) "Morah," stool. (**a**, **b**, **c**, and **h**) captured from the Boro tribal community of Kamrup, Assam, and (**d**, **e**, **f**, and **g**) from Nyishi tribal community of Senki view, Itanagar, Arunachal Pradesh, India

Table 14.5 Ethnobotanical usages of bamboos for making baskets and furnitures by ethnic communities. The available local names are indicated after species names within third brackets

Name of species	Place from where reported	Tribe/ethnic people	Specific use	Reference
A. falcata	Bhagirathi valley (Western Himalayas), India	N.A.	Baskets	Uniyal et al. (2002)
Arundinaria sp.; *Chimonobambusa falcata*; *Thamnocalamus spathiflorus*	Upper Kedarnath valley, Garhwal, India	N.A.	Baskets	Dhyani and Dhyani (2016)
B. pallida [Sho]; *B. tulda*; *D. hamiltonii*	Arunachal Pradesh, India	Yak pastoralists	Baskets (Frokpa)	Bora et al. (2013)

(continued)

Table 14.5 (continued)

Name of species	Place from where reported	Tribe/ethnic people	Specific use	Reference
B. blumeana; *B. lapidea*; *B. sinospinosa*; *B. vulgaris*	Yunnan, China	N.A.	Furniture	Yuming et al. (2004)
B. textiles; *Cephalostachyum scandens*; *Neosinocalamus affinis*; *Schizostachyum funghomii*			Baskets	
F. papyrifera; *F. utilis*; *F. yunnanensis*			Bed rests	
N.A.	Kandi belt, Jammu, India	N.A.	Baskets, chairs	Slathia and Paul (2012)
N.A.	Meghalaya, India	Biates	Baskets	Karolia and Ladia (2012)
N.A.	Manipur, India	Nagas and Kukis	Basketry	Yumkham and Singh (2013)
N.A.	Kashmir, India	Shaakhsaazi	Wicker handicraft	Islam and Sheikh Shah (2017)
N.A.	Southwest Ethiopia	N.A.	Chairs, baskets, shelf, bed	Worku (2015)
N.A.	Milin County, Nanyi, Tibet	Lhoba	Baskets	Li et al. (2015)
N.A.	Assam, India	Mikir (Karbis)	Special basket (Hak-chili)	Jain and Borthakur (1980)
N.A.	Tamil Nadu, India	N.A.	Baskets used for hatching and brooding the ducklings	Gajendran and Karthickeyan (2011)
N.A.	India	N.A.	Traditional baskets (pala, Olia/Doli)	Sethi et al. (2011)
N.A.	Southern Assam, India	N.A.	Baskets (Dol/Tukre)	Das and Das (2014)
N.A.	Manipur, Northeast India	Meitei	Baskets used to prepare Kum-dye	Ningombam et al. (2012)
N.A.	Eastern Lateritic Part of India	Santals	Baskets (Khanchi) used to collect mushrooms	Manna et al. (2014)
N.A.	Nandurbar district, Maharashtra, India	Pawra	Basket is used in preparing Malkangni and Behada oil	Mukherjee et al. (2013)

(continued)

Table 14.5 (continued)

Name of species	Place from where reported	Tribe/ethnic people	Specific use	Reference
N.A.	Andaman Islands, India	Jarawa	Planks making for sleeping and sitting	Sharief and Panda (2018)
N.A.	Gujarat, India	Kotwalia	Baskets (Dalu, Shiku)	Patel (2005)

Abbreviations used: *N.A.* not available

14.2.6 Atypical Ethnobotanical Uses of Bamboos by Various Ethnic Communities

14.2.6.1 Reports from India

Bamboos are used in different rituals by the tribal people of India. To name a few are "Jamur Awas" ritual celebrated in Mendha village, Maharashtra (Heda 2012); "Rongali Bihu" festival by "Sonowal Kachari" tribe of Assam (Fig. 14.2b–d, Sonowal 2016); "Mosomao Kheda Utsav" to drive mosquitoes away by "Hajong" community of Assam (Sharma et al. 2012); festivals celebrated by "Chothe" tribe, Manipur (Sanglakpam et al. 2012); "Monpa" community of Arunachal Pradesh (Singh 2013); and "Kurichya" tribes of Wayanad district in Kerala (Pramod et al. 2003). Bamboos sticks are also used in making plates from leaves of "Sal" (*Shorea robusta*) plants in Jharkhand (Islam et al. 2015). The ethnic communities of Assam and "Gond" tribe in Andhra Pradesh use bamboo structures to harbor and scare birds, respectively (Majumder et al. 2013; Sivaraj et al. 2012). The "Kani" tribe of Tamil Nadu (Kumar et al. 2012) and ethnic people of Nagaland (Singh 2013) use bamboo items in their bee-keeping practices. The "Seri cultural" community of West Bengal use bamboo structures to protect silkworms from parasite flies (Guha and Roy Choudhury 2001). Due to their resistance to earthquakes, bamboo-made constructions are frequently used in different parts of Assam (Fig. 14.5a–j, Pareek and Trivedi 2011). Bamboo-made ladders are popularly used in south of Tamil Nadu (Fig. 14.5f, Sundaramari et al. 2011). Bamboos are widely used to make channels for drip irrigation in different states of India such as Udhampur district of Jammu and Kashmir (Slathia et al. 2016), Khasi and Jaintia hills of Meghalaya (Singh and Gupta 2002), and Banderdewa forest range of Arunachal Pradesh (Fig. 14.5h, Pangging et al. 2011) and also in Jharkhand (Dey and Sarkar 2011). People of Kikruma village in Nagaland use bamboo channels for transporting dung and urine from cattle enclosures to rice fields (Das et al. 2012).

Bamboos are used as fuels in the Upper Kedarnath valley of Garhwal (Dhyani and Dhyani 2016, Table 14.6) and even used to ignite fire by the people residing in Bhagirathi valley of Western Himalayas (Uniyal et al. 2002). In some parts of Assam, bamboo items are used to rear and protect muga silkworms (Borthakur

Fig. 14.5 Uses of bamboos for diverse constructional and allied purposes useful for ethnic lifestyle. (**a**) Community stage; (**b**) wall material ready to be fixed in a room; (**c**) bridge; (**d**) wall, fencing pillar of bamboo house; (**e**) bamboo float; (**f**) ladder; (**g**) bamboo-made house; (**h**) bamboo channel used for transporting water; (**i**) bamboo gate; (**j**) bamboo poles used as supporting pillars of a hanging house. (**a**, **b**, **c**, **e**, **f**, **g**, **h**, **i**, and **j**) were obtained from the Nyishi tribal community of Senki view, Itanagar, Arunachal Pradesh, and (**b**) from Boro tribal community of Kamrup, Assam, India

Table 14.6 Multiple, atypical ethnobotanical utilities of bamboos by various ethnic communities. The available local names are indicated after species names within third brackets

Name of species	Reported from	Tribe/ethnic people	Details of use	Reference
Arundinaria falcata	Bhagirathi Valley (Western Himalayas), India	N.A.	Fire ignition in lower altitude, household articles, support seedlings	Uniyal (2002)
Arundinaria sp.; *Chimnobambusa falcata*; *Thamnocalamus spathiflorus*	Upper Kedarnath valley, Garhwal, India	N.A.	Minor fuels	Dhyani and Dhyani (2016)
B. arundinacea	Mendha (Lekha) village, Maharashtra, India	Gond	Used in Jamur Awas ritual	Heda (2012)
B. arundinacea (Retz.) Willd. [Tomalbah]; *B. balcooa* Roxb. [Bholukabah]; *B. tulda* Roxb. [Jatibah]	Assam, India	N.A.	Muga silkworm rearing, duck inside a bamboo cage used to drive away nocturnal predators of muga silkworms	Borthakur (2003)
B. arundinacea Willd.	Nilgiri Biosphere Reserve, Western Ghats, India	Aboriginal inhabitants	Brooms for grain collection	Rasingam and Jeeva (2013)
B. bambos (L.) Voss [Mullukayali]	Wayanad District, Kerala, India	Kurichya	Used in ceremonies "Thulappathu," "Kumbham," and funeral rites, frame of bows, striking surface of "Mottambu" (an arrow), water pots used in "Kumbham" and funerals	Pramod et al. (2003)
B. bambos (L.) Voss.; *D. strictus* (Roxb.) Nees.	Jharkhand, India	N.A.	Raw plates (Pattals)	Islam et al. (2015)
B. blumeana	Balang, China	Miao people	High ethnic values in birth rituals	Mao et al. (2018)
B. longispiculata gamble. ex brandis. [Mrittinga-lakla]	Bangladesh	N.A.	Fencing of gardens	Partha (2014)

(continued)

Table 14.6 (continued)

Name of species	Reported from	Tribe/ethnic people	Details of use	Reference
B. multiplex; *B. ventricosa*	Yunnan, China	N.A.	In Buddhist rituals	Yuming et al. (2004)
Dendrocalamus giganteus			Sign of future success	
Fargesia spp.			Considered holy	
Phyllostachys nigra		Yi people	Worshipped	
B. pallida Munro. [Watang]	Bishnupur district (Manipur), India	Chothe	In the festival of Chothe God	Sanglakpam et al. (2012)
B. pallida Munro. [Dibang]; *Dendrocalamus hamiltonii* Nees.	Arunachal Pradesh, India	Adi	Water conservation structures like Yetbung Lingang and Linkum	Pattanaaik et al. (2012)
B. tulda Roxb. [Bans]	Assam, India	Hajong	Used in "Mosomao Kheda Utsav" as torches to drive away mosquitoes	Sharma et al. (2012)
B. tulda Roxb. [Bans]	Nagaon district, Assam, India	Teagarden and ex-teagarden communities	Live fencing, checking soil erosion and construction	Borkataki et al. (2008)
B. tulda Roxb.	North bank plain zone, Assam, India	N.A.	For bird sitting	Majumder et al. (2013)
B. tulda Roxb. [Jati Bah]	Assam, India	Sonowal Kachari	In the Rongali Bihu festival, Toka (a bamboo made musical instrument) accompanied by Toka-mari (small bamboo sticks for striking)	Sonowal (2016)
B. vulgaris	Trinidad and Tobago	N.A.	Dogs grooming	Lans et al. (2000)
Dendrocalamus sp.	Jharkhand, India	N.A.	Brush, hollow boxes, ivory black color can be prepared	Kumar (2014)
D. hamiltonii Nees. and Arn.	Assam, India	Karbis	Making Kido (used for communications)	Teron and Gogoi (2004)
D. strictus (Roxb.) Nees.	Udhampur district, Jammu and Kashmir, India	N.A.	Construction of Kools (traditional irrigation channels)	Slathia et al. (2016)

(continued)

Table 14.6 (continued)

Name of species	Reported from	Tribe/ethnic people	Details of use	Reference
D. strictus Roxb.	Uttarakhand Himalayas, India	N.A.	Topare or Doke (storage bins)	Mehta et al. (2012)
Phyllostachys mannii [Almal]	Yunnan, Southwest China	Hani Communities, Mengsong, Xishuangbanna	Flutes, fertilizer for tobacco	Kanglin et al. (2000)
Schizostachyum dulloa (Gamble) R. B. Majumder	Cachar district, Assam, India	N.A.	Dala (bamboo trays), mats, carving and weaving of slats	Dattagupta and Gupta (2014)
Thamnocalamus spathiflorus	Bhagirathi Valley (Western Himalayas), India	N.A.	Fire ignition in higher altitudes	Uniyal (2002)
N.A.	Cachar district, Assam, India	N.A.	Racks for drying, trays used in smoking, storage rooms	Dattagupta and Gupta (2014)
N.A.	Tamil Nadu, Western Ghats, India	Kani	Bamboo pole beehives	Kumar et al. (2012)
N.A.	Mostly in districts of North Bihar, India	N.A.	In harvesting and seed collection of Makhana, Kaara (a bamboo pole), Larna (bamboo stick), bamboo containers	Mandal et al. (2010)
N.A.	North eastern hill region of India	N.A.	Traps (Vaithang, Chepthang, Hnawhtawt, Chehrap), Bamboo bait stations	Thakur et al. (2013)
N.A.	Orissa, India	Kandha	Tools for hunting and protection	Panda et al. (2005)
N.A.	Adilabad district, Andhra Pradesh, India	Gond	Making bird scare structures	Sivaraj et al. (2012)
N.A.	Khasi and Jaintia hills, Meghalaya, India	Tribal farmers	Prefabricated open channel is made using split bamboo lines, bamboo stands support the pipelines	Singh and Gupta (2002)
N.A.	Phunying, Bumdeling Wildlife	N.A.	Used for benediction	Wangyal (2012)

(continued)

Table 14.6 (continued)

Name of species	Reported from	Tribe/ethnic people	Details of use	Reference
	Sanctuary, Trashiyangtse, Bhutan			
N.A.	Fringe villages in Banderdewa forest range, Arunachal Pradesh, India	N.A.	In slope water harvesting, house construction, and commercial purposes	Pangging et al. (2011)
N.A.	Nagaland, India	N.A.	Making of queen cage in underground bee hives	Singh (2013)
N.A.	Tambrauw, West Papua, Indonesia	Ireres	Fire triggering during cooking	Marwa et al. (2013)
N.A.	Dhemaji district, Assam, India	N.A.	Upcoming devastating flood is thought to be indicated by massive bamboo flowering	Bordoloi and Muzaddadi (2015)
N.A.	India	N.A.	Protect against soil erosion in dykes, bhasa (bhela)	Sethi et al. (2011)
N.A.	Barak Valley, Assam, India	Manipuri rice farmers	In the control of *Leptocorisa* sp.	Bhattacharjee and Ray (2010)
N.A.	Bangladesh	N.A.	Floating platform protection, bamboo cage/raft	Irfanullah et al. (2011)
N.A.	Jharkhand, India	N.A.	Rain water storage, drip irrigation, bamboo pipes	Dey and Sarkar (2011)
N.A.	Northeast India	N.A.	Orchid bed protection, mite accumulation (pest)	Meena et al. (2018)
N.A.	South Tamil Nadu, India	N.A.	Bamboo ladder	Sundaramari et al. (2011)
N.A.	Assam, India	N.A.	Ekra, construction (prevent collapse during seismic activity).	Pareek and Trivedi (2011)
N.A.	Kerala, India	N.A.	Making brush handle	Mini (2010)
N.A.	Bangladesh	Khasia	Single bamboo pole made ladder (Lo-u)	Haider et al. (2013)
N.A.	Manipur, Northeast India	Meitei	Ladles used to prepare lime solution	Ningombam et al. (2012)

(continued)

Table 14.6 (continued)

Name of species	Reported from	Tribe/ethnic people	Details of use	Reference
			which is used in making Kum-dye	
N.A.	Andaman and Nicobar, India	Mongoloid	Ornaments	Senthilkumar et al. (2014)
N.A.	Ambo and Dandi districts, west-central Ethiopia	N.A.	Gotara (used for the storage of grains)	Geleta and Grausgruber (2013)
N.A.	Marquesas Islands (French Polynesia)	N.A.	Used in supercision process	Jost et al. (2016)
N.A.	Southern Nigeria	Yoruba-speaking people	Used as an item in prayers	Omonhinmin (2012)
N.A.	Nandurbar district, Maharashtra, India	Pawra	Pipes are used in preparation of balm oil	Mukherjee et al. (2013)
N.A.	Kikruma village, Nagaland, India	N.A.	Split bamboo channels are used to channel dung and urine from cattle enclosures to rice fields	Das et al. (2012)
N.A.	Jammu and Kashmir, Gujarat, Theni district of Tamil Nadu, India	N.A.	Bamboo strips oiled with ONGC oil wells prevent termite attacks	Mahapatro et al. (2017)
N.A.	West Bengal, India	Seri cultural community	Protection from parasite flies	Guha and Roy Choudhury (2001)
N.A.	India	N.A.	Bamboo poles	Tesfaye and Gautam (2003)
N.A.	Upper Magdalena River Valley. Colombia	N.A.	Tall palm trees climbing poles	Bernal et al. (2010)
N.A.	Subtropical regions of India	N.A.	Barejas (for betel vine cultivation)	Pradhan et al. (2013)
N.A.	Raigad district, Maharashtra, India	Mahadeo Koli	Food storage containers	Mishra et al. (2013)
N.A. [Arundo]	Fiche, Ethiopia	N.A.	Pens	d'Avigdor et al. (2014)

(continued)

Table 14.6 (continued)

Name of species	Reported from	Tribe/ethnic people	Details of use	Reference
N.A.	Arunachal Pradesh, India	Monpa	Fried dry bamboo shoots (tenga) taken as prasad; idols of animals and birds are hung on bamboo sticks and prasad kept on bamboo containers	Singh (2013)
N.A.	Northern Lao PDR	N.A.	Fans	Delang (2008)
N.A.	Northeast India	Karbis	Bamboo charcoal used in making blue/indigo dyes, twig used in ritual, bamboo culm with the node open at one end and split bamboo used in preparing red/pink dyes	Teron and Borthakur (2012)
N.A.	Central Bangladesh	N.A.	Winnower	Oakley and Momsen (2007)
N.A.	Andaman Islands, India	Jarawa	Arrows (patho)	Sharief and Panda (2018)
N.A.	Lombok Island, Indonesia	Sasak	Firewood for cooking	Sukenti et al. (2016)
N.A.	Gujarat, India	Kotwalia	Marriage pandals, symbol of long life and good luck	Patel (2005)
N.A.	Siang belt, Arunachal Pradesh, India	Adi and Galo	Fish cooking	Hussain et al. (2016)
N.A.	Lanten Yao, Northern Lao PDR	N.A.	Paper making, religious purposes	Delang (2006)

Abbreviations used: *N.A.* not available

2003). The "Mongoloid" tribes and settlers of Andaman and Nicobar islands use bamboos in making ornaments (Fig. 14.1j–m, Senthilkumar et al. 2014). Bamboos are used in making dyes and also in communication purposes by the ethnic "Karbis" tribe of Northeast India (Teron and Borthakur 2012; Teron and Gogoi 2004). In the Western Ghats, bamboo brooms are used for collection of food grains (Fig. 14.3n–p, Rasingam and Jeeva 2013). The Manipuri rice farmers of Barak valley in Assam use bamboos to control pests, e.g., *Leptocorisa* sp. (Bhattacharjee and Ray 2010, Table 14.6). In the Cachar district of Assam, ethnic groups make items for daily uses like trays and racks (Fig. 14.4f, Dattagupta and Gupta 2014). In Uttarakhand, many ethnic communities use bamboo bins for seed storage (Mehta et al. 2012), whereas “Meitei” community of Manipur and “Mahadeo Koli” tribes of Maharashtra make ladles and storage containers from bamboo (Ningombam et al. 2012; Mishra et al. 2013, Table 14.6). Bamboo items are used in the “Pytkar” and “Jadopatia” folk arts of Jharkhand (Kumar 2014). Rodent-catching traps are often made of bamboos in some parts of northeastern hill region of India (Thakur et al. 2013). The “Kandha” tribes of Orissa make bamboo-based tools for hunting and protection (Panda et al. 2005). The possible occurrence of devastating floods can be predicted by observing massive bamboo flowering by some ethnic groups residing in Dhemaji district, Assam (Bordoloi and Muzaddadi 2015). In Kerala, the handles of brushes used for traditional mural paintings are made of bamboos (Fig. 14.2h, Mini 2010, Table 14.6). In different parts of India, bamboos are used for making float “Bhela” (Fig. 14.5e, Sethi et al. 2011). Sometimes, bamboo is used to support climber plants like the betel vine (Pradhan et al. 2013) and also in the management and protection of orchid beds in Northeast India (Meena et al. 2018). Bamboo items are used in harvesting Makhana (*Euryale ferox*) seeds in North Bihar (Mandal et al. 2010).

14.2.6.2 Reports from Other Countries

In Bangladesh, different ethnic groups use bamboos for winnowing (Oakley and Momsen 2007), fencing of homestead gardens (Fig. 14.5b, d, Partha 2014), and management of floating platforms (Irfanullah et al. 2011). The “Khasia” community of Bangladesh uses bamboo to make a distinct single-pole ladder (Haider et al. 2013, Table 14.6). Climbing poles are made from bamboo in the upper Magdalena river valley of Colombia (Bernal et al. 2010). The “Miao” people in Balang and different ethnic communities of Yunnan province of China use bamboo in their holy rituals (Yuming et al. 2004; Mao et al. 2018). Again, in Yunnan of China, flutes and fertilizers are made from bamboos by “Hani” communities of Mengsong, Xishuangbanna (Kanglin et al. 2000). In Fiche of Ethiopia, bamboo pens are used by ethnic people during rituals (d’Avigdor et al. 2014). The tribes of Ambo and Dandi districts in west-central Ethiopia use bamboos to construct “Gotara” (grain storage sack, Geleta and Grausgruber 2013). Bamboos are used for benediction in Phunying of Bumdeling Wildlife Sanctuary, Trashiyangtse, Bhutan (Wangyal 2012). In Indonesia, the “Ireres” tribe of Tambrauw, West Papua, and “Sasak” tribe of Lombok island use bamboo to trigger fire for cooking (Marwa et al. 2013; Sukenti et al. 2016). In Northern Lao PDR, bamboos are used in rituals, as firewood and in paper making (Delang 2006, Table 14.6), and as fans for welcoming gestures (Fig. 14.2a, Delang 2008). The “Yoruba”-speaking people of Southern Nigeria use bamboo items in prayers (Omonhinmin 2012). Some tribes in Trinidad and Tobago use bamboo sticks to groom dogs (Lans et al. 2000). Bamboos are also used for supercisions in Marquesas islands of French Polynesia (Jost et al. 2016, Table 14.6).

14.2.7 Recent Progress on Bamboo Genomics and Its Future Implication on "Ethnobamboology"

Although a significant body of literature is available on diverse ethnobotanical uses of different bamboo species, genomic intervention strategies to exploit the existing knowledge has not been thought of. This is due to many reasons such as unavailability of genomic information on bamboos, the complexity of bamboo genomes, lack of collaboration among researchers working on ethnobamboology and those working on molecular aspects, and lack of funding opportunities. Due to the presence of high ploidy levels, genome sequencing of bamboos remained a great challenge over decades (Biswas et al., 2016). The draft sequences of the first bamboo genome *P. edulis* ($2n = 4x = 48$) published in 2013 were able to discover different molecular pathways responsible for growth and development. In addition, the draft genomes of four bamboo species belonging to tribes Olyreae and Bambuseae got sequenced (Guo et al. 2019). Interestingly, these four species, *Bonia amplexicaulis* (Bambuseae, $2n = 6x = 72$), *Guadua angustifolia* (Bambuseae, $2n = 4x = 46$), *Olyra latifolia* (Olyreae, $2n = 2x = 22$), and *Raddia guianensis* (Olyreae, $2n = 2x = 22$), have different ploidy level and also cover a large geographic distribution (Table 14.7). In the meantime, the Genome Atlas of Bamboo and Rattan (GABR) project announced the sequencing of more than 300 species from 37 different bamboo genera in 2017 (Zhao et al. 2017). This kind of large-scale multi-omics project is expected to generate data that can be used by the larger bamboo research community for diverse purposes (Zhao et al. 2017). In addition to genome sequencing, targeted sequencing of a limited number of genes (Lin et al. 2009; Dutta et al. 2018, Chakraborty et al. 2019, Yang et al. 2019) and characterization of transcriptomes have also been undertaken (Zhang et al. 2012; Gao et al. 2014; Zhao et al. 2016; Li et al. 2019). Therefore, it can be concluded that it is high time now to build on the available genomic resources to reshape the future of ethnobamboology research.

Table 14.7 Summary of the genome sequence data available on different bamboo species till to date

Bamboo species	Tribe	Ploidy level	Chromo some number	Genome size (Mb)	Number of protein-coding genes	References
Phyllostachys edulis	Arundinarieae	Tetraploid	$2n = 48$	2050	31,987	Peng et al. (2013)
Bonia amplexicaulis	Bambuseae	Hexaploid	$2n = 72$	848	47,056	Guo et al. (2019)
Guadua angustifolia	Bambuseae	Tetraploid	$2n = 46$	1614	38,575	Guo et al. (2019)
Olyra latifolia	Olyreae	Diploid	$2n = 22$	646	36,578	Guo et al. (2019)
Raddia guianensis	Olyreae	Diploid	$2n = 22$	626	24,275	Guo et al. (2019)

14.2.8 Conclusions and Future Perspective

In spite of enormous prospects, the full economic potential of bamboos has not been harvested, yet. One of the major reasons is the lack of translational research that can be useful for major industries such as pharmaceuticals and food. Only depending on government agencies for seeking research support will be too minimal, and private funding needs to be poured in. Initiatives like the National Bamboo Mission, Government of India (https://nbm.nic.in/#:~:text=About%20Us,under%20bamboo%20cultivation%20and%20marketing), are a good model that brings stakeholders from various levels, backgrounds, and expertise in one platform. In the meantime, bamboo researchers may consider shifting "publication-driven research" to "product-driven research" to be able to invent products and to make the academy-industry partnership viable. One potential area for future research would be to heavily invest on the findings of medicinally important compounds from different bamboo species, and this chapter compiles all available ethnobotanical knowledge to identify the target species and their medicinal applications. We also propose a pipeline on how conventional ethnobotanical knowledge can be connected to modern analytical tools to improve commercial utilities of bamboos in the future (Fig. 14.6). In parallel, bamboo-based handicraft industry can be supported to promote small-scale cottage industry. Taken together, this article summarizes important ethnobotanical uses of

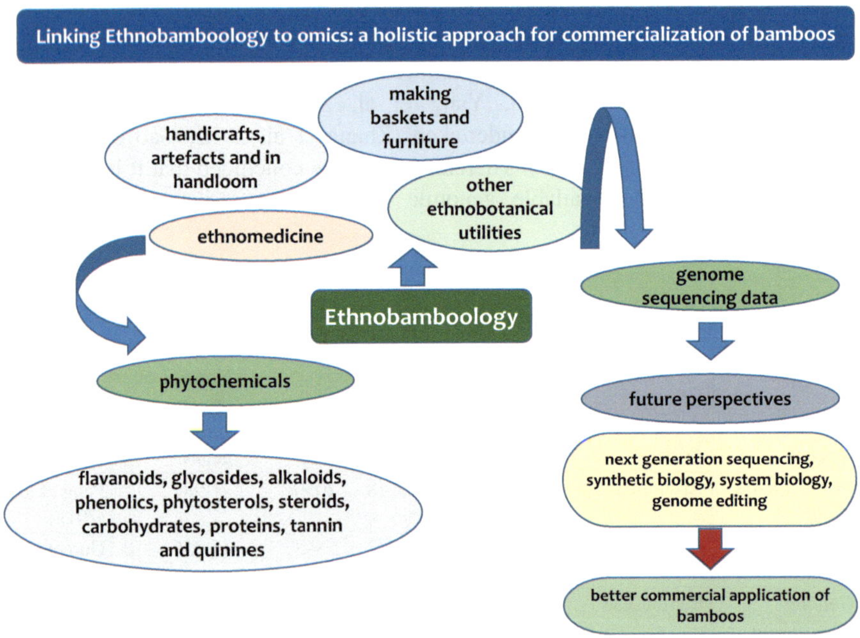

Fig. 14.6 Linking ethnobamboology to omics: a holistic approach for commercialization of bamboos

bamboos and provides new directions of research that are relevant in the post-genome and post-COVID era.

Acknowledgements MD lab also acknowledge funding from the Council of Scientific and Industrial Research, India [38(1386)/14/EMR-II], [38(1493)/19/EMR-II]; Department of Biotechnology, Govt. of India (BT/PR10778/PBD/16/1070/2014, BT/PR28859/FCB/125/3/2018); and Alexander von Humboldt Foundation, Germany. MB and SD acknowledges fellowship from CSIR, India. The authors thank Suparna Bhattacharyaand Jiten Kalita for providing information on ethnobotanical uses of bamboos from Assam, India. CC thanks the Director, Botanical Survey of India, for providing necessary support.

References

Albuquerque UP, Monteiro JM, Ramos MA, Amorim ELC (2007) Medicinal and magic plants from a public market in northeastern Brazil. J Ethnopharmacol 110:76–91

Baddu VD, Ouano NB (2018) Ethnobotanical survey of medicinal plants used by the Y'Apayaos of Sta. Praxedes in the province of Cagayan, Philippines. Mindanao J Sci Technol 16:128–153

Bam J, Rai S, Bhattacharya D, Maiti S, Islam S, Pathak P, Bera AK, Deb SM (2015) Indigenous curative and prophylactic traditional practices used against haematophagous leeches in Arunachal Pradesh and Sikkim. Indian J Tradit Knowl 14(3):493–497

Bernal R, Galeano G, Garcia N, Olivares IL, Cocoma C (2010) Uses and commercial prospects for the wine palm, Attaleabutyracea, in Colombia. Ethnobot Res Appl 8:255–268

Bharati KA, Sharma BL (2012) Plants used as ethnoveterinary medicines in Sikkim Himalayas. Ethnobot Res Appl 10:339–356

Bhattacharjee PP, Ray DC (2010) Pest management beliefs and practices of Manipuri rice farmers in Barak Valley, Assam. Indian J Tradit Knowl 9(4):673–676

Biswas P, Chakraborty S, Dutta S, Pal A, Das M (2016) Bamboo Flowering from the Perspective of Comparative Genomics and Transcriptomics. Front Plant Sci 7:1900

Bora L, Paul V, Bam J, Saikia A, Hazarika D (2013) Handicraft skills of Yak Pastoralists in Arunachal Pradesh. Indian J Tradit Knowl 12(4):718–724

Bordoloi R, Muzaddadi AU (2015) Indigenous technical knowledge associated with disaster management and fisheries related activities in the highest flood affected district (Dhemaji) of Assam, India. Indian J Tradit Knowl 14(3):407–415

Borkataki S, Chutia M, Borthakur SK (2008) Ethnobotany of biofencing among teagarden and ex-teagarden communities of Nagaon district of Assam. Indian J Tradit Knowl 7(4):666–668

Borthakur SK (1979) Traditional weaving implements among the Mikirs (Karbis) of Assam. Indian Mus Bull 11:46–50

Borthakur SK (2003) Ethnobotanical wisdom behind the traditional muga silk industry in Assam. Indian J Tradit Knowl 2(3):230–235

Bystriakova N, Kapos V, Lysenko I, Stapleton C (2003) Distribution and conservation status of forest bamboo biodiversity in the Asia-Pacific region. Biodivers Conserv 12:1833–1841

Chakraborty S, Dutta S, Biswas P, Das M (2019) Identification of candidate reference genes in tropical bamboos stable across tissues, developmental stages and species barrier. Biol Plant 63:253–261

Chandra R, Mahato M, Mandal SC, Kumar K, Kumar J (2007) Ethnomedicinal formulations used by traditional herbal practitioners of Ranchi, Jharkhand. Indian J Tradit Knowl 6(4):599–601

Chaveerach A, Mokkamul P, Sudmoon R, Tanee T (2006) Ethnobotany of genus *Piper* (Piperaceae) in Thailand. Ethnobot Res Appl 4:223–231

Chettri D, Chowdhury M (2018) Ethnomedicinal approach against bone fracture among the tea garden workers of Darjeeling Hills. Indian J Tradit Knowl 17(3):512–517 d'Avigdor E, Wohlmuth H, Asfaw Z, Awas T (2014) The current status of knowledge of herbal medicine and medicinal plants in Fiche, Ethiopia. J Ethnobiol Ethnomed 10:38

Das T, Das AK (2014) Inventory of the traditional rice varieties in farming system of southern Assam: A case study. Indian J Tradit Knowl 13(1):157–163

Das S, Kumar B (2019) Indian military involvement in the 1971 crisis of East Pakistan: a justification of level of analysis. Am J Theor Appl Bus 5(4):84–89

Das AJ, Dutta BK, Sharma GD (2008a) Medicinal plants used by different tribes of Cachar district, Assam. Indian J Tradit Knowl 7(3):446–456

Das M, Bhattacharya S, Singh P, Filgueiras TS, Pal A (2008b) Bamboo taxonomy and diversity in the Era of molecular markers. Adv Bot Res 47:225–268

Das A, Ramkrushna GI, Choudhury BU, Munda GC, Patel DP, Ngachan SV, Ghosh PK, Tripathi AK, Das S, Kumar M (2012) Natural resource conservation through indigenous farming systems: Wisdoms alive in North East India. Indian J Tradit Knowl 11(3):505–513

Daswad AK, Shendarkar GR, Shelke DP, Pohare JG, Wangawar SN, Dhadwe AK (2017) Phytochemicalevaluationandantioxidantstudyof *Dendrocalamusstrictus*leafextracts. World J Pharma Med Res 3(3):147–152

Dattagupta S, Gupta A (2014) Traditional processing of non-timber forest products in Cachar, Assam, India. Indian J Tradiit Knowl 13(2):427–433

de Boer HJ, Lamxay V, Bjork L (2012) Comparing medicinal plant knowledge using similarity indices: A case of the Brou, Saek and Kry in Lao PDR. J Ethnopharmacol 141:481–500

Delang CO (2006) Bamboo paper production and religious use by the Lanten Yao in Northern Lao PDR. Ethnobot Res Appl 4:097–102

Delang CO (2008) Keeping the Spirit Alive: Rice whiskey production in Northern Lao P.D.R. Ethnobot Res Appl 6:459–470

Dey P, Sarkar AK (2011) Revisiting indigenous farming knowledge of Jharkhand (India) for conservation of natural resources and combating climate change. Indian J Tradit Knowl 10 (1):71–19

Dhyani S, Dhyani D (2016) Significance of provisioning ecosystem services from moist temperate forest ecosystems: lessons from upper Kedarnath valley, Garhwal, India. Energy Ecol Environ 1 (2):109–121

Dixit U, Goyal VC (2011) Traditional knowledge from and for elderly. Indian J Tradit Knowl 10 (3):429–438

Dutta S, Biswas P, Chakraborty S, Mitra D, Pal A, Das M (2018) Identification, characterization and gene expression analyses of important flowering genes related to photoperiodic pathway in bamboo. BMC Genomics 19:190. https://doi.org/10.1186/s12864-018-4571-7

Gajendran K, Karthickeyan SMK (2011) Indigenous Technical knowledge in duck production in Tamil Nadu. Indian J Tradit Knowl 10(2):307–310

Gao J, Zhang Y, Zhang C, Qi F, Li X, Mu S, Peng Z (2014) Characterization of the floral transcriptome of Moso bamboo (*Phyllostachys edulis*) at different flowering developmental stages by transcriptome sequencing and RNA-seq analysis. PLoS One 9(6):e98910

Geleta N, Grausgruber H (2013) Homemade products and socio-cultural values of wheat seed production in Ambo and Dandi districts of West Central Ethiopia. Sci Technol Arts Res J 2 (4):62–70

Guha SC, Roy Choudhury TK (2001) Sericulture in India: industrial application from seventeenth to nineteenth century. J Sci Ind Res 60:369–377

Guo ZH, Ma PF, Yang GQ, Hu JY, Liu YL, Xia EH, Zhong MC, Zhao L, Sun GL, Xu YX, Zhao YJ, Zhang YC, Zhang YX, Zhang XM, Zhou MY, Guo Y, Guo C, Liu JX, Ye XY, Chen YM, Yang Y, Han B, Lin CS, Lu Y, Li DZ (2019) Genome sequences provide insights into the reticulate origin and unique traits of woody bamboos. Mol Plant 12:1353–1365

Gupta R, Vairale MG, Deshmukh RR, Chaudhary PR, Wate SR (2010) Ethnomedicinal uses of plants by Gond tribes of Maharashtra. Indian J Tradit Knowl 9(4):713–717

Gurrapu S, Mamidala E (2016) Medicinal plants used by traditional medicine practitioners in the management of HIV/AIDS-related diseases in tribal areas of Adilabad district, Telangana region. Am J Sci Med Res 2(1):239–245

Gutiérrez J (1993) Structural behavior and seismic resistance of bamboo housing. In: Proceedings of the 1st World Congress on Bamboo/Guadua, Pereira, pp 161–171

Haider MR, Khair A, Rahman MM, Alam MK (2013) Indigenous management practices of betel-leaf (*Piperbetle* L.) cultivation by the Khasia community in Bangladesh. Indian J Tradit Knowl 12(2):231–239

Hariyadi B, Ticktin T (2012) Uras: Medicinal and ritual plants of Serampas, Jambi Indonesia. Ethnobot Res Appl 10:133–149

He C, Cui K, Zhang J, Duan A, Zeng Y (2013) Next-generation sequencing-based mRNA and microRNA expression profiling analysis revealed pathways involved in the rapid growth of developing culms in Moso bamboo. BMC Plant Biol 13:119

Heda N (2012) Folk conservation practices of the Gond tribal of Mendha (Lekha) village of Central India. Indian J Tradit Knowl 11(4):727–732

Hussain A, Khan MN, Iqbal Z, Sajid MS (2008) An account of the botanical anthelmintics used in traditional veterinary practices in Sahiwal district of Punjab, Pakistan. J Ethnopharmacol 119:185–190

Hussain SM, Sen D, Riba T, Pathak M, Singh RK (2016) Fishing in the Siang belt of Arunachal Pradesh, India: learning traditional ecological knowledge of Adi and Galo communities. Indian J Trad Knowl 15(4):685–692

Irfanullah HM, Azad MAK, Kamruzzaman MD, Wahed MA (2011) Floating gardening in Bangladesh: a means to rebuild lives after devastating flood. Indian J Tradit Knowl 10(1):31–38

Islam MA, Sheikh Shah SSG (2017) Traditional wicker handicraft by the Shaakhsaazi communities of rural Kashmir. Indian J Tradit Knowl 16(2):350–355

Islam MA, Quli SMS, Rai R, Singh PK (2015) Livelihood promotion through value addition to household traditional Sal (*Shorearobusta* Gaertn.) leaf plate making in Jharkhand, India. Indian J Nat Prod Resour 6(4):320–325

Jagtap SD, Deokule SS, Bhosle SV (2006) Some unique ethnomedicinal uses of plants used by the Korku tribe of Amravati district of Maharashtra, India. J Ethnopharmacol 107:463–469

Jain SK, Borthakur SK (1980) Ethnobotany of the Mikirs of India. Econ Bot 34(3):264–272

Jawaid T, Awasthi A, Kamal M (2015) Estrogenic activity of a hydro-alcoholic extract of *Bambusaarundinaceae* leaves on female wistar rats. J Adv Pharm Technol Res 6(1):19–24

Jin K, Xue C, Wu X, Qian J, Zhu Y, Yang Z, Yonezawa T, Crabbe MJC, Cao Y, Hasegawa M, Zhong Y, Zheng Y (2011) Why Does the Giant Panda Eat Bamboo? A Comparative Analysis of Appetite-Reward-Related Genes among Mammals. PLoS One 6(7):e22602

Jost X, Ansel JL, Lecellier G, Raharivelomanana P, Butaud JF (2016) Ethnobotanical survey of cosmetic plants used in Marquesas Islands (French Polynesia). J Ethnobiol Ethnomed 12:55

Ju Y, Zhuo J, Liu B, Long C (2013) Eating from the wild: diversity of wild edible plants used by Tibetans in Shangri-la region, Yunnan, China. J Ethnobiol Ethnomed 9:28

Kanglin W, Jianchu X, Shengji P, Sanyang C (2000) Folk classification and conservation of bamboo in Xishuangbanna, Yunnan, Southwest China. J Ethnobiol 20(1):113–127

Karolia A, Ladia B (2012) Traditional textiles and costumes of Karbi and Biate tribes of Meghalaya. Indian J Tradit Knowl 11(2):309–316

Karolia A, Prakash R (2014) Design and development of fashion accessories inspired from the hand woven shawls of Nagaland. Indian J Tradit Knowl 13(2):416–426

Kaur P (2018) Bamboo availability and utilization potential as a building material. Environ Sci. https://doi.org/10.15406/freij.2018.02.00056

Khadda BS, Singh B, Singh DV, Singh JL, Singh SK, Singh CB, Singh D (2018) Inventory of traditional ethno-veterinary practices followed by goatkeepers in Uttarakhand. Indian J Tradit Knowl 17(1):155–161

Khatoon R, Das AK, Dutta BK, Singh PK (2014) Study of Traditional handloom weaving by Kom tribe of Manipur. Indian J Tradit Knowl 13(3):596–599

Kumar S (2014) Use of plants as Colour material in Pytkar and Jadopatia folk arts of Jharkhand. Indian J Tradit Knowl 13(1):202–207
Kumar R, Bharati KA (2012a) Folk veterinary medicine in Sitapur district of Uttar Pradesh, India. Indian J Nat Prod Resour 3(2):267–277
Kumar R, Bharati KA (2012b) Folk veterinary medicines in Jalaun district of Uttar Pradesh, India. Indian J Tradit Knowl 11(2):288–295
Kumar R, Bharati KA (2013) Folk veterinary medicines in the Bareilly district of Uttar Pradesh, India. Indian J Tradit Knowl 12(1):40–46
Kumar MS, Singh AJAR, Alagumuthu G (2012) Traditional beekeeping of stingless bee (*Trigona*sp) by Kani tribes of Western Ghats, Tamil Nadu, India. Indian J Tradit Knowl 11 (2):342–345
Lans C, Harper T, Georges K, Bridgewater E (2000) Medicinal plants used for dogs in Trinidad and Tobago. Prev Vet Med 45:201–220
Laxmay V, de Bour HJ, Bjork L (2011) Traditions and plant use during pregnancy, childbirth and postpartum recovery by the Kry ethnic group in Lao PDR. J Ethnobiol Ethnomed 7:14
Lepcha SR, Gurung R, Arrawatia ML (2012) Traditional Lepcha craft Sumok-thyaktuk (Lepcha hat) and its conservation in Dzongu Tribal Reserved Area (DTRA), Sikkim, India. Indian J Tradit Knowl 11(3):537–541
Li F, Zhuo J, Liu B, Jarvis D, Long C (2015) Ethnobotanical study on wild plants used by Lhoba people in Milin County, Tibet. J Ethnobiol Ethnomed 11:23
Li Y, Zhang C, Yang K, Shi J, Ding Y, Gao Z (2019) *De novo* sequencing of the transcriptome reveals regulators of the floral transition in *Fargesia macclureana* (Poaceae). BMC Genomics 20:1035
Lin E, Peng H, Jin Q, Deng M, Li T, Xiao X, Hua X, Wang K, Bian H, Han N, Zhu M (2009) Identification and characterization of two Bamboo (*Phyllostachys praecox*) AP1/SQUA-like MADS-box genes during floral transition. Planta 231:109–120
Liu W, Hui C, Wang F, Wang M, Liu G (2018) Review of the resources and utilization of bamboo in China. Bamboo - current and future prospects, Abdul Khalil HPS, IntechOpen. https://doi.org/10.5772/intechopen.76485
Magel E, Kruse S, Lütje G, Liese W (2005) Soluble carbohydrates and acid invertases involved in the rapid growth of developing culms in *Sasa palmata* (Bean) Camus. Bamboo Sci Cult 19:23–29
Mahapatro GK, Debajyoti C, Gautam RD (2017) India indigenous traditional knowledge (ITK) on termites: eco-friendly approaches to sustainable management. Indian J Trad Knowl 16(2):333–340
Majumder D, Deka SN, Pujari D, Das PK (2013) Traditional knowledge adopted by the farmers for management of rice pests in North bank plain zone of Assam. Indian J Tradit Knowl 12 (4):725–729
Mandal SK, Rahaman CH (2014) Determination of informants' consensus and documentation of ethnoveterinary practices from Birbhum district of West Bengal, India. Indian J Tradit Knowl 13 (4):742–751
Mandal RN, Saha GS, Sarangi N (2010) Harvest and processing of Makhana (*Euryale ferox* Salisb.)-an unique assemblage of traditional knowledge. Indian J Tradit Knowl 9(4):684–688
Manna S, Ray D, Roy A (2014) Tribal relation to spatiotemporal variation of wild mushrooms in Eastern lateritic part of India. Ethnobot Res Appl 12:015–024
Mao S, Shen Y, Deng H (2018) Multipurpose plant utilization in ethnic areas of Guizhou, southwest China. Ecol Indic 90:547–553
Marwa J, Cabuy RL, Tawer AY (2013) Potential and pattern of utilization of renewable energy sources from vegetation based on local knowledge of Ireres tribe in Tambrauw, West Papua, Indonesia. Indian J Tradit Knowl 12(3):411–417
Meena NK, Pal R, Barman D, Pant RP (2018) Indigenous approaches of orchid pest management in North East India. Indian J Tradit Knowl 17(1):209–214

Mehta PS, Negi KS, Rathi RS, Ojha SN (2012) Indigenous methods of seed conservation and protection in Uttarakhand Himalayas. Indian J Tradit Knowl 11(2):279–282

Mei J, Wang P, Chen K, Wang L, Wang Y, Liu Y (2015) Archaeometallurgical studies in China: some recent developmentsand challenging issues. J Arch Sci 56:221–232

Mini PV (2010) Preparation techniques of pigments for traditional mural paintings of Kerala. Indian J Tradit Knowl 9(4):635–639

Mishra S, Chaudhury SS (2012) Ethnobotanical flora used by four major tribes of Koraput, Odisha, India. Genet Resour Crop Evol 59:793–804

Mishra M, Jadhav PA, Sinha R (2013) Ethnobotanical knowledge of Mahadeo Koli tribes of Raigad District of Maharashtra: Plant used during famine. Int J Integrat Sci Innov Technol 2(5):27–32

Mistry N, Silori CS, Gupta L, Dixit AM (2003) Indigenous knowledge on animal healthcare practices in district Kachchh, Gujarat. Indian J Tradit Knowl 2(3):240–254

Mondal S, Rahaman CH (2012) Medicinal plants used by the tribal people of Birbhum district of West Bengal and Dumka district of Jharkhand in India. Indian J Tradit Knowl 11(4):674–679

Mukherjee S, Kulkarni O, Deokule S, Gadgil S, Harsulkar A, Jagtap S (2013) Oil extraction from medicinal plants by Pawra tribe of Nandurbar district (Maharashtra): Value addition and sustainable utilization with aid of Ayurved. Indian J Tradit Knowl 12(2):272–276

Muniappan M, Jagadeesh JS, Inbaraj SD (2014) Evaluation of *Bambusaarundinacea*leaves for wound healing in albino rats. Glob J Pharmacol 8(1):84–86

Nazreen S, Kaur G, Alam MM, Haider S, Hamid H, Alam MS (2011) Hypoglycemic activity of *Bambusaarundinacea*leaf ethanolic extract in streptozotocininduced diabetic rats. Pharmacol Online 1:964–972

Ningombam DS, Ningthoujam SS, Singh PK, Singh OB (2012) Legacy of kum dye: a case study with the Meitei community in Manipur, Northeast India. Ethnobot Res Appl 10:561–570

Oakley E, Momsen JH (2007) Women and seed management: A study of two villages in Bangladesh. Singap J Trop Geogr 28:90–106

Oladele AT, Elem JA (2018) Medicinal plants used to treat sexual diseases by Ogba/Egbema/Ndoni ethnics of Rivers State, Nigeria. World News Nat Sci 17:16–38

Omonhinmin AC (2012) Ethnobotany of *Dacryodesedulis* (G. Don) H.J. Lam in Southern Nigeria 1: Practices and applications among the Yoruba speaking people. Ethnobot Res Appl 10:175–184

Owolabi MS, Lajide L (2015) Preliminary phytochemical screening and antimicrobial activity of crude extracts of *Bambusa vulgaris* Schrad. Ex J.C. Wendl. (Poaceae) from southwestern Nigeria. Am J Essent 3(1):42–45

Panda T, Panigrahi SK, Padhy RN (2005) A sustainable use of phytodiversity by the Kandha tribe of Orissa. Indian J Tradit Knowl 4(2):173–178

Pandya A, Thoudam J (2010) Handloom weaving, the traditional craft of Manipur. Indian J Tradit Knowl 9(4):651–655

Pangging G, Arunachalam A, Mawphlang ISL, Biswas S (2011) Traditional management practices of natural resources of forest dependent communities in Arunachal Pradesh – a case study of fringe villages in Banderdewa forest range. Indian J Tradit Knowl 10(2):269–275

Pareek A, Trivedi PC (2011) Cultural values and indigenous knowledge of climate change and disaster prediction in Rajasthan, India. Indian J Tradit Knowl 10(1):183–189

Partha (2014) Ethnobotany of the Laleng (Patra) Community in Bangladesh. J Pharmacogn Phytochem 2(6):173–184

Patel AB (2005) Traditional bamboo uses by the tribes of Gujarat. Indian J Tradit Knowl 4 (2):179–184

Patil JU, Biradar SD (2011) Folkloric medicinal plants of Hingoli district, Maharashtra. Indian J Tradit Knowl 2(1):97–101

Pattanaaik SK, Sen D, Kumar N, Moyong O, Debnath P (2012) Traditional system of water management in watersheds of Arunachal Pradesh. Indian J Tradit Knowl 11(4):719–723

Peng Z, Lu Y, Li L, Zhao Q, Feng Q, Gao Z, Lu H, Hu T, Yao N, Liu K, Li Y, Fan D, Guo Y, Li W, Lu Y, Weng Q, Zhou CC, Zhang L, Huang T, Zhao Y, Zhu C, Liu X, Yang X, Wang T, Miao K,

Zhuang C, Cao X, Tang W, Liu G, Liu Y, Chen J, Liu Z, Yuan L, Liu Z, Huang X, Lu T, Fei B, Ning Z, Han B, Jiang Z (2013) The draft genome of the fast- growing non-timber forest species moso bamboo (Phyllostachys heterocycla). Nat Genet 45:456–461

Pradhan D, Suri KA, Pradhan DK, Biswasroy P (2013) Golden heart of the nature: *Piperbetle* L. J Pharmacogn Phytochem 1(6):147–167

Pramod C, Sivadasan M, Anilkumar N (2003) Ethnobotany of religious and supernatural beliefs of Kurichya of Wayanad District, Kerala, India. Ethnobotany 15:11–19

Rai PK, Lalramnghinglova H (2011) Ethnomedicinal plants of India with special reference to an Indo-Burma hotspot region: an overview. Ethnobot Res Appl 9:379–420

Rao JK, Reddi TVVS, Kumar OA (2011) Stem bark of certain plants of Vishakhapatnam district, Andhra Pradesh. Curr Bot 2(5):01–06

Rasingam L, Jeeva S (2013) Indigenous brooms used by the aboriginal inhabitants of Nilgiri Biosphere Reserve, Western Ghats, India. Indian J Nat Prod Resour 4(3):312–316

Rathod JD, Pathak NL, Patel RG, Jivani NP, Bhatt NM (2011) Phytopharmacological properties of *Bambusa arundinacea* as a potential medicinal tree: an overview. J Appl Pharm 01(10):27–31

Rautray AK, Sahoo R, Sardar KK, Patra RC, Sahoo A (2015) Ethnoveterinary practices for small ruminants followed by the rural folks in southern Odisha. Indian J Tradit Knowl 14(2):319–324

Ravikumar T, Krishnan P, Ram N, Sankar RK, Roy SD, Ahmed SKZ, Neethiselvan N (2015) Hodi, a traditional craft of Nicobari tribe. Indian J Tradit Knowl 14(1):161–168

Roosita K, Kusharto CM, Sekiyama M, Fachrurozi Y, Ohtsuka R (2008) Medicinal plants used by the villagers of a Sundanese community in West Java, Indonesia. J Ethnopharmacol 115:72–81

Saha MR, Sarker DD, Sen A (2014) Ethnoveterinary practices among the tribal community of Malda district of West Bengal, India. Indian J Tradit Knowl 13(2):359–367

Sahu SC, Pattnaik SK, Dash SS, Dhal NK (2013) Fibre-yielding plant resources of Odisha and traditional fibre preparation knowledge – an overview. Indian J Nat Prod Resour 4(4):339–347

Sanglakpam P, Mathur RR, Pandey AK (2012) Ethnobotany of *Chothe* tribe of Bishnupur district (Manipur). Indian J Nat Prod Resour 3(3):414–425

Saran PL, Choudhary R, Solanki IS, Devi G (2015) Traditional medicaments through papaya in North eastern plains zone of India. Indian J Tradit Knowl 14(4):537–543

Senthilkumar MK, Sivakumar P, Changanakkattil F, Rajesh V, Perumal P (2011) Evaluation of anti-diabetic activity of *Bambusa vulgaris* leaves in Streptozotocin induced diabetic rats. Int J Pharm Sci Drug Res 3(3):208–210

Senthilkumar U, Choudhary RK, Sanjappa M, Narasimhan D, Uma Shaanker R, Ravikanth G (2014) Livelihood and revenue: role of rattans among Mongoloid tribes and settlers of Andaman and Nicobar Islands, India. Ethnobot Res Appl 12:141–154

Sethi SN, Sundaray JK, Panigrahi A, Chand S (2011) Prediction and management of natual disasters through indigenous technical knowledge, with special reference to fisheries. Indian J Tradit Knowl 10(1):167–172

Shah B, Sheth F, Parabia M (2012) Folk herbal knowledge on the management of respiratory disorders prevailing in ethnic society of Valsad district, Gujarat. Indian J Nat Prod Resour 3 (3):438–447

Shankar R, Rawat MS (2008) Medicinal plants used in traditional medicine in Lohit and Dibang valley districts of Arunachal Pradesh. Indian J Tradit Knowl 7(2):288–295

Sharief MU, Panda SP (2018) Material culture of the ethnic aboriginal Jarawa folk in Andaman Islands, India. Genet Resour Crop Evol 65:1369–1384

Sharma UK, Pegu S, Hazarika D, Das A (2012) Medico-religious plants used by the Hajong community of Assam, India. J Ethnopharmacol 143:787–800

Singh RK (2013) Ecoculture and subsistence living of Monpa community in the eastern Himalayas: an ethnological study in Arunachal Pradesh. Indian J Tradit Knowl 12(3):441–453

Singh RA, Gupta RC (2002) Traditional land and water management systems of North-East hill region. Indian J Tradit Knowl 1(1):32–39

Singh VN, Chanu LI, Chiru Community, Baruah MK (2011) An ethnobotanical study of Chirus- a less known tribe of Assam. Indian J Tradit Knowl 10(3):572–574

Situmorang ROP, Harianja AH, Silalahi J (2015) Karo'local wisdom: the use of woody plants for traditional diabetic medicines. Indones J For Res 2(2):121–113
Sivaraj N, Pandravada SR, Jairam R, Sunil N, Varaprasad KS, Ramesh P, Bisht IS, Pareek SK (2012) An indigenous cost effective innovation for bird scares by the Gond tribe in Adilabad district, Andhra Pradesh. Indian J Tradit Knowl 11(4):714–718
Sivasankari B, Anandharaj M, Gunasekaran P (2014) An ethnobotanical study of indigenous knowledge on medicinal plants used by the village peoples of Thoppampatti, Dindigul district, Tamilnadu, India. J Ethnopharmacol 153:408–423
Slathia PS, Paul N (2012) Traditional practices for sustainable livelihood in Kandi belt of Jammu. Indian J Tradit Knowl 11(3):548–552
Slathia PS, Sharma BC, Paul N, Kumar R, Gupta SK (2016) Traditional water conveyance modes, their management and use in Udhampur district of Jammu and Kashmir. Indian J Tradit Knowl 15(3):503–508
Sonowal R (2016) Sacred plants of the Sonowal Kachari tribe of Assam, Northeast India. Dimensions of multidisciplinary research studies, 1st edn. Kasturi Prakaxaan, Guwahati
Sukenti K, Hakim L, Indriyani S, Purwanto Y, Matthews PJ (2016) Ethnobotanical study on local cuisine of the Sasak tribe in Lombok Island, Indonesia. J Ethn Foods 3(3):1–12
Sundaramari M, Ganesh S, Kannan GS, Seethalakshmi M, Gopalsamy K (2011) Indigenous grain storage structures of south Tamil Nadu. Indian J Tradit Knowl 10(2):380–383
Tamuli P, Sharma P (2010) Ethno-medico-botany of Dimasa Kachari of North Cachar Hills district of Assam. Indian J Tradit Knowl 9(4):718–720
Tantengco OAG, Condes MLC, Estadilla HHT, Ragragio EM (2018) Ethnobotanical survey of medicinal plants used by Ayta Communities in Dinalupihan, Bataan, Philippines. Pharm J 10 (5):859–870
Teron R, Borthakur SK (2012) Biological motifs and designs on traditional costumes among Karbis of Assam. Indian J Tradit Knowl 11(2):305–308
Teron R, Gogoi P (2004) Indigenous means of communication amongst social dignitaries of hill Karbis of Assam. Indian J Tradit Knowl 3(3):294–298
Tesfaye A, Gautam RD (2003) Traditional pest management practices and lesser exploited natural products in Ethiopia and India: Appraisal and revalidation. Indian J Tradit Knowl 2(2):189–201
Thakur NSA, Firake DM, Kumar D (2013) Indigenous traps for the management of rodent outbreak in North Eastern Hill region of India. Indian J Tradit Knowl 12(4):730–735
Thamizharasan S, Umamaheswari S, Rajeswari H, Ulagaratchagan V (2015) Quantitative Phytochemical Analysis of *Bambusa arundinacea* Seeds. Int J Pharmacogn Phytochem Res 7 (5):980–983
Uddin MZ, Arefin MK, Alam MF, Kibria MG, Podder SL, Hassan MA (2017) Knowledge of ethnomedical plants and informant consensus in and around Lawachara National Park. J Asiat Soc Bangladesh Sci 43(1):101–123
Uniyal SK, Awasthi A, Rawat GS (2002) Traditional and ethnobotanical uses of plants in Bhagirathi Valley (Western Himalayas). Indian J Tradit Knowl 1(1):7–19
Upadhya V, Hedge HV, Bhat S, Hurkadale PJ, Kholkute SD, Hedge GR (2012) Ethnomedicinal plants used to treat bone fracture from North-Central Western Ghats of India. J Ethnopharmacol 142:557–562
Verma RK (2014) An ethnobotanical study of plants used for the treatment of livestock diseases in Tikamgarh District of Bundelkhand, Central India. Asian Pac J Trop Biomed 4(Suppl 1):S460–S467
Wangyal JT (2012) Ethnobotanical knowledge of local communities of Bumdeling Wildlife Sanctuary, Trashiyangtse, Bhutan. Indian J Tradit Knowl 11(3):447–452
Worku M (2015) Traditional uses of non-timber forest products in southwest Ethiopia: opportunities and challenges for sustainable forest management. J Agric Environ Sci 1(1):42–63
Yakubu MT, Bukoye BB (2009) Abortifacient potentials of the aqueous extract of *Bambusa vulgaris* leaves in pregnant Dutch Rabbits. Contraception 80:308–313

Yang Z, Chen L, Kohnen MV et al (2019) Identification and Characterization of the PEBP Family Genes in Moso Bamboo (*Phyllostachys heterocycla*). Sci Rep 9:14998

Yeasmin L, Ali MN, Gantait S, Chakraborty S (2015) Bamboo: an overview on its genetic diversity and characterization. 3. Biotech 5:1–11

Yigezu Y, Haile DB, Ayen WY (2014) Ethnoveterinary medicines in four districts of Jimma zone, Ethiopia: cross sectional survey for plant species and mode of use. BMC Vet Res 10(1):76

Yuming Y, Kanglin W, Shengji P, Jiming H (2004) Bamboo diversity and traditional uses in Yunnan, China. Mt Res Dev 24(2):157–165

Yumkham SD, Singh PK (2013) Study on Uses and Trading of *Huperzia squarrosa* (G. Forst.) Trevis. (Lycopodiaceae) in Manipur, India. Ethnobot Res Appl 11:153–161

Zhang X, Zhao L, Larson-rabin Z, Li D, Guo Z (2012) *De Novo* Sequencing and Characterization of the Floral Transcriptome of *Dendrocalamus latiflorus* (Poaceae: Bambusoideae). PLoS One 7: e42082

Zhao H et al (2016) Comprehensive analysis of multi- tissue transcriptome data and the genome-wide investigation of GRAS family in *Phyllostachys edulis*. Sci Rep 6:1–15

Zhao H, Zhao S, International Network for Bamboo and Rattan, Fei B, Liu H, Yang H, Dai H, Wang D, Jin W, Tang F, Gao Q, Xun H, Wang Y, Qi L, Yue X, Lin S, Gu L, Li L, Zhu T, Wei Q, Su Z, Wan TBWA, Ofori DA, Muthike GM, Mengesha YM, de Castro E, Silva RM, Beraldo AL, Gao Z, Liu X, Jiang Z (2017) Announcing the genome atlas of bamboo and rattan (GABR) project: promoting research in evolution and in economically and ecologically beneficial plants. Gigascience 6:1–7

Zubair M, Bibi Z, Rizwan K, Rasool N, Zahoor AF, Riaz M (2013) In Vitro Antimicrobial and Haemolytic Studies of *Bambusa arundinaceae* leaves. J Appl Pharm Sci 3(04):111–115

Chapter 15
Bamboo Flowering in South America: What the Past Tells about the Future

Carolina Guerreiro and Andrea S. Vega

Abstract A key characteristic of most woody bamboos is the process of cyclical flowering after a long vegetative period. These flowering events are generally gregarious, affecting an entire region, and are followed by the death of individuals. The duration of the life cycle has been determined in very few species. After reconstructing the history of massive reproductive episodes of South American woody bamboo species, the life span of many species was estimated. For most of the species considered, mean flowering period multiples of 15–16 years were found, with an ca. 30-year cycle being the most usual. Evidence of a certain level of reproductive synchrony among different species was found in South America. This survey had also led to several predictions about probable dates of future flowering events. The existence of a relationship between a mass flowering event and climate factors is preliminarily assessed, yielding original results. Although historical records and climate data series are fragmentary, there are some hints that point out a trend that should be taken into account in future studies. Proposed environmental and genetic causes of bamboo flowering are discussed. Finally, the environmental and social consequences of bamboo flowering are listed.

Keywords Climatic factors · Life cycles · Phenology · South America · Woody bamboos

C. Guerreiro
Instituto de Botánica Darwinion (CONICET-ANCEFN), Buenos Aires, Argentina

A. S. Vega (✉)
Facultad de Agronomía, Departamento de Recursos Naturales y Ambiente, Cátedra de Botánica General, Universidad de Buenos Aires, Buenos Aires, Argentina
e-mail: avega@agro.uba.ar

Z. Ahmad et al. (eds.), *Biotechnological Advances in Bamboo*,
https://doi.org/10.1007/978-981-16-1310-4_15

15.1 Introduction

The subfamily Bambusoideae comprises ca. 125 genera and 1680 species within the tribes Arundinarieae (temperate woody bamboos), Bambuseae (tropical woody bamboos), and Olyreae (herbaceous bamboos). The tribe Bambuseae consists of two major clades: the Paleotropical (Old World or Asian) woody bamboos and the Neotropical (New World or American) woody bamboos (Soreng et al. 2017). According to recent phylogenetic worldwide classifications of the Poaceae, the Neotropical woody bamboo clade is composed of the subtribes Arthrostyliidinae, Guaduinae, and Chusqueinae, with 17, 5, and 1 genera, respectively, and is found from Mexico along Central America and the Caribbean Islands to South America (Clark et al. 2015; Soreng et al. 2017; Saarela et al. 2018). These subtribes are composed of the genera *Actinocladum* McClure ex Soderstr.; *Alvimia* Calderón ex Soderstr. and Londoño; *Apoclada* McClure; *Arthrostylidium* Rupr.; *Athroostachys* Benth.; *Atractantha* McClure; *Aulonemia* Goudot; *Aulonemiella* L.G. Clark, Londoño, C.D. Tyrrell, and Judz.; *Cambajuva* P.L. Viana, L.G. Clark, and Filg.; *Chusquea* Kunth; *Colanthelia* McClure and E.W. Sm.; *Didymogonyx* (L.G. Clark and Londoño) C.D. Tyrrell, L.G. Clark, and Londoño; *Elytrostachys* McClure; *Eremocaulon* Soderstr. and Londoño; *Filgueirasia* Guala; *Glaziophyton* Franch.; *Guadua* Kunth; *Merostachys* Spreng.; *Myriocladus* Swallen; *Olmeca* Soderstr.; *Otatea* (McClure and E.W. Sm.) C.E. Calderón and Soderstr.; *Rhipidocladum* McClure; and *Tibisia* C.D. Tyrrell, Londoño, and L.G. Clark (Vorontsova et al. 2016; Tyrrell et al. 2018; Clark et al. 2020).

Bamboos are as American as potatoes, maize, and chili peppers. Though it may be surprising for some, nearly half of the world's tropical woody bamboo species are native to America (Judziewicz et al. 1999). Particularly, South America is the part of the Americas with the highest diversity and number of woody bamboo species. They occur in a broad range of environments in many South American landscapes. Woody bamboos occupy extensive areas of humid, lowland tropical evergreen forests of the Amazonian basin and the Atlantic forests of Argentina, Brazil, and Paraguay and are understory dominants in several different types of forests throughout South America. They are also characteristic of tropical and subtropical montane forests along the Andes mountain range, extending southward into cold temperate beech forests of southern Argentina and Chile. Above tree line, woody bamboos are also the main component of the unique Andean vegetation formations known as *páramos*, which are humid, rainy, windswept, open high-altitude grasslands. Similar bamboo dominated grassland vegetation is also found on the higher peaks, between 1700 and 2800 m, in southeastern Brazil referred to by locals as *campos de altitude* (Judziewicz et al. 1999; Clark et al. 2015).

The complex morphology and unusual flowering behavior shown by most bamboo species are probably the consequence of adaptations to these habitats (BPG 2012). This article brings together all that is known about flowering in bamboos native to the New World. No single work can answer all the questions concerning this interesting topic, but the objective here is to provide an informative sample of current knowledge regarding bamboo flowering in South America.

15.2 Flowering in Bamboo

The most intriguing characteristic of bamboos is their flowering behavior. The sexual reproduction pattern of bamboo species differs in the degree or proportion of individuals flowered, as in the time elapsed between flowering events and the area in which these occur. This variability ranges from continuous flowerings, where all individuals flower annually or seasonally without the subsequent death, up to mass flowering events, in which all individuals of a certain species in a certain area flower, seed, and die synchronously (Janzen 1976; Judziewicz et al. 1999; Franklin 2004; Guerreiro 2014, 2016; Zheng et al. 2020). Given this variability, four main flowering habits have been described (Zheng et al. 2020). Isolated or sporadic flowering at irregular intervals involves only a few individuals of one or more populations and does not necessarily imply the death of the plant (Zheng et al. 2020). Mass or gregarious flowering involves the majority of individuals in a single or several populations, which flower synchronously (Janzen 1976; Campbell 1985; Judziewicz et al. 1999; Zheng et al. 2020). Other species exhibit isolated flowering together with mass flowering events. Finally, partial flowering generally occurs in a patchy distribution, and the degree of flowering is between isolated and gregarious flowering (Zheng et al. 2020).

Within the plant kingdom, the phenomenon of mass flowering, and in some cases at regular intervals, is not unusual, as seen in some species of Acanthaceae (Ramanayake 2006) and cycads (Jones 1993) and several species of Dipterocarpaceae (Janzen 1976; Sakai 2002). Monocarpic plant species are those that manifest a sexual reproduction strategy that is characterized by a single reproductive episode before death. This characteristic is found in several species of the families Arecaceae, Asteraceae, Bromeliaceae, Fabaceae, and Poaceae, among others (Kitajima and Augspurger 1989). Therefore, this feature is not unusual either. However, it is interesting to highlight that only certain species of woody bamboos exhibit both flowering behaviors, massive and monocarpic.

The species of bamboo that show a massive and cyclical flowering pattern, after a long period of vegetative growth, are the most captivating from the biological point of view. Mass flowering events occur at approximately regular intervals and are followed by the death of flowering individuals (McClure 1996; Alam 2008). Moreover, flowering can be synchronous on a quite wide range of spatial scales (Janzen 1976; Gielis et al. 1999). In South America, the massive flowering of *Chusquea culeou* E. Desv. during the spring-summer of 2000–2001 spread over 2000 km^2 of temperate beech forests in the Andean-Patagonian region of southern Argentina and Chile (Sanguinetti and García 2001; Giordano et al. 2009; Marchesini et al. 2009). Furthermore, the massive flowering of *Chusquea valdiviensis* E. Desv. in 1990–1992 spanned 12,000 km^2 in southern Chile as reported by Gallardo and Mercado (1999). Occasionally, a geographically directed progression of the flowering phenomenon along the range of a species has been observed (Stern et al. 1999; Franklin 2004; Alam 2008).

The period of time between two mass flowering events of a given species is generally regarded as its life cycle or flowering period and is different for each species (Alam 2008). Many bamboos grow vegetatively for a few years before flowering, while others surpass 100 years (Janzen 1976). The duration of the life cycle has been determined in very few species. The clearest example of periodicity has been documented by Kawamura (1927), who described seven mass flowering events of Japanese bamboo, *Phyllostachys reticulata* (Rupr.) K. Koch., since the year 813 at intervals of 120 years.

There are only very few people who have been able to observe the complete life cycle of bamboo, from the germination of the seed until its flowering and death. For example, Parodi (1955) reported that after the mass flowering of *Guadua trinii* (Nees) Nees ex Rupr. in 1922–1923, numerous seedlings were collected and cultivated at the Lucien Hauman Botanical Garden of the University of Buenos Aires (Argentina), where they developed vigorously. In 1952, the flowering of these cultivated individuals began and lasted until the summer of 1954, and then the plants died. Parodi (1955) states: "From the end of 1922, when these seedlings emerged, until the end of 1952, when they flowered, exactly 30 years elapsed, which is, therefore, the longevity of this species." Similarly, Vega and Cámara Hernández (2008) reported the flowering of specimens of *Guadua chacoensis* (Rojas Acosta) Londoño and P.M. Peterson, which had been cultivated in the abovementioned institution for 28 years, after seedlings were collected and taken to the Botanical Garden.

The estimation of flowering cycles is mainly based on past flowering records (Alam 2008). Herbarium specimens can solve the lack of direct field observations since they act as valuable testimonies for the reconstruction of flowering dates (Pohl 1991; Lavoie and Lanchance 2006). In this way, the length of the life cycle of a woody bamboo species may be estimated based on records that give evidence of past flowering events, which are readily available in botanical collections kept in the herbaria of museums, universities, and other institutions around the world (Thiers 2020). A reproductive herbarium specimen may be regarded as evidence of a gregarious flowering only if they bear definite information on the nature of the flowering event or multiple collections from the same populations were made (Clark 1989).

Regarding South American woody bamboos particularly, little was known about their flowering periods. With the information gathered from the study of herbarium specimens kept in botanical collections of the most important institutions in Argentina and the world, plus an extensive literature search, the occurrence of mass flowering episodes of South American bamboos was determined (Guerreiro 2014, 2016). Furthermore, based on this data, the life span of each species was calculated by recording the time periods between reported flowering events. This method, originally used by Kawamura (1927), consists of calculating the time lapse between the reported mass flowering episodes of a certain species in a certain location and obtaining a recurrent lapse. However, when working with historical records, absence of thorough chronological data is usually found, which may result in time intervals of very different lengths. In those cases, it is valid to assume that there were other intermediate mass flowering episodes that were not reported. When different time periods are found, these are reduced to a recurrent multiple which leads to obtaining an approximately regular interval (Kawamura 1927; Pohl 1991; Guerreiro and Lizarazu 2010). Following these steps, it was possible to reconstruct the history of mass flowerings of 16 South American woody bamboo species, since the nineteenth century up to our days, and, thus, estimate the duration of the life cycle of each species. Table 15.1 shows results of the authors' work for the past 10 years regarding bamboo flowering, along with information for other species native to South America available in literature. Clearly, these data are not definitive. Bamboo flowering is such an exceptional situation that proper confirmation is yet to be achieved.

It is worth highlighting the importance of historical records for this kind of work. While the vast majority of the data collected corresponds to bamboo mass flowering events occurring throughout the twentieth century and the first decade of the present century, a small percent of the data retrieved give evidence of flowering events occurring during the eighteenth and nineteenth century. The oldest flowering record obtained from a herbarium specimen is from the year 1818, found in the botanical collection of the Muséum National D'Histoire Naturelle in Paris, France. It was a specimen of *Chusquea ramosissima* Lindm., gathered in Minas Gerais, Brazil, by Auguste de Saint-Hilaire. The oldest bibliographic record found gives evidence of the mass flowering event of *Chusquea quila* Kunth in southern Chile in 1795 (Gunckel 1948).

Sometimes, bamboo species may show isolated flowering of a few individuals in the time period between mass flowering events. In those species, at any given moment it is possible to find an individual of a certain population in flower. The proportion of reproductive specimens is low and varies annually (McClure 1996; Clark 1989). This is the case, for example, of *Chusquea argentina* Parodi (Fig. 15.1), *C. culeou* (Guerreiro 2014, 2016; Fig. 15.2), and *C. ramosissima* (Montti et al. 2011, 2014). However, the opposite may also be the case. On certain occasions, during a gregarious flowering episode, few individuals may persist in their vegetative state for unknown reasons (Ramanayake 2006). This was reported in *C. ramosissima* by Montti et al. (2011, 2014) and in *C. argentina* (Fig. 15.1). It is clear that flowering behavior varies to a great degree even in phylogenetically related species (Judziewicz et al. 1999).

Various bamboo species are considered to be facultatively monocarpic, which means that after setting seeds, the plants may die or otherwise may resume their vegetative state (Judziewicz et al. 1999). Montti et al. (2011) documented that not all individuals of *C. ramosissima* died after the reproductive event. The isolated flowering of a *C. culeou* specimen, which surprisingly did not die afterward, was reported by Guerreiro (2014). Also, a *G. chacoensis* specimen flowered in the last gregarious flowering episode of 2004–2008, but, unlike every other clump in the population, it did not die. This specimen is now part of the current cohort (Fig. 15.3a; author's personal observation). The same situation was recorded in *Bambusa tuldoides* Munro (Guerreiro and Lizarazu 2010).

Campbell (1985) examined the flowering periods of 20 bamboo species, mainly from southeastern Asia but also from South America. The author found an outstanding concentration of mean flowering cycle multiples of 15–16. This singular pattern was also found in South American bamboo species (Guerreiro 2014, 2016). In most

Table 15.1 Mass flowering events, estimated flowering cycles, and approximate date of the next mass flowering event of Neotropical bamboo species. In some species, according to their estimated flowering cycles, there are flowering events that should have already happened, although no evidence was found by the authors. Nevertheless, these putative mass flowering events were added to the table and are indicated with a "?"

Species	Mass flowering events	Estimated flowering cycle (years)	Approximate date of the next mass flowering event	Reference
Actinocladum verticillatum (Nees) McClure ex Soderstr.	1953; 1984–1986; 2017?	32	2049	Filgueiras and Pereira (1988)
Arthrostylidium venezuelae (Steud.) McClure	1989			Pohl (1991)
Aulonemia radiata (Rupr.) McClure and L.B. Sm. (under *Aulonemia fimbriatifolia* L.G. Clark)		20		Clark (2004)
A. patriae R.W.Pohl	1982			Pohl (1991)
Chusquea abietifolia Griseb.	1884–1886; 1918–1919; 1948–1949; 1978?; 2008?	30	2038	Seifriz (1920, 1950)
C. argentina Parodi	1935; 2010–2013	77	2087	Guerreiro and Rúgolo (2020)
C. caparaoensis L.G. Clark	2015			Pianissola et al. (2018)
C. capituliflora Trin.		16		Schmidt and Longhi-Wagner (2009)
C. culeou E. Desv.	1938–1939; 2000–2001	62	2062	Guerreiro (2014)
C. foliosa L.G. Clark	1987–1990			Widmer (1997, 1998)
C. gamarrae Fadrique and L.G. Clark	2016–2017			Fadrique et al. (2019)
C. longispiculata L.G. Clark		32		Clark (2004)
C. lorentziana Griseb.	1874; 1941–1942; 1971–1974; 2003–2005	32	2036	Guerreiro (2014)
C. magnifolia L.G. Clark (under *Neurolepis pittieri* McClure)	1978; 1983?; 1988?; 1993?; 1998?; 2003?; 2008?; 2013?; 2018?	5	2023	Davidse and Huber (1979)

C. meyeriana Rupr. ex Döll		35		Schmidt and Longhi-Wagner (2009)
C. mollis (Swallen) L.G. Clark		5–7		Ely and González (2020)
C. montana Phil.	1858–1862; 1944–1945; 1985	41	2026	Guerreiro (2014)
C. multiramea L.G. Clark and Ely	2009			Clark and Ely (2013)
C. patens L.G. Clark	1987–1990			Widmer (1997, 1998)
C. purdieana Munro	1945; 1989	44	2033	Clark and Ely (2013)
C. quila Kunth	1795; 1886–1888; 1929; 1947; 1992	45	2037	Guerreiro (2014)
C. ramosissima Lindm.	1916–1919; 1945–1949; 1977–1981; 2003–2005	29	2033	Guerreiro (2014)
C. riosaltensis L.G. Clark	1987–1988; 2006–2007	19	2025	Moreira et al. (2008)
C. scabra Soderstr. and C.E. Calderón	1979–1980			Pohl (1991)
C. simpliciflora Munro	1990			Pohl (1991)
C. subtilis Widmer and L.G. Clark	1987–1990			Widmer (1997, 1998)
C. talamancensis Widmer and L.G. Clark	1987–1990			Pohl (1991); Widmer (1997, 1998)
C. tenella Nees	1901–1905; 1916; 1932–1933; 1948–1949; 1963–1964; 1979–1982; 2000–2001; 2016	16	2032	Guerreiro (2014)
C. tomentosa Widmer and L.G. Clark	1987–1990			Widmer (1997, 1998)
C. tonduzii Hack.	1966			Pohl (1991)
C. tovarii L.G. Clark (under *Neurolepis weberbaueri* Pilg.)	1992			Judziewicz et al. (1999)
C. uliginosa Phil.	1962			Matthei (1997)
C. valdiviensis E. Desv.	1992			Guerreiro (2014)
C. virgata Hack.	1989–1990			Pohl (1991)
Colanthelia intermedia (McClure and L.B. Sm.) McClure	2003			Santos-Gonçalves et al. (2018)

(continued)

Table 15.1 (continued)

Species	Mass flowering events	Estimated flowering cycle (years)	Approximate date of the next mass flowering event	Reference
C. rhizantha (Hack.) McClure	1992–1996			Guerreiro (2014)
Elytrostachys clavigera McClure	1978; 1985?; 1992?; 1999?; 2006?; 2013?; 2020?	7	2027	Pohl (1991)
Guadua amplexifolia J.Presl	1953–1954			Kennard (1955)
G. chacoensis (Rojas) Londoño and P.M. Peterson	1883; 1914–1916; 1943–1946; 1974–1977; 2004–2008	31	2037	Guerreiro (2014)
G. paraguayana Döll	1936–1939; 1974–1976; 2013	38	2051	Guerreiro (2014)
G. sarcocarpa Londoño and Peterson	1904; 1933; 1959; 1987–1990; 2016?	26–29	2043	Nelson (1994)
G. tagoara (Nees) Kunth		30		Lizarazu et al. (2012)
G. trinii (Nees) Nees ex Rupr.	1920–1923; 1952–1953; 1984; 2018	31	2049	Guerreiro (2014); Guerreiro et al. (2020)
G. weberbaueri Pilg.	1976; 2004	28	2032	Carvalho et al. (2013)
Merostachys burmanii Send.	1915; 1944–1948; 1972–1978; 2003?	28–31	2033	Sendulsky (1992)
M. clausenii Munro	1874–1877; 1906–1910; 1940–1946; 1974–1976; 2004–2006	32	2037	Guerreiro (2014)
M. filgueraisii Send.	1980–1982			Sendulsky (1995)
M. fistulosa Döll		30–34		Janzen (1976)
M. magellanica Send.	1891; 1921–1922; 1956; 1988–1990	30–34	2020	Sendulsky (1995)
M. multiramea Hack.	1875; 1906–1909; 1937–1943; 1971–1975; 2003–2007	32	2037	Guerreiro (2014)
M. scandens Send.	1941; 1972–1974; 2005?	31–33	2037	Sendulsky (1995)

M. skvortzovii Send.	1941; 1972–1974; 2003–2006	30–34	2036	Sendulsky (1995); Liebsch and Reginato (2009)
M. tatianae Santos-Gonç., Carv.-Okano and Filg.	2002			Santos-Gonçalves et al. (2012)
Otatea acuminata (Munro) C.E. Calderón and Soderstr.		17–30		Ruiz-Sanchez et al. (2011)
O. fimbriata Soderstr.	2007	8–20	2015–2027	Ruiz-Sanchez et al. (2011)
O. glauca L.G.Clark and G.Cortés	2003			Ruiz-Sanchez et al. (2011)
O. ramirezii Ruiz-Sanchez	2012			Ruiz-Sanchez (2013)
Rhipidocladum bartlettii (McClure) McClure		15–20		Tyrrell (2008)
R. martinezii Davidse and R.W. Pohl	1987; 2003; 2019?	16	2035	Tyrrell (2008)
R. neumannii Sulekic, Rúgolo, and L.G. Clark	1944–1946; 1988–1989; 2008–2010	21	2030	Guerreiro (2014)
R. pacuarense R.W. Pohl	1982			Pohl (1991)
R. parviflorum (Trin.) McClure	1835; 1856; 1960; 1978–1979; 1998?; 2018?	20	2038	Tyrrell (2008)
R. pittieri (Hack.) McClure	1973; 1989; 2005?	16	2021	Pohl (1991)
R. racemiflorum (Steud.) McClure	1973; 1989; 2005?	16	2021	Pohl (1991)
R. sibilans Davidse, Judz., and L.G. Clark	1981			Tyrrell (2008)
R. zoqueorum Ruiz-Sanchez, C.D. Tyrrell, and Sosa	1986; 2018	32	2050	Ruiz-Sanchez et al. (2019)

Fig. 15.1 *Chusquea argentina* Parodi: (**a**) isolated vegetative clump in a mass flowering event, (**b**–**c**) dead clumps after flowering, (**d**) flowering of an isolated clump

Fig. 15.2 *Chusquea culeou* E. Desv. flowering clump

of the species with some evidence of regularity, the duration of their estimated life cycles is a multiple of 15–16, with an ca. 30-year flowering period being the most usual (Table 15.1).

Fig. 15.3 Mass flowering events of (**a**) *Guadua chacoensis* (Rojas Acosta) Londoño and P.M. Peterson, (**b–c**) *Merostachys clausenii* Munro

Thus, flowering cycles seem to be narrowly grouped in numbers that may be factored into small prime numbers. Now, that is definitively a situation not resulting from chance. A simple mathematical model of the evolution of bamboo mass flowering intervals was proposed a few years ago by Veller et al. (2015). First, in an initial phase, natural synchronization on a small multiyear interval (3 or 5 years) would take place in an annual flowering population. Later, multiple-interval lineages might arise, be selected for and fix in the population. Thus, the flowering cycle evolves to a multiple of its original synchronization lapse. Then, consequent small multiplications of the original synchronization lapse would result in the current dramatically extended flowering cycles (Veller et al. 2015).

Estimating the life cycle length of a bamboo species leads to being able to predict probable dates of future flowering events. Some of the predictions made by Guerreiro (2014) have already been fulfilled! For example, *C. tenella* flowered in 2016 in northeastern Argentina and southern Brazil (in herb.; J. I. Areta, pers. comm.; H. Keller, pers. comm.). Guerreiro et al. (2020) reported the flowering of natural and cultivated stands of *G. trinii* and verified its flowering period. For the rest of the predictions made by Guerreiro (2014), only time will tell.

Bamboo mass flowering events have been documented around the world, leading many to hypothesize about the evolutionary forces that drove bamboo to flower and die cyclically after decades. There are still conflicting ideas regarding the ecological and evolutionary meaning of bamboo flowering patterns and many conjectures to explain bamboo flowering synchronism. Currently, the predator-escape or consumer satiation hypothesis is the most generally regarded. Janzen (1976) proposed that gregarious flowering is a strategy by which bamboo saturates its seed predators with more seeds (caryopsis, to be botanically precise) than they can consume, so some seeds manage to escape predation to germinate and give rise to a new generation that will manifest the same interval of vegetative growth as its parents, thus perpetuating the synchrony. He also hypothesized flowering synchrony to be the result of stabilizing selection, where severe seed predation in sporadic or isolated reproductive events and satiation of predators during gregarious flowering episodes strongly supports synchrony. Another hypothesis proposed that in gregariously flowering, wind-pollinated species, synchrony may be a positive adaptive trait, due to higher rates of pollination (Koenig and Ashley 2003). Reproductive synchronization has proven benefits, such as reduced seed predation and increased seed disperser attraction and cross-pollination rates (Crone et al. 2011). Gregarious, cyclical flowerings of woody bamboos portrait an ultimate example of said reproductive synchrony.

Clearly, many species of bamboos exhibit regular flowering cycles. Even more interesting is that the flowering cycles of different species seem to be synchronized. Gregarious reproductive episodes of different species have been recorded concurrently on multiple opportunities. This was documented by Campbell (1985), who noted "some synchrony between species, with a general periodicity of ca. 30 years in East Asia." Pohl (1991) recorded the simultaneous flowering of *Rhipidocladum pittieri* (Hack.) McClure and *R. racemiflorum* (Steud.) McClure on two occasions, 1973 and 1989. Also, *Chusquea foliosa* L.G. Clark, *C. patens* L.G. Clark, *C. subtilis* Widmer and L.G. Clark, *C. talamancensis* Widmer and L.G. Clark, and *C. tomentosa* Widmer and L.G. Clark mass flowered in synchrony from 1987 to 1990 (Widmer 1997, 1998). Furthermore, *Chusquea lorentziana* Griseb., *C. ramosissima*, *G. chacoensis*, *Merostachys clausenii* Munro (Fig. 15.3), and *M. multiramea* Hack. show synchronous life cycles of ca. 30 years. These species, native to southern South America, flowered gregariously and simultaneously on several occasions over the past 75 years (Guerreiro 2014). This may be interpreted, according to Campbell (1985), as "evidence of a general synchronous tendency over large continental regions."

It is clear that bamboo flowering is still one of the largest botanical mysteries. The selective forces that explain the mass flowering pattern are still little known and even unsatisfactory for many of the bamboo species, which is not surprising given the enormous diversity of flowering patterns they present. Reproductive synchrony between individuals of a species is a phenomenon widespread in both plants and animals. In general, this synchrony is given by a particular season being more favorable for breeding and the survival of the new generation (Ims 1990). Massive flowering to long intervals exhibited by some bamboo species represents an extreme case of such synchrony.

15.3 Environmental Factors

One of the fundamental biological problems that have kept botanists very occupied is to what extent biological processes are influenced by external factors. Assessing which environmental factor results in a plant flowering becomes much more interesting and difficult when it comes to plants that flower after a long vegetative period and, even more, when all the plants in an area flower simultaneously, such as woody bamboos. The complexity of flowering behavior shown by many species has sparked much discussion about the causes of the sudden onset of this process. Mass flowering bamboos remain in a vegetative state for many years before, as Janzen (1976) defined it, "embarking on a single suicidal bout of sexual reproduction."

Certain environmental cues, such as temperature or photoperiod, are the most efficient at inducing flowering in most plants (Ramanayake 2006). However, in bamboos, the factors that induce flowering are still uncertain. Many have been proposed but none has general acceptance. One of the oldest theories proposes drought as the cause of mass flowering (Morris 1886). The occasional but striking coincidence of a bamboo flowering event with conditions of extreme drought in certain areas of southeastern Asia gave rise to the idea that such flowering is prompted by low rainfall (Gadgil and Prasad 1984). This coincidence has also been recorded in southern South America, in the Andean Patagonian area, where *C. culeou* and *C. montana* Phil., among others, inhabit. When analyzing annual climate data sets of the areas where mass flowering events of these species have been reported, a period of severe drought was identified 2 years before the flowering events of each species (Guerreiro 2013). According to Campbell (1985), under drought conditions, growth would decline and then may be the optimal time to produce seeds. Also, seedlings may be more tolerant to drought than the parental generation, due to lower transpiration rates.

However, the opposite phenomenon has also been recorded. This is the existence of a period of excess rainfall prior to the occurrence of the mass flowering event of a species. This was observed before the flowering events of *C. abietifolia*, which occurred in Jamaica in 1918 and 1948–1949 (Seifriz 1920, 1950). And in southern South America, the same situation has been recorded in northeastern Argentina, prior to the mass reproductive episodes of *C. ramosissima*, *G. chacoensis*, and *M. clausenii* which occurred in the 1970s and the first years of the present century. Mean annual maximum temperature and mean annual minimum temperature data sets were also considered, but no anomaly was found prior to the recorded mass flowering events (Guerreiro 2013). Although several events of mass flowering of many species have been recorded throughout the twentieth century in South America (Guerreiro 2014), detailed and reliable meteorological data sets are available since the 1960s. Thus, the chances of linking flowering events with meteorological extremes are limited due to the lack of long-term detailed climate records. This is a common limitation in descriptive phenological studies (Forrest and Miller-Rushing 2010).

It is natural to wonder why bamboo flowers massively and periodically. Then, it is only logical to wonder why different species of bamboo flower at the same time over and over again. If an environmental factor is to be considered as a possible cause of this phenomenon, one must speculate with a climatic factor that would act cyclically at a continental scale. It has been widely known for many years now the existence of patterns of climate variability that act on a hemispheric scale, meaning their influence extends to the entire southern hemisphere. The best known example of this type of pattern is El Niño-Southern Oscillation (ENSO). This phenomenon is a recurring climate pattern involving changes in sea surface temperature and the air pressure of the overlying atmosphere across the equatorial Pacific Ocean. This oscillating pattern is one of the most important climate phenomena on Earth due to its ability to change the global atmospheric circulation, which in turn influences temperature and precipitation across the globe. It mainly affects South America, both on the Atlantic and the Pacific coasts, causing climatic anomalies. The intensity of this phenomenon can be measured using various methods. In recent years, the most widely used is the Southern Oscillation Index (SOI), which is calculated based on the observed sea-level atmospheric pressure differences between the island of Tahiti and the city of Darwin, in Australia (Trenberth 1997; Trenberth and Stepaniak 2001; Trenberth and Fasullo 2013; Johnson 2013; Meng et al. 2018).

Another lesser known, but very much studied in recent years, example of a pattern of climate variability is the Antarctic Oscillation (AAO) or Southern Annular Mode (SAM). This phenomenon refers to a large-scale alternation of atmospheric mass between the middle and high latitude surface pressure. The effects of this phenomenon range from different aspects of the climate of the southern hemisphere, including temperatures and rainfall (Silvestri and Vera 2003; Bettolli et al. 2012), to the abundance of phytoplankton and disturbances in biogeochemical cycles by altering ocean circulation patterns (Lovenduski and Gruber 2005; Butler et al. 2007). An objective index that measures the intensity of the Antarctic Oscillation is defined as the difference of zonal mean sea-level atmospheric pressure between 40°S and 65°S (Gong and Wang 1999).

In order to evaluate the existence of a possible relationship between an environmental factor that acts on a hemispheric scale and the massive and simultaneous flowering of various species of woody bamboos, a preliminary comparison of the occurrence of these flowering events to the Southern Oscillation Index and the Antarctic Oscillation Index was carried out. In the first case, the data was obtained from the National Climate Center of the Australian Bureau of Meteorology (available at http://www.bom.gov.au/climate/current/soihtm1.shtml). In the case of the Antarctic Oscillation Index, data was obtained from the British Antarctic Survey (available at http://www.nerc-bas.ac.uk/icd/gjma/sam.html, calculated according to Marshall (2003)).

After a preliminary analysis, the occurrence of negative values of the Antarctic Oscillation Index preceded the massive and simultaneous flowering of five different species of woody bamboo in two occasions, in 1972 and 2002, but only in 1972 this variable reached an extreme value (below percentile 5; Guerreiro 2013). Of course, these results are not conclusive, given that the historical and the climate data sets are fragmentary. However, they may be considered as hints pointing to a situation that should be followed in time and taken in consideration in future studies. Nevertheless, caution is strongly advised since it is relatively easy to find a correlation between a climatic variable and a phenological response, but this does not show that the climatic variable is the ultimate cause of the observed response. In other words, correlation does not imply causality.

Anecdotally, other possible environmental "causes" of bamboo flowering that have been proposed, but have not reached great adhesion, are sunspots cycles, wildfire cycles, and depletion of soil nutrients. In the first case, Suessenguth (1925: cited in Kawamura 1927) suggested the existence of a relationship between flowering and the increase of sunspots, and Gunckel (1948) stated that "the sun is the quintessential cause of all the phenomena that occur in nature with astonishing regularity." Keeley and Bond (1999) hypothesized that massive death of bamboos after seeding produces a widespread fuel load that greatly increases the risk for potential wildfires. This strategy enhances the availability of resources for seedling recruitment and resumes the successional cycle to support persistence of the new generation. However, there is no evidence to support neither of these hypotheses. On the other hand, the hypothesis of lack of nutrients in the soil as a flowering trigger could not be applied to species that show gregarious flowering. It is practically impossible for all the individuals occurring in an area of hundreds of km^2 to exhaust simultaneously the resources of the soil in which they are growing.

15.4 Genetic Factors

In literature, there are multiple reports of bamboo specimens collected from a natural stand and taken to different parts of the world, which flower in synchrony with the parental population as well as individuals occurring in other distant places. For instance, specimens of *C. abietifolia* that were taken from Jamaica to the Royal Botanic Gardens in Kew, England, in 1883 flowered and died in 1884–1885 simultaneously with their former companions in the mountains of the Caribbean island (Morris 1886; Bean 1907; Seifriz 1923, 1950; Tucker 1988).

Guerreiro et al. (2020) tell a similar tale to describe the flowering of specimens of *G. trinii* cultivated at the Lucien Hauman Botanical Garden in Argentina. Nearly 100 years ago, after a gregarious flowering episode, seedlings were brought to the Botanical Garden. Since then, it has perpetuated itself by flowering and dying cyclically in a 30–33-year period, displaying a quite striking phenomenon in bamboo nature unfolding before us.

Other examples are those of *Pseudosasa japonica* (Siebold & Zucc. ex Steud.) Makino ex Nakai, originally from Japan and cultivated in Argentina, which flowered simultaneously in Africa, Europe, and Japan (Rúgolo de Agrasar 1991); *Fargesia murielae* (Gamble) T.P. Yi and *F. nitida* (Mitford) Keng f. ex T.P.Yi, both native to temperate Asia, flowered in the early 1990s in China, Europe, and North America (Gielis et al. 1999; Ramanayake 2006; Saarela 2007).

Apparently, at least in these cases, synchronous flowering occurs independently of the size of the plant, soil fertility or moisture, sun exposure, climatic factors, etc. These events of synchronous flowering in various environments would indicate that flowering is not caused by environmental factors but would be determined by some type of "internal clock," typical of each species of bamboo (Kawamura 1927; Janzen 1976; Simmonds 1980). To explain the fascinating phenomenon of periodic and gregarious flowering, some authors have proposed the existence of an endogenous mechanism of genetic bases, relatively immune to environmental influences, which would be related to the passage of time, regardless of the history, the state of the plant, or its environment. That is, after a certain time, the plant goes into a particular developmental stage where it is capable of flowering (Kawamura 1927; Janzen 1976; Simmonds 1980). This internal clock would determine when to flower, for example, by beginning hormone synthesis when a specific minimum value of a certain variable is attained.

Flowering transition is controlled by a large number of genes. A number of bamboo genes homologous to those of *Oryza sativa* L. and *Brachypodium distachyon* (L.) P. Beauv., implicated in well-known flowering pathways, have been identified in different species of *Bambusa* Schreb., *Dendrocalamus* Nees, and *Phyllostachys* Siebold and Zucc. (Biswas et al. 2016). Also, Wysocki et al. (2016) identified expressed MADS-box genes in several bamboo species. The function of MADS-box genes in floral organ development has been extensively studied using model species. In addition, the expression of some MADS-box genes can affect the timing of flowering (Shih et al. 2014; Ge et al. 2017).

Flowering after a certain time and regularly cyclical reproductive episodes of many species point to a time-related gene activity (Ramanayake 2006). However, the molecular mechanisms responsible for age-related flowering have not been discovered yet. Research on several perennial plant species suggests a major role of microRNAs in remaining in a vegetative stage for a long time (Huijser and Schmid 2011). There is growing evidence that microRNAs manage the ability to flower, since their amount depends on shoot age (Hyun et al. 2017). Regarding this, six microRNAs may play fundamental controlling roles in flower development and floral transition in *Phyllostachys edulis* (Carrière) J. Houz. (Ge et al. 2017). These studies generated partial yet important evidence of gene regulation. Overall description of genes implicated in a specific flowering pathway in bamboos is yet to be accomplished. Many other further elements may govern flowering timing in the end (Dutta et al. 2018).

15.5 Both?

Whether an internal clock or the influence of the environment, there is not enough experimental data so far. And the existing evidence for both positions remains, as Simmonds (1980) noted, "anecdotal and scrappy" but nevertheless persuasive. On the other hand, these theories may not be opposite but complementary (Zheng et al.

2020). In many plant species, it has been found that physiological stress caused by environmental disturbances (such as drought, excess rainfall, depletion of nutrients, fire, physical damage, etc.) can trigger the internal flowering mechanism (Ramanayake 2006). Almost a century ago, Seifriz (1923) raised the idea that "when bamboo is close to the moment of its reproduction, an unusually dry (or wet) season can accelerate flower bud formation." Also, the existence of a number of genes involved in integrating endogenous and environmental signals to induce flowering in some plant species has been shown (Alam 2008; Hyun et al. 2017).

Likely, the minimum amount of physiological stress necessary to stimulate the flowering of a bamboo declined as the "programmed" moment of reproducing approaches. So, if any environmental disturbance occurred shortly before the "programmed" moment, it could affect flowering time. This may be because the internal clock could be "counting" environmental regularities (e.g., annual changes in temperature), and therefore, they are susceptible to "miscalculations" if climatic irregularities or disturbances take place (Franklin 2004). A certain stage of development may be triggered when the internal clock reaches a particular biological or chemical threshold or state, and the progression rate toward this threshold may depend on external, metabolic, or chemical factors (Rensing et al. 2001).

In a range of evolutionary and environmental contexts, natural selection could favor the interactions between exogenous and endogenous signals and genetic predetermination of development (Sakai et al. 2006). In conclusion, reaching sexual maturity is innate in any species of bamboo but the time when sexual maturity is reached may be sensitive to external influences. One thing's for sure, though. Finding the ultimate cause of bamboo flowering will require a few generations of scientists (Zimmer 2015).

15.6 Consequences of Bamboo Flowering

Bamboos are the only great lineage of the Poaceae family to occur in temperate and tropical forests. Nowadays, after a long process of diversification and adaptation, there are more than a thousand bamboo species worldwide, with different habits and morphology. Some species of bamboo may be part of the dominant vegetation in the ecosystem in which they inhabit and therefore play a fundamental biological role (Judziewicz et al. 1999; Lima et al. 2012). The integrated response of the ecosystem to such events is not adequately known due to the infrequent and diverse nature of the flowering events and the multiple and complex interactions among ecosystem processes (Montti et al. 2011).

In any place where bamboos are a major part of the understory (Fig. 15.4), their flowering and death have an important impact on plant community structure and dynamics (Holz and Veblen 2006; Campanello et al. 2007; Caccia et al. 2009). Furthermore, many studies have shown that bamboo flowering and death result in substantial changes in environmental conditions such as the increase in understory

Fig. 15.4 Typical Andean Patagonian forest with *Chusquea*-dominated understory

light availability, changes in the daily pattern of temperature and thermal amplitude of the understory, and altered biogeochemical cycles in forest ecosystems. These changes in resource availability enhanced canopy tree germination and regeneration but also promote the growth of other understory plants (Giordano et al. 2009; Marchesini et al. 2009; Montti et al. 2011; Austin and Marchesini 2012). These studies demonstrated that this infrequent event is an important component controlling forest regeneration in diverse ecosystems. However, vegetation response to these extraordinary events depends on environmental characteristics and also on the flowering patterns of the species involved and its synchronicity (Holz and Veblen 2006).

On the other hand, population dynamics of animal species that obtain food and shelter from bamboos are definitively altered after flowering events. Worldwide, many mammals take advantage of the changing food and habitat resources provided by bamboo species, the best known (and most beautiful) example being that of giant pandas (Nie et al. 2015, 2019). In South America, bamboo-seed specialist birds undergo outstanding variability in their food and habitat sources (Areta et al. 2009, 2013, 2016; Areta and Cockle 2012; Cockle and Areta 2013; Milesi et al. 2017).

In cultures associated with bamboo, flowering is regarded as a bad omen resulting in famine, massive death, and natural disasters (Ramanayake 2006). Reference to this can be found in Indian poetry written more than 5000 years ago (John and Nadgauda 2002). The unpredictable nature of the flowering event of bamboos, and the consequent death, has turned into disaster mainly in southeastern Asia. Land becomes bare, causing landslides in mountainous areas; flowering of economically important species, such as *Dendrocalamus asper* (Schult. and Schult. f.) Backer ex

K. Heyne, causes heavy losses in the bamboo processing industry; flowering of bamboo in home gardens polluted water due to the great number of fallen flowers; dying culms tended to collapse over houses and electric wires and are also prone to wildfires (Janzen 1976; John and Nadgauda 2002; Ramanayake 2006; Zaitinvawra and Kanagaraj 2013).

Particularly, massive flowering and fruiting of a bamboo species produces large quantities of seeds that remain available for the native granivorous fauna, mostly rodents. They relish this extraordinary supply of food increasing their population density very quickly, triggering a worldwide known phenomenon, which in South America is given the name *ratadas* (Gallardo and Mercado 1999; Piudo et al. 2005; Piudo and Monteverde 2016; Milesi et al. 2017). *Ratadas* stand for the very rapid increase in rodent density in a few weeks' periods. Once the seeds are exhausted due to their germination, rodents move to nearby farms and devour crops, grains, and basically whatever they encounter, hence bringing about famine as well (John and Nadgauda 2002). Furthermore, many rodent species involved in *ratadas* are reservoirs of emerging diseases that affect human populations (Jaksic and Lima 2003). For example, in the Andean Patagonian beech forests of southwestern South America, the long-tailed mouse (*Oligoryzomys longicaudatus*) is of special interest for public health since this rodent is the main reservoir of hantavirus, which causes the disease known as hantavirus pulmonary syndrome (HPS) in southern Argentina and Chile (Piudo and Monteverde 2016).

These rodent outbreaks have been historically associated with massive invasion into homes and subsequent calamities of different sorts such as famine, diseases, and even "bad luck" (Gallardo and Mercado 1999; John and Nadgauda 2002; Ramanayake 2006; Zaitinvawra and Kanagaraj 2013). However, in northern South America, there is a saying that goes: "he or she who finds a flowering *Guadua* (native bamboo) on a Friday midnight is destined to be a millionaire. . . ."

Wherever man has come into contact with bamboo, he has found multiple uses for it, from food, biochar, and forage to construction and papermaking. Apart from such practical uses, many bamboo species are cultivated as ornamentals. On the basis of its growth habits and biological characteristics, bamboos constitute a multipurpose economic investment and also have many applications on environmental problems. Bamboo ecological uses on water remediation, soil erosion control, carbon sequestration, and land rehabilitation have been reported around the world (Soderstrom and Calderón 1979; Zhou et al. 2005; Alchouron et al. 2019; Paudyal et al. 2019).

In any place where bamboo is the dominant component of the flora, it plays a fundamental role in the development of human societies, meeting economic, ecological, and spiritual needs. Bamboo has become a resource that stimulates creativity, nourishes the spirit, restores landscapes and provides economic benefits. In South America, bamboo is not an important asset in forestry activity. Its use is usually restricted to local communities nearby native bamboo forests. However, it could represent a greater economic value if its cultivation was promoted and its applications were widely communicated.

Bamboo flowering research is hampered by the sporadic, infrequent, usually unpredictable nature of the phenomenon itself. However, more careful observation reveals that it is not a whimsical event but that even these phenomena so exceptional, in some cases, are subject to regular cycles. This peculiar behavior has fascinated people for centuries and still persists as one of the greatest botanical mysteries. This review provided the state-of-the-art of bamboo flowering knowledge in South America. There was no attempt here to explain bamboo flowering process, or discover the ultimate cause of these massive events, but simply to explore data in order to find possible patterns. Since the first understanding of a pattern in nature and its description provides the inferences that guide theoretical development and lead subsequent experimental work (Werner 1998), this review is intended to be a contribution to the development of a general framework of knowledge that will eventually allow proposing empirical, inductive hypotheses, which may be put to the test in the future.

Author's Contribution

The authors contributed equally to this work.

Funding Sources This work received financial support through grants PICT 2015-0727 and UBACyT 20020160100061BA.

Conflict of Interest The authors have no conflict of interest to declare.

References

Alam MK (2008) Recent flowering of Muli Bamboo (*Melocanna baccifera*) in Chittagong hill tracts and rat infestation: eco-environmental aspects. International Resources Group, USAID

Alchouron J, Navarathna C, Chludil HD, Dewage NB, Perez F, Hassan EB, Pittman CU Jr, Vega AS, Mlsna TE (2019) Assessing South American *Guadua chacoensis* bamboo biochar and Fe_3O_4 nanoparticle dispersed analogues for aqueous arsenic (V) remediation. Sci Total Environ 706:135943. https://doi.org/10.1016/j.scitotenv.2019.135943

Areta JI, Bodrati A, Cockle K (2009) Specialization on *Guadua* bamboo seeds by three bird species in the Atlantic forest of Argentina. Biotropica 41(1):66–73

Areta JI, Bodrati A, Cockle K (2016) Un panorama de las aves especialistas en bambúes de la Argentina. In: Rúgolo ZE (ed) Bambúes leñosos nativos y exóticos de la Argentina. Trama S.A, Buenos Aires, pp 221–236

Areta JI, Bodrati A, Silva GT, Rupp AE, Velazquez M, Holzmann I, Carrano E, Laps RR, Zimmerman CE (2013) Natural history, distribution and conservation of two nomadic *Sporophila* seedeaters specializing on bamboo in the Atlantic Forest. Condor 115:237–252

Areta JI, Cockle K (2012) A theoretical framework for understanding the ecology and conservation of bamboo-specialist birds. J Ornithol 153:163–170

Austin AT, Marchesini VA (2012) Gregarious flowering and death of understory bamboo slow litter decomposition and nitrogen turnover in a southern temperate forest in Patagonia, Argentina. Funct Ecol 26(1):265–273. https://doi.org/10.1111/j.1365-2435.2011.01910.x

Bean WJ (1907) The flowering of cultivated bamboos. Bull Misc Inform 6:228–233

Bettolli ML, Krieger PA, Penalba OC (2012) Exploración de predictores climáticos de la circulación diaria de verano en el sur de Sudamérica. Paper presented at the XI Congreso Argentino de Meteorología, Mendoza, Argentina, 28 May–1 June 2012

Biswas P, Chakraborty S, Dutta S, Pal A, Das M (2016) Bamboo flowering from the perspective of comparative genomics and transcriptomics. Front Plant Sci 7:1900. https://doi.org/10.3389/fpls.2016.01900

BPG (2012) An updated tribal and subtribal classification of the bamboos (Poaceae: Bambusoideae). Bamboo Sci Cult: J Am Bamboo Soc 25(1):1–10

Butler AH, Thompson DWJ, Gurney KR (2007) Observed relationships between the southern annular mode and atmospheric carbon dioxide. Global Biogeochem Cycles 21:GB4014. https://doi.org/10.1029/2006GB002796

Caccia FD, Chaneton EJ, Kitzberger T (2009) Direct and indirect effects of understory bamboo shape tree regeneration niches in a mixed temperate forest. Oecologia 161(4):771–780

Campanello PI, Gatti MG, Ares A, Montti L, Goldstein G (2007) Tree regeneration and microclimate in a liana and bamboo-dominated semideciduous Atlantic forest. For Ecol Manag 252 (1–3):108–117

Campbell JJN (1985) Bamboo flowering patterns: a global view with special reference to East Asia. Bamboo Sci Cult: J Am Bamboo Soc 6:17–35

Carvalho AL, Nelson BW, Bianchini MC, Plagnol D, Kuplich TM, Daly DC (2013) Bamboo-dominated forests of the Southwest Amazon: detection, spatial extent, life cycle length and flowering waves. PLoS One 8(1):e54852. https://doi.org/10.1371/journal.pone.0054852

Clark LG (1989) Systematics of *Chusquea* section *Swallenochloa*, section *Verticillatae*, section *Serpentes* and section *Longifoliae* (Poaceae, Bambusoideae). Syst Bot Monogr 27:1–127

Clark LG (2004) New species of *Aulonemia* and *Chusquea* (Poaceae: Bambusoideae: Bambuseae) from southeastern Brazil. Revista Brasil Bot 27(1):31–36

Clark LG, Ely F (2013) Two new species of *Chusquea* (Poaceae: Bambusoideae: Bambuseae) from Venezuela and a redescription of *Chusquea purdieana*. Syst Bot 38(4):1087–1095. https://doi.org/10.1600/036364413X674922

Clark LG, Londoño X, Ruiz-Sanchez E (2015) Bamboo taxonomy and habitat. In: Liese W, Köhl M (eds) Bamboo. The plant and its uses. Springer, Switzerland, pp 1–30

Clark LG, Londoño X, Tyrrell CD, Judziewicz EJ (2020) Convergence strikes again in the Neotropical woody bamboos (Poaceae: Bambusoideae: Bambuseae): a new Andean genus and a new species. Bot J Linn Soc 192(1):21–33. https://doi.org/10.1093/botlinnean/boz047

Cockle K, Areta JI (2013) Specialization on bamboo by Neotropical birds. Condor 115:217–220

Crone EE, McIntire EJB, Brodie J (2011) What defines mast seeding? Spatio-temporal patterns of cone production by whitebark pine. J Ecol 99(2):438–444

Davidse G, Huber O (1979) Notes on the flowering and life history of *Neurolepis pittieri* (Gramineae, Bambusoideae). Ann Missouri Bot Gard 66:900–902

Dutta S, Biswas P, Chakraborty S, Mitra D, Pal A, Das M (2018) Identification, characterization and gene expression analyses of important flowering genes related to photoperiodic pathway in bamboo. BMC Genomics 19:190. https://doi.org/10.1186/s12864-018-4571-7

Ely F, González JV (2020) Ciclo de floración de *Chusquea mollis* L.G. Clark (Poaceae, Bambusoideae, Bambuseae, Chusqueinae) en los Andes venezolanos. Ciencia 22:78–90. https://doi.org/10.24133/ciencia.v22i1.1411

Fadrique B, Pianissola EM, Feeley KJ, Clark LG (2019) A preliminary revision of *Chusquea* sect. *Swallenochloa* (Bambuseae, Bambusoideae, Poaceae) in Peru including the description of two new species and the resurrection of two other species. Phytotaxa 418(2):171–194

Filgueiras TS, Pereira BAS (1988) On the flowering of *Actinocladum verticillatum* (Gramineae: Bambusoideae). Biotropica 20(2):164–166

Forrest J, Miller-Rushing AJ (2010) Towards a synthetic understanding of the role of phenology in ecology and evolution. Philos Trans R Soc B – Biol Sci 365:3101–3112

Franklin DC (2004) Synchrony and asynchrony: observations and hypotheses for the flowering wave in a long lived semelparous bamboo. J Biogeogr 31:773–786

Gadgil M, Prasad SN (1984) Ecological determinants of life history evolution of two Indian bamboo species. Biotropica 16:161–172

Gallardo MH, Mercado CL (1999) Mast seeding of bamboo shrubs and mouse outbreaks in southern Chile. Mastozool Neotrop 6(2):103–111

Ge W, Zhang Y, Cheng Z, Hou D, Li X, Gao J (2017) Main regulatory pathways, key genes and microRNAs involved in flower formation and development of moso bamboo (*Phyllostachys edulis*). Plant Biotechnol J 15:82–96. https://doi.org/10.1111/pbi.12593

Gielis J, Goetghebeur P, Debergh P (1999) Physiological aspects and experimental reversion of flowering in *Fargesia murieliae* (Poaceae, Bambusoideae). Syst Geogr Plants 68:147–158

Giordano CV, Sánchez RA, Austin AT (2009) Gregarious bamboo flowering opens a window of opportunity for regeneration in a temperate forest of Patagonia. New Phytol 181:880–889

Gong D, Wang S (1999) Definition of Antarctic oscillation index. Geophys Res Lett 26(4):459–462

Guerreiro C (2013) Análisis fenológico de los bambúes leñosos (Poaceae, Bambusoideae, Bambuseae) nativos y exóticos de América austral, con la aplicación de estudios ecológicos, sistemáticos, morfológicos y anatómicos. Dissertation, Universidad de Buenos Aires

Guerreiro C (2014) Flowering cycles of woody bamboos native to southern South America. J Plant Res 127:307–313

Guerreiro C (2016) Fenología. In: Rúgolo ZE (ed) Bambúes leñosos nativos y exóticos de la Argentina. Trama S.A, Buenos Aires, pp 167–178

Guerreiro C, Lizarazu MA (2010) Flowering of *Bambusa tuldoides* (Poaceae, Bambusoideae, Bambuseae) in southern South America. Darwiniana 48(1):25–31

Guerreiro C, Rúgolo Z (2020) Un evento de floración en el género Chusquea (Poaceae, Bambusoideae, Bambuseae). Darwiniana nueva serie 8(2):576–582

Guerreiro C, Peichoto MC, Vega AS (2020) Flowering of *Guadua trinii* (Poaceae, Bambusoideae, Bambuseae). Rodriguésia 71:e03352018. https://doi.org/10.1590/2175-7860202071033

Gunckel H (1948) La floración de la quila y del colihue en la Araucania. Ciencia e Invest 4:91–95

Holz A, Veblen TT (2006) Tree regeneration responses to *Chusquea montana* bamboo die-off in a subalpine *Nothofagus* forest in the southern Andes. J Veg Sci 17(1):19–28

Huijser P, Schmid M (2011) The control of developmental phase transitions in plants. Development 138:4117–4129

Hyun Y, Richter R, Coupland G (2017) Competence to flower: age-controlled sensitivity to environmental cues. Plant Physiol 173:36–46. https://doi.org/10.1104/pp.16.01523

Ims RA (1990) The ecology and evolution of reproductive synchrony. Trends Ecol Evol 5:135–140

Jaksic FM, Lima M (2003) Myths and facts on ratadas: bamboo blooms, rainfall peaks and rodent outbreaks in South America. Austral Ecol 28:237–251

Janzen DH (1976) Why bamboos wait so long to flower. Annu Rev Ecol Syst 7:347–391

John CK, Nadgauda RS (2002) Bamboo flowering and famine. Curr Sci 82(3):261–262

Johnson NC (2013) How many ENSO flavors can we distinguish? J Clim 26:4816–4827. https://doi.org/10.1175/JCLI-D-12-00649.1

Jones DL (1993) Cycads of the world. Smithsonian Institution Press, Washington D.C

Judziewicz EJ, Clark LG, Londoño X, Stern MJ (1999) American bamboos. Smithsonian Institution Press, Washington D.C

Kawamura S (1927) On the periodical flowering of the bamboo. Jpn J Bot 3:335–342

Keeley JE, Bond WJ (1999) Mast flowering and semelparity in bamboos: the bamboo fire cycle hypothesis. Am Nat 154(3):383–391

Kennard WC (1955) Flowering of the bamboo *Guadua amplexifolia* Presl in Puerto Rico. Lloydia 18(4):193–196

Kitajima K, Augspurger CK (1989) Seed and seedling ecology of a monocarpic tropical tree, *Tachigalia versicolor*. Ecology 70(4):1102–1114

Koenig WD, Ashley MV (2003) Is pollen limited? The answer is blowing in the wind. Trends Ecol Evol 18:157–159

Lavoie C, Lanchance D (2006) A new herbarium-based method for reconstructing the phenology of plant species across large areas. Am J Bot 93(4):512–516

Liebsch D, Reginato M (2009) Florescimento e frutificação de *Merostachys skvortzovii* Sendulsky (taquara-lixa) no estado do Paraná. Iheringia Sér Bot 64(1):53–56

Lima RAF, Rother DC, Muler AE, Lepsch IF, Rodrigues RR (2012) Bamboo overabundance alters forest structure and dynamics in the Atlantic Forest hotspot. Biol Conserv 147:32–39

Lizarazu MA, Rúgolo de Agrasar ZE, Vega AS (2012) Nueva cita de *Guadua tagoara* (Poaceae, Bambusoideae, Bambuseae) en la Argentina. Darwiniana 50(1):162–169

Lovenduski NS, Gruber N (2005) Impact of the southern annular mode on Southern Ocean circulation and biology. Geophys Res Lett 32:L11603. https://doi.org/10.1029/2005GL022727

Marchesini VA, Sala OE, Austin AT (2009) Ecological consequences of a massive flowering event of bamboo (*Chusquea culeou*) in a temperate forest of Patagonia, Argentina. J Veg Sci 20:424–432

Marshall GJ (2003) Trends in the southern annular mode from observations and reanalyses. J Climatol 16:4134–4143

Matthei O (1997) Las especies del género *Chusquea* Kunth (Poaceae: Bambusoideae), que crecen en la X Región, Chile. Gay Bot 54(2):199–220

McClure FA (1996) The bamboos. Smithsonian Institution Press, Washington, D.C.

Meng J, Fan J, Ashkenazy Y, Bunde A, Havlin S (2018) Forecasting the magnitude and onset of El Niño based on climate network. New J Phys 20:043036. https://doi.org/10.1088/1367-2630/aabb25

Milesi FA, Guichón ML, Monteverde MJ, Piudo L, Sanguinetti J (2017) Ecological consequences of an unusual simultaneous masting of *Araucaria araucana* and *Chusquea culeou* in North-West Patagonia, Argentina. Austral Ecol 42(6):711–722. https://doi.org/10.1111/aec.12489

Montti L, Campanello PI, Goldstein G (2011) Flowering, die-back and recovery of a semelparous woody bamboo in the Atlantic Forest. Acta Oecol 37:361–368. https://doi.org/10.1016/j.actao.2011.04.004

Montti L, Villagra M, Campanello PI, Gatti MG, Goldstein G (2014) Functional traits enhance invasiveness of bamboos over co-occurring tree saplings in the semideciduous Atlantic Forest. Acta Oecol 54(1):36–44. https://doi.org/10.1016/j.actao.2013.03.004

Moreira F, Viana PL, Clark LG (2008) Flowering in *Chusquea riosaltensis* (Poaceae: Bambusoideae: Bambuseae: Chusqueinae). Bamboo Sci Cult: J Am Bamboo Soc 21(1):48–52

Morris D (1886) Chusquea abietifolia. Gard Chron 669(26):524

Nelson BW (1994) Natural forest disturbance and change in the Brazilian Amazon. Remote Sens Rev 10:105–125

Nie Y, Speakman JR, Wu Q, Zhang C, Hu Y, Xia M, Yan L, Hambly C, Wang L, Wei W, Zhang J, Wei F (2015) Exceptionally low daily energy expenditure in the bamboo-eating giant panda. Science 349(6244):171–174

Nie Y, Wei F, Zhou W, Hu Y, Senior AM, Wu Q, Yan L, Raubenheimer D (2019) Giant pandas are macronutritional carnivores. Curr Biol 29:1677–1682. https://doi.org/10.1016/j.cub.2019.03.067

Parodi LR (1955) La floración de tacuara brava (*Guadua trinii*). Revista Argent Agron 22 (3):134–136

Paudyal K, Adhikari S, Sharma S, Samsudin YB, Paudyal BR, Bhandari A, Birhane E, Darcha G, Trinh TL, Baral H (2019) Framework for assessing ecosystem services from bamboo forests: Lessons from Asia and Africa. Working Paper 255. CIFOR, Bogor. doi:https://doi.org/10.17528/cifor/007433

Pianissola EM, Parma DF, Santos-Gonçalves AP, Clark LG (2018) Two new species of *Chusquea* subg. *Swallenochloa* (Poaceae: Bambusoideae: Bambuseae) from Minas Gerais, Brazil, and complete description of *C. caparaoensis*. Phytotaxa 358(3):235–250

Piudo L, Monteverde M (2016) Efectos ecoepidemiológicos de la floración masiva de la caña colihue. In: Rúgolo ZE (ed) Bambúes leñosos nativos y exóticos de la Argentina. Trama S.A, Buenos Aires, pp 179–188

Piudo L, Monteverde M, González Capria S, Padula P, Carmanchahi P (2005) Distribution and abundance of sigmodontine rodents in relation to hantavirus in Neuquén, Argentina. J Vector Ecol 30:119–125

Pohl RW (1991) Blooming history of the Costa Rican bamboos. Revista Biol Trop 39(1):111–124

Ramanayake SMSD (2006) Flowering in bamboo: An enigma! Ceylon J Sci (Biol Sci) 35 (2):95–105

Rensing L, Meyer-Grahle U, Ruoff P (2001) Biological timing and the clock metaphor: oscillatory and hourglass mechanisms. Chronobiol Int 18(3):329–369

Rúgolo de Agrasar ZE (1991) La floración de *Arundinaria japonica* (Gramineae - Bambusoideae). Biol Soc Argent Bot 27(1–2):49–57

Ruiz-Sanchez E (2013) *Otatea ramirezii* (Poaceae: Bambusoideae: Bambuseae) flower description and the importance of the Mexican national living bamboo collection. Phytotaxa 150(1):54–60

Ruiz-Sanchez E, Sosa V, Mejía-Saules MT, Londoño X, Clark LG (2011) A taxonomic revision of *Otatea* (Poaceae: Bambusoideae: Bambuseae) including four new species. Syst Bot 36 (2):314–336. https://doi.org/10.1600/036364411X56M516

Ruiz-Sanchez E, Tyrrell CD, Ortiz-Rodriguez AE, Sosa V, Gómez-Domínguez H (2019) A new species of *Rhipidocladum* (Poaceae: Bambusoideae: Arthrostylidiinae) from Mexico. Phytotaxa 420(4):255–263

Saarela JM (2007) The north American flowering of the cultivated fountain bamboo, *Fargesia nitida* (Poaceae: Bambusoideae), in Vancouver, British Columbia, Canada. Davidsonia 18 (2):43–55

Saarela JM, Burke SV, Wysocki WP, Barrett MD, Clark LG, Craine JM, Peterson PM, Soreng RJ, Vorontsova MS, Duvall MR (2018) A 250 plastome phylogeny of the grass family (Poaceae): topological support under different data partitions. PeerJ 6:e4299. https://doi.org/10.7717/peerj.4299

Sakai S (2002) General flowering in lowland mixed dipterocarp forests of South-East Asia. Biol J Linn Soc 75:233–247

Sakai S, Harrison RD, Momose K, Kuraji K, Nagamasu H, Yasunari T, Chong L, Nakashizuka T (2006) Irregular droughts trigger mass flowering in asesonal tropical forests in Asia. Am J Bot 93(8):1134–1139

Sanguinetti J, García L (2001) Floración masiva de *Chusquea culeou* en el Parque Nacional Lanín. Eventuales consecuencias ecológicas, su vinculación con las actividades humanas y necesidad de manejo y monitoreo. Informe de la Administración de Parques Nacionales, Argentina

Santos-Gonçalves AP, Carvalho-Okano RM, Filgueiras TS (2012) A new species of *Merostachys* (Poaceae: Bambusoideae) from southeastern Brazil. Syst Bot 37(4):938–940. https://doi.org/10.1600/036364412X656400

Santos-Gonçalves AP, Jesus-Costa C, Filgueiras TS, Clark LG (2018) A revision of *Colanthelia* (Poaceae: Bambusoideae: Bambuseae: Arthrostylidiinae) and new species for the Atlantic forest. Syst Bot 43(4):956–974. https://doi.org/10.1600/036364418X697652

Schmidt R, Longhi-Wagner HM (2009) Flora ilustrada do Rio Grande do Sul. A tribo Bambuseae (Poaceae, Bambusoideae) no Rio Grande do Sul, Brasil. Revista Brasil Biociências 7(1):71–128

Seifriz W (1920) The length of the life cycle of a climbing bamboo. A striking case of sexual periodicity in *Chusquea abietifolia* Griseb. Am J Bot 7(3):83–94

Seifriz W (1923) Observations on the causes of gregarious flowering in plants. Am J Bot 10 (2):93–112

Seifriz W (1950) Gregarious flowering of *Chusquea*. Nature 22:635–636

Sendulsky T (1992) *Merostachys burmanii* (Poaceae: Bambusoideae: Bambuseae), a new species from Brazil. Novon 2:111–113

Sendulsky T (1995) *Merostachys multiramea* (Poaceae: Bambusoideae: Bambuseae) and similar species from Brazil. Novon 5:76–96

Shih M, Chou M, Yue J, Hsu C, Chang W, Ko S, Liao D, Huang Y, Chen JJW, Yuan J, Gu X, Lin C (2014) BeMADS1 is a key to delivery MADSs into nucleus in reproductive tissues-de novo characterization of *Bambusa edulis* transcriptome and study of MADS genes in bamboo floral development. BMC Plant Biol 14:179. https://doi.org/10.1186/1471-2229-14-179

Silvestri GE, Vera CS (2003) Antarctic oscillation signal on precipitation anomalies over south-eastern South America. Geophys Res Lett 30(21):2115. https://doi.org/10.1029/2003GL018277

Simmonds NW (1980) Monocarpy, calendars and flowering cycles in angiosperms. Kew Bull 35 (2):235–245

Soderstrom TR, Calderón CE (1979) A commentary on the bamboos. Biotropica 11(3):161–172

Soreng RJ, Peterson PM, Romaschenko K, Davidse G, Teisher JK, Clark LG, Barbera P, Gillespie LJ, Zuloaga FO (2017) A worldwide phylogenetic classification of the Poaceae (Gramineae) II: an update and a comparison of two 2015 classifications. J Syst Evol 55(4):259–290. https://doi.org/10.1111/jse.12262

Stern MJ, Goodell K, Kennard DK (1999) Local distribution of *Chusquea tomentosa* before and after a flowering event. Biotropica 31(2):365–368

Thiers B (2020) Index Herbariorum: A global directory of public herbaria and associated staff, New York Botanical Garden. http://sweetgum.nybg.org/ih. Accessed 20 Sept 2020

Trenberth KE (1997) The definition of El Niño. Bull Am Meteorol Soc 78(12):2771–2777

Trenberth KE, Fasullo JT (2013) An apparent hiatus in global warming? Earth's Future 1:19–32. https://doi.org/10.1002/2013EF000165

Trenberth KE, Stepaniak DP (2001) Indices of El Niño evolution. J Clim 14:1697–1701

Tucker GC (1988) The genera of Bambusoideae (Gramineae) in the southeastern United States. J Arnold Arbor 69:239–273

Tyrrell CD (2008) Systematics of the neotropical woody bamboo genus *Rhipidocladum* (Poaceae: Bambusoideae). Dissertation, Iowa State University

Tyrrell CD, Londoño X, Oviedo Prieto R, Attigala L, McDonald K, Clark LG (2018) Molecular phylogeny and cryptic morphology reveal a new genus of West Indian woody bamboo (Poaceae: Bambusoideae: Bambuseae) hidden by convergent character evolution. Taxon 67 (5):916–930. https://doi.org/10.12705/675.5

Vega AS, Cámara Hernández J (2008) La floración de *Guadua chacoensis* (Poaceae, Bambusoideae, Bambuseae). Rev Fac Agronomía UBA 28(2–3):107–110

Veller C, Nowak MA, Davis CC (2015) Extended flowering intervals of bamboos evolved by discrete multiplication. Ecol Lett 18(7):653–659. https://doi.org/10.1111/ele.12442

Vorontsova MS, Clark LG, Dransfield J, Govaerts R, Baker WJ (2016) World checklist of bamboos and rattans. Int Network Bamboo Rattan, Beijing

Werner EE (1998) Ecological experiments and a research program in community ecology. In: Resetarits WJ, Bernardo J (eds) Experimental ecology. Issues and perspectives. Oxford University Press, New York, pp 5–26

Widmer Y (1997) Life history of some *Chusquea* species in old-growth oak forest in Costa Rica. In: Chapman GP (ed) The bamboos. Academic Press, London, pp 17–31

Widmer Y (1998) Flowering phenology of *Chusquea* bamboos with special reference to *Chusquea talamancensis* in Costa Rica. J Am Bamboo Soc 12(1):1–20

Wysocki WP, Ruiz-Sanchez E, Yin Y, Duvall MR (2016) The floral transcriptomes of four bamboo species (Bambusoideae; Poaceae): support for common ancestry among woody bamboos. BMC Genomics 17:384. https://doi.org/10.1186/s12864-016-2707-1

Zaitinvawra D, Kanagaraj E (2013) Social dynamics of bamboo flowering in Mizoram. Soc Work J 4(1):31–42

Zheng X, Lin S, Fu H, Wan Y, Ding Y (2020) The bamboo flowering cycle sheds light on flowering diversity. Front Plant Sci. https://doi.org/10.3389/fpls.2020.00381

Zhou B, Fu M, Xie J, Yang X, Li Z (2005) Ecological functions of bamboo forest: research and application. J For Res 16(2):143–147

Zimmer C (2015) Bamboo mathematicians. National Geographic. https://www.nationalgeographic.com/science/phenomena/2015/05/15/bamboo-mathematicians. Accessed 10 Sept 2020

Chapter 16
Molecular Markers in Bamboo Genotyping: Prospects for Conservation and Breeding

Lucina Yeasmin and Md. Nasim Ali

Abstract Bamboo (family: Poaceae) is a cash crop, and due to its economic benefit, this gift of nature is considered as "green gold." This widely distributed plant group is facing concern regarding its conservation due to its continually increasing demand. The successful conservation of plant species lies in proper identification and characterization. In the case of bamboo, as the flowering cycle is long, the identification and taxonomical classification is dependent on mainly its vegetative features like culm and culm-sheath characters. Due to its inappropriate flowering cycle as well as widespread polyploidization of the genome, the taxonomy of bamboo is highly unstable. The molecular techniques in taxonomic classification have been employed since its discovery as it is not influenced by external factors. Molecular taxonomy can resolve many discrepancies regarding the classification and identification of genotypes which are long-standing and could not be solved based on phenotypic characters. Molecular descriptors such as hybridization-based marker like restriction fragment length polymorphism (RFLP) or polymerase chain reaction (PCR)-based markers like random amplified polymorphic DNA (RAPD), inter simple sequence repeat (ISSR), simple sequence repeat (SSR), sequence characterized amplified regions (SCARs), and amplified fragment length polymorphism (AFLP) or sequence-based marker like single-nucleotide polymorphism (SNP), diversity array technology (DArT), etc. are used to evaluate the genetic diversity as well as for accurate identification of the bamboo species. The present study will elaborate on the utility of different molecular markers for identification and taxonomic classification in bamboos.

Keywords Bamboo · Molecular breeding · Molecular markers · Molecular taxonomy · Molecular phylogeny

---

L. Yeasmin
Department of Plant and Microbial Technology, Institute of Life Sciences, Bhubaneswar, Odisha, India

Md. Nasim Ali (✉)
Department of Agricultural Biotechnology, Bidhan Chandra KrishiViswavidyalaya, Mohanpur, Nadia, West Bengal, India

Z. Ahmad et al. (eds.), *Biotechnological Advances in Bamboo*,
https://doi.org/10.1007/978-981-16-1310-4_16

16.1 Introduction

Bamboo, an economically important plant species, belongs to the grass family (Poaceae). This diverse plant group constitutes a single subfamily Bambusoideae distributed under 121 genera and 1662 species (Canavan et al. 2017). Subfamily Bambusoideae is further grouped into three major tribes, viz., temperate woody bamboos consisting of 546 species, tropical woody bamboos having 812 species, and herbaceous bamboos consisting of 124 species (Clark et al. 2015). It is distributed globally except Europe which has no native species (Liese and Hamburg 1987); the Asia-Pacific and South America have the maximum species diversity, whereas Africa has the minimum bamboo species diversity (Bystriakova et al. 2003). The highest bamboo-producing country in the world is China which produces 164 million tons of bamboo yearly with a productivity of 30 tons/ha (Lobovikov et al. 2007). Due to its multifarious utility and versatile benefit in day-to-day life of human beings, it is named as "green gold" or in someplace "poor man's timber." It has cultural (Kurz 1876), artistic (McNeely 1995), religious (Skeat 1900), as well as economic value. According to the 2015–2016 National Institution for Transforming India (NITI Ayog) report, 136 million bamboo cultivators in India fetch approximately 1750 USD annual income from bamboo (NITI Policy Paper 2017). Bamboo plays a manifold role in environmental protection such as preventing soil erosion and conserving soil moisture (Christanty et al. 1996, 1997; Mailly et al. 1997). It can produce 30% more oxygen than an equivalent stand of other trees (Prieto et al. 2013). It provides food for many wild animals and thus plays a great role in the forest ecosystem. Research has proven that bamboo is nine times stronger than that of commercial geocell material and thus can be useful in soft ground engineering (Hegde and Sitharam 2014). Many parts of bamboo plants have therapeutic values also, and it has manifold use in Ayurvedic medicine (Das et al. 2012; Nirmala et al. 2018; Ren et al. 2019).

Keeping aside the aforesaid importance of bamboo in human life, it is exploited irrationally for a long time. Due to irrational utilization or exploitation and genetic erosion of bamboo species, collection and preservation of germplasms is a necessary task now (Thomas et al. 1988; Loh et al. 2000; Nilkanta et al. 2017), along with the classification and identification of the species which becomes the need of the hour (Rao and Rao 1995; Bahadur 1979; Soderstrom and Calderon 1979). To protect the species, conservation and utilization of the balance of germplasms, characterization, and identification are a special need (Nayak et al. 2003; Liu et al. 2013; Mei et al. 2014; Desai et al. 2015; Migicovsky et al. 2019). During the recent past, good numbers of articles have been published discriminately using different methods of molecular genotyping in bamboo. In the present article, the aim is to summarize all the information related to molecular genotyping as evident from the published literatures, which ultimately will help in breeding and conservation of the bamboo species across the globe.

16.2 Morphometric Taxonomy in Bamboo

Traditionally classification and identification of bamboo are based on morphological characters. In modern times, technologies like biochemical, anatomical, and physiological features are also included for the identification of bamboo species (Stapleton 1997). The flowering cycle in the bamboo species ranged between 3 and 120 years (Janzen 1976). So, the morphology-based classification of bamboo is mainly dependent on nonreproductive characters like culm and culm-sheath characters. The vegetative features are often influenced by the environment (Wu 1962) and thus less trustworthy. As reported by Shalini et al. (2013), vegetative features that demarcate species may be more delicate and inaccessible for study which is also a great cause of concern regarding bamboo identification.

In the year 1896, Prof. Gamble identified most of the old world bamboos based on flowers and vegetative descriptors. Later on, many workers have suggested different morphological parameters such as culm-sheath characters (Chatterjee and Raizada 1963), young vegetative shoot, and branching pattern (Bennet et al. 1990). Later Bhattacharya et al. (2006) and Das et al. (2007) utilized 32 key morphological parameters for phylogenetics relationship study in bamboo species. The dendrogram generated using key morphological parameters was not in compliance with the species as reported by Gamble (1896). A gregarious flowering bamboo, *Thamnocalamus spathiflorus* subsp. *spathiflorus*, was studied by Bhattacharya et al. (2009). The vegetative and floral morphology illustrated was in conformity with the previous report (Naithani et al. 2003; Clayton et al. 2006). Attigala et al. (2016) studied *Kuruna*, a new temperate woody bamboo genus, and it included seven species distributed in Sri Lanka and South India. This study provides a reorganized report on the morphology of the genus *Bamboo*, a comprehensive description of seven bamboo species, detailed phylogenetics of entire species, and a morphological descriptor useful for their identification. This study includes *Arundinaria wightiana* in *Kuruna* based on its morphology and which is the only prevalent *Kuruna* species in India.

16.3 Limitations of Morphometric Taxonomy

There are several instances of taxonomical discrepancies regarding the classification of bamboo species based on vegetative characters. Soderstrom and Ellis (1987) have transferred one species of *Oxytenanthera* (*Oxytenanthera monadelpha*) to a new genus *Pseudoxytenanthera* based on its vegetative and floral characteristics, but later *Pseudoxytenanthera* was merged with *Oxytenanthera* by Majumdar (1989) due to its similarity with type specimen, *Oxytenanthera abyssinica*. As opined by Sharma (1996), morphological features are not consistent to distinguish the two genera, viz., *Pseudoxytenanthera* and *Oxytenanthera munro*, and as a result, the genus name *Oxytenanthera* persisted. Das et al. (2007) reported vegetative parameters alone are incapable of distinguishing strongly associated species. The cluster pattern of 15 bamboo species was not in accordance with the classification given by Gamble (1896). In this context, a need for a different approach to classification and identification of bamboo was raised. In taxonomical classification, a new concept of integrative taxonomy appeared which is a multidimensional approach and considers multiple lines of evidence related to its development, ecology, and behavior (Dayrat 2005).

For the taxonomic classification of plants, molecular data can support valuable information (Das et al. 2008). The use of DNA markers can solve the limitations of classical taxonomy and the unavailability of reproductive characters. It can be applied at any stage and cannot be influenced by the environment. As reported by Zhu et al. (2014), the sequence-related amplified polymorphism (SRAP) markers are more efficient to differentiate 13 accessions of bamboo in comparison with 22 key morphological characters studied.

16.4 Application of Different Molecular Markers for Bamboo Taxonomy

To overcome the limitations of morphology-based bamboo taxonomy and identification, the inclusion of molecular marker-based technology is the latest solution. There are different types of DNA markers for identification and diversity study. Broadly three types of molecular markers are available: (1) hybridization-based marker (restriction fragment length polymorphism (RFLP)), (2) PCR-based marker (random amplified polymorphic DNA (RAPD), inter simple sequence repeat (ISSR), simple sequence repeat (SSR), amplified fragment length polymorphism (AFLP)), and (3) sequence-based marker (single-nucleotide polymorphism (SNP), diversity array technology (DArT)).

From time to time with the available resources, different bambusiasts have included the molecular markers for bamboo taxonomy. As discussed earlier, Attigala et al. (2016) included *Arundinaria wightiana*, a South Indian woody bamboo under the genus *Kuruna* based on morphology. Because of the similarities between the climatic condition of the Western Ghats of India (original habitat of the species) to Sri Lanka and its similar habitation to other species under the genus, reassignment of *Arundinaria wightiana* to *Kuruna* was supported. But, as opined by the author, this species needs further study based on the molecular analysis for confirmation. This section deals with the application of various molecular markers to explore the bamboo systematics.

16.4.1 Restriction Fragment Length Polymorphism (RFLP)

Friar and Kochert included RFLP in 81 species of bamboo during two consecutive studies in (1991) and (1994). In the first study, they constructed *the PstI* library of random probes from *Phyllostachys nigra* genome and screened 61 accessions of temperate bamboo. In another study, they evaluated genetic variation among 20 *Phyllostachys* species. Their study showed that RFLP can be fruitfully used for the identification of species and species demarcation. A hybrid approach was taken by Sen and Goyal (2014) for a diversity study of 29 accessions of bamboo from.

North Bengal, India. They did the study through PCR-RFLP based on the *trnL-F* region. The *trnL*-F region was amplified using specific primer pair and then digested with *Taq*I, *Alu*I, and *Hinf*I. Their study suggested that PCR- RFLP can be used as a tool for phylogenetics screening. Konzen et al. (2017) employed a modified technology of RFLP, i.e., RAPD-RFLP, to study variation among four genera of bamboos. Their study involved digestion of RAPD products (PCR products with RAPD primers) with three different combinations of restriction enzymes. The result showed significant variation among the genera studied and suggested the usefulness of this alternative technique for diversity study at generic level. The study of RFLP in bamboo is limited to date due to its hazardous and time-consuming nature.

16.4.2 Random Amplified Polymorphic DNA (RAPD)

RAPD is a low-cost, easy, and fast marker (Belaj et al. 2001; Deshwall et al. 2005) utilized in the analysis of the phylogenetic relationship among different species, since its discovery by Williams et al. (1990). It requires a high level of polymorphism among species with less quantity of genomic DNA (Williams et al. 1990). Different research conducted by Ko et al. (1998) exhibited RAPD is a fast and responsive technique for polymorphism. As reported by several other workers, RAPD in plant diversity study has significance (Kapteyn and Simon 2002; Welsh and McClelland 1990). In this method, the genomic DNA is amplified with a random decamer primer which binds to DNA on two different sites in opposite direction and amplified if priming sites are within amplifiable distance. The polymorphisms between two species resulted from differences in sequences on one or both of the primer annealing sites and are represented as the presence or absence of a particular RAPD band. It does not require any prior knowledge of sequence information as a result of which it was a marker of choice to many scientists until the draft genome of bamboo came in 2013 (Peng et al. 2013).

Many bambusiasts have employed RAPD for phylogenetic relationship study and identification of bamboo. Lai and Hsaio (1997) employed 13 RAPD markers to characterize and distribute different clones of the *Phyllostachys pubescens* grown in Taiwan. Out of 170 samples collected around the island, 9 clones were identified and the findings suggested very inadequate genetic variation, and the center of variation happened to be the first region on the island of the successful introduction of the species from China mainland. In the year 2003, Nayak et al. utilized 30 RAPD primers to estimate genetic diversity among 12 bamboo species. Bhattacharya et al. (2006) used 30 decamer RAPD primers from Operon technology for characterization of 17 different geographically different accessions of *B. tulda* in West Bengal. RAPD was also employed to explore phylogenetic relationships among 15 South-eastern China species of *Bambusa* which were placed under several other subgenera (Sun et al. 2006). Their findings suggested that homoplasious floral character-based traditional classifications of woody bamboos require comprehensive evaluation. In the next year, Das et al. (2007) evaluated the phylogenetic relationship among 15 bamboo species from the Botanical Survey of India using morphological and molecular markers. Their study based on molecular parameters (RAPD) was in full compliance with the classical bamboo taxonomist Gamble (1896), though the cluster developed based on qualitative and quantitative morphological traits varied greatly. Ramanayake et al. (2007) estimated genetic diversity among 9 species of bamboo that belong to 4 genera from Sri Lanka by using 41 RAPD primers. Their study showed that RAPD analysis is a potential technique determining genetic diversity as well as solving problematic generic assignment. A gregarious flowering bamboo, *Thamnocalamus spathiflorus* subsp. *Spathiflorus*, was characterized using 30 RAPD primers (Bhattacharya et al. 2009). Lalhruaitluanga and Prasad (2009) estimated genetic diversity among 12 *Melocanna baccifera* accessions from Mizoram by using RAPD markers.

Shalini and his coworkers evaluated the genetic diversity of ten bamboo species using morphological traits along with DNA markers (Shalini et al. 2013). They applied 21 RAPD primers to distinguish the genotypes and found a wide range of genetic variability among the species. Thirty RAPD primers were utilized for the evaluation of phylogenetic relationships among 13 genotypes of Indian bamboo (Desai et al. 2015). Goyal and Sen (2015) studied the phylogenetic relationship using 30 RAPD primers among the 29 bamboo species growing in North Bengal which they have already encountered and studied their distribution pattern (Goyal et al. 2012). Hafzari et al. (2019) evaluated RAPD markers for the identification of five bamboo genera from Indonesia. A total of 25 species from 5 genera were collected for analysis, and their findings suggested RAPD as a useful marker for the diversity study of bamboo. Makmur et al. (2020) very recently conducted a study on eight different types of bamboos from Indonesia. They utilized 20 RAPD primers for genetic diversity evaluation of collected bamboo species. Their result showed RAPD markers successfully estimated genetic diversity among the studied bamboo groups.

16.4.3 Sequence Characterized Amplified Region (SCAR)

SCAR is a fragment of genomic DNA present on a defined genetic locus which is amplified using a specific pair of oligonucleotide primers. Paran and Michelmore

(1993) derived SCAR by sequencing the polymorphic RAPD band of interest and used two ends of the amplified products as primers. SCARs have advantages over RAPD primers as they can amplify a single locus and developed into a codominant marker (Das et al. 2005). SCAR marker with all its advantage is useful for the identification of bamboo species and can resolve taxonomic discrepancies, but to date, the number of SCARs is limited in bamboo.

Das et al. (2005) developed SCAR markers specific for two bamboo species. Their work involved initial screening of 30 random decamer primers to identify loci that are species explicit. Bb_{836} for *B. balcooa* were obtained from random primer PW-02, and Bt_{609} for *B. tulda* were obtained from OPA-08. They validated those two markers with a large number of samples collected from different ecogeographical regions of West Bengal. Another work was reported by Rangsiruji et al. (2018), on the development of SCAR markers in *the Dendrocalamus*. They initially screened 50 RAPD primers in 8 different species of *Dendrocalamus*, and 5 primers showed species-specific loci. Finally, they derived five species-specific SCAR markers which showed potential for identification of five species of *Dendrocalamus*.

16.4.4 Inter Simple Sequence Repeat (ISSR)

ISSRs are regions between two microsatellites or simple sequence repeat regions. The primer is designed for a microsatellite sequence along with two to four arbitrary bases, often degenerate nucleotides at 3′ or 5′ end to amplify the ISSR regions (Qian et al. 2001). The length of the primer is usually 16–25 nucleotides, and thus the annealing temperature is higher than RAPD (decamer primer) making the marker higher reproducible than RAPD. Also, the primer designed from microsatellite regions is unique and distributed across the entire genome which amplifies more molecular allele in comparison with RAPD (Saha et al. 2016). ISSR is considered as the choice of marker for bamboo diversity and phylogenetics relationship study as it does not require any prior knowledge of gene sequence information. Many bambusiasts employed ISSR along with RAPD for genetic diversity assessment in bamboo (Lalhruaitluanga and Prasad 2009; Shalini et al. 2013; Desai et al. 2015; Goyal and Sen 2015).

Lin et al. (2009) utilized 16 ISSR primers along with 15 pairs of AFLP primers for genetic similarity assessment among *Phyllostachys pubescens* cultivars. Statistical analysis showed that a significant correlation existed between the two similarity matrixes obtained from two molecular marker systems. Their study also showed these two molecular markers could prominently identify ten cultivars of *P. pubescens*. In the next year, Lin et al. (2010) utilized ISSR markers for the identification of hybrids produced by crossbreeding of *Phyllostachys* species. Mukherjee et al. (2010) evaluated the genetic relationship with 12 ISSR primers among 22 bamboo taxa. The dendrogram and principal coordinate analysis showed conformity with earlier reports with few exceptions. Their findings revealed that species of one genus were clustered with members of another genus, which suggests correct delineation of genus and species based on morphology (both vegetative and reproductive features) and inclusion of molecular data along with morphology for classification. Genetic diversity was evaluated among 12 populations of *Dendrocalamus membranaceus* in Yunnan, China, using 10 ISSR primers (Yang et al. 2012). Their findings revealed that the genetic differentiation among the population is significant and no significant correlation exists between genetic and geographical differences among the population tested. Tian et al. (2012) investigated seven populations of *Dendrocalamus giganteus* using seven inter simple sequence repeat primers for the assessment of genetic diversity as the prologue to an effective breeding program.

Seven ISSR markers were applied to analyze genetic variability among six cultivated bamboo species in the Gujrat region (Chaudhary et al. 2015). Nilkanta et al. (2017) investigated the genetic diversity of *Melocanna baccifera*, an economically important bamboo species of Northeast India, by using five ISSR markers. They have conducted the study in seven populations sampled from five districts of Manipur. ISSR marker analysis revealed high within-population genetic variation but low genetic diversity between populations. The result revealed the urge for preservation and protection of all the natural bamboo population in the Northeast region. Amom et al. (2018) employed 10 ISSR primers to investigate the phylogenetic relationship among 15 different bamboo species. Dendrogram analysis and principal component analysis revealed the genetic relationship of 15 bamboo species was in agreement with traditional classification with minor deviations. Ely et al. (2019) studied ecophysiology and genetic diversity of *Chusquea* bamboo species from Venezuela. Genetic diversity study includes ISSR and RAPD markers. Both the marker system showed genetic variation, but ISSR showed higher genetic diversity than RAPD markers, thus suggesting better marker system for genetic diversity study in bamboo. Rajput et al. (2020) applied ISSR along with start codon targeted (SCoT) marker for clonal identification of the tissue culture raised plantlets of *Bambusa balcooa*. In a study by Oumer et al. (2020) in Ethiopia genetic diversity, population structure and gene flow analysis of lowland bamboo (*Oxytenanthera abyssinica*) was conducted using ISSR markers. Their study successfully evidenced that the genetic diversity of the lowland bamboo is associated with geographic locations. A recent report of Amom et al. (2020) witnessed the potential of molecular markers to study the bamboo systematics wherein ISSR and three other DNA markers were used for five native and economically significant bamboo species. Their study revealed that the cluster resulting from phytochemical analysis coordinated strongly with the dendrogram generated from DNA markers suggesting the possibility of combining molecular and phytochemical approaches for genetic relationship study.

16.4.5 Simple Sequence Repeat (SSR) or Microsatellite

Microsatellites are a repeat sequence of one to six nucleotides and abundantly distributed across the eukaryotic genome (Morgante et al. 2002). Microsatellite or SSR markers are characterized by codominant inheritance, reproducibility, high genome coverage, and random dispersion and with a provision to automation through high-throughput genotyping (Lin et al. 2014). The number of SSR markers in bamboo was limited due to the lack of genome information until 2013. Development of SSR in bamboo was both time- and cost-consuming (Chen et al. 2010b). Nayak and Rout (2005) characterized six microsatellites in *Bambusa arundinacea* which were cross-amplified in other bamboo species. Kaneko et al. (2008) isolated and characterized nine SSR markers from *Bambusa arnhemica*. They have suggested that the SSR markers will be useful for investigation of gene flow, evolution, clump characters, and biogeographic history of endemic *B. arnhemica*. Kitamura et al. (2009) isolated ten polymorphic SSR markers from dwarf bamboo *Sasa cernua* and *Sasa kurilensis* and confirmed their applicability in open-pollinated seeds and leaf samples from the natural population. Later Kitamura and Kawahara (2009) investigated the distribution of clones to determine the genetic nature of sporadic flowering in a flowering patch of *Sasa cernua* using eight microsatellite markers developed in their earlier study. Miyazaki and coworkers in the same year developed eight polymorphic SSR markers from *Sasa senanensis* and investigated cross transferability in other dwarf bamboos. Moreover, the cross transferability of rice SSR markers was tested in bamboo, and 120 SSR markers from rice were evaluated in 21 species of bamboo (Chen et al. 2010b). Out of 120 markers, 82 amplified successfully in bamboo genotype, and SSR markers positioned on rice chromosome 7 and chromosome 1 exhibited, respectively, the highest and the lowest transferability. Their study demonstrated rice SSR can be successfully utilized for diversity study in bamboo. Tang et al. (2010) examined the available public domain sequence database and analyzed 1532 *Phyllostachys pubescens* sequences. They discovered 3241 SSR loci of di- or more nucleotide repeat sequences in 920 genomic and 68 cDNA sequences. They developed a total of 19 microsatellite markers and checked for cross transferability in 6 other *Phyllostachys* species. All the markers were transferred successfully in all six species of *Phyllostachys* and showed high polymorphism. Sixteen novel microsatellite markers were developed in the strongest woody bamboo, *Dendrocalamus sinicus*, using Fast Isolation by AFLP of Sequences COntaining Repeats (FIASCO) protocol (Dong et al. 2012). The SSRs developed were successfully cross-amplified in other *Dendrocalamus* species and thus signified that the markers can be employed in diversity analysis of the *Dendrocalamus*.

Twenty microsatellite markers were developed by Jiang et al. (2013) for *Phyllostachys edulis*, an economically important bamboo species of China. They have tested those 20 markers on 71 samples collected from 3 geographically isolated regions. Each marker produced between two and ten amplicons and will support in future studies on different aspects of *P. edulis* like molecular ecology, conservation, etc. Full-length cDNA (FL-cDNAs) databases give a prosperous resource for developing potential FL-cDNA SSR markers. Lin et al. (2014) screened 10,608 cDNAs of *Phyllostachys pubescens* and discovered 1614 SSRs in 1382 SSR-containing FL-cDNAs. The applicability was confirmed by using the FL-cDNA SSR markers to spot the parental stock in interspecific hybrids of bamboo. Later, Zhao et al. (2015) physically mapped 1098 microsatellites on the Moso bamboo (*Phyllostachys edulis*) genome and validated 917 markers in 9 accessions with ~39.8% polymorphisms. They have implemented a database for bamboo microsatellite (http://www.bamboogdb.org/ssr). The markers developed are valuable for the study of molecular marker-based taxonomy in bamboo. Very recently Rossarolla et al. (2020) have characterized ten polymorphic SSR markers of *Guadua chacoensis* and also successfully cross-amplified in other species of bamboo.

16.4.6 Expressed Sequence Tagged-Simple Sequence Repeat (EST-SSR)

Nowadays online databases are a great source of sequence information. The EST sequence project for the discovery of genes in several plants generated a huge amount of DNA sequence data deposited in online databases (Rudd 2003). The public domain databases can be accessed with some specific computer programs and can be searched for SSR motifs, which are known as genic or EST-SSR microsatellites. The genic SSRs or EST-SSRs are limited to those species or closely related species for which large amounts of ESTs are available and submitted (Varshney et al. 2005).

At the beginning of the twenty-first century, gene sequence data was scarce in bamboo, so EST-SSR in bamboo was unavailable. However, few bambusiasts employed EST-SSR study in bamboo from other cereal crops such as Barkley et al. (2005) and Sharma et al. (2008). Barkley et al. (2005) used 25 EST-SSR markers from maize, wheat, sorghum, and rice and checked transferability in bamboo. Ninety-two accessions of bamboo belonging to 11 genera and 44 species were studied for genetic diversity study. Sharma et al. (2008) utilized 20 EST-SSRs, developed from the sugarcane genome, in assessing genetic distances among 23 bamboo species. Their major findings indicated EST-SSRs from cereal crops can be successfully utilized in bamboo for diversity study.

In the subsequent year, Sharma et al. (2009) mined 329 ESTs of *Bambusa oldhamii* and *Phyllostachys edulis* in public sequence databases and using different computer programs identified 10 successful EST-SSR markers from *B. oldhamii* ESTs. Their cross transferability level was higher and amplified consistently in other species suggesting their usefulness in diversity study as well as in genetic analyses of bamboo species. Dong et al. (2011) searched 3406 publically available ESTs of *Bambusa oldhamii* and *Phyllostachys edulis* and discovered 245 nonredundant SSR markers in 205 EST contigs and developed 15 EST-SSR markers. The transferability of those markers was checked in 14 caespitose bamboo species and 2 markers, viz., BOM01 and BOM02, transferred to almost all the caespitose bamboos producing species-specific alleles which could be utilized for identification of caespitose bamboo interspecies hybrids. Bhandawat et al. (2019) mined 8121 EST-SSR markers from *Dendrocalamus hamiltonii* transcriptome data, and they developed a set of 114 polymorphic markers which are linked with several biogenic factors such as transcription factors, cell cycle regulators, signaling, etc. Genetic diversity and population structure were evaluated among 72 accessions belonging to three populations of *D. hamiltonii*. Cai et al. (2019) identified 18,356 EST-SSR loci from *Phyllostachys violascens* transcriptomic data. A total of 11,264 primer pairs were designed, and a total of 96 primer pairs were selected randomly and synthesized. Out of 96 primers synthesized, 54 were used to study variation among 16 *P. violascens* bamboo and 10 other species of *Phyllostachys*. Their study generated rich EST-SSR markers for genetic diversity study in bamboo.

16.4.7 Amplified Fragment Length Polymorphism (AFLP)

AFLP is a robust, reliable genetic marker discovered by Vos et al. (1995) and shows a significant level of DNA polymorphism. Loh et al. (2000) conducted a study using AFLP markers for genetic variation and relationship study in four bamboo genera under the Bambusinae subtribe. AFLPs discriminated against different species understudy with a unique banding profile. Unique AFLPs were detected in 13 out of the 15 bamboo species studied. To explore the clonal structure of a dense population of dwarf bamboo, *Sasa senanensis* AFLP profiling was used (Suyama et al. 2000). AFLP fingerprinting of 51 *S. senanensis* population from a study plot in Japan suggested that the plot was consisted of at least 22 clones. A comparative study on two contrasting molecular techniques, viz., AFLP and ITS-nrDNA, for phylogenetic relationship assessment of *Phyllostachys* bamboo was conducted (Hodkinson et al. 2000). Twenty-two species of *Phyllostachys* were considered for the study and the 5S spacer region of nrDNA investigated along with two selective AFLP markers. Their result showed that AFLP analysis exhibited a higher degree of discrimination, and the 5S spacer region is unsuitable for this purpose. AFLP was suggested as the better choice of marker for phylogenetics relation study. Marulanda and coworker utilized AFLP markers to describe the association between accessions and biotypes of *Guadua angustifolia* and compared them with other *Guadua* species of Columbia (Marulanda et al. 2002). Fifty-five accessions were studied using three combinations of primer, and a clear genetic difference was observed between the different species of the *Guadua* genus.

Molecular marker-based identification of clonal plants is superior to other techniques, but in this process also there are two limitations: first, wrong identification of genetically similar seedlings as clones, and secondly, wrong identification of two same clones having different fingerprints as genetically different individuals. The problem was addressed by Douhovnikoff and Dodd (2003) in *Salix exigua* using the higher precision of differentiation of AFLP fingerprint and developed the threshold value of Jaccard's similarity index (0.983) for assigning individuals to clones. The result showed the approach was useful in the precise identification of clones. The flowering incident in bamboo is itself an interesting phenomenon as the flowering cycle varies greatly in different species of bamboo. The flowering cycle of *Phyllostachys pubescens* was calculated precisely as 67 years. In a community of *P. pubescens*, the flowering and nonflowering culms were mixed, and flowering episodes lasted as long as 3 years in the population. AFLP analysis by Isagi et al. (2004) confirmed separate stands or genets of *Phyllostachys pubescens* that originated from an earlier flowering incident. AFLP fingerprint technology was utilized by Mathews et al. (2009) for the clonal diversity assessment of *Arundinaria gigantea* in Western North Carolina. Their study helped in the restoration project of *A. gigantea* by identifying the clonal diversity in different stands and cause of culm loss. The genetic structure of *Sasa pubiculmis* was identified, and the flowering pattern along with the seed set was investigated using the AFLP technique (Miyazaki et al. 2009).

Phylogenetic relationship as well as the genetic variability based on AFLP markers among edible bamboos from Northeast India (Ghosh et al. 2011) and landraces of *Dendrocalamus hamiltonii* (Waikhom et al. 2012) were evaluated. In a different study by Lin et al. (2011a, b), the efficiency of the AFLP marker for genetic diversity assessment was established in *Phyllostachys violascens*. ISSR, sequence-related amplified polymorphism (SRAP), and AFLP techniques were used for the evaluation of phylogenetic relationships within different cultivars of *Phyllostachys violascens*. Their findings demonstrated that all three marker system were useful for genetic diversity estimation in *P. violascens*, though AFLP was the most resourceful marker. Waikhom et al. (2012) further tested the four pairs of markers through multiple regression analysis for marker-trait association identification, and a positive correlation was found between AFLP data and biochemical attributes, i.e., antioxidant activity and total cyanide content. Eight AFLP primers along with 42 RAPD primers were utilized for genetic diversity assessment of industrially important red bamboo, *Ochlandra travancorica*, from Kerala, India (Nag et al. 2013). A relatively high amount of polymorphism was observed which could be useful in the selection of elite germplasm for improvement.

16.4.8 DNA Barcoding and Molecular Phylogeny in Bamboo

A short stretch of DNA sequence from a standardized region of the genome which can be utilized uniquely for identification of the species is termed as DNA barcode. The concept of using DNA barcode for species identification was proposed by Hebert et al. (2003) using mitochondrial cytochrome *c* oxidase (*COI*) barcode region in the animal. In the case of plants, the *COI* gene and other mitochondrial regions are not useful barcode region for identification of species due to its low genetic variation and variable structure of the mitochondrial genome (Kress et al. 2005; Chase et al.

2005; Pennisi 2007; Chase et al. 2007; Fazekas et al. 2008). Consortium for the Barcode of Life's (CBOL) Plant Working Group (2009) suggested two locus combinations of *rbcL+ matK* as a potential DNA barcode for plant species identification. The nuclear ribosomal -*ITS*2 region is recommended as a potential tool for the identification of plant taxa (Chen et al. 2010a). Several studies reported different regions of plastid DNA alone or in combination as potential barcode regions in plants, viz., *trnH-psbA* (Kress et al. 2005), *rpoC1* + *rpoB* + *matK* or *rpoC1* + *matK* + *trnH-psbA* (Chase et al. 2007), and *rbcL* + *trnH-psbA* (Kress and Erickson 2007). DNA barcoding is a helpful tool for taxonomic classification and recently gaining preference over classical taxonomy due to its accuracy in identification (Sijimol et al. 2014).

Sequence-based phylogenetic relationship study in bamboo was conducted as early as in (2005) by Qiang et al. They applied nrDNA *ITS* region and cpDNA *trnL-F* intergenic spacer for genetic relationship study of *Arundinaria* and related genera. The sequence-based phylogenetic tree was inconsistent with the morphological character-based tree. In the same year, Sun et al. (2005) employed a nuclear rDNA *ITS* sequence for phylogenetic analysis of *Bambusa*. Their study raised the question about the monophyly of the different subgenera under the *Bambusa* genus. Yang et al. (2008) employed nuclear rDNA *ITS* and GBBSI gene along with a plastid *trnL-F* spacer sequence to assess the phylogenetic relationship and fruit evolutionary analysis. Their study suggested reorientation at subtribe level and fruit characters are not reliable for phylogeny. The study further reveals bacoid caryopsis may represent particular ecological condition-based specialization. Ruiz-Sanchez and Sosa (2010) utilized molecular data along with morphological and ecological data for delimiting species boundaries within the Neotropical bamboo *Otatea*. They employed cpDNA regions *atpF-atpH*, *psbK-psbI*, and *trnL-rpl32* in combination and nrDNA internal transcribed spacer (*ITS*) region. Their result assigned seven species under *Otatea* instead of three as previously described.

Cai et al. (2012) conducted a study for testing four candidate barcode markers, viz., *matK*, *rbcL*, *psbA-trnH*, and *ITS2*, in temperate woody bamboos. The study revealed *rbcL* + *ITS2* as the potential marker combination for the identification of temperate woody bamboos. The core barcode *matK* failed to identify *Bambusa* species due to polyploidization, interspecies hybridization, and introgression (Das et al. 2013). Sosa et al. (2013) worked on 36 species of bamboo and evaluated the efficiency of *rbcL*, *matK*, and *psbI-K* spacer region individually and in combination. Their study revealed that *matK* + *psbI-K* were able to identify the woody bamboos at the generic level. Ghosh et al. (2017) worked on 21 tropical bamboo species and reconstructed phylogeny based on *ITS1* and *ITS2* sequence and their consensus secondary structure. Their findings reveal that the sequence alone could reconstruct the traditional phylogeny, but few inconsistencies were found and they integrated the secondary structure of the *ITS* sequence which helped in the resolution of the tree. Their study suggested a combination of the structure along with sequence for phylogenetic relationship assessment of bamboo. Tyrrell et al. (2018) employed four plastid markers, viz., *ndhF*, *trnC-rpoB*, *trnD-trnT*, and *rps16-trnQ*, for phylogenetic analysis of 31 Neotropical woody bamboos under the genus *Arthrostylidium*. Their molecular phylogenetic study along with leaf anatomy study revealed that three species under the abovementioned genus do not belong to them, but rather they represent a different genus under the subtribe Guaduinae and they erected a new genus *Tibisia*. CBOL-recommended seven standard barcode regions were evaluated for the detection of potential barcode for commercially important bamboo species identification (Dev et al. 2020). Their study suggests *psbA-trnH* DNA barcode region can be utilized to spot the species and identify the planting materials of their bamboo.

16.5 Conclusions

Bamboo classification based on traditional taxonomic analysis is tedious and sometimes misleading. The classification based on floral characters is limited due to long flowering cycles, and that may also be homoplasious. So, for taxonomic classification, bambusiasts depend on vegetative characters which are frequently influenced by the environment. The molecular marker-based taxonomic classification is a step forward for confirmation and solving taxonomic discrepancies as well as for delineation and identification of bamboo species. As evident, various molecular markers have been successfully utilized for the characterization of bamboo species. For solving problematic generic assignment, the utilization of RAPD, ISSR, SSR, and AFLP, as a sole and/or in combination, was very successful in this species. The results, so obtained, may also be very useful for preserving and protecting the natural bamboo populations. Besides, SSR and/or EST-SSRs have the potential for the investigation of clump structure, the evolution of the bamboo flowering signal, models of gene flow, ecology, population structure, the biogeographic history, and conservation of endemic bamboo species. The high level of cross transferability of SSRs and reliable amplification in other species suggested their utility in diversity study as well as in functional and genetic analyses of bamboo species. SCAR marker with all its advantage is useful for the identification of bamboo species and can resolve taxonomic discrepancies. DNA barcoding is a robust and reliable tool for taxonomic classification and recently gaining preference over classical taxonomy due to its accuracy in the identification of candidate barcode markers. Among the different available barcode region, nrDNA *ITS* in combination with cpDNA *rbcL* and *psbA-trnH* spacer proved as a potential barcode for bamboo identification. DNA markers combining with morphological and phytochemical approaches are suggested to use for proper characterization, exploring the genetic property, and studying genetic relationship which can reconstruct the phylogenetic tree with further resolution among bamboo species.

Funding Sources: No funding was received for this publication.

Author Contributions: Ali MN conceptualized the idea of the article and finalized the initial drafts. Yeasmin L reviewed and prepared the initial draft of the manuscript.

Conflict of Interest There is no conflict of interest.

References

Amom T, Tikendra L, Rahaman H, Potshangbam A, Nongdam P (2018) Evaluation of genetic relationship between 15 bamboo species of North-East India based on ISSR marker analysis. Mol Biol Res Commun 7(1):7

Amom T, Tikendra L, Apana N et al (2020) Efficiency of RAPD, ISSR, iPBS, SCoT and phytochemical markers in the genetic relationship study of five native and economical important bamboos of North-East India. Phytochemistry 174:112330

Attigala L, Kathriarachchi HS, Clark LG (2016) Taxonomic revision of the temperate woody bamboo genus Kuruna (Poaceae: Bambusoideae: Arundinarieae). Syst Bot 41(1):174–196

Bahadur KN (1979) Taxonomy of bamboos. Indian J Forest 2(3):222–241

Barkley NA, Newman ML, Wang ML, Hotchkiss MW, Pederson GA (2005) Assessment of the genetic diversity and phylogenetic relationships of a temperate bamboo collection by using transferred EST-SSR markers. Genome 48(4):731–737

Belaj A, Trujillo I, De la Rosa R, Rallo L, Gimenez MJ (2001) Polymorphism and discrimination capacity of randomly amplified polymorphic markers in an olive germplasm bank. J Am Soc Hortic Sci 126(1):64–71

Bennet SSR, Gaur RC, Sharma PN (1990) Thirty seven bamboos growing in India. Forest Research Institute, Dehradun

Bhandawat A, Sharma V, Singh P et al (2019) Discovery and utilization of EST-SSR marker resource for genetic diversity and population structure analyses of a subtropical bamboo, *Dendrocalamus hamiltonii*. Biochem Genet 57(5):652–672

Bhattacharya S, Das M, Bar R, Pal A (2006) Morphological and molecular characterization of *Bambusa tulda* with a note on flowering. Ann Bot 98(3):529–535

Bhattacharya S, Ghosh JS, Das M, Pal A (2009) Morphological and molecular characterization of *Thamnocalamus spathiflorus* subsp. *spathiflorus* at population level. Plant Syst Evol 282 (1–2):13–20

Bystriakova N, Kapos V, Lysenko I, Stapleton C (2003) Distribution and conservation status of forest bamboo biodiversity in the Asia-Pacific region. Biodivers Conserv 12(9):1833–1841

Cai ZM, Zhang YX, Zhang LN, Gao LM, Li DZ (2012) Testing four candidate barcoding markers in temperate woody bamboos (Poaceae: Bambusoideae). J Syst Evol 50(6):527–539

Cai K, Zhu L, Zhang K, Li L et al (2019) Development and characterization of EST-SSR markers from RNA-Seq data in *Phyllostachys violascens*. Front Plant Sci 10:50

Canavan S, Richardson DM, Visser V et al (2017) The global distribution of bamboos: assessing correlates of introduction and invasion. AoB Plants 9(1):plw078

CBOL Group, Hollingsworth PM, Forrest LL, Spouge JL et al (2009) A DNA barcode for land plants. Proc Natl Acad Sci 106(31):12794–12797

Chase MW, Salamin N, Wilkinson M et al (2005) Land plants and DNA barcodes: short-term and long-term goals. Philos Trans R Soc Lond Ser B Biol Sci 360:1889–1895

Chase MW, Cowan RS, Hollingsworth PM et al (2007) A proposal for a standardised protocol to barcode all land plants. Taxon 56:295–299

Chatterjee RN, Raizada MB (1963) Culm sheaths as an aid to identification of bamboos. Indian Forester 89(11):744–756

Chaudhary S, Patel E, Kharadi A et al (2015) Evaluation of genetic variability of bamboo varieties cultivated in Gujarat region using ISSR markers. Indian J Biotech Pharm Res 3(1):1–6

Chen S, Yao H, Han J et al (2010a) Validation of the ITS2 region as a novel DNA barcode for identifying medicinal plant species. PLoS One 5(1):e8613. https://doi.org/10.1371/journal.pone.0008613

Chen SY, Lin YT, Lin CW et al (2010b) Transferability of rice SSR markers to bamboo. Euphytica 175(1):23–33

Christanty L, Mailly D, Kimmins JP (1996) "Without bamboo, the land dies": biomass, litterfall, and soil organic matter dynamics of a Javanese bamboo talun-kebun system. Forest Ecol Manag 87(1–3):75–88

Christanty L, Kimmins JP, Mailly D (1997) "Without bamboo, the land dies": a conceptual model of the biogeochemical role of bamboo in an Indonesian agroforestry system. Forest Ecol Manag 91(1):83–91

Clark LG, Londoño X, Ruiz-Sanchez E (2015) Bamboo taxonomy and habitat. In: Liese W, Kohl M (eds) Bamboo, tropical forestry, vol 10. Springer, Cham, pp 1–30

Clayton WD, Harman KT, Williamson H (2006) Grass Base- the online world grass flora. http://www.kew.org/data/grasses-db.html

Das M, Bhattacharya S, Pal A (2005) Generation and characterization of SCARs by cloning and sequencing of RAPD products: a strategy for species- specific marker development in bamboo. Ann Bot 95(5):835–841

Das M, Bhattacharya S, Basak J, Pal A (2007) Phylogenetic relationships among the bamboo species as revealed by morphological characters and polymorphism analyses. Biol Plant 51 (4):667–672

Das M, Bhattacharya S, Singh P et al (2008) Bamboo taxonomy and diversity in the era of molecular markers. Adv Bot Res 47:225–268

Das S, Md R, Basu SP, Das S (2012) Therapeutic potentials of *Bambusa bambos* Druce. Indo Glob J Pharm Sci 2(1):85–87

Das MM, Mahadani P, Singh R et al (2013) MatK sequence based plant DNA Barcoding failed to identify Bambusa (Family: Poaceae) species from Northeast India. J Environ Sociobiol 10 (1):49–54

Dayrat B (2005) Towards integrative taxonomy. Biol J Linn Soc 85(3):407–417

Desai P, Gajera B, Mankad M et al (2015) Comparative assessment of genetic diversity among Indian bamboo genotypes using RAPD and ISSR markers. Mol Biol Reports 42(8):1265–1273

Deshwall RPS, Singh R, Malik K, Randhawa GJ (2005) Assessment of genetic diversity and genetic relationships among 29 populations of *Azadirachta indica* A Juss. using RAPD markers. Genet Resour Crop Evol 52(3):285–292

Dev SA, Sijimol K, Prathibha PS et al (2020) DNA barcoding as a valuable molecular tool for the certification of planting materials in bamboo. 3 Biotech 10(2):59

Dong WJ, Wu MD, Lin Y et al (2011) Evaluation of 15 caespitose bamboo EST-SSR markers for cross-species/genera transferability and ability to identify interspecies hybrids. Plant Breed 130 (5):596–600

Dong YR, Zhang ZR, Yang HQ (2012) Sixteen novel microsatellite markers developed for *Dendrocalamus sinicus* (Poaceae), the strongest woody bamboo in the world. Am J Bot 99 (9):e347–e349

Douhovnikoff V, Dodd RS (2003) Intra-clonal variation and a similarity threshold for identification of clones: application to *Salix exigua* using AFLP molecular markers. Theor Appl Genet 106 (7):1307–1315

Ely F, Rada F, Fermin G, Clark LG (2019) Ecophysiology and genetic diversity in species of the bamboo Chusquea in the high Andes, Venezuela. Plant Ecol Divers 12(6):555–572

Fazekas AJ, Burgess KS, Kesanakurti PR et al (2008) Multiple multilocus DNA barcodes from the plastid genome discriminate plant species equally well. PLoS One 3:e2802

Friar E, Kochert G (1991) Bamboo germplasm screening with nuclear restriction fragment length polymorphisms. Theor Appl Genet 82(6):697–703

Friar E, Kochert G (1994) A study of genetic variation and evolution of *Phyllostachys* (Bambusoideae: Poaceae) using nuclear restriction fragment length polymorphisms. Theor Appl Genet 89(2):265–270

Gamble JS (1896) The Bambuseae of British India. Ann R Bot Gard Calcutta 7:1–133

Ghosh S, Devi WS, Mandi S, Talukdar NC (2011) Amplified fragment length polymorphism based study of phylogenetic relationship and genetic variability among some edible bamboo species of north-East India. J Plant Mol Biol Biotechnol 2(2):8–15

Ghosh JS, Bhattacharya S, Pal A (2017) Molecular phylogeny of 21 tropical bamboo species reconstructed by integrating non-coding internal transcribed spacer (ITS1 and 2) sequences and their consensus secondary structure. Genetica 145(3):319–333

Goyal AK, Sen A (2015) Phylogenetic relationships among accessions of bamboos encountered in North Bengal, India based on RAPD and ISSR markers. Indian J Biotechnol 14(4):495–503

Goyal AK, Ghosh PK, Dubey PK, Sen A (2012) Inventorying bamboo biodiversity of North Bengal: a case study. Int J Fund Appl Sci 1(1):5–8

Hafzari R, Setiawati T, Irawan B, Kusmoro J (2019) Evaluation of RAPD markers for molecular identification of five bamboo genera from Indonesia. Folia For Pol 61(4):255–266

Hebert PDN, Cywinska A, Ball SL, deWaard JR (2003) Biological identifications through DNA barcodes. Proc R Soc Biol Sci 270:313–321

Hegde A, Sitharam TG (2014) Use of bamboo in soft-ground engineering and its performance comparison with Geosynthetics: experimental studies. J Mater Civ Eng 27(9):1–9

Hodkinson TR, Renvoize SA, Chonghaile GN et al (2000) A comparison of ITS nuclear rDNA sequence data and AFLP markers for phylogenetic studies in *Phyllostachys* (Bambusoideae, Poaceae). J Plant Res 113(3):259–269

Isagi Y, Shimada K, Kushima H et al (2004) Clonal structure and flowering traits of a bamboo [*Phyllostachys pubescens* (Mazel) Ohwi] stand grown from a simultaneous flowering as revealed by AFLP analysis. Mol Ecol 13(7):2017–2021

Janzen DH (1976) Why bamboos wait so long to flower. Ann Rev Ecol Syst 7(1):347–391

Jiang WX, Zhang WJ, Ding YL (2013) Development of polymorphic microsatellite markers for *Phyllostachys edulis* (Poaceae), an important bamboo species in China. Appl Plant Sci 1 (7):1200012

Kaneko S, Franklin DC, Yamasaki N, Isagi Y (2008) Development of microsatellite markers for *Bambusa arnhemica* (Poaceae: Bambuseae), a bamboo endemic to northern Australia. Conserv Genet 9(5):1311–1313

Kapteyn J, Simon JE (2002) The use of RAPDs for assessment of identity, diversity, and quality of Echinacea. In: Janick J, Whipkey A (eds) Trends in new crops and new uses. ASHS Press, Alexandria, VA, pp 509–513

Kitamura K, Kawahara T (2009) Clonal identification by microsatellite loci in sporadic flowering of a dwarf bamboo species, *Sasa cernua*. J Plant Res 122(3):299–304

Kitamura K, Saitoh T, Matsuo A, Suyama Y (2009) Development of microsatellite markers for the dwarf bamboo species *Sasa cernua* and *Sasa kurilensis* (Poaceae) in northern Japan. Mol Ecol Resour 9(6):1470–1472

Ko MK, Yang J, Jin YH et al (1998) Genetic relationships of *Viola* species evaluated by random amplified polymorphic DNA analysis. J Hort Sci Biotech 73(5):601–605

Konzen ER, Peron R, Ito MA, Brondani GE, Tsai SM (2017) Molecular identification of bamboo genera and species based on RAPD-RFLP markers. Silva Fenn 51(4):1–16. https://doi.org/10.14214/sf.1691

Kress WJ, Erickson DL (2007) A two-locus global DNA barcode for land plants: the coding rbcL gene complements the non-coding trnH-psbA apacer region. PLoS One 2:e508

Kress WJ, Wurdack KJ, Zimmer EA et al (2005) Use of DNA barcodes to identify flowering plants. Proc Natl Acad Sci U S A 102:8369–8374

Kurz S (1876) Bamboo and its use. Indian For 1:219–269

Lai CC, Hsaio JY (1997) Genetic variation of *Phyllostachys pubescens* (Bambusoideae, Poaceae) in Taiwan based on DNA polymorphism. Bot Bull Acad Sinica 38:145–152

Lalhruaitluanga H, Prasad MNV (2009) Comparative results of RAPD and ISSR markers for genetic diversity assessment in *Melocanna baccifera* Roxb. Growing in Mizoram state of India. Afr J Biotechnol 8(22):6053–6062
Liese W, Hamburg FRG (1987) Research on bamboo. Wood Sci Technol 21(3):189–209
Lin XC, Ruan XS, Lou YF et al (2009) Genetic similarity among cultivars of *Phyllostachys pubescens*. Plant Syst Evol 277(1–2):67–73
Lin XC, Lou YF, Liu J et al (2010) Cross breeding of *Phyllostachys* species (Poaceae) and identification of their hybrids using ISSR markers. Genet Mol Res 9(3):1398–1404
Lin X, Lou Y, Zhang Y et al (2011a) Identification of genetic diversity among cultivars of *Phyllostachys violascens* using ISSR, SRAP and AFLP markers. Bot Rev 77(3):223–232
Lin Y, Lu JJ, Wu MD et al (2011b) Identification, cross-taxon transferability and application of full-length cDNA SSR markers in *Phyllostachys pubescens*. Springerplus 3(1):486
Lin Y, Lu JJ, Wu MD et al (2014) Identification, cross-taxon transferability and application of full-length cDNA SSR markers in *Phyllostachys pubescens*. Springerplus 3(1):486
Liu J, Shi S, Chang E et al (2013) Genetic diversity of the critically endangered *Thujasutchuenensis* revealed by ISSR markers and the implications for conservation. Intl J Mol Sci 14 (7):14860–14871
Lobovikov M, Paudel S, Ball L, Piazza M, Guardia M, Ren H, Russo L, Wu J (2007) World bamboo resources: a thematic study prepared in the framework of the global forest resources assessment 2005. Food & Agriculture Org of United Nations, Rome
Loh JP, Kiew R, Set O et al (2000) A study of genetic variation and relationship within the bamboo subtribe Bambusinae usng amplified fragment length polymorphism. Ann Bot 85(5):607–612
Mailly D, Christanty L, Kimmins JP (1997) "Without bamboo, the land dies": nutrient cycling and biogeochemistry of a Javanese bamboo talun-kebun system. For Ecol Manag 91(2–3):155–173
Majumdar R (1989) Bambusoideae. In: Karthikeyan S, Jain SK, Nayar MP, Sanjappa M (eds) Florae Indicae Enumeratio: Monocotyledonae. Botanical Survey of India, Calcutta, pp 254–274
Makmur MF, Larekeng SH, Restu M (2020) Genetic diversity of eight types of bamboo based on random amplified polymorphic DNA (RAPD) markers. Plant Arch 20(2):2333–2337
Marulanda ML, Márquez P, Londoño X (2002) AFLP analysis of *Guadua angustifolia* (Poaceae: Bambusoideae) in Columbia with emphasis on the coffee region. J Am Bamboo Soc 16 (1):32–42
Mathews KG, Huguelet J, Lanning M et al (2009) Clonal diversity of *Arundinaria gigantea* (Poaceae; Bambusoideae) in western North Carolina and its relationship to sexual reproduction: an assessment using AFLP fingerprints. Castanea 74(3):213–223
McNeely AJ (1995) Bamboo, biodiversity and conservation in Asia. Bamboo, people and the environment. In: The proceedings of Vth international bamboo workshop and the IV international bamboo congress, Ubud, Bali, 19 June 1995
Mei Z, Fu S, Yu H et al (2014) Genetic characterization and authentication of *Dimocarpus longan* Lour. Using an improved RAPD technique. Genet Mol Res 13(1):1447–1455
Migicovsky Z, Warschefsky E, Klein LL, Miller AJ (2019) Using living germplasm collections to characterize, improve, and conserve woody perennials. Crop Sci 59(6):2365–2380
Miyazaki Y, Ohnishi N, Takafumi H, Hiura T (2009) Genets of dwarf bamboo do not die after one flowering event: evidence from genetic structure and flowering pattern. J Plant Res 122 (5):523–528
Morgante M, Hanafey M, Powell W (2002) Microsatellites are preferentially associated with non repetitive DNA in plant genomes. Nat Genet 30:194–200
Mukherjee AK, Ratha S, Dhar S et al (2010) Genetic relationships among 22 taxa of bamboo revealed by ISSR and EST-based random primers. Biochem Genet 48(11–12):1015–1025
Nag A, Gupta P, Sharma V et al (2013) AFLP and RAPD based genetic diversity assessment of industrially important reed bamboo (*Ochlandra travancorica* Benth). J Plant Biochem Biotechnol 22(1):144–149
Naithani HB, Pal M, Lepcha STS (2003) Gregarious flowering of *Thamnocalamus spathiflorus* and *T. falconeri*, bamboos from Uttaranchal, India. Ind For 129(4):517–526

Nayak S, Rout GR (2005) Characterization of microsatellites in *Bambusa arundinacea* and cross species amplification in other bamboos. Zeitschrift für Naturforschung C 60(7–8):605–610
Nayak S, Rout GR, Das P (2003) Evaluation of genetic variability in bamboo using RAPD markers. Plant Soil Environ 49:24–28
Nilkanta H, Amom T, Tikendra L et al (2017) ISSR marker based populationg genetic study of *Melocanna baccifera* (Roxb.) Kurz: a commercially important bamboo of Manipur, North-East India. Scientifica 2017:1–9
Nirmala C, Bisht MS, Bajwa HK, Santosh O (2018) Bamboo: a rich source of natural antioxidants and its applications in the food and pharmaceutical industry. Trends Food Sci Technol 77:91–99
NITI Policy Paper 1 (2017). http://agricoop.nic.in/sites/default/files/NITI%20Aayog%20Policy%20Paper.pdf
Oumer OA, Dagne K, Feyissa T, Tesfaye K, Durai J, Hyder MZ (2020) Genetic diversity, population structure, and gene flow analysis of lowland bamboo [Oxytenanthera abyssinica (A. Rich.) Munro] in Ethiopia. Ecol Evol 10(20):11217–11236
Paran I, Michelmore RW (1993) Development of reliable PCR based markers linked to downy mildew resistance genes in lettuce. Theor Appl Genet 85:985–993
Peng Z, Lu Y, Li L et al (2013) The draft genome of the fast-growing non-timber forest species moso bamboo (*Phyllostachys heterocycla*). Nat Genet 45(4):456–461
Pennisi E (2007) Wanted: a barcode for plants. Science 318:190–191
Prieto MG, Vidal NJ, Rojas T, Eduardo J (2013) Bamboo as a sustainable material for several city uses. In: 2013 international conference on new concepts in smart cities: fostering public and private alliances (SmartMILE), 11 December 2013. IEEE, Gijon
Qian W, Ge S, Hong DY (2001) Genetic variation within and among populations of a wild rice *Oryza granulate* from China detected by RAPD and ISSR markers. Theor Appl Genet 102 (2):440–449
Qiang Z, Yu-long D, Chen XU et al (2005) A preliminary analysis of phylogenetic relationships of *Arundinaria* and related genera based on nucleotide sequences of nrDNA (ITS region) and cpDNA (trnL-F intergenic spacer). J For Res 16(1):5–8
Rajput BS, Jani M, Ramesh K, Manokari M, Jogam P, Allini VR, Kher MM, Shekhawat MS (2020) Large-scale clonal propagation of Bambusa balcooa Roxb.: an industrially important bamboo species. Ind Crop Prod 157:112905
Ramanayake SMSD, Meemaduma VN, Weerawardene TE (2007) Genetic diversity and relationships between nine species of bamboo in Sri Lanka, using random amplified polymorphic DNA. Plant Syst Evol 269(1):55–61
Rangsiruji A, Binchai S, Pringsulaka O (2018) Species identification of economic bamboos in the genus *Dendrocalamus* using SCAR and multiplex PCR. Songklanakarin J Sci Technol 40 (3):640–647
Rao AN, Rao VR (1995) Patterns of variation in bamboo. In: Williams JT, IVR R, Rao AN (eds) Genetic enhancement of bamboos and rattan. International Network for Bamboo and Rattan, New Delhi. Appendix 5
Ren Y, Ma Y, Zhang Z, Qiu L, Zhai H, Gu R, Xie Y (2019) Total alkaloids from bamboo shoots and bamboo shoot shells of Pleioblastus amarus (Keng) Keng f. and their anti-inflammatory activities. Molecules 24(15):2699
Rossarolla MD, Tomazetti TC, Vieira LN et al (2020) Identification and characterization of SSR markers of *Guadua chacoensis* (Rojas) Londoño & PM Peterson and transferability to other bamboo species. 3 Biotech 10(6):273
Rudd S (2003) Expressed sequence tags: alternative or complement to whole genome sequences? Trends Plant Sci 8:321–329
Ruiz-Sanchez E, Sosa V (2010) Delimiting species boundaries within the Neotropical bamboo Otatea (Poaceae: Bambusoideae) using molecular, morphological and ecological data. Mol Phylogenet Evol 54(2):344–356
Saha S, Adhikari S, Dey T, Ghosh P (2016) RAPD and ISSR based evaluation of genetic stability of micropropagated plantlets of *Morus alba* L. variety S-1. Meta Gene 7:7–15

Sen A, Goyal A (2014) Molecular phylogeny of North Bengal bamboos inferred through PCR-RFLP based on the trnL-trnF region. The Experiment 19(1):1304–1315

Shalini A, Meena RK, Tarafdar S, Thakur S (2013) Evaluation of genetic diversity in bamboo through DNA marker and study of association with morphological traits. Bull Environ Pharm Life Sci 2(8):78–83

Sharma ML (1996) The genus *Oxytenanthera* in Asia. Indian For 122(2):187–188

Sharma RK, Gupta P, Sharma V et al (2008) Evaluation of rice and sugarcane SSR markers for phylogenetic and genetic diversity analyses in bamboo. Genome 51(2):91–103

Sharma V, Bhardwaj P, Kumar R et al (2009) Identification and cross-species amplification of EST derived SSR markers in different bamboo species. Conserv Genet 10(3):721–724

Sijimol K, Dev SA, Muralidharan EM, Sreekumar VB (2014) DNA barcoding: An emerging tool for precise identification and certification of planting stock in taxonomically challenging bamboo species. J Bamboo Rattan 13(1/2):29–43

Skeat WW (1900) Malay magic: being an introduction to the folklore in popular religion of the Malay peninsula. MacMillan and Co. Ltd., London

Soderstrom TR, Calderon CE (1979) A commentary on bamboos (Poaceae: Bambusoideae). Biotropica 11:161–172

Soderstrom TR, Ellis RP (1987) The position of bamboo genera and allies in a system of grass classification. In: Soderstrom TR, Hilu KW, Campbell CS, Barkworth ME (eds) Grass systematics and evolution. Smithsonian Institution Press, Washington DC; London

Sosa V, Mejía-Saules T, Cuéllar MA, Vovides AP (2013) DNA barcoding in endangered Mesoamerican groups of plants. Bot Rev 79(4):469–482

Stapleton CMA (1997) Morphology of Woody bamboos. In: Chapman GP (ed) The Bamboos. Academic Press, New York, pp 251–267

Sun Y, Xia N, Lin R (2005) Phylogenetic analysis of Bambusa (Poaceae: Bambusoideae) based on internal transcribed spacer sequences of nuclear ribosomal DNA. Biochem Genet 43 (11–12):603–612

Sun Y, Xia N, Stapleton CM (2006) Relationships between Bambusa species (Poaceae, Bambusoideae) revealed by random amplified polymorphic DNA. Biochem Syst Ecol 34 (5):417–423

Suyama Y, Obayashi K, Hayashi I (2000) Clonal structure in a dwarf bamboo (*Sasa senanensis*) population inferred from amplified fragment length polymorphism (AFLP) fingerprints. Mol Ecol 9(7):901–906

Tang DQ, Lu JJ, Fang W et al (2010) Development, characterization and utilization of GenBank microsatellite markers in *Phyllostachys pubescens* and related species. Mol Breed 25 (2):299–311

Thomas TA, Arora RK, Singh R (1988) Genetic wealth of bamboos in India and their conservation strategies. In: Proceedings of international bamboo workshop, bamboos current research, Cochin, 14–18 November 1988

Tian B, Yang HQ, Wong KM et al (2012) ISSR analysis shows low genetic diversity versus high genetic differentiation for giant bamboo, *Dendrocalamus giganteus* (Poaceae: Bambusoideae), in China populations. Genet Res Crop Evol 59(5):901–908

Tyrrell CD, Londoño X, Prieto RO et al (2018) Molecular phylogeny and cryptic morphology reveal a new genus of west Indian woody bamboo (Poaceae: Bambusoideae: Bambuseae) hidden by convergent character evolution. Taxon 67(5):916–930

Varshney RK, Graner A, Sorrells ME (2005) Genic microsatellite markers in plants: features and applications. Trends Biotechnol 23(1):48–55

Vos P, Hogers R, Bleeker M et al (1995) AFLP: a new technique for DNA fingerprinting. Nucleic Acids Res 23:4407–4414

Waikhom SD, Ghosh S, Talukdar NC, Mandi SS (2012) Assessment of genetic diversity of landraces of *Dendrocalamus hamiltonii* using AFLP markers and association with biochemical traits. Genet Mol Res 11(3):2107–2121

Welsh J, McClelland M (1990) Fingerprinting genomes using PCR with arbitrary primers. Nucleic Acids Res 18(24):7213–7218

Williams JGK, Kubelik KJ, Livak KJ et al (1990) DNA polymorphisms amplified by arbitrary primers are useful genetic markers. Nucleic Acids Res 18(22):6531–6535

Wu MCY (1962) The classification of Bambuseae based on leaf anatomy. Bot Bull Acad Sinica 3:83–108

Yang HQ, Yang JB, Peng ZH et al (2008) A molecular phylogenetic and fruit evolutionary analysis of the major groups of the paleotropical woody bamboos (Gramineae: Bambusoideae) based on nuclear ITS, GBSSI gene and plastid trnL-F DNA sequences. Mol Phylogenet Evol 48 (3):809–824

Yang HQ, An MY, Gu ZJ, Tian B (2012) Genetic diversity and differentiation of *Dendrocalamus membranaceus* (Poaceae: Bambusoideae), a declining bamboo species in Yunnan, China, as based on inter-simple sequence repeat (ISSR) analysis. Int J Mol Sci 13(4):4446–4457

Zhao H, Yang L, Peng Z et al (2015) Developing genome-wide microsatellite markers of bamboo and their applications on molecular marker assisted taxonomy for accessions in the genus *Phyllostachys*. Sci Rep 5:8018

Zhu S, Liu T, Tang Q et al (2014) Evaluation of bamboo genetic diversity using morphological and SRAP analyses. Russ J Genet 50(3):268–273

Chapter 17
Application of Bamboo in the Food and Pharmaceutical Industry

Oinam Santosh, Harjit Kaur Bajwa, Madho Singh Bisht, and Nirmala Chongtham

Abstract Bamboo is an enduring plant blessed with a plethora of functional components that make it incredibly useful for the development of functional foods. It has served as food and medicine since ancient times. Bamboo-based products commercially available in the markets of different countries are canned and fermented shoots, pickle, shoot powder, bamboo juice and water, beverages, and bamboo shoot fortified food products. The bioactive compounds such as phenols, flavonoids, and phytosterols in bamboo shoot and leaves provide youthful feeling, athletic energy, and longevity to regular consumers. Recent studies have also shown that bamboo shoot is a good source of acetylcholine, which has preventive effects against Alzheimer's disease. Bamboo salt, bamboo vinegar, bamboo extracts, and bamboo silica are some important bamboo-based pharmaceutical preparations that are now gaining importance. Although some chemical compounds in shoots are labeled as antinutrients, their role as potential healthy biochemical components for the prevention of several health problems has been scientifically elucidated. The importance of bamboo shoots linked to functional health-modulating functions are anti-oxidation, antidiabetic, anticancer, cardiovascular, anti-inflammatory, antimicrobial, antiviral, and antihypertensive. Bamboo being rich in nutrients, antioxidants, and bioactive compounds has attracted significant research and commercial interest and is gaining popularity worldwide. This chapter discusses the nutrients and bioactive compounds in bamboo and their potential role in developing novel food and pharmaceutical products to improve health globally.

Keywords Bamboo · Nutrients · Bioactive compounds · Antioxidants · Health food

O. Santosh · H. K. Bajwa · N. Chongtham (✉)
Department of Botany, Panjab University, Chandigarh, India

M. S. Bisht
Department of Environmental Studies, North Eastern Hill University, Shillong, Meghalaya, India

Z. Ahmad et al. (eds.), *Biotechnological Advances in Bamboo*,
https://doi.org/10.1007/978-981-16-1310-4_17

17.1 Introduction

Bamboo is one such precious resource that has played a significant role in human civilization since time immemorial for subsistence to entertainment and emotional support of people living in tropical and subtropical parts of the world, in particular regions like East and South Asia (Chongtham and Bisht 2021). Due to extensive use and application of bamboo for each and every aspect of life, these regions are in fact known as "Bamboo Civilization," and countries like China are called "Bamboo Civilized Country" (Jiayan 2014). It serves as construction material, in making of various agricultural and household items, and as fiber and fabrics for food and medicine (Chongtham and Bisht 2021). Bamboo is rightly called "plant with thousand faces due to its multiple uses." In the present time too, bamboo is ahead of steel in tensile strength, and at the same time, it is proving better than plastic in plasticity and multiple applications. For centuries bamboo has been a favorite food for many animals and for humans as well (Nirmala et al. 2011; Chongtham and Bisht 2021). Bamboo is a rare example that has become a staple food for animals like the giant panda, basically a carnivorous animal (Xue et al. 2015). In many societies and cultures also, bamboo is revered as high valued food. Like in Japan the young juvenile shoots of bamboo are called "king of forest vegetables," and in the Tang Dynasty of China, it was said, "there is no banquet without bamboo" (Nirmala et al. 2011).

Bamboo is also considered valued medicine and health rejuvenating tonic particularly in countries like China, Japan, Korea, and India, and every part of the plant like leaves, shoots, sap, nodes, internodes, and roots was used for treating various ailments and for various health benefits (Wróblewska et al. 2019). The use of "banslochan" (an amorphous substance collected from the culms of some bamboos) as an important ingredient in one of the earliest formulated health tonic called "Chyawanprash" in India is there for centuries (Nirmala et al. 2011). Similarly, the use of bamboo leaves, shavings, and sap in the Chinese traditional medicine is very old (Chongtham and Bisht 2021). Old works of literature like Compendium Materia Medica (a pharmaceutical text of Ming Dynasty, 1368–1644), Ayurveda, and Bhavprakash Nighantu (old medical compilations in India) have mentioned a number of medicinal uses of bamboo (Nirmala et al. 2011). However, for the last 100–200 years the use of bamboo for food and medicine has been pushed back due to industrialization, change in food habits, and the popularity of allopathic medicines. But now bamboo is being seen as a solution for various new-age diseases like diabetes, obesity, cancer, and various cardiovascular diseases and reported to be endowed with antioxidants, antibacterial, antifungal, anti-fatigue, cholesterol-lowering, and neuron protecting properties (Hu et al. 2000; Fujimura et al. 2005; Zhang et al. 2006). Bamboo is no more a neglected crop or poor man's food but a rich man's delicacy and superfood (Chongtham and Bisht 2021). Recent studies revealed the potential of bamboo, rich in nutrition, presence of various bioactive compounds, and several health benefits, that can prevent many chronic diseases that has led to increasing interests among researcher for its application in the food, beverages, and pharmaceutical industries (Nirmala et al. 2011, 2018; Nirmala and Bisht 2017; Bajwa et al. 2015; Rawat et al. 2016; Saini et al. 2017; Chongtham and Bisht 2021). Bamboo leaves are being used for tea and alcoholic drinks, young and old culms, shavings, and the sap is being used for the formulation of various medicinal products, and young juvenile shoots of many species of bamboo have taken the status of health food and are being consumed as fresh, dried, paste, and fermented and in fortified forms (Chongtham and Bisht 2021).

In recent years, global attention has increased for a sustainable and long-term solution for micronutrient deficiency supporting the relation between nutrition and advancement in agriculture and food system (Black et al. 2013). There is also a need of integrated approaches in food sector regarding the rising situation for noncommunicable diseases due to the fast-changing lifestyle of the present century (Roos et al. 2019). According to the report of the World Health Organization (WHO 2010), the fast-changing lifestyle with poor choices of nutritious food and several other critical factors have increased the risk of noncommunicable diseases like cardiovascular diseases, obesity, diabetes, osteoporosis, cancer, gastrointestinal diseases, and respiratory diseases. Thus, there is a significant change in the food industries due to shifting of consumer's choice from healthy foods to food that prevents nutrition-related diseases and improves the physical and mental well-being of consumers leading to the development of modern functional foods. Recent studies revealed the potential of bamboo in food sector being rich in nutrition, presence of good quantity of bioactive compounds, and several health benefits that can prevent many chronic diseases that has led to increasing interest among researcher for its application in the food industry (Nirmala et al. 2011; Nirmala and Bisht 2017; Bajwa et al. 2015; Rawat et al. 2016; Saini et al. 2017). There have been recent reports in the food sectors for the uses of bamboo, such as shoots in bakery products (Santosh et al. 2018, 2019), leaf extract for treatment of ailments (Das 2019), source of natural bioactive compounds and antioxidants (Nirmala et al. 2018), dried bamboo culm for the physical characteristics and proximate composition of fortified products in relation to sugar and fat substitute formulation (Felisberto et al. 2019), and bamboo fiber for improving the sensory quality of food products (Silva et al. 2020). Phenol, phytosterol, and dietary fiber are extensively used in industrial purposes for their several health benefits; however, the enhancement of such bioactive compounds in fortified products is not studied in detail so far. Bamboo is rich in mineral (Chongtham et al. 2020), but the potential in improving the content in the product is still to be explored. This chapter discusses the importance of bamboo in the food sector and its application in the food and pharmaceutical industries.

17.2 Bamboo as Food

Young juvenile shoots, leaves, culms, and seeds of many bamboo species are food to many animals and birds and are, for some, exclusive food (giant panda and golden lemurs) and seasonal delicacy for animals like mountain gorilla (Xue et al. 2015).

For humans, young juvenile shoots of some 100–200 species and seeds of few species are food in parts of the world like in East and Southeast Asian countries. The consumption of bamboo seeds is known in few regions particularly in South India (Prasad et al. 1985). The seeds of *Bambusa arundinacea* and *Phyllostachys bambusoides* are called "bamboo rice" in some parts of South India and cooked and eaten as rice occasionally (Prasad et al. 1985). There are reports that during the famine of 1883, *B. arundinacea* produced seeds in abundance which were consumed at many places of South India for sustenance (Prasad et al. 1985).

Young juvenile shoots are consumed in many parts of East and Southeast Asia (Collins and Keilar 2005; Chongtham and Bisht 2021). China, Japan, Korea, and the northeastern regions of India are the main bamboo shoot producers as well as consumers. Fresh shoots have a crisp and crunchy taste and are used in making soups, stir-fries, snacks, salads, fried rice, spring rolls, and several other dishes (Nirmala et al. 2011). The most common species preferred as fresh or canned are *Phyllostachys pubescens*, *Dendrocalamus hamiltonii*, *D. giganteus*, *Bambusa balcooa*, *B. bambos*, *Thyrsostachys siamensis* (Chongtham and Bisht 2021). Shoots are also consumed as fermented dried, pickled, and processed/canned (Bashir 2010). In North-East India, bamboo shoots are mainly consumed as fermented and there are a number of fermentation processes that are developed according to bamboo species as well as the taste of the people (Giri and Janmejay 2000). Fermentation has two major roles in bamboo shoot processing, one is to increase the shelf life of shoots for more than 1 year, which is otherwise just 2–3 days, and secondly decrease the content of taxiphylin and other antinutrients in the shoots (Sarangthem and Singh 2013; Choudhury et al. 2012). Fermentation also increases the food value of shoots by producing additional bioactive compounds and increasing the content of proteins, amino acids, and mineral elements (Nirmala et al. 2011). Bamboo shoots being seasonal crops with very fast growth and very short shelf life after harvest need proper processing techniques for future use, easy transportation, and an increase of the shelf life. Some bamboo species have a very high amount of antinutrients and bitterness and need proper processing to make shoots fit for consumption. Simple washing, boiling, soaking, sun drying, and fermentation are the traditional methods of processing bamboo shoots particularly in India and many other South-East Asian countries (Chongtham and Bisht 2021). Now bamboo shoots are also being processed with various modern methods with high-tech machines and tools particularly in countries like China, Taiwan, Korea, and Thailand. Methods like freeze-drying, hot air drying, oven drying, solar drying, osmotic dehydration, and canning are the most common ones (Wongsakpairod 2000; Madamba 2003; Xu et al. 2005; Cheng 2006; Chongtham and Bisht 2021).

Bamboo shoots are consumed in different forms (fresh, dried, fermented, powder, paste) and in different ways, like just boiled in Japan, stir-fry and soups in China and Korea, heavily spiced in Thailand, Indonesia and India, or pickled in Myanmar, Nepal and in many other South-East Asian countries (Table 17.1). The juvenile bamboo shoots are delicious as well as rich in nutrient components mainly proteins, carbohydrates, minerals, vitamins, dietary fiber, and various bioactive compounds like phenols and phytosterols which exhibit a great potential of bamboo shoots as a

Table 17.1 Traditional bamboo shoot dishes of different countries

Country	Local name of bamboo dish	Reference
India	*Khorisa, Tuaithur, Byapu, Papu sududanjii, Usoi-Ooti, Usoi-kangsu, Soibum thongba, Soibum eronba, Tenga, Lung-seij, Jhur, Jingtah, Rawtui-bai, Rhuchak, Voyen, Handua, Pu-erh, Sabji, Mia-gudhog*	Tripathi (2011); Tamang et al. (2012); Bhatt et al. (2005); Singh et al. (2007); Jeyaram et al. (2009); Bisht et al. (2015); Kithan et al. (2015); Kumar et al. (2017); Thomas et al. (2014)
China	*Ulanzi*	Qing et al. (2008)
Japan	*Menma, Takenoko gohan*	Tripathi (2011)
Thailand	*Ma khua proh, Dom jud naomi, Naw-mai-dong, Kaeng kae*	Phithakpol et al. (1995); Tangkanakul et al. (2006); Kumar et al. (2017)
Philippines	*Ginataang labong, Diendeng na labong*	Phithakpol et al. (1995)
Indonesia	*Gulai rebung, Sayur ladeh, Lumpia*	Bhatt et al. (2003); Tripathi (2011)
Vietnam	*Sup mang cua*	Avieli (2005)
Korea	*Jooksun*	Kim et al. (2007)
Nepal and Bhutan	*Alu tama, Mesu*	Tamang (2005)

food resource (Bhargava et al. 1996; Chen et al. 1999; Kumbhare and Bhargava 2007; Nirmala et al. 2007, 2008, 2018; Satya et al. 2010; Choudhury et al. 2012). Bamboo shoots are quite rich in some dietary components like potassium, silica, selenium, manganese, dietary fiber, phenols, phytosterols, amino acids, and vitamin (Satya et al. 2010; Chongtham and Bisht 2021). This neglected vegetable is richer in many nutritional components compared to common vegetables we consume regularly (Nirmala et al. 2011). Moreover, bamboo shoot is low in calories and have fewer carbohydrates, nearly negligible fats, and cholesterols and have a high amount of phytosterols, which is quite good for the present-day sedentary lifestyle of the majority of people particularly in industrial and well-developed countries. The amount of phytosterols in bamboo shoots ranges from 91–265 mg/100 g dry weight which is quite a good amount compared to many other food items.

Recent studies revealed the potential of shoots being rich in nutrition, presence of a good quantity of bioactive compounds, and several health benefits that can prevent many chronic diseases that have led to increasing interest among researchers for its application in the food industry (Nirmala et al. 2011; Nirmala and Bisht 2017; Bajwa et al. 2015; Rawat et al. 2016; Saini et al. 2017). Bamboo shoot in fresh and fermented forms is an important ingredient in the cuisines across the Himalayas. For centuries, young edible bamboo shoots have remained one of the highly palatable dishes in delicacies (Satya et al. 2012) and an important forest vegetable in the traditional culinary preparation of China for more than 2500 years. In India, the use of bamboo shoots is still limited to the Northeast region and some other hilly parts in South and North-West India. Though, bamboo shoots remain a neglected crop, consumed usually by the local people, the delicacy of the vegetables is in high demand in up-scale markets and standard restaurant, which is why shoots are no longer considered as "poor man's timber" but are considered as "rich man's delicacy" (Nirmala et al. 2011). Fresh shoots have a crisp and crunchy taste and are used as an ingredient in making soups, stir-fries, snacks, salads, fried rice, spring rolls, and several other fried dishes (Nirmala et al. 2011). In countries like China, Japan, and India, they are sold in various processed forms like dried, fermented, pickled, and canned (Bashir 2010). Popular fresh and fermented bamboo shoot products of Northeast India include *Usoi*, *Soibum*, and *Soidon* of Manipur; *Hirring*, *Ekung*, and *Eup* of Arunachal Pradesh, *Rep* of Mizoram; *Kardi* or *amil* of Assam, and *Lung-Siej* of Meghalaya. Some of the famous local dishes include *Usoi-Ooti*, *Soibum eromba* of Manipur, *Rawtui-bai* of Mizoram, and *Mia-gudhog* of Tripura (Thomas et al. 2014; Bisht et al. 2015). Young tender shoots are used for the preparation of fermented products *Soibum* and *Soijin*, whereas apical meristem is used for the preparation of *Soidon* in Manipur, India. Some of the local delicious dishes of bamboo shoots of other countries include *Gulai rebung*, *Sayur ladeh* of Indonesia, *Ulanzi* of China, *Mesu* of Nepal and Bhutan, *Takenoko gohan* of Japan, *Naw-mai-dong* of Thailand, *Ginatang labong* of the Philippines, *Sup mang cua* of Vietnam, and *Jooksun* of Korea (Table 17.1) (Phithakpol et al. 1995; Avieli 2005; Tamang 2005; Kim et al. 2007; Qing et al. 2008).

17.3 Nutritional Properties of Bamboo

Bamboo in food sectors is known mostly for young shoots and leaves which are used for feeding humans and animals (Halvorson et al. 2011). In many Asian and African countries, bamboo leaves are used as fodder for many animals like the giant panda, golden bamboo lemur, elephants, as well as cattle, sheep, and goat, and they are also considered of highly medicinal value in many Asian countries (Chongtham and Bisht 2021). Singhal et al. (2011) analyzed the nutritional content in the leaves of 27 bamboo species and reported rich crude protein and low crude fiber content with 70% silica of the total ash and other insoluble mineral matters. Andriarimalala et al. (2019) analyzed the chemical composition and nutritive values of the leaves of nine bamboo species, *Bambusa balcooa*, *Bambusa bambos*, *Bambusa vulgaris*, *Bambusa tulda*, *Dendrocalamus asper*, *Dendrocalamus giganteus*, *Dendrocalamus strictus*, *Phyllostachys aurea*, and *Gigantochloa pseudoarundinacea*, which are used as cattle fodder and did not affect the ruminant's diet and the milk production. Due to higher content of water, crude protein, phosphorus, and less tannin in the leaf of bamboo species *Bonia saxatilis*, the Assamese macaques (*Macaca assamensis*) predominantly consume bamboo leaf (Li et al. 2020). Bamboo leaves are also rich in antioxidant, and the polyphenols and the extract from the leaves of *Phyllostachys* Sieb. have been certified for use in edible oil, meat product, aquatic product, and various other food additive (Lu et al. 2006). The antioxidant extract from bamboo leaves is also used to improve the storage stability and extend the shelf life of seafood (Xie et al. 2020). Bamboo leaf extracts are used in traditional medicine due to content of phenolic acids and flavonoids such as cryptochlorogenic acid, chlorogenic acid and neo-chlorogenic acid, caffeic acid, ferulic acid, luteolin and tricin, isoorientin, orientin, vitexin, and isovitexin (Ma et al. 2020).

The juvenile bamboo shoots are delicious as well as rich in nutrient components mainly proteins, carbohydrates, minerals, vitamins, and dietary fiber which exhibit great potential as a food resource (Table 17.2). Nutritional analysis of bamboo shoots has been conducted by many researchers which showed a high amount of dietary fiber, vitamins, minerals, protein, antioxidants, and polyphenols and low amount of fat (Bhargava et al. 1996; Chen et al. 1999; Bhatt et al. 2005; Kumbhare and Bhargava 2007; Nirmala et al. 2007, 2008, 2018; Satya et al. 2010; Choudhury et al. 2012). In 1953, the US Bureau of Human Nutrition and Home Economics reported the average food value of many species of bamboo shoot (Young 1954). Xia (1989) analyzed the nutritional profile of *Phyllostachys pubescens* locally known as Moso bamboo from Guangdong province, China, and reported the content of reducing sugar, protein, crude fat, fatty acids, vitamins, minerals, and amino acids. Tripathi (1998) analyzed the nutritional value of edible shoots of *Bambusa vulgaris*, *B. bambos*, and *Melocanna baccifera* and observed 88.8% moisture, 3.9% protein, 0.5% fat, 5.7% carbohydrate, and 1.1% minerals, particularly in *B. bambos*. Young shoots are a good source of dietary fiber due to which the calorie content of shoot is very low (Shi and Yang 1992; Nirmala et al. 2011). A high amount of dietary fiber in bamboo shoot is associated with several health benefits that include reducing the risk of cardiovascular diseases, hypertension, obesity, cancer, and certain gastrointestinal disorders (Anderson et al. 2009; Lattimer and Haub 2010; Brennan et al. 2012). It also controls or lowers the level of sugar in the blood, promotes regularity and prevents constipation, lowers blood cholesterol levels, and helps in weight control (Behall 1997). Shoots are also a good source of health-promoting bioactive compounds such as phytosterols, flavonoids, and phenolic acids which show effectiveness in decreasing blood pressure and cholesterol, increasing appetite, and having anticancerous and antidiabetic properties (Park and Jhon 2009; Hong et al. 2010; Nirmala et al. 2011; Singhal et al. 2013). Bamboo shoot has a good profile of minerals mainly potassium, calcium, manganese, zinc, chromium, copper, and iron, plus lower amounts of phosphorus and selenium (Shi and Yang 1992; Saini et al. 2017; Bajwa et al. 2019). Fresh bamboo shoots are also a good source of thiamine, niacin, vitamin A, vitamin B6, and vitamin E (Visuphaka 1985; Xia 1989; Shi and Yang 1992). Shoots contain 17 amino acids, 8 of which are essential for the human body (Qiu 1992; Ferreira et al. 1995). It is rich in amino acid tyrosine comprising about 57–67% of total amino acids (Kozukue et al. 1983). Fat content is comparatively low (0.26–0.94%), and the total sugar content, 2.5% on average, is lower than that of other vegetables, whereas the water content is 90% or more (Nirmala et al. 2011). Lignans in bamboo shoots have anticancer, antibacterial, and antiviral activities (Fujimura et al. 2005). High content of cellulose in bamboo shoot promotes digestion by increasing the peristaltic movement of the intestines (Fujimura et al. 2005; Shi and Yang 1992). Recent studies have also shown a good source of acetylcholine in the upper portion of *Phyllostachys bambusoides*, which is an important neurotransmitter in the cholinergic nervous systems of vertebrates and insects, which has preventive effects against Alzheimer's disease (Singhal et al.

Table 17.2 Nutritional content in fresh juvenile shoots of some edible bamboo species (macronutrients g/100 g fresh weight, vitamins mg/100 g fresh weight, and minerals mg/100 g dry weight) (Adapted from Chongtham and Bisht 2021)

Species	*B. balcooa*	*B. bambos*	*B. nutans*	*C. callosa*	*D. giganteus*	*D. hamiltonii*	*M. baccifera*	*P. mannii*
Carbohydrate	3.22 ± 0.13	1.72 ± 0.08	2.76 ± 0.10	1.26 ± 0.01	5.65 ± 0.06	3.33 ± 0.04	2.22 ± 0.01	2.73 ± 0.02
Amino acids	2.13 ± 0.03	2.31 ± 0.01	2.21 ± 0.02	4.61 ± 0.02	2.26 ± 0.04	2.33 ± 0.02	2.43 ± 0.05	2.36 ± 0.06
Starch	1.21 ± 0.02	1.24 ± 0.08	1.36 ± 0.08	0.71 ± 0.02	2.38 ± 0.04	1.74 ± 0.02	0.85 ± 0.02	1.09 ± 0.02
Protein	3.70 ± 0.09	5.87 ± 0.39	3.47 ± 0.23	4.57 ± 0.03	3.64 ± 0.05	3.37 ± 0.03	3.22 ± 0.02	3.24 ± 0.03
Fat	0.47 ± 0.01	0.53 ± 0.04	0.70 ± 0.05	0.43 ± 0.02	0.49 ± 0.02	0.42 ± 0.02	0.34 ± 0.02	0.44 ± 0.01
Vitamin C	2.63 ± 0.02	1.63 ± 0.03	1.52 ± 0.03	2.59 ± 0.03	2.21 ± 0.02	2.48 ± 0.07	1.44 ± 0.04	3.23 ± 0.05
Vitamin E	0.42 ± 0.03	0.60 ± 0.04	0.49 ± 0.02	0.81 ± 0.02	0.56 ± 0.03	0.68 ± 0.03	0.40 ± 0.07	0.53 ± 0.04
Potassium	4230 ± 60	5980 ± 60	5230 ± 60	6570 ± 60	4590 ± 50	5230 ± 60	6480 ± 60	6660 ± 70
Phosphorus	560 ± 30	750 ± 50	580 ± 30	750 ± 50	540 ± 30	560 ± 40	620 ± 40	930 ± 60
Magnesium	210 ± 10	230 ± 20	200 ± 10	220 ± 10	190 ± 10	200 ± 10	300 ± 20	230 ± 20
Calcium	180 ± 10	190 ± 10	180 ± 10	220 ± 20	210 ± 20	150 ± 10	210 ± 20	130 ± 10
Silicon	150 ± 2.8	130 ± 1.8	160 ± 2.8	100 ± 1.8	120 ± 1.6	190 ± 2.8	120 ± 2.4	70 ± 1.2
Iron	8.2 ± 0.8	8.0 ± 0.8	8.8 ± 0.8	6.5 ± 0.6	6.9 ± 0.6	7.4 ± 0.3	7.2 ± 0.8	9.1 ± 0.8
Zinc	6.8 ± 0.4	10 ± 0.6	9.5 ± 0.6	8.0 ± 0.8	6.1 ± 0.4	6.8 ± 0.4	10 ± 0.8	10 ± 0.8
Copper	2.6 ± 0.4	2.5 ± 0.4	1.9 ± 0.2	3.4 ± 0.6	5.1 ± 0.8	2.6 ± 0.4	2.8 ± 0.4	2.6 ± 0.4
Manganese	2.5 ± 0.2	3.6 ± 0.2	9.7 ± 0.8	3.5 ± 0.4	1.3 ± 0.1	1.2 ± 0.1	5.5 ± 0.6	9.0 ± 0.8
Nickel	0.9 ± 0.1	0.7 ± 0.1	0.8 ± 0.1	0.9 ± 0.1	0.8 ± 0.1	0.7 ± 0.1	1.0 ± 0.1	0.8 ± 0.1

Note: *B., Bambusa; C., Chimonobambusa; D., Dendrocalamus; M., Melocanna.* Data presented in mean values ± standard deviation ($n = 3$)

2013). Bamboo salt, bamboo vinegar, bamboo extracts, and bamboo silica are some important bamboo-based pharmaceutical preparations which are now gaining importance. Akakabe et al. (2006) mentioned that bamboo vinegar can act as an insecticide and bactericide and can be used as a deodorant. Bamboo salt is known to have many therapeutic effects on diseases such as inflammations, viral diseases, diabetes, circulation, organ disorders, and cancer (Hwang et al. 2008). In the traditional medicine of the Republic of Korea, it is used for treating cancer patients (Singhal et al. 2013). With major advancement in technology and reports of the nutritional profile, bamboo shoot is recommended as a healthy food.

17.4 Bamboo as Medicine

The medicinal properties of bamboo have been recognized for centuries particularly in Indian and Chinese traditional medicine systems. Leaves, seeds, bamboo shavings, sap, culms, rhizomes, and shoots are all reported to have medicinal properties (Nirmala et al. 2011; Wróblewska et al. 2019; Chongtham and Bisht 2021). In China, the medicinal properties of bamboo are compiled in the Compendium of Materia Medica (during the Ming Dynasty, 1368–1644). Similarly, in ancient India, bamboo is defined as, “Bamboo by nature is laxative, frigid, seminal, curative, palatable, bladder purifier and full of astringent juice. It splits cough, subsides bile and cures leprosy, bloody flux, wounds and swelling” (Tewari 1992). In India, the use of bamboo for medicinal and health benefits goes back to nearly 8000 to 10,000 years. Banslochan (also called Tabasheer), an amorphous substance collected from the culms of some species of bamboo, is being used in the health tonic called “Chyawanprash” since the time of Chyawanrishi (a sage named Chyawan) who lived nearly 10,000 years ago (Chongtham and Bisht 2021). Banslochan is used in various Ayurvedic prescriptions like for asthma, cough, etc. and considered as an astringent, stimulant, febrifuge, cooling tonic, antispasmodic, and aphrodisiac (Nirmala and Bisht 2017). Banslochan (Tabasheer) is also used in the preparation of various Chinese traditional medicines like Chenjin-wan, Quinghua-ditan-tang, and Xiaoer-qizhen-dan along with other bamboo parts like shaving, leaves, and other plant herbs (Chongtham and Bisht 2021). In Chinese medicines, bamboo sap is also extensively used which is considered to help to treat cold and fever and resolves phlegm or loss of consciousness associated with phlegm heat (Sangeetha et al. 2015). Bamboo shoots for the treatment of ailments such as chickenpox, skin diseases, infections, ulcers, etc. are also reported since ancient times (Sangtam et al. 2012).

The leaves of bamboo are also endowed with various medicinal properties and are a rich source of antioxidants and various bioactive compounds. The leaves of bamboo are given for the treatment of cough, fever, leprosy, and hematemesis in the Ayurvedic and Chinese traditional medicine systems (Das 2019). The leaves also have anti-inflammatory, antiulcer, antimicrobial, and hypoglycemic activities (Das 2019). The aqueous and ethanolic extract of bamboo leaves proved effective against

Staphylococcus aureus, *Escherichia coli*, *Pseudomonas aeruginosa*, and *Bacillus* sp. (Das 2019). The BLE (bamboo leaf extract) is mainly composed of flavonoids, lactones, and phenolic acids which have preventive measures to the reactive oxygen species (ROS) in the body and protects DNA, proteins, and lipids from the highly reactive ROS (Wróblewska et al. 2019; Chongtham and Bisht 2021). In South India, the *Bambusa* leaf extract is reported to be administered for the strengthening of cartilage and for treating osteoarthritis and osteoporosis (Sangeetha et al. 2015). This leaf extract has a vital role in the integrity of the bones, arterial walls, skin, teeth, gums, hair, and nails and is used to alleviate eczema and psoriasis (Vanithakumari et al. 1989). In eastern Asia, particularly in Japan, the leaves of *Sasa senanensis* (Kumaizasa) are called "Sasa Health" for centuries due to various health benefits of the *S. senanensis* leaf extract (Sangeetha et al. 2015). The extract of Kumaizasa leaves is also reported to have antitumor activity and immune potentiating efficiency.

The seeds of bamboo are very nutritious and also reported to have medicinal and aphrodisiac properties. The Kani tribes of Tamil Nadu in South India believe that the seeds of *Bambusa arundinacea* enhance the fertility (Sangeetha et al. 2015) and the explosive increase of rodent population in the northeastern region of India is also considered due to the fertility-increasing property of the seeds of *Melocanna baccifera* on which they feed and then multiply (Biswas et al. 2016).

17.5 Pharmaceutical Properties of Bioactive Compounds in Bamboo Shoot

Pharmaceutical properties of the bamboo shoot were recorded dating back to the Ming Dynasty (1368 to 1644), with a statement as: "It's slightly cold, sweet, non-toxic, and it quenches thirst, benefits the liquid circulatory system and can be served as a daily dish" (Yuming et al. 1999). The medicinal benefits of bamboo shoot in human health were proclaimed for more than 2000 years in archaic Chinese medicinal books, such as "Ben Chao Qui Zheng," "Ben Jing Feng Yuan," "Yao Pin Hua Yi," and "Jing Yue". It mentioned promoting digestion through the peristalsis motion of the intestine and also to be effective in cardiovascular disease prevention (Nirmala et al. 2011). Bamboo leaves are also used as herbal medicines in different areas of the world due to major bioactive compounds such as phenolic acids including cryptochlorogenic acid, chlorogenic acid and neo-chlorogenic acid, caffeic acid, ferulic acid, luteolin, and tricin (Zhu et al. 2018). Leaves are also rich in flavonoids such as isoorientin, orientin, vitexin, and isovitexin (Yang et al. 2014). The phenolic extracts from bamboo leaves have the properties of treating inflammation, hypertension, cardiovascular disease, arteriosclerosis, and cancer (Ma et al. 2020). The extract from *Phyllostachys nigra* leaves which is rich in isoorientin, orientin, and isoitexin is reported to have increased coronary blood flow and prevent myocardial ischemia in rabbits (Fu et al. 2005). Leaves of *Sasa quelpaertensis* Nakai are used for tea as a therapeutic purpose with antidiabetic, diuretic, and anti-

Table 17.3 Health benefits of bamboo shoots

No.	Potential activities	Reference
1	Antioxidant and anti-inflammatory effects of bamboo shoot extracts	Hu et al. (2000); Lu et al. (2005); Bajwa et al. (2019)
2	Antimicrobial and antifungal activities of bamboo shoot	Fujimura et al. (2005)
3	Cholesterol and body weight lowering effects of bamboo leaves extract	Ryou et al. (2012)
4	Antioxidant and phenolic extract of bamboo leaves promotes digestion	Ma et al. (2020)
5	Anti-inflammatory and anti-obesity effects of bamboo leaf extract	Moon-Hee et al. (2017)
6	Anti-obesity activities of bamboo shoot	Li et al. (2016)
7	Antiapoptotic activities of bamboo shoot	Hong et al. (2010)
8	Anticancer, antibacterial, antiviral activity of bamboo shoot fiber	Shi and Yang (1992); Hiromichi (2007)
9	Antidiabetic properties of bamboo leaves	Yang et al. (2010)
10	Anti-fatigue activity of bamboo shavings	Zhang et al. (2006)
11	Cholesterol-lowering properties of bamboo shoot	Park and Jhon (2009)
12	Antihypertension effects of bamboo shoot extract	Liu et al. (2013)
13	Bamboo lignin protects neurons from oxidative stress	Akao et al. (2004)
14	Bamboo shoot extract enhances the antioxidant activities	Bajwa et al. (2019)

inflammatory effects (Ryou et al. 2012). The old-age practice of bamboo for numerous health benefits is now authenticated with modern research in terms of preventing cancer, weight control, maintaining cholesterol level, and improving appetite and digestion, and there are many bamboo-based nutraceutical products available in the market (Shi and Yang 1992; Fujimura et al. 2005; Park and Jhon 2009; Nirmala et al. 2011; Ryou et al. 2012; Bajwa et al. 2015, 2019; Yang et al. 2010; Ma et al. 2020) (Tables 17.3 and 17.4).

17.5.1 Phytosterol

Bioactive compounds are phytochemicals that are typically present in small quantities in foods that promote health benefits by modulating the metabolic process of the human body system such as antioxidant activity, enzyme activity, receptor activity, and also gene expression (Correia and Beirao-da-Costa 2012). Bamboo shoots are a rich source of bioactive compounds with various dietary fiber components, phytosterols which are a precursor of many pharmaceutical steroids and phenols that act as free radical terminators, metal chelators, and singlet oxygen quenchers (Srivastava 1990; Kris-Etherton et al. 2002; Nirmala et al. 2011). There are several health

Table 17.4 Nutraceutical products of bamboo

No.	Product name	Health benefits
1	Biotin bamboo extract	Promotes and maintains skin tissue
2	Swanson bamboo extract	Silica supplement for hair, skin, and nails
3	BioFinest bamboo extract	Weight control, improves digestion, boosts immune system
4	NutriStart bamboo silica	Skin, ligament, tendon, and bone supplement
5	Boo bamboo Suncare natural sunscreen	Protection from broad-spectrum UVA/UVB
6	Enerex bamboo silica	Antiaging; strengthens the arteries, joints, nail, hair, skin, and bones
7	Silicon mix bamboo extract	Hair and skin supplement
8	Shudhanta herbal bamboo capsule	Aids digestion, immune booster, antibiotic and anti-inflammatory
9	Herbal papaya bamboo leaf extract liquid	Improving blood circulation
10	Bamboo Nutra	Antiaging, anti-obesity
11	Bamboo flex	Anti-inflammatory, remineralization, and development of bone structure
12	Bonusan forte	Anti-fatigue, supports energy metabolism, good for nervous system
13	Guozen bamboo leaf essence	Purifies blood and strengthens bones
14	Hawlik Cappillary capsules	Improves hair health
15	Lambert silica capsules	Contributes to structure and resilience of connective tissue, synthesis of bone collagen and cartilage
16	Sanacel	Improves digestion
17	Silice de Bambou	Prevents premature aging, preserves skin youthfulness, and promotes strong hair and healthy bones and teeth
18	Solaray bamboo capsules	Stimulates collagen synthesis in bone and connective tissue

benefits of phytosterols such as anticancer, cholesterol-lowering, anti-inflammatory, and anti-atherogenicity properties (Shi and Yang 1992; Hu et al. 2000; Lu et al. 2005; Hiromichi 2007; Park and Jhon 2009) (Table 17.3). In the pharmaceutical and nutraceutical industry, bamboo shoot can be a good source of phytosterol which is used for manufacturing steroids (Nirmala et al. 2011). The concentration of phytosterol in edible bamboo species, *Bambusa tulda* and *Dendrocalamus giganteus*, was estimated in fresh and fermented shoots (Srivastava 1990). The study reported higher phytosterol content in fermented shoots (1.6–2.8%) which was higher than the fresh shoot (0.21–0.39%). Lachance and He (1998) studied the phytosterol composition from the crude extract of various bamboo species, *Bambusa oldhami*, *B. edulis*, *Pseudosasa usawai*, *Dendrocalamus latiflorus*, *Phyllostachys edulis*, *P. pubescens*, and *P. makinoi*. The crude extract of bamboo shoots was analyzed using gas chromatography-mass spectrometry (GC-MS) and liquid chromatography-mass spectrometry (LC-MS), and several mixtures of phytosterol were reported including sitosterol, sitastanol, stigmasterol, beta-sitosterol, stigmasta-3,5-dien-7-one, stigmast-4-en-3-on, stigmasta-5,22-dien-3-ol, campesterol, and derivatives. The study also reports the lowering of cholesterol levels in mammals that include LDL (low-density lipoprotein) cholesterol, serum cholesterol, and total liver lipids with the administration of crude extract of bamboo shoot containing phytosterol. Analysis of phytosterol composition in the bamboo shoot extract from *Phyllostachys pubescens* using gas chromatography-mass spectrometry (GC/MS) led to the identification of 17 compounds with main components of 26% β-sitosterol, 10.5% of 9,12-octadecadienoic acid, and 9.83% of 9,12,15-octadecatrienoic acid (Lu et al. 2009a). The total phytosterol content in shoots of *Bambusa balcooa* and *Dendrocalamus strictus* was also studied in fresh and fermented shoots (Sarangthem and Singh 2003). Phytosterol concentration was higher in fermented bamboo shoots of *B. balcooa* (0.61%) and *D. strictus* (0.42%) as compared to the fresh shoot which ranges from 0.14 to 0.18%. Lu et al. (2009b) analyzed the phytosterol composition of bamboo species *Pleioblastus amarus*, *Dendrocalamus latiflorus*, *Phyllostachys pubescens*, and *P. praecox* using a UPLC-APCI-MS method and identified the presence of phytosterol compounds, 6-ketocholestanol, desmosterol, ergosterol, cholesterol, lanosterol, cholestanol, stigmasterol, campesterol, β-sitosterol, and stigmastanol (Fig. 17.1.).

Nirmala et al. (2014) reported the total phytosterol content of fresh juvenile shoots of *Bambusa balcooa*, *B. tulda*, *B. nutans*, *Dendrocalamus giganteus*, *D. hamiltonii*, *D. membranaceus*, and *D. strictus* which ranges from 0.19 g/100 g to 0.13 g/100 g with maximum content in *B. balcooa* and *D. hamiltonii*. Similarly, Ingudam and Sarangthem (2016) also reported maximum content of phytosterol with 0.29 g/100 g dry weight in *D. hamiltonii* during analysis in four different portions of shoots, i.e., apex, middle, the base, and the culm sheath covering the soft shoots of twelve bamboo species *Dendrocalamus brandisii*, *D. strictus*, *D. giganteus*, *D. flagellifer*, *D. hamiltonii*, *D. sericeus*, *Bambusa tulda*, *B. balcooa*, *B. nutans*, *B. kingiana*, *B. khasiana*, and *Cephalostachyum pergracile*. Apex portion of the shoot was observed with maximum content of phytosterol in *D. hamiltonii*, whereas minimum content was observed in culm sheath with 0.03 g/100 g dry weight in *D. strictus*. The total phytosterol content in *Dendrocalamus latiflorus* was reported as 0.16 g/100 g dry weight in the fresh shoot (Thounaojam et al. 2017). Santosh et al. (2019) evaluated the total phytosterol content in *D. hamiltonii* shoot paste used for fortification and reported a maximum content of 0.47 g/100 g dry weight which was higher than previously reported 0.19 g/100 g dry weight (Nirmala et al. 2014) and 0.29 g/100 g dry weight (Ingudam and Sarangthem 2016) (Table 17.5).

Fig. 17.1 Phytosterols isolated from bamboo shoot

Table 17.5 Phenol and phytosterol (mg/100 g) content in some species of fresh bamboo shoot

No.	Species	Phenol (mg/100 g)	Phytosterol (mg/100 g)	Reference
1	*Bambusa balcooa*	191.37 ± 2.62	190.00 ± 0.01	Nirmala et al. (2014)
		–	190.0 ± 10.5	Ingudam and Sarangthem (2016)
		101.65 ± 2.75	–	Badwaik et al. (2015)
2	*B. bambos*	360.00 ± 0.05	–	Pandey and Ojha (2014)
		760.00 ± 0.47		Pandey et al. (2012)
3	*B. kingiana*	–	222.00 ± 10.8	Ingudam and Sarangthem (2016)
4	*B. khasiana*	–	289.00 ± 5.8	Ingudam and Sarangthem (2016)
5	*B. nutans*	275.36 ± 2.04	–	Nirmala et al. (2014)
		–	94.00 ± 7.4	Ingudam and Sarangthem (2016)
		489.83 ± 5.08	164.20 ± 0.30	Bajwa et al. (2015)
6	*B. pallida*	79.85 ± 3.98	–	Badwaik et al. (2015)
7	*B. tulda*	443.97 ± 6.09	130.00 ± 0.01	Nirmala et al. (2014)
		390.00 ± 0.07	–	Pandey and Ojha (2014)
		960.00 ± 0.56		Pandey et al. (2012)
		–	190.00 ± 15.7	Ingudam and Sarangthem (2016)
		80.54 ± 3.21	–	Badwaik et al. (2015)
8	*Cephalostachyum pergracile*	–	215.00 ± 31.8	Ingudam and Sarangthem (2016)
9	*Dendrocalamus asper*	580.00 ± 0.07	–	Pandey and Ojha (2014)
		840.00 ± 0.25		Pandey et al. (2012)
10	*D. brandisii*	–	276.60 ± 1.60	Ingudam and Sarangthem (2016)
11	*D. flagellifer*	–	144.30 ± 15.8	Ingudam and Sarangthem (2016)
12	*D. giganteus*	222.40 ± 6.26	150.00 ± 0.08	Nirmala et al. (2014)
		336.56 ± 9.3	136.23 ± 2.40	Bajwa et al. (2015)
		–	198.70 ± 1.3	Ingudam and Sarangthem (2016)
13	*D. hamiltonii*	264.83 ± 6.75	190.00 ± 0.04	Nirmala et al. (2014)
		–	293.80 ± 16.1	Ingudam and Sarangthem (2016)
		586.36 ± 4.3	198.27 ± 2.30	Bajwa et al. (2015)
		88.23 ± 4.38	–	Badwaik et al. (2015)
		610.00 ± 0.01	470.00 ± 0.02	Santosh et al. (2019)
14	*D. latiflorus*	414.43 ± 6.26	160.76 ± 0.08	Thounaojam et al. (2017)
		612.24 ± 1.8	146.33 ± 3.10	Bajwa et al. (2015)
15	*D. membranaceus*	302.73 ± 8.53	–	Nirmala et al. (2014)

(continued)

Table 17.5 (continued)

No.	Species	Phenol (mg/100 g)	Phytosterol (mg/100 g)	Reference
16	*D. sericeus*	–	266.00 ± 14.5	Ingudam and Sarangthem (2016)
17	*D. strictus*	271.23 ± 5.64	140.00 ± 0.03	Nirmala et al. (2014)
		0.63 ± 0.08	–	Pandey and Ojha (2014)
		1250 ± 0.68		Pandey et al. (2012)
		–	212.40 ± 12.1	Ingudam and Sarangthem (2016)

Data presented in mean values ± standard deviation ($n = 3$)

17.5.2 Phenol

Plant phenols or polyphenols are secondary metabolites of plants that act as free radical terminator, metal chelators, and singlet oxygen quenchers (Kris-Etherton et al. 2002). There are several health-promoting properties of phenol such as antioxidant and antimicrobial activity, due to which bamboo shoot, rich in phenol, is gaining importance in the food industry (Park and Jhon, 2010). Satya et al. (2009) reported the total phenolic content of bamboo shoots ranging from 153.91 to 222.81 GAE (gallic acid equivalents)/100 g dry weight in four species, *Bambusa balcooa*, *B. tulda*, *B. vulgaris*, and *Dendrocalamus hamiltonii*. Park and Jhon (2010) identified eight phenolic compounds from the shoot extracts of *P. pubescens* and *P. nigra* which include protocatechuic acid, p-hydroxybenzoic acid, catechin, caffeic acid, chlorogenic acid, syringic acid, p-coumaric acid, and ferulic acid (Fig. 17.2).

Nirmala et al. (2014) investigated the bioactive compounds in bamboo species *Bambusa balcooa*, *B. tulda*, *B. nutans*, *Dendrocalamus giganteus*, *D. hamiltonii*, *D. membranaceus*, and *D. strictus* (Table 17.5). Phenol content in the fresh shoots of all the seven selected species ranged from 191.37 mg/100 g to 443.97 mg/100 g fresh weight which was highest in *B. tulda* and minimum in *B. balcooa*. Pandey and Ojha (2013) studied total phenol content in the shoots of *Bambusa tulda*, *Dendrocalamus asper*, and *D. strictus* at different optimum harvesting ages of fresh shoots which ranged from 0.57 to 2.97 g/100 g with maximum content in *D. strictus*, whereas total phenol content in fresh shoots of *B. bambos*, *B. tulda*, *D. asper*, and *D. strictus* was reported with the range from 0.36 to 0.63 g/100 g (Pandey and Ojha, 2014).

Total phenol content in the fresh bamboo shoot of *Dendrocalamus latiflorus* was reported as 414.43 mg/100 g fresh weight (Thounaojam et al. 2017). Bajwa et al. (2015) analyzed the physicochemical and nutritional qualities of *Dendrocalamus hamiltonii* shoots in which the phenol content was reported to be 0.59 g/100 g fresh weight, whereas Santosh et al. (2019) also reported the total phenol content of the same species with 0.61 g/100 g fresh weight.

Fig. 17.2 Phenolic compounds isolated from bamboo shoot

17.5.3 Dietary Fiber

Dietary fiber comprises a unique blend of bioactive components that are indigestible parts of plant food that cannot be digested by the human digestive enzyme. They are composed of straight chains of carbohydrate molecules that have the potential to bind and remove harmful toxins and carcinogens in the digestive tract (Lattimer and Haub, 2010). There are two types of dietary fiber: soluble and insoluble. Soluble fiber can dissolve in or absorb water in the large intestine passing undigested from the small intestine to produce short-chain fatty acids that are effective in binding toxins and cholesterol in the intestinal tract (Young et al. 2005). On the other hand, insoluble fiber cannot dissolve in water, therefore increasing the fecal bulk and viscosity which remove potential toxins and carcinogens from the intestinal tract in less contact time by speeding out from the body (Adlercreutz et al. 1987). Bamboo shoots are a rich source of dietary fiber which has been investigated by several researchers (Rajyalakshmi and Geervani 1994; Bhatt et al. 2005; Kumbhare and Bhargava 2007; Nirmala et al. 2008, 2011, 2014; Bajwa et al. 2015; Rawat et al. 2016; Thounaojam et al. 2017; Santosh et al. 2019). Several health benefits of dietary fiber of bamboo shoot have been reported including treatment and prevention of obesity and diabetes, reduced cardiovascular diseases, and decreased incidence of certain types of cancer (Nirmala et al. 2009; Tucker and Thomas 2009). Park and Jhon (2009) investigated the health benefits of bamboo shoot dietary fiber on humans by the administration of a fiber-free diet and a diet with bamboo shoot fiber which confirmed the beneficial effects of bamboo shoot dietary fiber in lowering blood cholesterol levels and improving bowel functions. Lignans are an important component of fiber present in bamboo shoots which are reported to have anticancer, antibacterial, and antiviral activity (Shi and Yang 1992; Akao et al. 2004). Nirmala et al. (2009) reported a comparative account on the dietary fiber components of *Dendrocalamus giganteus* in fresh, canned, and boiled shoots. Shoots have a high amount of neutral detergent fiber (NDF), ranging from 2.23 g/ 100 g to 4.18 g/100 g which also showed an increase in other fiber components including acid detergent fiber (ADF), lignin, hemicellulose, and cellulose after fermentation. A high content of dietary fiber was also reported from fresh bamboo shoots of *Bambusa bambos*, *B. kingiana*, *B. nutans*, *B. polymorpha*, *B. tulda*, *B. vulgaris*, *Dendrocalamus asper*, *D. brandisii*, *D. giganteus*, *D. hamiltonii*, *D. membranaceus*, *D. strictus*, *Gigantochloa albociliata*, and *G. rostrate*, ranging from 2.26 to 4.49 g/100 g fresh weight with maximum in the shoots of *B. kingiana* (Nirmala et al. 2011). Bhatt et al. (2005) determined crude fiber for 11 bamboo species, *Bambusa balcooa*, *B. nutans*, *B. tulda*, *Dendrocalamus giganteus*, *D. hamiltonii*, *D. hookerii*, *D. longispathus*, *D. sikkimensis*, *Melocanna baccifera*, *Phyllostachys bambusoides*, and *Teinostachyum wightii*. The estimation was carried out using acid and alkaline digestion methods (Maynard 1970) in a dry matter which ranged from minimum of 23.1 g/100 g dry weight in *P. bambusoides* to maximum35.5 g/100 g dry weight in *M. baccifera*. Dietary fiber in the shoots of *Bambusa arundinacea*, *D. hamiltonii*, *D. latiflorus*, *B. nutans*, and *D. strictus* has also been reported (Rajyalakshmi and Geervani 1994; Kumbhare and Bhargava 2007; Bajwa et al. 2015; Rawat et al. 2016; Thounaojam et al. 2017; Santosh et al. 2019) (Table 17.6).

17.6 Application of Bamboo Shoot in Food Fortification

An adequate diet is one of the important factors influencing growth and immunity as it supplies the proper amount of nutrients, minerals, and vitamins that are critical for the human body. Research suggests that a plant-based diet provides us with almost all macronutrients (proteins, lipids, carbohydrates), micronutrients (minerals, trace elements, vitamins), and bioactive compounds in sufficient amount. In addition, people who eat primarily plant-based diets tend to have a lower body mass index and lower rates of obesity, diabetes, and heart diseases. There is a significant change in the food industries due to the shifting of consumer's choice from healthy foods to food that prevents nutrition-related diseases and improves the physical and mental well-being of consumers leading to the development of modern functional foods. Fortification of foods is an effective means to prevent micronutrient deficiencies. Foods which have already been used successfully as food vehicle include wheat, rice, milk, salt, cooking oils, sugar, and condiments. Basic foods such as bread, biscuits, dairy products, packaged cereals, flours, and ready-to-eat products are a convenient means for the maximum intake of micronutrients in the population. Over the years, several important and nutritious plants remain neglected in the world of food and nutrition diversity due to the emphasis given only on few major crop plants

Table 17.6 Dietary fiber % (crude fiber/NDF) content of fresh bamboo shoots

Bamboo species	Dietary fiber	Reference
Bambusa arundinacea	6.9	Rajyalakshmi and Geervani (1994)
B. bamboos	3.54	Nirmala et al. (2011)
B. balcooa	26.4[a]	Bhatt et al. (2005)
	6.75	Nirmala et al. (2014)
B. kingiana	4.49	Nirmala et al. (2011)
B. nutans	28.5[a]	Bhatt et al. (2005)
	2.28	Nirmala et al. (2011, 2014)
	0.76	Kumbhare and Bhargava (2007)
B. polymorpha	3.82	Nirmala et al. (2011)
B tulda	24.6[a]	Bhatt et al. (2005)
	3.97	Nirmala et al. (2011, 2014)
B. vulgaris	0.97	Kumbhare and Bhargava (2007)
	4.24	Nirmala et al. (2011)
D. asper	0.71	Kumbhare and Bhargava (2007)
	3.54	Nirmala et al. (2011)
D. brandisii	4.03	Nirmala et al. (2011)
D. giganteus	27.6[a]	Bhatt et al. (2005)
	2.65	Nirmala et al. (2008, 2011, 2014)
D. hamiltonii	25.4[a]	Bhatt et al. (2005)
	3.9	Nirmala et al. (2011)
	8.52	Bajwa et al. (2015)
	5.5	Santosh et al. (2019)
D. hookeri	34.7[a]	Bhatt et al. (2005)
D. latiflorus	5.88	Rawat et al. (2016)
	5.39	Thounaojam et al. (2017)
D. longispathus	26.7[a]	Bhatt et al. (2005)
D. membranaceus	2.91	Nirmala et al. (2011, 2014)
D. sikkimensis	23.5[a]	Bhatt et al. (2005)
D. strictus	2.26	Nirmala et al. (2011, 2014)
	0.98	Kumbhare and Bhargava (2007)
G. albociliata	4.15	Nirmala et al. (2011)
G. rostrata	4.2	Nirmala et al. (2011)
Melocanna baccifera	35.5[a]	Bhatt et al. (2005)
Phyllostachys bambusoides	23.1[a]	Bhatt et al. (2005)
Teinostachyum wightii	23.7[a]	Bhatt et al. (2005)

[a]Crude fiber % dr. wt

(Haq 2007). Neglected plants have been explored and received much attention recently for several biologically active substances such as dietary fibers, amino acids, proteins, minerals, vitamins, phytochemicals, and antioxidant properties which are used in the development of functional foods through fortification in the food industry (Rawat and Indrani 2015). Bamboo is one such neglected plant as food which has the potential of being used for food fortification.

Nowadays, the importance of food fortification is gaining momentum with the increasing urbanization and changing lifestyle for the prevention of several nutrition-related diseases. The present generation is aware of the relationship between healthy living and a healthy diet and also the cost-effectiveness of healthcare. The development of nutritious and health-promoting functional foods is very important in the food industries with the identification of new sources of nutraceuticals (Kris-Etherton et al. 2002). Fortification of widely consumed foods has been practiced in many developed countries as an effective strategy to address several nutrition-related diseases through existing food delivery systems, without requiring major changes in existing consumption (Serdula 2010). Interest in utilizing bamboo shoot which is rich in nutrients, bioactive compounds, and minerals for the production of natural functional food is gaining popularity in the food industries (Nirmala et al. 2011; Santosh et al. 2019). For long-term preservation and removal of antinutrient from bamboo shoots, many processing techniques are reported for the application in the production of value-added food products (Choudhury et al. 2012; Santosh et al. 2019). Recently, bamboo shoot has been used for several value-added products such as pickles, candies, nuggets, crackers, chutney, chips, cookies, chapatis, and buns (Farris and Piergiovanni 2008; Bisht et al. 2012; Choudhury et al. 2012; Pandey et al. 2012; Sood et al. 2013; Das et al. 2013; Thomas et al. 2014; Chavhan et al. 2015; Maroma 2015; Nimisha et al. 2015; Mustafa et al. 2016; Zhang et al. 2017; Felisberto et al. 2019) (Table 17.7).

Bamboo shoot fortified products hold great potential as a health food and a good source for nutraceutical and pharmaceutical products. Bamboo fiber was used in the preparation of the well-known Italian food "Amaretti" cookies which shows improvement in the texture and shelf life of the product (Farris and Piergiovanni 2008). Choudhury et al. (2015) investigated the influence of fortifying biscuits with the shoots of *Bambusa balcooa* for physicochemical, texture, and organoleptic characteristics. The study observed a decrease in the gluten content and an increase in moisture, fiber, protein, fat, ash, and phenolic and antioxidant properties with an increase in the fortification level. Sensory observation recommended a 10% level of bamboo shoot incorporation without affecting the overall quality. Mustafa et al. (2016) also analyzed the physical characteristics and sensory acceptance of cookies fortified with dried bamboo shoot powder and recommended a 6% level of fortification. The enhancement of nutritional and organoleptic properties of biscuit fortified with a processed form of bamboo shoot paste and freeze-dried powder of *Dendrocalamus hamiltonii* was also reported (Santosh et al. 2018, 2019) (Table 17.8). The study observed an increase in the nutritional and bioactive compounds and mineral content in a fortified biscuit; however, the sensory acceptability for aroma, texture, taste, and overall quality was maximum in the biscuits fortified with boiled shoots. Cookies fortified with the flour obtained from dried bamboo culm were reported to increase its crispness (Felisberto et al. 2019). Pandey et al. (2012) evaluated the nutritional profile in several value-added products such as nuggets, crackers, and pickles from different bamboo species, viz., *Dendrocalamus asper*, *D. strictus*, *Bambusa bambos*, and *B. tulda*. Sood et al. (2013) prepared products such as candy, chutney, nuggets, crackers, and chukh from shoots of

Table 17.7 Products fortified with bamboo shoots

Sl.	Fortified products	Bamboo species	Processed form	References
1	Amaretti cookies	*Not mentioned*	Bamboo fiber	Farris and Piergiovanni (2008)
2	Crackers, nugget, pickle	*Bambusa bambos*, *B. tulda*, *Dendrocalamus asper*, *D. strictus*	Brine-treated boiled shoot	Pandey et al. (2012)
3	Chicken nuggets	*B. auriculata*	Shoot fermented for 2 months	Das et al. (2013)
4	Candy, chutney, chukh, cracker, nugget	*D. hamiltonii*	Boiled shoot	Sood et al. (2013)
5	Pork nuggets	*B. polymorpha*	Brine-treated, boiled, and fermented for 6 months	Thomas et al. (2014)
6	Biscuit	*B. balcooa*	Boiled, dried, and powdered	Choudhury et al. (2015)
7	Chips	*B. vulgaris*	Shoot boiled for 2 h	Maroma (2015)
8	Pork pickles	*Not mentioned*	Minced shoot exposed to sun and fermented for 21 days, dried, and powdered	Chavhan et al. (2015)
9	Candies	*Not mentioned*	Boiled shoot	Nimisha et al. (2015)
10	Cookies	*Not mentioned*	Boiled shoot, dried, and powdered	Mustafa et al. (2016)
11	Pork nuggets	*B. polymorpha*	Brine-treated boiled shoot extract	Thomas et al. (2016)
12	Battered and breaded fish balls	*Not mentioned*	Bamboo shoot fiber of Hubei Ruifa biological engineering co., LTD	Zeng et al. (2016)
13	Fried potato chips	*Bambusa balcooa*	Bamboo shoot powder and bamboo shoot extract	Shanmugam et al. (2016)
14	Frozen dough	*Not mentioned*	Bamboo shoot fiber of Zhejiang Geng sheng tang ecological agriculture co., ltd.	Zhang et al. (2017)
15	Milk pudding	*D. latiflorus*	Shoot fiber extracted with cellulase and papain enzyme method	Zheng et al. (2017)
16	Biscuit	*D. hamiltonii*	Freeze-dried powder of fresh, boiled, and soaked shoots	Santosh et al. (2018)
17	Biscuit	*D. hamiltonii*	Fresh, boiled, and soaked shoot paste	Santosh et al. (2019)
18	Cookies	*D. asper*	Bamboo culm treated with metabisulfite, dried, and powdered	Felisberto et al. (2019)

Table 17.8 Phenol and dietary fiber/ADF (g/100 g) of bamboo shoot fortified products

Product	Shoot species	Processed form of shoot	Phenol	Fiber/ADF	Reference
Nugget	*D. strictus*	Boiled shoots and green gram	2.43 ± 0.03	–	Pandey et al. (2012)
	D. hamiltonii	Boiled bamboo shoot	–	6.40 ± 0.26	Sood et al. (2013)
Papad/ crackers	*D. asper*	15-min boiled shoots	1.02 ± 0.06	–	Pandey et al. (2012)
	D. hamiltonii	Boiled bamboo shoot	–	3.90 ± 0.31	Sood et al. (2013)
Pickle	*D. asper*	15-min boiled shoots	0.5 ± 0.03	–	Pandey et al. (2012)
Biscuit	*B. balcooa*	10% fortification of 30-min balanced dried powder	0.02 ± 0.01	1.85 ± 0.09	Choudhury et al. (2015)
	D. hamiltonii	10% fortification of 20-min boiled freeze-dried powder	0.20 ± 0.01	3.82 ± 0.01	Santosh et al. (2018)
	D. hamiltonii	10% fortification of 20-min boiled shoot paste	0.14 ± 0.01	3.58 ± 0.01	Santosh et al. (2019)

Note: *D., Dendrocalamus; B., Bambusa.* Data presented in mean values ± standard deviation ($n = 3$)

Dendrocalamus hamiltonii and analyzed their nutritional and sensory attributes. The study revealed good sensory acceptability of all the products in terms of color, flavor, aroma, taste, and texture. A good profile of moisture content, protein, ash, fiber, and total carbohydrates was also reported in shoot fortified nuggets and crackers. Nimisha et al. (2015) observed that the sensory acceptability of bamboo shoot candy flavored with pineapple was higher compared to the candy flavored with ginger with a stable storage period of 6 months under the normal condition without any microbial contamination during storage. Maroma (2015) utilized shoots of *Bambusa vulgaris* for the preparation of bamboo shoot chips, and the study revealed that the product was safe for consumption in terms of microbiological analysis and the sensory score was good for aroma, flavor, texture, and audible crispness. Shanmugam et al. (2016) investigated the reduction of acrylamide levels in fried potato chips with the application of bamboo shoot powder and bamboo shoot extract of *Bambusa balcooa*. Reduction in acrylamide level of potato chips was reported upto 50% in 1 g/L bamboo extract treatment and 25% in 50 g/L bamboo shoot powder treatment. Das et al. (2013) studied the effect of the fermented bamboo shoot of *Bambusa auriculata* in the fortified chicken nuggets and observed improvement in the emulsion stability, cooking yield, and sensory attributes of the fortified products. Improved sensory and microbial qualities of pork nuggets fortified with fermented bamboo shoots of *Bambusa polymorpha* were reported which also increased the shelf life of the nuggets for 2 weeks (Thomas et al. 2014). Thomas et al. (2016) also compared the antioxidant and antimicrobial effects of boiled bamboo shoot extract and *Averrhoa carambola* extract in the preparation of pork nuggets. Improvement in the sensory and increased shelf life of pork nuggets with the 6% level of the extract incorporated were observed from 21 days to 35 days compared to control samples. The increase in the shelf life of pork pickles fortified with fermented bamboo shoot was reported by Chavhan et al. 2015.

Bamboo shoots are a good source of dietary fiber, and the utilization of bamboo fiber which has several health benefits is reported in fortified bakery products, meat, sausage, beverages, spices, pasta, and ketchup (Nirmala et al. 2011). The inclusion of bamboo shoot dietary fiber in the diet has a beneficial effect on healthy digestion and lowering of lipid profile (Park and Jhon 2009). Fortification of dietary fiber in food lowers the fat content in deep-fried products which will also solve the problems of obesity and various cardiovascular diseases due to the over-ingestion of high fat-containing food items (Mellema 2003). Sensory acceptability of deep-fried fish balls was improved with a 6% level of dried bamboo shoot dietary fiber fortification and a decreased fat content of the crust and the core from 25.5% to 17.7% and 2.4% to 1.3%, respectively (Zeng et al. 2016). The importance of dietary fiber in the dairy industries for the improvement of rheological and texture properties is also gaining interest. Improvement in the mechanical properties, freezable water content, and thermal stability of the dough with the incorporation of bamboo shoot dietary fiber was reported (Zhang et al. 2017). Zheng et al. (2017) extracted fiber from the shoots of *Dendrocalamus latiflorus* through the compound enzyme method of cellulase and papain and observed that milk pudding fortified with 2 g/100 g of bamboo fiber was observed to have better rheological and texture properties due to the improved elasticity where the system stability was attained. Hemicellulose components of bamboo, a mixture of xylose and xylo-oligosaccharides (XOS) isolated from *Sasa senanensis* by steaming and subsequent water extraction, are reported to be a potential raw material of functional food and pharmaceutical industries (Peng and She 2014). Miura et al. (2013) reported the presence of xylitol in *Phyllostachys pubescens* which is converted from hemicelluloses through microbial activity. Since the compound has several health benefits such as anticaries, anti-inflammatory, and sweetening properties, it is of great interest in the food industries (Mäki-Arvela et al. 2011).

17.7 Conclusion

Bamboo, the "plant with a thousand faces," has indeed etched out a place in the food and pharmaceutical sector. Bamboo shoots in particular are gaining worldwide importance as health food being a rich repository of nutrients and health-promoting bioactive compounds. It has always been a herbal component of the traditional medicinal system since ancient times for the treatment of several diseases. Scientific reports of nutrient richness and proven health benefits have led to the emergence of bamboo as a highly potent ingredient for the development of novel functional foods and pharmaceutical products. Thus, bamboo shoots with a good source of nutrients and natural bioactive compounds are aptly considered as a superfood and can play a vital role in the food and pharmaceutical industries.

Acknowledgments The authors are grateful to the Ministry of Food Processing Industries (V45/MFPI/R&D/2000 Vol.IV), Department of Biotechnology, New Delhi (BT/475/NE/TBP/20132), DST PURSE Grant, Govt. of India, American Bamboo Society, and Ned Jaquith Foundation, USA, for providing financial assistance to conduct this research work.

Declaration of Interest Statement The authors declare none.

References

Adlercreutz H, Hamalainen E, Gorbach SL et al (1987) Association of diet and sex-hormones in relation to breast-cancer. Eur J Cancer Clin Oncol 23:1725–1726

Akakabe Y, Tamura Y, Iwamoto S et al (2006) Volatile organic compounds with characteristic odor in bamboo vinegar. Biosci Biotechnol Biochem 70:2797–2799

Akao Y, Seki N, Nakagawa Y et al (2004) A highly bioactive lignophenol derivative from bamboo lignin exhibit a potent activity to suppress apoptosis induced by oxidative stress in human neuroblastoma SH-SY5Y cells. Bioorg Med Chem 12:4791–4801

Anderson JW, Baird P, Davis RH et al (2009) Health benefits of dietary fiber. Nutr Rev 67 (4):188–205

Andriarimalala JH, Kpomasse CC, Salgado P et al (2019) Nutritional potential of bamboo leaves for feeding dairy cattle. Pesquisa Agropecuária Tropical 49

Avieli N (2005) Roasted pigs and bao dumplings: festive food and imagined transnational identity in Chinese–Vietnamese festivals. Asia Pac Viewp 46(3):281–293

Badwaik LS, Gautam G, Deka SC (2015) Influence of blanching on antioxidant, nutritional and physical properties of bamboo shoot. J Agric Sci 10(3):140–150

Bajwa HK, Nirmala C, Koul A et al (2015) Effects of processing and preservation on phenols and phytosterols content in bamboo shoots. In: 10th world bamboo congress, 2015, Korea

Bajwa HK, Santosh O, Koul A et al (2019) Phytomodulatory effects of fresh and processed shoots of an edible bamboo *Dendrocalamus hamiltonii* Nees & Arn. Ex Munro on antioxidant defense system in mouse liver. J Food Meas Charact 13(4):3250–3256

Bashir T (2010) Prospects of bamboo shoot processing in north-East India. Curr Sci 98(3):288–289

Behall KM (1997) Dietary fiber: nutritional lessons for macronutrient substitutes. Ann N Y Acad Sci 819:142–154

Bhargava A, Kumbhare V, Srivastava A et al (1996) Bamboo parts and seeds for additional source of nutrition. J Food Sci Technol 33(2):145–146

Bhatt BP, Singha LB, Singh K et al (2003) Some commercial edible bamboo species of north East India: production, indigenous uses, cost-benefit and management strategies. Bamboo Sci Cult 17(1):4–20

Bhatt BP, Singha LB, Sachan MS et al (2005) Commercial edible bamboo species of north eastern Himalayan region, India, part II: fermented, roasted and boiled bamboo shoots sale. J Bamboo Rattan 4:13–31

Bisht MS, Nirmala C, Vyas P (2012) Bamboo shoot a neglected natural resource: a source of food and prosperity for North-East India. In: Recent advances in natural products research, pp 18–22

Bisht MS, Nirmala C, Meetei OS (2015) Bamboo shoots for food in North-East India: conventional and contemporary. In: Proceedings of 10th world bamboo congress, Damyang

Biswas S, Kumar S, Mittal V (2016) A note on rodent migration following gregarious bamboo flowering in North-Eastern Hill region with particular reference to Mizoram (India) and its consequences. J Commun Dis 48(3):1–12

Black RE, Victora CG, Walker SP et al (2013) Maternal and child undernutrition and overweight in low-income and middle-income countries. Lancet 382(9890):427–451

Brennan MA, Derbyshire EJ, Brennan CS et al (2012) Impact of dietary fibre-enriched ready-to-eat extruded snacks on the postprandial glycaemic response of non-diabetic patients. Mol Nutr Food Res 56(5):834–837

Chavhan DM, Hazarika M, Brahma ML et al (2015) Effect of incorporation of fermented bamboo shoot on physicochemical and microbial quality of pork pickle. J Food Sci Technol 52 (2):1223–1227

Chen CJ, Qiu EF, Huang RZ et al (1999) Study on the spring shoot nutrient content of *Phyllostachys pubescens* of different provenances. J Bamboo Res 18(1):6–11

Cheng HP (2006) Vacuum cooling combined with hydrocooling and vacuum drying on bamboo shoots. Appl Therm Eng 26(17–18):2168–2175

Chongtham N, Bisht MS (2021) Bamboo shoot: superfood for nutrition, health and medicine. CRC Press, New York

Chongtham N, Bisht MS, Bajwa HK et al (2020) Mineral elements in bamboo shoots and potential role in food fortification. J Food Compos Anal 95:103662

Choudhury D, Sahu JK, Sharma GD (2012) Value addition to bamboo shoots: a review. J Food Sci Technol 49(4):407–414

Choudhury M, Badwaik LS, Borah PK et al (2015) Influence of bamboo shoot powder fortification on physico-chemical, textural and organoleptic characteristics of biscuits. J Food Sci Technol 52 (10):6742–6748

Collins RJ, Keilar S (2005) The Australian bamboo shoots industry: a supply chain approach. A report for rural industries research and development corporation, Australia. Community Med 33:9–10

Correia P, Beirao-da-Costa ML (2012) Effect of drying temperatures on starch-related functional and thermal properties of chestnut flours. Food Bioprod Process 90(2):284–294

Das M (2019) Bamboo: inherent source of nutrition and medicine. J Pharmacogn Phytochem 8 (2):1338–1344

Das A, Nath DR, Kumari S et al (2013) Effect of fermented bamboo shoot on the quality and shelf life of nuggets prepared from desi spent hen. Vet World 6(7):419–423

Farris S, Piergiovanni L (2008) Effects of ingredients and process conditions on ‘Amaretti’cookies characteristics. Int J Food Sci Technol 43(8):1395–1403

Felisberto MHF, Beraldo AL, Costa MS et al (2019) Characterization of young bamboo culm starch from *Dendrocalamus asper*. Food Res Int 124:222–229

Ferreira VLP, Yotsuyanagi K, Carvalho CRL (1995) Elimination of cyanogenic compounds from bamboo shoots *Dendrocalamus giganteus* Munro. Trop Sci 35:342–346

Fu XC, Wang MW, Li SP et al (2005) Vasodilatation produced by orientin and its mechanism study. Biol Pharm Bull 28(1):37–41

Fujimura M, Ideguchi M, Minami Y et al (2005) Amino acid sequence and antimicrobial activity of chitin binding peptides, Pp-AMP 1 and Pp-AMP2, from Japanese bamboo shoots (*Phyllostachys pubescens*). Biosci Biotechnol Biochem 69:642–645

Giri SS, Janmejay L (2000) Effect of bamboo shoot fermentation and aging on nutritional and sensory qualities of soibum. J Food Sci Technol 37(4):423–426

Halvorson JJ, Cassida KA, Turner KE et al (2011) Nutritive value of bamboo as browse for livestock. Renewable Agric Food Syst 26:161–170

Haq N (2007) Underutilized fruits: a resource for sustainable livelihoods. Science Publishers, An imprint of Edenbridge Ltd., British Isles, pp 1–28

Hiromichi H (2007) Use of an antitumor agent European Patent EP1413208

Hong EJ, Jung EM, Lee GS et al (2010) Protective effects of the pyrolyzates derived from bamboo against neuronal damage and haemato aggregation. J Ethnopharmacol 128:594–599

Hu CH, Zhang Y, David DK (2000) Evaluation of antioxidant and prooxidant activities of bamboo *Phyllostachys nigra* var. henonis leaf extract in vitro. J Agric Food Chem 48(8):3170–3176

Hwang KM, Jung KO, Song CH et al (2008) Increased antimutagenic and anticlas-togenic effects of doenjang (Korean fermented soybean paste) prepared with bamboo salt. J Med Food 11 (4):717–722

Ingudam M, Sarangthem K (2016) Phytosterols content in different bamboo species of Manipur, India. Int J Curr Microbiol Appl Sci 5(8):943–948

Jeyaram K, Singh TA, Romi W et al (2009) Traditional fermented foods of Manipur. Indian J Tradit Knowl 8:115–121

Jiayan Y (2014) A comparative study of bamboo culture and its applications in China, Japan and South Korea. In: Proceedings of SOCIOINT14 - international conference on social sciences and humanities, Istanbul, 8–10 September 2014. ISBN: 978-605-64453-1-6

Kim HJ, Lee HO, Min DB (2007) Effects and prooxidant mechanisms of oxidized α-tocopherol on the oxidative stability of soybean oil. J Food Sci 72(4):223–230

Kithan N, Goswami S, Biswas S (2015) Traditional techniques for processing bamboo shoots in Northeast India as food resource. Int J Curr Res 7(3):13524–13528

Kozukue E, Kozukue N, Kurosaki T (1983) Organic acid, sugar and amino acid composition of bamboo shoots. J Food Sci 48(3):935–938

Kris-Etherton PM, Hecker KD, Bonanome A et al (2002) Bioactive compounds in foods: their role in the prevention of cardiovascular disease and cancer. Am J Med 113(9):71–88

Kumar PS, Kumari U, Devi MP et al (2017) Bamboo shoot as a source of nutraceuticals and bioactive compounds: a review. Indian J Nat Prod Resour 8(1):32–46

Kumbhare V, Bhargava A (2007) Effect of processing on nutritional value of central Indian bamboo shoots. Part I. J Food Sci Technol 44(1):29–31

Lachance PA, He YH (1998) Hypocholesterolemic composition from bamboo shoots. PCT International Patent PCT/US98/12556

Lattimer JM, Haub MD (2010) Effects of dietary fiber and its components on metabolic health. Nutrients 2(12):1266–1289

Li X, Guo J, Ji K et al (2016) Bamboo shoot fiber prevents obesity in mice by modulating the gut microbiota. Sci Rep 6:32953

Li Y, Ma G, Zhou Q et al (2020) Nutrient contents predict the bamboo-leaf-based diet of Assamese macaques living in limestone forests of Southwest Guangxi, China. Ecol Evol 10:5570–5581

Liu L, Liu L, Lu B et al (2013) Evaluation of bamboo shoot peptide preparation with angiotensin converting enzyme inhibitory and antioxidant abilities from by-products of canned bamboo shoots. J Agric Food Chem 61(23):5526–5533

Lu B, Wu X, Tie X et al (2005) Toxicology and safety of antioxidant of bamboo leaves. Part I: acute and sub chronic toxicity studies on antioxidant of bamboo leaves. Food Chem Toxicol 43 (5):783–792

Lu B, Wu X, Shi J et al (2006) Toxicology and safety of antioxidant of bamboo leaves. Part 2: developmental toxicity test in rats with antioxidant of bamboo leaves. Food Chem Toxicol 44 (10):1739–1743

Lu B, Ren Y, Zhang Y et al (2009a) Effects of genetic variability, parts and seasons on the sterol content and composition in bamboo shoots. Food Chem 112:1016–1021

Lu B, Bao J, Shan L et al (2009b) Technology for supercritical CO_2 extraction of bamboo shoot oil and components of product. Trans Chin Soc Agric Eng 25(8):312–316

Ma Y, Yang Y, Gao J et al (2020) Phenolics and antioxidant activity of bamboo leaves soup as affected by in vitro digestion. Food Chem Toxicol 135:110941

Madamba PS (2003) Physical changes in bamboo (*Bambusa phyllostachys*) shoot during hot air drying: shrinkage, density, and porosity. Dry Technol 21(3):555–568

Mäki-Arvela PI, Salmi T, Holmbom B et al (2011) Synthesis of sugars by hydrolysis of hemicelluloses: a review. Chem Rev 111(9):5638–5666

Maroma DP (2015) Utilization of bamboo shoots (*Bambusa vulgaris*) in chips production. Open Access Libr J 2(05):1

Maynard AJ (1970) Methods in food analysis. Academic, New York, p 176
Mellema M (2003) Mechanism and reduction of fat uptake in deep-fat fried foods. Trends Food Sci Technol 14:364–373
Miura M, Watanabe I, Shimotori Y et al (2013) Microbial conversion of bamboo hemicellulose hydrolysate to xylitol. Wood Sci Technol 47(3):515–522
Moon-Hee CH, Yang JH, Yang SH, et al (2017) The anti-inflammatory and anti-Adipogenic properties of bamboo (*Sasa coreana* Nakai) leaves extract. 한국생물공학회 학술대회 10:228
Mustafa U, Naeem N, Masood S (2016) Effect of bamboo powder supplementation on physico-chemical and organoleptic characteristics of fortified cookies. Food Sci Technol 4(1):7–13
Nimisha SM, Chauhan AS, Rekha MN et al (2015) Composition of edible portion of tender bamboo shoot (TBS) and development of various candies with and without incorporation of ginger and pineapple flavours. Am J Nutr Food Sci 2(1):7–15
Nirmala C, Bisht MS (2017) Bamboo: a prospective ingredient for functional food and nutraceuticals. Bamboo J 30:82–99
Nirmala C, David E, Sharma ML (2007) Changes in nutrient components during ageing of emerging juvenile bamboo shoots. Int J Food Sci Nutr 58(8):612–618
Nirmala C, Sharma ML, David E (2008) A comparative study of nutrient components of freshly harvested, fermented and canned bamboo shoots of *Dendrocalamus giganteus* Munro. J Am Bamboo Society 21(1):33–39
Nirmala C, Sheena H, David E (2009) Bamboo shoots: a rich source of dietary fibres. In: Klein F, Moller G (eds) Dietary fibres, fruit and vegetable consumption and health. Nova Science Publisher, Inc., New York, pp 173–187
Nirmala C, Bisht MS, Sheena H (2011) Nutritional properties of bamboo shoots: potential and prospects for utilization as a health food. Comp Rev Food Sci Food Saf 10:153–169
Nirmala C, Bisht MS, Laishram M (2014) Bioactive compounds in bamboo shoots: health benefits and prospects for developing functional foods. Int J Food Sci Technol 49:1425–1431
Nirmala C, Bisht MS, Bajwa HK et al (2018) Bamboo: a rich source of natural antioxidants and its applications in the food and pharmaceutical industry. Trends Food Sci Technol 77:91–99
Pandey AK, Ojha V (2013) Standardization of harvesting age of bamboo shoots with respect to nutritional and anti-nutritional components. J For Res 24:83–90
Pandey AK, Ojha V (2014) Precooking processing of bamboo shoots for removal of antinutrients. J Food Sci Technol 51(1):43–50
Pandey AK, Ojha V, Choubey SK (2012) Development and shelf-life evaluation of value-added edible products from bamboo shoots. Am J Food Technol 7(6):363–371
Park EJ, Jhon DY (2009) Effects of bamboo shoot consumption on lipid profiles and bowel function in healthy young women. Nutrition 25:723–728
Park EJ, Jhon DY (2010) The antioxidant, angiotensin converting enzyme inhibition activity, and phenolic compounds of bamboo shoot extracts. LWT-Food Sci Technol 43(4):655–659
Peng P, She D (2014) Isolation, structural characterization, and potential applications of hemi-celluloses from bamboo: a review. Carbohydr Polym 112:701–720
Phithakpol B, Varanyanond W, Reungmaneepaitoon S et al (1995) The traditional fermented foods of Thailand. Institute of Food Research and development. Kasetsart University, Bangkok
Prasad SN, Hegde MS, Gadgil M (1985) An experiment in eco-development in Uttara Kannada district of Karnataka. South Asian Anthropol 6(1):73–83
Qing Y, Zhu BD, Zheng LW et al (2008) Bamboo resources, utilization and ex-situ conservation in Xishuangbanna, South-Eastern China. J For Res 19(1):79–83
Qiu FG (1992) The recent development of bamboo foods. In: Proceedings of the international symposium on industrial use of bamboo. International Tropical Timber Organization and Chinese Academy, Beijing, pp 333–337
Rajyalakshmi P, Geervani P (1994) Nutritive value of the foods cultivated and consumed by the tribal of South India. Plant Foods Hum Nutr 46(1):53–61
Rawat N, Indrani D (2015) Functional ingredients of wheat-based bakery, traditional, pasta, and other food products. Food Rev Int 31(2):125–146

Rawat K, Sharma V, Saini N et al (2016) Impact of different boiling and soaking treatments on the release and retention of Antinutrients and nutrients from the edible shoots of three bamboo species. Am J Food Sci Nutr Res 3(3):31–41

Roos N, Ponce MC, Doak CM et al (2019) Micronutrient status of populations and preventive nutrition interventions in South East Asia. Matern Child Health J 23(1):29–45

Ryou SH, Kang MS, Kim KI et al (2012) Effects of green tea or *Sasa quelpaertensis* bamboo leaves on plasma and liver lipids, erythrocyte Na efflux, and platelet aggregation in ovariectomized rats. Nutr Res Pract 6(2):106–112

Saini N, Rawat K, Bisht MS et al (2017) Qualitative and quantitative mineral element variances in shoots of two edible bamboo species after processing and storage evaluated by wavelength dispersion x-ray fluorescence spectrometry. Int J Innov Res Sci Eng Technol 6(5):8265–8270

Sangeetha R, Diea YKT, Chaitra C et al (2015) The amazing bamboo: a review on its medicinal and pharmacological potential. Indian J Nutr 2(1):1–7

Sangtam TL, Jamir NS, Deb CR et al (2012) A study on the medicinal plants used by the Sangtam Naga tribe in Kiphire District, Nagaland, India. Int J Ayurvedic Herbal Med 2(2):267–275

Santosh O, Bajwa HK, Bisht MS (2018) Freeze-dried bamboo shoot powder for food fortification: enrichment of nutritional content and organoleptic qualities of fortified biscuits. MOJ Food Process Technol 6(4):342–348

Santosh O, Bajwa HK, Bisht MS et al (2019) Functional biscuits from bamboo shoots: enrichment of nutrients, bioactive compounds and minerals in bamboo shoot paste fortified biscuits. Int J Food Sci Nutr 4(1):89–94

Sarangthem K, Singh TN (2003) Transformation of fermented bamboo (*Dendrocalamus hamiltonii*) shoots into phytosterols by microorganisms. J Food Sci Technol 40:622–625

Sarangthem K, Singh TN (2013) Fermentation decreases the antinutritional content in bamboo shoots. Int J Curr Microbiol Appl Sci 2:361–369

Satya S, Singhal P, Prabhu VG et al (2009) Exploring the nutraceutical potential and food safety aspect of bamboo shoot of some Indian species. In: VIII World bamboo conference, Bangkok

Satya S, Bal LM, Singhal P et al (2010) Bamboo shoot processing: food quality and safety aspect (a review). Trends Food Sci Technol 21:181–189

Satya S, Singhal P, Bal LM et al (2012) Bamboo shoot: a potential source of food security. Mediterr J Nutr Metabol 5(1):1–10

Serdula M (2010) Maximizing the impact of flour fortification to improve vitamin and mineral nutrition in populations. Food Nutr Bull 31(1):S86–S93

Shanmugam S, Bora A, Raju A et al (2016) Application of antioxidant from bamboo shoot (*Bambusa balcooa*) as an effective way to reduce the formation of acrylamide in fried potato chips. Curr Nutr Food Sci 12(4):256–263

Shi QT, Yang KS (1992) Study on relationship between nutrients in bamboo shoots and human health. In: Bamboo and its use. Proceedings of the international symposium on industrial use of bamboo. International Tropical Timber Organization and Chinese Academy, Beijing, pp 338–346

Silva MF, Menis-Henrique MEC, Felisberto MHF et al (2020) Bamboo as an eco-friendly material for food and biotechnology industries. Curr Opin Food Sci 33:124–130

Singh A, Singh RK, Sureja AK (2007) Cultural significance and diversities of ethnic foods of Northeast India. Indian J Trad Knowl 6(1):79–94

Singhal P, Satya S, Sudhakar P (2011) Antioxidant and pharmaceutical potential of bamboo leaves. Bamboo Sci Cult 24(1):19–28

Singhal P, Bal LM, Satya S et al (2013) Bamboo shoots: a novel source of nutrition and medicine. Crit Rev Food Sci Nutr 53(5):517–534

Sood S, Walia S, Gupta M et al (2013) Nutritional characterization of shoots and other edible products of an edible bamboo–*Dendrocalamus hamiltonii*. Curr Res Nutr Food Sci J 1 (2):169–176

Srivastava RC (1990) Bamboo: new raw materials for phytosterols. Curr Sci 59:1333–1334

Tamang JP (2005) Food culture of Sikkim. In: Sikkim study series, vol IV. Department of Information and Public Relations, Govt. of Sikkim, Gangtok

Tamang JP, Tamang N, Thapa S et al (2012) Microorganisms and nutritional value of ethnic fermented foods and alcoholic beverages of north East India. Indian J Tradi Knowl 11(1):7–25

Tangkanakul P, Trakoontivakorn G, Auttaviboonkul P et al (2006) Antioxidant activity of northern and northeastern Thai foods containing indigenous vegetables. Kasetsart J (Nat Sci) 40:47–58

Tewari DN (1992) A monograph on bamboo. International Book Distributors, Dehradun, Uttaranchal

Thomas R, Jebin N, Barman K et al (2014) Quality and shelf life evaluation of pork nuggets incorporated with fermented bamboo shoot (*Bambusa polymorpha*) mince. Meat Sci 96 (3):1210–1218

Thomas R, Jebin N, Saha R et al (2016) Antioxidant and antimicrobial effects of kordoi (*Averrhoa carambola*) fruit juice and bamboo (*Bambusa polymorpha*) shoot extract in pork nuggets. Food Chem 190:41–49

Thounaojam P, Bisht MS, Nirmala C (2017) Effect of processing on nutritional and phytochemical contents in shoots of an edible bamboo *Dendrocalamus latiflorus* Munro. Int J Agric Sci 8 (1):79–87

Tripathi YC (1998) Food and nutrition potential of bamboo. MFP News 8(1):10–11

Tripathi YC (2011) Edible bamboo shoots - a nutritional appraisal. In: Singh S, Das R (eds) Productivity enhancement and value addition of bamboos. Excel India Publishers, New Delhi, pp 272–281

Tucker LA, Thomas KS (2009) Increasing total fiber intake reduces risk of weight and fat gains in women. J Nutr 139:576–581

Vanithakumari G, Manonayagi S, Padma S et al (1989) Antifertility effect of *Bambusa arundinacea* shoot extracts in male rats. J Ethnopharmacol 25(2):173–180

Visuphaka K (1985) The role of bamboo as a potential food source in Thailand. Recent research on bamboo. In: Proceedings of the international bamboo workshop, 6–14 October 1985

WHO (2010) Global status report on noncommunicable diseases. ISBN 978 92 4 156422 9

Wongsakpairod T (2000) Bamboo shoot drying using superheated steam. Meng Thesis, King Mongkut's University of Technology, Thonburi, Bangkok

Wróblewska KB, Baby AR, Guaratini MTG et al (2019) In-vitro antioxidant and photoprotective activity of five native Brazilian bamboo species. Ind Crop Prod 130:208–215

Xia NH (1989) Analysis of nutritive constituents of bamboo shoots in Guangdong. Acta Botanica Austro Sinica 4:199–206

Xie HK, Li A, Zhao MT et al (2020) Effects of antioxidants of bamboo leaves (AOB) on the oxidative susceptibility of glycerophosphocholine and glycerophosphoethanolamine in dried scallop (*Argopecten irradians*) adductor muscle during storage. LWT Food Sci Technol 134:110214

Xu Y, Zhang M, Tu D et al (2005) A two-stage convective air and vacuum freeze-drying technique for bamboo shoots. Int J Food Sci Technol 40(6):589–595

Xue Z, Zhang W, Wang L et al (2015) The bamboo-eating giant panda harbors a carnivore-like gut microbiota, with excessive seasonal variations. M Bio 6(3). https://doi.org/10.1128/mBio.00022-15

Yang JH, Lim HS, Heo YR (2010) *Sasa borealis* leaves extract improves insulin resistance by modulating inflammatory cytokine secretion in high fat diet-induced obese C57/BL6J Mice. Nutr Res Pract 4(2):99–105

Yang JP, He H, Lu YH (2014) Four flavonoid compounds from *Phyllostachys edulis* leaf extract retard the digestion of starch and its working mechanisms. J Agric Food Chem 62 (31):7760–7770

Young RA (1954) Flavor qualities of some edible oriental bamboos. Econ Bot 8(4):377–386

Young GP, Hu Y, Leu L et al (2005) Dietary fibre and colorectal cancer: a model for environment—gene interactions. Mol Nutr Food Res 49(6):571–584

Yuming Y et al (1999) Bamboo resources and their utilization in China. In: Rao AN, Ramanatha Rao V (eds) Proceedings of training course cum workshop, 10–17 May 1998

Zeng H, Chen J, Zhai J et al (2016) Reduction of the fat content of battered and breaded fish balls during deep-fat frying using fermented bamboo shoot dietary fiber. LWT Food Sci Technol 73:425–431

Zhang Y, Yao X, Bao B et al (2006) Anti-fatigue activity of atriterpenoid-rich extract from Chinese bamboo shavings *(Caulis bamfusae* in Taeniam). Phytother Res 20(10):872–876

Zhang H, Zhang Y, Wang X et al (2017) Effects of bamboo shoot dietary fiber on mechanical properties, moisture distribution, and microstructure of frozen dough. J Chem 2017:Article ID 4513410. https://doi.org/10.1155/2017/4513410

Zheng J, Wu J, Dai Y et al (2017) Influence of bamboo shoot dietary fiber on the rheological and textural properties of milk pudding. LWT Food Sci Technol 84:364–369

Zhu D, Wang C, Zhang Y et al (2018) Insight into solvent effects on phenolic content and antioxidant activity of bamboo leaves extracts by HPLC analysis. J Food Meas Charac 12 (3):2240–2246

Chapter 18
Functional Pasta: A Comparative Study of the Use of Bamboo Fibers and White Fibers

Amanda Rios Ferreira, Felipe Gianasi, Thaisa de Menezes Alves Moro, Mária Herminia Ferrari Felisberto, Elisa Cristina Andrade Neves, and Maria Teresa Pedrosa Silva Clerici

Abstract Fiber consumption is related to health benefits, and considering this approach, the production of main meals with high fiber content may increase the daily fiber intake and provide income to communities depending on bamboo. This chapter presents the study of *fettuccine* formulation elaborated with bamboo fibers (BF) (60 μm and 145 μm, respectively BFA and BFB), in comparison with white fibers: *Psyllium* and cellulose fiber (PCF) (80:20 and 50:50, respectively PCFA and PCFB), wheat stalk fiber (WSF) (60 μm and 145 μm, respectively WSFA and WSFB), and a mixture of *Psyllium* and BF (PBF). *Fettuccine* was produced and evaluated by technological characteristics of replacing *Triticum durum* semolina by 3.5% and 7% of each fiber, totaling 14 trials plus the standard formulation (SF), 100% semolina, and the obtained data were analyzed by analysis of variance

A. R. Ferreira · F. Gianasi · M. T. P. S. Clerici (✉)
Department of Food Science and Nutrition, University of Campinas (UNICAMP), Campinas, SP, Brazil
e-mail: mclerici@unicamp.br

T. de Menezes Alves Moro
Department of Food Science and Nutrition, University of Campinas (UNICAMP), Campinas, SP, Brazil

School of Nutrition, Santa Ursula University (USU), Rio de Janeiro, RJ, Brazil

M. H. F. Felisberto (✉)
Department of Food Science and Nutrition, University of Campinas (UNICAMP), Campinas, SP, Brazil

Department of Food Technology (DTA), Federal University of Viçosa (UFV), Viçosa, MG, Brazil
e-mail: maria.felisberto@ufv.br

E. C. A. Neves
Department of Food Science and Nutrition, University of Campinas (UNICAMP), Campinas, SP, Brazil

School of Food Engineering, Institute of Technology, Federal University of Pará (UFPA), Belém, PA, Brazil

Z. Ahmad et al. (eds.), *Biotechnological Advances in Bamboo*,
https://doi.org/10.1007/978-981-16-1310-4_18

($p < 0.05$). All formulations showed a lighter color; pasta with BFA and WSFA presented the most comparable parameters to the standard formulation. The results suggest that it is possible to use up to 7% of BF in pasta with no prejudice to color, texture, and flavor. Thus, pasta could be a gateway to introduce its essential raw material as an ingredient in food industries and provide regional economic benefits.

Keywords Bamboo · Healthiness · Nondigestible carbohydrates · Sustainability

Abbreviations

BF	Bamboo fibers
BFA	Commercial dietary fiber composed of bamboo fiber with a granulometry of 60 micrometers, known commercially as QC90
BFB	Commercial dietary fiber composed of bamboo fiber with a granulometry of 145 micrometers, known commercially as QC200
LS	Loss of soluble solids in cooking
OCT	Optimal cooking time
PBF	Commercial dietary fiber composed of 20:80 *Psyllium* fiber and bamboo fiber, known commercially as Q80
PCF	*Psyllium* and cellulose fiber
PCFA	Commercial dietary fiber composed of 80:20 *Psyllium* fiber and cellulose fiber, known commercially as S20
PCFB	Commercial dietary fiber composed of 50:50 *Psyllium* fiber and cellulose fiber, known commercially as S50
SF	Standard formulation
SI	Swelling index
WG	Weight gain
WSF	Wheat stalk fiber
WSFA	Commercial dietary fiber of wheat stalk with a granulometry of 60 micrometers, known commercially as WC90
WSFB	Commercial dietary fiber of wheat stalk with a granulometry of 145 micrometers, known commercially as WC200

18.1 Introduction

The International Pasta Organization shows that the world produced 16.5 million tons of pasta in 2019 (IPO International Pasta Organisation 2020). Despite this, the statistics on consumption and sale of whole-grain pasta is not yet established. The main ingredient for pasta production is *Triticum durum* semolina due to its protein composition and more crystalline structure, which allows obtaining a firm and yellowish food matrix (Kill and Turnbull 2001).

For the production of whole-grain pasta, whole wheat flour or wheat bran is used, which can both negatively affect not only the technical characteristics (Kill and

Turnbull 2001) but also the sensory characteristics of the product, being perceived by the consumer with a grassy and bitter taste, limiting the concentration of its incorporation (West et al. 2013). To improve the sensory quality of pasta, fibers of another nature have been tested, as shown in studies by Foschia et al. (2015). They evaluated the incorporation of *Psyllium* sp. fiber, wheat fiber, and hemicellulose in pasta formulations and observed superior technological characteristics for whole pasta with bran and wheat.

Ajila et al. (2015) evaluated a partial substitution of semolina (3%, 5%, and 7%) by powdered mango peel in pasta production, seeking to increase the fiber content and improve nutritional quality. They found that concerning cooking characteristics, the highest level of substitution presented the highest values of loss of soluble solids, having increased from 5.84% to 8.71%. The authors conducted the pasta sensory analysis for color, texture, flavor, and overall impression and observed the lowest note of flavor (5.1) for the highest level of substitution, while the control presented 8.1 and the overall evaluation presented 5.9 against 7.8 of the control.

Other ingredients, such as broad bean flour (Tazrart et al. 2016), tomatoes (Pasqualone et al. 2016), legumes (Pasqualone et al. 2017), cereals (Kosović et al. 2018), wheat bran (Alzuwaid et al. 2020), cereal coffee (Biernacka et al. 2020), oat bran (Levent et al. 2020), and inulin (Aravind et al. 2012), have already been used in the preparation of pasta, in order to delay the release of postprandial glucose with consequent reduction of the glycemic index (GI) and nutritional improvements.

Considering that most of the fiber consumption comes from the main meals and also from cereals or other minimally processed products obtained from them, as is the case with pasta, the incorporation of fibers in this kind of food whose consumption habit is already established can be a way to increase the daily fiber intake. The World Health Organization recommends the intake of 25 g of dietary fiber per day for a healthy adult, which unfortunately is not adopted in all countries related to unhealthy parameters for preventing diet-related chronic diseases (WHO 2009). The consumption of more than 25–29 g of dietary fiber/day is related to a lower risk of diabetes, breast cancer, and other comorbidities (Reynolds et al. 2019).

Bamboo clumps are scattered worldwide in regions with a tropical climate, where a large part of the world population lives and where hunger and food security plague populations. Despite this ability to multiply without management and live for over a hundred years, the incorporation of bamboo in food habits is strong mostly in Asian countries with the consumption of bamboo shoots (Nongdam and Tikendra 2014), but not yet as a food ingredient and in several countries probably due to the need to boil and prepare the bamboo shoot to remove cyanogenic compounds (Wang et al. 2020). It was only recently that Ferreira et al. (2020) evaluates bamboo fiber incorporation in whole-grain cookie formulations.

Assessing the viability of pasta production from commercial bamboo fibers could present a way to encourage the production of bamboo flours in communities or regions where this plant has an impact, which can favor not only trade but also quality improvement of food and the establishment of a new source of income and maintenance of habits, considering the occurrence of pandemics, such as that of Covid19 (Kissler et al. 2020). The present study aimed to evaluate the performance of two different bamboo fibers (60 and 145 micrometers) in partial replacement of

Triticum durum semolina for *fettuccine* pasta production and the comparison of this performance with other white and pale-yellow commercial fibers such as *Psyllium* sp. fiber, cellulose fiber, and wheat stalk fiber, in pasta formulations.

18.2 Material and Methods

18.2.1 Raw Materials

Durum wheat semolina was donated by Pastificio Selmi S/A (Sumaré, Brazil). The commercial fibers donated by Nutrassim Indústria, Comércio, Importação e Exportação LTDA (Extrema, Brazil) were bamboo fibers (BF) (60 μm and 145 μm, respectively BFA and BFB), *Psyllium* and cellulose fiber (PCF) (80:20 and 50:50, respectively PCFA and PCFB), wheat stalk fiber (WSF) (60 μm and 145 μm, respectively WSFA and WSFB), and a mixture of *Psyllium* and BF (PBF).

18.2.2 Characterization of Semolina and Commercial Fibers

Semolina was characterized by moisture content, ether extract, proteins, ash, (methods 44–15.02, 30–25.01, 46–13.01, 08–01.01, respectively), and total fiber through the Megazyme kit (K-TDFR 05/12), based on the American Association Cereal Chemists International (2010) (method 32–05.01) and Association Official Analytical Chemists (2007) (method 985.29). The total carbohydrate content was calculated by difference (100 – moisture – ether extract – proteins – ash –total fibers) according to AACCI methodology (2010). The rheological characteristics of the semolina were evaluated using the farinograph method 54–21.02 (AACCI 2010).

The commercial fibers were evaluated for their moisture content (AACCI 2010), and the other characteristics were obtained through a report provided by the manufacturer.

The color parameters of the raw materials were analyzed on an opaque white background, by the CIELab system, using a MiniScan XE portable colorimeter from Hunter Associates Laboratory, Inc. (Reston, USA). Six dough strips were analyzed before and after cooking for parameters L*, a*, and b* with D65 illuminant and 10° observation angle.

18.2.3 Pasta Production

Fourteen pasta formulations were produced with partial replacement of wheat semolina by 3.5% and 7% of selected commercial fiber and the standard formulation (SF) with wheat semolina and water only, as can be seen in Table 18.1.

Table 18.1 Formulations used in the preparation of pasta with fibers and standard formulation (without commercial fibers)

Formulations[a]	Formulation (%)			Fiber
	Fiber	Semolina	Water	
BFA1	3.5	96.5	45	BFA
BFA2	7	93	45	
BFB1	3.5	96.5	45	BFB
BFB2	7	93	45	
PCFA1	3.5	96.5	45	PCFA
PCFA2	7	93	45	
PCFB1	3.5	96.5	45	PCFB
PCFB2	7	93	45	
WSFA1	3.5	96.5	45	WSFA
WSFA2	7	93	45	
WFSB1	3.5	96.5	45	WSFB
WFSB2	7	93	45	
PBF1	3.5	96.5	45	PBF
PBF2	7	93	45	
SF	–	100	33	–

[a]BFA (bamboo fiber with 60 μm), BFB (bamboo fiber with 145 μm), PCFA (80% *Psyllium* and 20% cellulose fiber), PCFB (50% *Psyllium* and 50% cellulose fiber), WSFA (wheat stalk fiber with 60 μm), WSFB (wheat stalk fiber with 145 μm), PBF (a mixture of *Psyllium* and bamboo fiber) and SF (standard formulation)

The manufacture of the pasta consisted of mixing the semolina with water for the standard formulation and to the pasta with fibers (Table 18.1) consisted in the hydration of the fibers for 24 h with 45% water, followed by incorporating the semolina into the hydrated fiber in Pastaia (Italvisa, Brazil); after mixing and standing for 5 min, the dough was cold extruded in the *fettuccine* format. The pasta was dried in an oven (at 55 °C) until moisture content is below 13%, and after being cooled to room temperature (25 °C), it was packaged, sealed, and stored (Fig. 18.1).

18.2.4 Cooking Performance of Pasta

The optimal cooking time (OCT), weight gain (WG), swelling index (SI), and loss of soluble solids (LS) were determined according to the AACCI methodology n°16–50 (2010), as well as the cutting force of the pasta, 10 min after cooking stopped at its optimum time, on Texturometer Stable Micro-System, model TA-XT2i (Surrey, United Kingdom), using a knife blade probe (A/LKB) as described by Lemes et al. (2012).

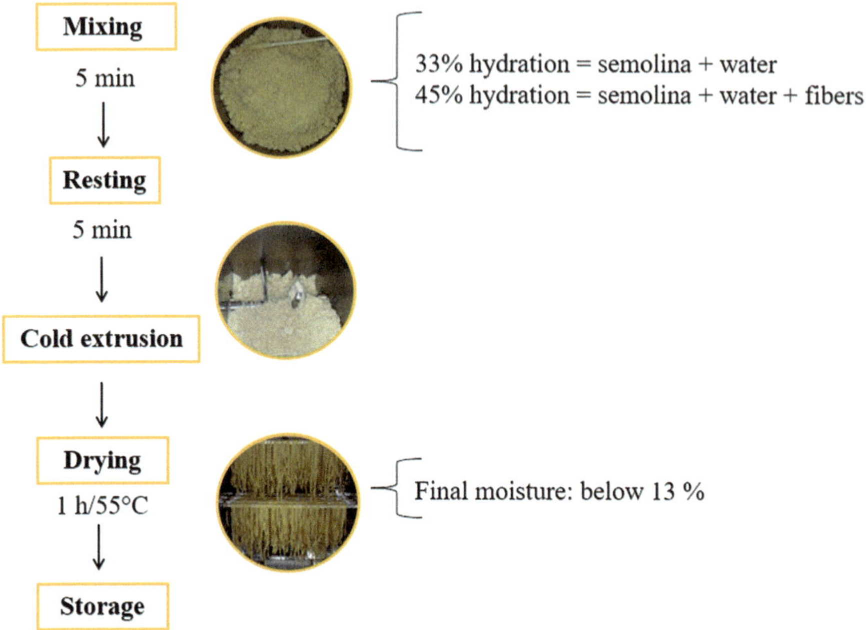

Fig. 18.1 Process flow diagram for the production of dry pasta with or without fibers

18.2.5 Technological Assessment of Pasta

Color parameters of dry and cooked pasta were obtained according to item 18.2.2.

The total dietary fiber content of the pasta was obtained through theoretical calculation considering the fiber content of the raw materials and the ready-to-eat product.

18.2.6 Statistical Analysis

All analyses were performed in triplicate, except for the cutting force performed in ten replicates. The data obtained were submitted to analysis of variance (ANOVA) followed by the Scott-Knott test when necessary, with a significance level of $p < 0.05$ using the SISVAR 5.6 program.

18.3 Results and Discussion

18.3.1 Characterization of Semolina and Commercial Fibers

As the pasta consists of a protein network that involves the starch granules, the incorporation of fibers in pasta causes the dilution of the gluten network and can cause it to break, increasing the loss of water-soluble solids, for example (Vignola et al. 2018). In this sense, to have considerable protein content is essential. The physicochemical characterization of the raw materials is presented in Table 18.2. Semolina showed moisture content of 13.47%, total ash content of 0.75 ± 0.01%, protein content of 12.87 ± 0.07%, ether extract of 1.2 ± 0.01%, total carbohydrate content of 65.18 ± 0.08%, water absorption on the farinograph of 53.65 ± 0.07 mL, and stability of 18.25 ± 0.07 min. The proximate composition is aligned with the results obtained by Gull et al. (2015).

Color parameters of semolina presented high L* parameter, on a scale where 100 is white and 0 is black; the parameter a* varies from –a, which is green, and + a to which is red; finally, parameter b* ranges from –b, which is blue, to +b, which is yellow. Considering the obtained values for semolina (L* 87.17, a* 1.58, and b* 20.86), we observed a light-yellow color, as can also be seen in Fig. 18.2. The behavior of increasing the L* value in enriched pasta is uncommon. Gull et al. (2015) evaluated durum wheat semolina replacement by finger millet flour, pearl millet flour, and carrot pomace powder in pasta making and observed a slight decrease in L* value for all raw and cooked pasta samples.

Knowing the moisture content of the fibers is critical regarding the stability of the pasta after drying since the sorption kinetics and the thermal characteristics of the doughs are altered depending on the presence of the fibers and the type of fibers incorporated into the formulation (Witczaka and Gałkowska 2021). Concerning the physicochemical characteristics of the commercial fibers, the moisture content of the samples varied from 3.66% for BFA to 11.05% for BFB. The fiber content was the lowest for WSFA, with 84.23%, and the highest for BFA. All fibers presented a white or light-yellow color, as shown in the color parameters and the visual aspect (Fig. 18.2).

18.3.2 Cooking Performance of Pasta

Regarding the pasta quality assessment (Table 18.3), only one-third of the pasta produced and evaluated in this study showed a homogeneous appearance, which indicates that the addition of some fibers, especially in higher concentrations, could be perceived by consumers and cause the rejection of the product. The color and ability of the fibers to mix with the other ingredients of the dough is an important parameter when considering food incorporation both for the bran used

Table 18.2 Physicochemical characteristics of raw materials used for pasta production

Raw material[a]	Moisture (%)	Fiber (%)[b]	Color parameters			Aspect
			L*	a*	b*	
BFA	3.66 ± 0.15	95.44	96.85 ± 0.40	−0.53 ± 0.015	4.93 ± 0.10	White
BFB	11.05 ± 0.40	85.75	95.28 ± 1.43	−0.58 ± 0.060	4.82 ± 0.03	White
PCFA	6.90 ± 0.11	92.54	87.96 ± 0.63	−1.91 ± 0.040	11.04 ± 0.23	Light beige
PCFB	9.65 ± 0.12	78.15	90.91 ± 1.63	−1.13 ± 0.120	8.97 ± 0.20	Light beige
WFSA	7.70 ± 0.24	84.23	95.77 ± 1.19	−0.81 ± 0.040	4.27 ± 0.46	White
WFSB	4.63 ± 0.18	89.91	94.11 ± 0.32	−0.017 ± 0.040	4.79 ± 0.05	White
PBF	6.19 ± 0.18	92.94	93.64 ± 0.65	−0.03 ± 0.020	6.9 ± 0.16	Light beige
Semolina	13.47 ± 0.09	5.79 ± 0.65	87.17 ± 0.80	−1.58 ± 0.040	20.86 ± 0.59	Light beige

[a]Where BFA (bamboo fiber with 60 μm), BFB (bamboo fiber with 145 μm), PCFA (80% *Psyllium* and 20% cellulose fiber), PCFB (50% *Psyllium* and 50% cellulose fiber), WSFA (wheat stalk fiber with 60 μm), WSFB (wheat stalk fiber with 145 μm), and PBF (a mixture of *Psyllium* and bamboo fiber)

[b]Semolina fiber content was determined by total fiber (AACCI, 2010 - method 32–05.01 and AOAC, 2007 - method 985.29), and the commercial fiber data of dietary was provided by the manufacturer, all of them on a dry basis

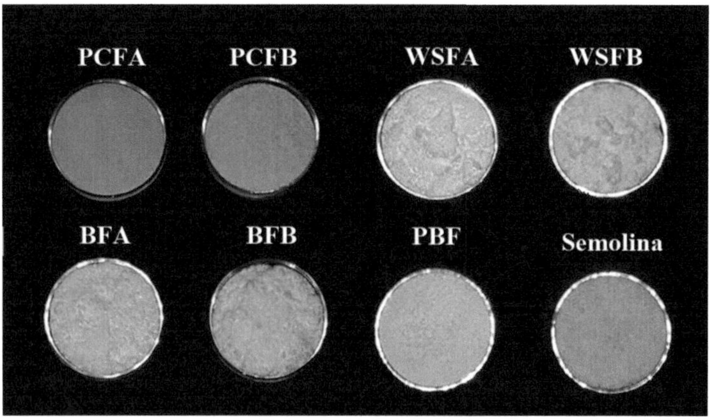

Fig. 18.2 Commercial fibers and semolina used in the formulation of pasta where BFA (bamboo fiber with 60 μm), BFB (bamboo fiber with 145 μm), PCFA (80% *Psyllium* and 20% cellulose fiber), PCFB (50% *Psyllium* and 50% cellulose fiber), WSFA (wheat stalk fiber with 60 μm), WSFB (wheat stalk fiber with 145 μm), and PBF (a mixture of *Psyllium* and bamboo fiber)

commercially, which has a brownish color (Stevenson et al. 2012), and for the white fibers of this study.

We also observed that the presence of non-wheat ingredients such as fibers in pasta formulation results in a different microstructure, cooking performance (Table 18.3), and quality of the cooked product, a tendency that was well established in a review by Mercier et al. (2016). All pasta presented a higher OCT than the standard formulation. The relationship between protein coagulation and starch gelatinization in the presence of each fiber may justify this behavior. This same trend was observed in the study of Foschia et al. (2015), who replaced semolina with short- and long-chain inulin, Glucagel, *Psyllium*, and oat derivatives. The authors observed that pasta produced from the replacement of semolina by all fibers presented a higher OCT than the standard pasta (only semolina); moreover, all pasta formulations presented increased LS than the standard, which proportionally increased with the amount of substitution.

The incorporation of fibers or even bran in pasta dough can result in changes in the water absorption of the dough which is directly related to the characteristics of each fiber, its water solubility, and particle size (Cho and Dreher 2001). The weight gains and swelling index (SI) are essential parameters to the perception of a consumer: Pasta that shows less weight gain could cause rejection. On the other hand, if the consumer could choose healthier pasta, with high fiber content, but with no changes in the appearance, weight gain, color parameters, and volume increase, it could cause less rejection.

The weight gains in this study varied between 2.21 and 3.54 times the weight of uncooked pasta, the SI varied between 3.14 and 7.25, and for neither of the two parameters, a significant difference was observed. Brennan and Tudorica (2007) evaluated pasta produced by replacing bread wheat flour with resistant starch II

Table 18.3 Results of technological characterization through the cooking test of pasta produced with substitution of wheat semolina in concentrations of 3.5% and 7% with white commercial fibers and standard formulation (without commercial fibers)

Pasta[a]	Aspect	Cooking performance of pasta[b]				
		OCT (min)[c]	Weight gain	Swelling index	Loss of solids[d] (g/100 g)	Force to cut (N)
BFA1	Homogeneous	13	2.62 ± 0.31^{ns}	3.43 ± 0.42^{ns}	6.70 ± 0.09^{c}	5.56 ± 0.41^{d}
BFA2	Homogeneous	13	2.82 ± 0.29^{ns}	3.57 ± 0.40^{ns}	7.65 ± 0.64^{b}	5.58 ± 0.35^{d}
BFB1	Heterogeneous	12.5	3.54 ± 0.44^{ns}	3.71 ± 0.38^{ns}	8.32 ± 1.58^{b}	5.86 ± 0.38^{d}
BFB2	Heterogeneous	13	2.85 ± 0.09^{ns}	4.33 ± 0.38^{ns}	7.14 ± 0.17^{b}	6.17 ± 0.42^{c}
PCFA1	Heterogeneous	12.5	2.96 ± 0.36^{ns}	7.25 ± 0.34^{ns}	8.81 ± 0.60^{b}	6.54 ± 0.63^{b}
PCFA2	Heterogeneous	12.5	2.96 ± 0.01^{ns}	3.50 ± 0.36^{ns}	5.47 ± 0.08^{c}	7.48 ± 0.42^{a}
PCFB1	Heterogeneous	13	2.35 ± 0.05^{ns}	4.29 ± 0.33^{ns}	6.07 ± 0.53^{c}	6.42 ± 0.24^{b}
PCFB2	Heterogeneous	13	2.91 ± 0.01^{ns}	3.86 ± 0.37^{ns}	11.80 ± 2.22^{a}	6.64 ± 0.37^{b}
WSFA1	Homogeneous	12.5	2.69 ± 0.15^{ns}	3.57 ± 0.40^{ns}	7.35 ± 1.23^{b}	6.52 ± 0.51^{b}
WSFA2	Homogeneous	14	2.21 ± 0.01^{ns}	3.14 ± 0.45^{ns}	6.68 ± 0.40^{c}	6.33 ± 0.38^{c}
WFSB1	Heterogeneous	12	3.03 ± 0.37^{ns}	4.67 ± 0.35^{ns}	5.80 ± 0.48^{c}	5.35 ± 0.31^{e}
WFSB2	Heterogeneous	12.5	2.89 ± 0.07^{ns}	3.38 ± 0.37^{ns}	6.52 ± 0.52^{c}	6.14 ± 0.37^{c}
PBF1	Heterogeneous	13	2.82 ± 0.26^{ns}	4.17 ± 0.40^{ns}	5.65 ± 0.19^{c}	5.92 ± 0.56^{d}
PBF2	Heterogeneous	13	2.79 ± 0.14^{ns}	5.20 ± 0.38^{ns}	7.88 ± 0.88^{b}	6.61 ± 0.43^{b}
SF	Homogeneous	11.5	3.13 ± 0.20^{ns}	3.38 ± 0.37^{ns}	6.33 ± 0.48^{c}	5.03 ± 0.37^{e}

[a]Where 1 and 2 = dry pasta formulation with partial replacement of wheat semolina by 3.5% and 7%, respectively, of each fiber; BFA (bamboo fiber with 60 μm), BFB (bamboo fiber with 145 μm), PCFA (80% *Psyllium* and 20% cellulose fiber), PCFB (50% *Psyllium* and 50% cellulose fiber), WSFA (wheat stalk fiber with 60 μm), WSFB (wheat stalk fiber with 145 μm -), PBF (a mixture of *Psyllium* and bamboo fiber) and SF (standard formulation)

[b]Results expressed as mean ± standard deviation, where data followed by different letters in the same column show a significant Scott-Knott difference ($p < 0.05$) and ns = not significant

[c]OCT: optimal cooking time

[d]LS: loss of soluble solids

(RSII), resistant starch IV (RSIV), oat bran, and inulin in different concentrations. They observed that water absorption and SI were higher in pasta with fiber when compared to the control pasta.

The loss of solids is also an important parameter related to consumer acceptance. The produced pasta presented values of loss of solids between 5.47 (g/100 g) and 11.80 (g/100 g). Our results are in line with the study of Tudorica et al. (2002) that used different fibers in pasta such as inulin (IN), guar gum, and pea fiber (PF) in the production of spaghetti. These authors found that pasta with PF (6.99 ± 0.3 g/100 g up to 7.41 ± 0.72 g/100 g) and IN (6.72 ± 0.3 g/100 g up to 8.06 ± 0.39 g/100 g) increased the loss of cooking solids (LS) of the pasta when compared to the standard formulation (5.06 ± 0.11 g/100 g). It was probably observed due to the great interference of fibers in the protein and starch matrix, in addition to competition from water between the fiber and these components, which can hinder the swelling of the starch granule. For our study, pasta with solid loss of up to 8% was considered ideal. Thus, pasta PCFA1 (08.81 ± 0.60), PCFB2 (05.47 ± 0.08), and BFB1 (08.32 ± 1.58) did not achieve this characteristic.

Although the effects and benefits of incorporating dietary fibers into pasta are well known, maintaining the cooking and texture characteristics is a challenge faced by the food industry. The results of the cutting force of our work varied between 5.03 and 7.48 N, with PCFA2 being a mixture of *Psyllium* fiber and cellulose fiber, which presented the highest cutting force. Regarding previous studies carried out with the enrichment of pasta with fibers, such as barley (Brennan and Simons 2004), the origin of fibers and its percentage of substitution in the pasta interfered with the cutting force of the product. The effect of this incorporation on texture parameters was deeper evaluated in a study that evaluates the differences between technological characteristics of refined wheat flour pasta and whole-grain wheat flour pasta, where the presence of bran in the whole-grain pasta caused an increase on the texture parameters (Vignola et al. 2018).

Remarkably, the fibers of this study were different in solubility and characteristics. PCFA, PCFB, and PBF are blends of soluble and insoluble dietary fibers while BFA, BFB, WSFA, and WSFB are versatile insoluble dietary fibers (CreaFill 2020), and this difference could play a role in the action of fibers in the human metabolism. On the other hand, concerning the technological parameters of pasta and not anymore the human metabolism, Rakhesh et al. (2014) evaluated the sensory and technical characteristics of pasta added with fibers and observed that the addition affected the gluten network, which could be perceived in aspects such as loss of cooking solids and firmness. However, this effect could not be generalized as a function of the solubility or source of the fibers since fibers of the same origin, when obtained in different ways, presented different effects.

18.3.3 Technological Assessment of Pasta

Table 18.4 shows the color parameters and the theoretical calculation of the fiber content of pasta. It is noticed that the color difference (ΔE) between the raw pasta

Table 18.4 Color parameters of dry and cooked pasta and dietary fiber content of pasta produced with the replacement of *T. durum* semolina by white commercial fibers and standard formulation (without commercial fibers)

Pasta[a]	Technological assessment of pasta[b]								
	Dry pasta				Cooked pasta				
	L*	a*	b*	ΔE	L*	a*	b*	Δ*E*	Dietary fiber (%)[c]
BFA1	74.69 ± 1.43^{c}	-0.23 ± 0.07^{d}	34.50 ± 1.68^{b}	3.90	77.90 ± 0.64^{b}	-3.24 ± 0.16^{d}	27.77 ± 0.39^{a}	9.77	8.92
BFA2	77.56 ± 1.01^{a}	-0.13 ± 0.24^{d}	28.21 ± 0.73^{d}	3.05	80.13 ± 0.44^{a}	-3.37 ± 0.12^{d}	23.28 ± 1.01^{c}	6.40	12.06
BFB1	79.43 ± 0.66^{a}	-0.19 ± 0.21^{d}	28.67 ± 0.81^{c}	4.01	79.64 ± 0.59^{a}	-3.50 ± 0.03^{d}	24.91 ± 1.43^{c}	7.52	8.58
BFB2	78.14 ± 0.95^{a}	-0.32 ± 0.14^{d}	30.65 ± 1.28^{d}	2.09	79.89 ± 1.13^{a}	-3.55 ± 0.01^{d}	24.45 ± 1.92^{b}	7.23	11.38
PCFA1	72.70 ± 0.03^{c}	-1.50 ± 0.12^{b}	32.99 ± 1.98^{b}	4.32	76.56 ± 0.68^{c}	-1.98 ± 0.15^{b}	25.92 ± 1.71^{b}	8.08	8.87
PCFA2	73.57 ± 0.72^{c}	-3.00 ± 0.18^{a}	29.58 ± 0.83^{d}	4.23	75.14 ± 0.50^{c}	-0.97 ± 0.12^{a}	22.61 ± 0.19^{c}	5.66	11.85
PCFB1	73.11 ± 0.61^{c}	-1.63 ± 0.33^{b}	33.72 ± 0.79^{b}	4.49	78.65 ± 0.99^{a}	-2.54 ± 0.21^{c}	23.69 ± 1.28^{c}	6.21	8.31
PCFB2	74.39 ± 1.01^{c}	-3.16 ± 0.16^{a}	31.32 ± 1.25^{c}	3.73	76.15 ± 2.05^{c}	-1.10 ± 0.11^{a}	24.32 ± 0.96^{c}	6.91	10.85
WSFA1	76.05 ± 0.72^{a}	-0.75 ± 0.35^{c}	36.81 ± 0.60^{a}	5.99	79.11 ± 0.19^{a}	-3.71 ± 0.02^{d}	24.11 ± 1.29^{c}	6.56	8.52
WSFA2	77.34 ± 1.11^{a}	-0.28 ± 0.16^{d}	33.21 ± 1.48^{b}	2.67	79.45 ± 0.28^{a}	-3.42 ± 0.06^{d}	24.87 ± 1.56^{b}	7.41	11.27
WFSB1	78.38 ± 0.25^{a}	-0.15 ± 0.17^{d}	32.14 ± 0.82^{b}	2.62	81.04 ± 0.44^{a}	-3.39 ± 0.06^{d}	22.75 ± 1.18^{c}	6.60	8.72
WFSB2	77.47 ± 0.78^{a}	-0.92 ± 0.17^{c}	35.46 ± 0.77^{c}	4.89	79.97 ± 0.21^{a}	-2.86 ± 026^{c}	22.50 ± 0.44^{c}	5.80	11.67
PBF1	73.75 ± 1.37^{c}	-0.66 ± 0.35^{c}	34.63 ± 1.85^{b}	4.50	77.88 ± 0.51^{b}	-3.09 ± 0.03^{b}	24.96 ± 1.10^{b}	7.05	8.83
PBF2	74.40 ± 0.49^{c}	-0.86 ± 0.13^{c}	33.49 ± 2.07^{b}	3.27	77.20 ± 0.10^{b}	-2.74 ± 0.05^{c}	25.67 ± 0.72^{b}	7.69	11.88
SF	76.10 ± 1.46^{b}	-0.13 ± 0.18^{d}	30.88 ± 0.87^{c}	0.00	76.36 ± 0.36^{c}	-4.28 ± 0.13^{d}	18.18 ± 1.18^{d}	0.00	5.79 ± 0.65^{d}

[a]Where 1 and 2 = dry pasta formulation with partial replacement of wheat semolina by 3.5% and 7%, respectively, of each fiber; BFA (bamboo fiber with 60 μm), BFB (bamboo fiber with 145 μm), PCFA (80% *Psyllium* and 20% cellulose fiber), PCFB (50% *Psyllium* and 50% cellulose fiber), WSFA (wheat stalk fiber with 60 μm), WSFB (wheat stalk fiber with 145 μm –), PBF (a mixture of *Psyllium* and bamboo fiber) and SF (standard formulation)
[b]Results expressed as mean ± standard deviation, on a dry basis, where data followed by different letters in the same column show a significant Scott-Knott difference ($p < 0.05$) and ns = not significant
[c]Theoretical calculation taking into account the characteristics of the raw materials presented in Table 18.1 and the formulation presented in Table 18.2
[d]Same fiber composition of semolina

with fibers and the standard pasta was classified as perceptible to the naked eye ($3.5 < \Delta E < 5$). For pasta after cooking, the variation was higher than 5; that is, any untrained consumer could perceive the difference between the semolina standard pasta and the pasta with fibers, according to the scale proposed by Mokrzycki and Tatol (2011).

The yellowish color of durum wheat products, derived from carotenoids, is an essential characteristic from a commercial point of view. According to Borelli et al. (2008), there is a relationship between the yellowish color of semolina and the color of the pasta produced, which can be explained by the degradation of carotenoids by the action of lipoxygenase and by the heat treatment during the processing (drying and cooking steps) of the pasta.

The pasta with a mixture of *Psyllium* fiber and cellulose fiber (PCFA1 and PCFA2) and *Psyllium* fiber and bamboo fiber (PBF1 and PBF2) presented lower luminosity than the SF, while the other tests with bamboo and wheat fibers showed higher luminosity than the SF. Studies with the incorporation of other fibers, such as wheat bran and oats, showed that the addition resulted in darker pasta compared to the standard (Bustos et al. 2015).

Concerning the theoretical content of fiber, while the standard semolina pasta presented a dietary fiber content of less than 6%, all pasta with fiber presented fiber content higher than 8.5%, contributing to the consumer's daily intake. In addition to the individual benefits of pasta with high fiber content for consumers, these could contribute to the companies, which could declare health claims on their labels depending on each regulation. For example, in the case of Europe, the specifications are 6 g per 100 g of a product to be considered "rich in fibers" (European Commission 2012).

Figure 18.3 shows the pasta that presented the best characteristics related to SF (BSFA1, BSFA2, WSFA1, and WSFA2) before and after cooking. It is possible to relate color and homogeneity with the cooking test presented before at Table 18.4. Only pastas produced with bamboo fiber and wheat stalk fiber were homogeneous, while the others had white dots, showing the presence of fibers. According to the theoretical calculation of the dietary fiber content tests, BFA1, BFA2, WSFA1, and

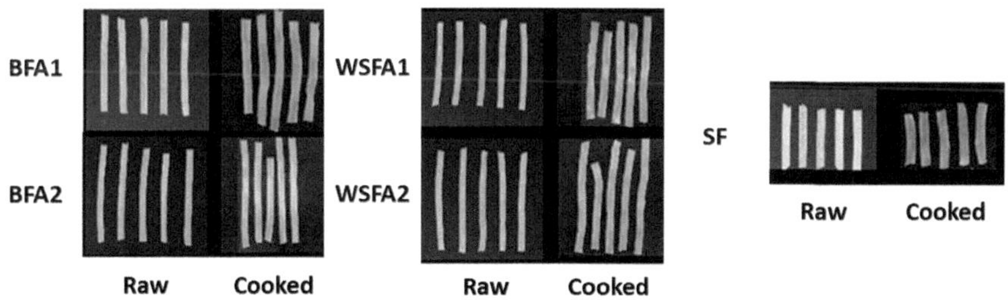

Fig. 18.3 Raw and cooked dry *fettuccine* pasta produced by replacing *Triticum durum* semolina with fibers, where BFA1 and BFA2 = 3.5% and 7%, respectively, of 60 μm bamboo fiber; WSFA1 and WSFA2 = wheat stalk fiber also in concentrations of 3.5% and 7% and standard formulation (without commercial fibers)

WSFA2, which presented the characteristics closest to the SF and are shown in Fig. 18.3, could fit in the European labeling claims as "high fiber content" and "fiber-rich," respectively.

18.4 Conclusion

The results demonstrated that it is possible to use bamboo fiber and wheat stalk fiber to increase dry pasta fiber content without significant changes in their technological characteristics. The development of a pasta formulation with fibers that presents similar characteristics to their standard can benefit consumers who would normally reject whole grain and/or with added fiber products. Therefore, they may have a high innovative potential in the pasta industry. Given the availability of these fibers in the market, future perspectives could include tests of this incorporation in pasta on an industrial scale.

Acknowledgments The authors are grateful to Conselho Nacional de Desenvolvimento Científico e Tecnológico (CNPq) for the scholarship of Amanda Rios Ferreira (132004/2016-0), and Elisa C. A. Neves (140807/2016-0 and 142287/2019-9) and Maria T.P.S.Clerici (Process 312786/2020-4); FAEPEX and FAPESP for the financial support (2015/19637-4); Coordenação de Aperfeiçoamento de Pessoal de Nível Superior (CAPES) financial code-001 and for the scholarship of Mária H. F. Felisberto and Thaísa de M. A. Moro; Selmi Alimentos, for the donation of the semolina of *Triticum durum*; and to the company Nutrassim for the donation of the commercial fibers.

Author Contributions All authors contributed equally to the work presented in this paper. Amanda Rios Ferreira: formal analysis – investigation; writing – original draft (equal). Felipe Gianasi: formal analysis – investigation (supporting). Thaisa de Menezes Alves Moro: investigation - writing – review and editing (equal). Mária Hermínia Ferrari Felisberto: conceptualization (supporting) - investigation - writing – review and editing. Elisa Cristina Andrade Neves: writing – review and editing (equal). Maria Teresa Pedrosa Silva Clerici: conceptualization - founding – resources - supervision (equal).

Conflict of Interest There is no conflict of interest.

References

Ajila CM, Aalami M, Leelavathi K, Rao UJSP (2015) Mango peel powder: a potential source of antioxidant and dietary fiber in macaroni preparations. Innov Food Sci Emerg Technol 11 (1):219–224

Alzuwaid NT, Fellows CM, Laddomada B, Sissons M (2020) Impact of wheat bran particle size on the technological and phytochemical properties of durum wheat pasta. J Cereal Sci 95:103–133

American Association Cereal Chemists International. AACCI (2010) American association cereal chemists, 11th edn. AACC International, St. Paul, MN. http://methods.aaccnet.org/toc.aspx. Accessed 30/07/2020

Aravind N, Sissons MJ, Fellows CM, Blazek J, Gilbert EP (2012) Effect of inulin soluble dietary fibre addition on technological, sensory, and structural properties of durum wheat spaghetti. Food Chem 132:993–1002
Association Official Analytical Chemists. AOAC (2007) Association official analytical chemists. 18th edn., Washington, DC., Method 985.29
Biernacka B, Dziki D, Gawlik-Dziki U, Różyło R (2020) Common wheat pasta enriched with cereal coffee: quality and physical and functional properties. LWT 128:142–161
Borelli GM, Leonardis AM, Platani C, Troccoli A (2008) Distribution along durum wheat kernel of the components involved in semolina color. J Cereal Sci 48:494–502
Brennan CS, Simons LJ (2004) The effect of barley beta-glucan fiber fractions on starch gelatinization and pasting properties. J Food Sci 69(4):257–261
Brennan CS, Tudorica CM (2007) Fresh pasta quality as affected by enrichment of non-starch polysaccharides. J Food Sci 72(9):659–665
Bustos MC, Perez GT, Leon AE (2015) Structure and quality of pasta enriched with functional ingredients. RSC Adv 39(5):30780–30792
Cho SS, Dreher ML (2001) Handbook of dietary fiber, vol 468. Marcel Dekker Incorporation, New York, NY
CreaFill (2020) Creafill Fibers Corporation—Food Ingredients. https://www.creafill.com/creafibe-and-creasplend-food/. Accessed 30/09/2020
European Commission (2012) Nutrition claims. http://ec.europa.eu/food/safety/labelling_nutrition/claims/nutrition_claims/index_en.htm. Accessed: 07/09/2020
Ferreira AR, Neves ECAN, Moro TMAM, Felisberto MHFF, Clerici MTPS (2020) Whole cookies with bamboo shoots: technological properties. Ciencia 22:72–78
Foschia M, Peressini D, Sensidoni A, Brennan MA, Brennan CS (2015) How combinations of dietary fibers can affect physicochemical characteristics of pasta. LWT 61:41–46
Gull A, Kamlesh P, Pradyuman K (2015) Effect of millet flours and carrot pomace on cooking qualities, color and texture of developed pasta. LWT 63:470–474
IPO International Pasta Organisation (2020) Annual report of the International Pasta Organisation. https://internationalpasta.org/annual-report/. Accessed 5/9/2020
Kill RC, Turnbull KK (2001) Pasta and semolina technology, vol 242. Blackwell Science, Oxford
Kissler SM, Tedijanto C, Goldstein E, Grad YH, Lipsitch M (2020) Projecting the transmission dynamics of SARS-CoV-2 through the postpandemic period. Science 368(6493):860–868
Kosović I, Benšić M, Ackar D, Jozinović A (2018) Microstructure and cooking quality of barley-enriched pasta produced at different process parameters. Foods Raw Mater 6(2):281–290
Lemes AC, Takeuchi KP, Carvalho JCM, Danesi EDG (2012) Fresh pasta production enriched with Spirulina platensis biomass. Braz Arch Biol Technol 55(5):741–750
Levent H, Koyuncu M, Bilgiçli N, Adıgüzel E, Dedeoğlu M (2020) Improvement of chemical properties of noodle and pasta using dephytinized cereal brans. LWT 128:109–125
Mercier S, Moresoli C, Mondor M, Villeneuve S, Marcos B (2016) A meta-analysis of enriched pasta: what are the effects of enrichment and process specifications on the quality attributes of pasta? Compr Rev Food Sci Food Saf 15:685–704
Mokrzycki W, Tatol M (2011) Color difference Delta E: a survey. Mach Graph Vis Warszawa 20:383–411
Nongdam P, Tikendra L (2014) The nutritional facts of bamboo shoots and their usage as important traditional foods of Northeast India. Int Sch Res Notices 14:679–683
Pasqualone A, Gambacorta G, Summo C et al (2016) Functional, textural and sensory properties of dry pasta supplemented with lyophilized tomato matrix or with durum wheat bran extracts produced by supercritical carbon dioxide or ultrasound. Food Chem 213:545–553
Pasqualone A, Punzi R, Trani A, Summo C, Paradiso VM, Caponio F, Gambacorta G (2017) Enrichment of fresh pasta with antioxidant extracts obtained from artichoke canning by-products by ultrasound-assisted technology and quality characterisation of the end product. Int J Food Sci Technol 52:2078–2087

Rakhesh N, Fellows CM, Sissons M (2014) Evaluation of the technological and sensory properties of durum wheat spaghetti enriched with different dietary fibres. J Sci Food Agric 95:2–11
Reynolds A, Mann J, Cummings J, Winter N, Mete E, Morenga LT (2019) Carbohydrate quality and human health: a series of systematic reviews and meta-analyses. Lancet 393:434–445
Stevenson L, Phillips F, O'Sullivan K, Walton J (2012) Wheat bran: its composition and benefits to health, a European perspective. Int J Food Sci Nutr 63:1001–1013
Tazrart K, Zaidi F, Lamacchia C (2016) Effect of durum wheat semolina substitution with broad bean flour (Vicia faba) on the Maccheronccini pasta quality. Eur Food Res Technol 242:477–485
Tudorica CM, Kuri V, Brennan CS (2002) Nutritional and physicochemical characteristics of pasta with dietary fiber. J Agric Food Chem 50:347–356
Vignola MB, Bustos MC, Perez GT (2018) Comparison of quality attributes of refined and whole wheat extruded pasta. LWT 89:329–335
Wang Y, Chein J, Wang D, Ye F, He Y, Hu Z, Zhao G (2020) A systematic review on the composition, storage, processing of bamboo shoots: focusing the nutritional and functional benefits. J Funct Food 71:104–115
West R, Seetharaman K, Duizer LM (2013) Effect of drying profile and whole grain content on flavour and texture of pasta. J Cereal Sci 58:82–88
WHO (2009) Population nutrient intake goals for preventing diet-related chronic diseases.http://www.who.int/dietphysicalactivity/publications/trs916/en/gsfao_overall.pdf. Accessed 1/8/2020
Witczaka T, Gałkowska D (2021) Sorption and thermal characteristics of ancient grain pasta of various compositions. LWT 137:110433

Chapter 19
Bamboo Fiber as a Substitute for Fat and/or Sugar in Cookies

Mária Herminia Ferrari Felisberto, Patricia Satie Endo Miyake, Amanda Rios Ferreira, Thaísa de Menezes Alves Moro, Elisa Cristina Andrade Neves, and Maria Teresa Pedrosa Silva Clerici

Abstract Bamboo is source of fiber, which has been used to produce isolated fiber as an ingredient for food products. This chapter will present the study of cookie formulations manufactured with bamboo fiber (3%). A linear factorial design with a central point and two independent variables (reduction of fat and/or sugar in the levels of 0%, 25%, and 50%) was used. The dependent responses were technological characteristics and nutritional composition and stability during storage of selected cookies, in addition to the control formulation (without bamboo fiber). The results showed color parameters in the slightly yellowish region, and we observed that it is possible to provide a caloric reduction in cookies, both concerning the content of added sugar and fat. Regarding stability over 28 days of storage, the formulations were stable, with low values for moisture content (5%) and water activity (0.5). Concerning texture, the formulation with reduced sugar content was the one that was closest to the control formulation. We conclude that it is possible to reduce three

M. H. F. Felisberto (✉)
Department of Food Science and Nutrition, School of Food Engineering, University of Campinas (UNICAMP), Campinas, SP, Brazil

Department of Food Technology (DTA), Federal University of Viçosa (UFV), Viçosa, MG, Brazil
e-mail: maria.felisberto@ufv.br; mclerici@unicamp.br

P. S. E. Miyake · A. R. Ferreira · M. T. P. S. Clerici
Department of Food Science and Nutrition, School of Food Engineering, University of Campinas (UNICAMP), Campinas, SP, Brazil

T. de Menezes Alves Moro
Department of Food Science and Nutrition, School of Food Engineering, University of Campinas (UNICAMP), Campinas, SP, Brazil

School of Nutrition, Santa Ursula University (USU), Rio de Janeiro, RJ, Brazil

E. C. A. Neves
Department of Food Science and Nutrition, School of Food Engineering, University of Campinas (UNICAMP), Campinas, SP, Brazil

School of Food Engineering, Institute of Technology, Federal University of Pará (UFPA), Belém, PA, Brazil

Z. Ahmad et al. (eds.), *Biotechnological Advances in Bamboo*,
https://doi.org/10.1007/978-981-16-1310-4_19

ingredients in biscuit formulations (refined wheat flour, fat, and sugar), adding 3% of bamboo fiber, which shows its potential as an ingredient capable of improving the healthiness of baking products.

Keywords Biscuits · Caloric reduction · Food ingredient · Healthiness · Sustainability

19.1 Introduction

Bamboo is a very fast-growing and versatile plant on Earth, which has gained increasing importance in the economic scenario as a substitute for wood, recently. It is a grass from the Poaceae family and Bambusoideae subfamily. Seventy-five different genera with approximately 1300 species is already documented, which cover around 25 million hectares globally (Liese and Kohl 2015).

Brazil is home to about 89% of the bamboo genera, 65% of the known species in the American continent, and one of the largest native bamboo reserves worldwide (180,000 km^2) (Judziewicz et al. 1999; Manhães 2008; Pereira and Beraldo 2016). Bamboo is a plant with more than 1000 uses, like renewable raw material, food, and regenerative energy. Culms have notable technological characteristics and are used for handicraft products, construction, furniture, and scaffolding and as material for secondary products such as flooring, boards, or bamboo mats (Akinlabi et al. 2017; Pereira and Beraldo 2016). In Brazil, the predominance of its use is in the manufacture of paper because of its long fibers, which give the product higher strength.

In addition to industrial applications, a young bamboo culm can be used for flour obtention as evaluated by Felisberto et al. (2017a, b). Furthermore, due to its composition, the flour can be used as a food ingredient and for starch extraction (Felisberto et al. 2020; Felisberto et al. 2019a, b). However, the starch extraction process presents as a by-product of the fibrous fraction in a substantial and significant amount, greater than 60 g/100 g, according to Felisberto et al. (2017b), which has not yet been studied. The same research group evaluated the wheat flour substitution (15%) by young bamboo culm flour in cookies, including an additional reduction in sugar and/or fat content (up to 50%), without significant changes in its technological characteristics. Nevertheless, the processing of young bamboo culm requires industrial equipment to transform it into the flour with specific granulometry after the drying process, limiting its use by local communities.

Bamboo shoot processing, on the other hand, is more straightforward and easier to adapt, despite the inconvenience of the presence of cyanogenic compounds that can be removed with adequate preparations, such as immersion and successive washing in hot water. Bamboo shoots have some nutrients and nutraceutical characteristics with additional benefits to the health of the consumer (Chongtham et al. 2011). Commercial fibers extracted from bamboo shoot already exist in the market, which is produced industrially and traded internationally (JELU-WERK 2020;

TIC-GUMS 2020; Nutrassim 2020). Although bamboo shoot fiber is commercialized internationally, few studies have evaluated its food product application.

The world is going through the Covid-19 pandemic this year (2020), which allows a reflection on the increase in the search for a still practical diet, due to the daily rush, but healthier, which will take industries to maintain this innovation strategy, meeting the demand of the consumer market.

Cookies are one of the most consumed bakery products, mainly by infants, due to their sweetness and crunchy texture, and by adults, due to their ease of consumption and long shelf life (Chevallier et al. 2000; Jacob and Leelavathi 2007; Moraes et al. 2010). According to Gökmen et al. (2008) and Pareyt et al. (2009), cookies are baked products, based on cereals with considerable levels of sugar and fat (20–50%), low water levels (1–5%), moisture content from 2% to 8%, and water activity between 0.1 and 0.3, whose characteristics of the final product depend on the quality of the ingredients used.

Recently Choudhury et al. (2015) produced cookies with whole bamboo shoot flour and obtained acceptable sensory acceptance. Farris and Piergiovanni (2008) and Farris et al. (2008) also evaluated the addition of bamboo shoot fiber in "Amaretti" cookies, a kind of cookie that is typical and traditional from Italy, and it is prepared in different ways with almond paste.

Considering greater consumer demand for products that contribute to healthier eating habits, the search for potential ingredients for food applications has been a substantial trend. Using bamboo fiber for the development of novel products may be interesting, in addition to being an alternative source of income for local communities. Thus, the objective of this study was to prepare cookies, with substitution of wheat flour by commercial bamboo fiber (3%), with a reduction of fat and/or sugar content, and to evaluate the nutritional and technological properties of the formulations and also the stability during storage.

19.2 Material and Methods

19.2.1 Raw Materials

Commercial bamboo fiber was obtained as a donation by Nutrassim Indústria, Comércio, Importação e Exportação Ltda (Extrema, Brazil), and all other ingredients were purchased at the local market in Campinas-SP, Brazil, taking care that they were from the same batch.

19.2.2 Wheat Flour and Commercial Bamboo Fiber Characterization

Wheat flour was evaluated for proximate composition, according to the American Association of Cereal Chemists International (2010) for moisture content (method 44–15.02), by the weight loss of the sample, when heated to 105 °C, until a constant mass. Protein content (method 46–13.01) was determined by the total nitrogen content in the sample, through acid digestion and subsequent titration and using 6.25 as the conversion factor of total nitrogen to protein. The total ash content (method 08–01.01) was determined as the inorganic residue after incinerating the sample in a muffle furnace (550–570 °C). Lipid content (method 30–25.01) was determined by ether extraction in Soxhlet. Total fiber (method 992–16) was evaluated by an enzymatic-gravimetric method from the Megazyme kit (K-TDFR 05/12), based on AACCI (2010) (method 32-05.01) and Association Official Analytical Chemists (2007) (method 985.29), according to protocols described in the kit. Digestible carbohydrates were calculated [100 – (moisture + protein + lipid + total fiber + ash)]. For the commercial bamboo fiber, a report provided by the manufacturer was used.

19.2.3 Experimental Design

A factorial design with two variables (fat and sugar reductions) was used (Table 19.1), and the substitution of wheat flour (3%) by commercial bamboo fiber was performed.

Table 19.1 Factorial design (2^2) with three central points of the cookie formulations

Formulations	Decoded levels (%)[a]	
	Sugar reduction	Fat reduction
F1	0	0
F2	50	0
F3	0	50
F4	50	50
F5	25	25
F6	25	25
F7	25	25

[a]Percentage of x_1 (sugar) and x_2 (margarine) reduction in each formulation

Table 19.2 Formulations of the cookies and design (F1–F7), with reduction of fat and/or sugar, and replacement of wheat flour (3%) by commercial bamboo fiber

Ingredients (%)[a]	F1	F2	F3	F4	F5	F6	F7
Wheat flour	97.00	97.00	97.00	97.00	97.00	97.00	97.00
Bamboo fiber	3.00	3.00	3.00	3.00	3.00	3.00	3.00
Sugar[a]	54.60	27.30	54.60	27.30	40.95	40.95	40.95
Margarine[a]	38.00	38.00	19.00	19.00	28.50	28.50	28.50
Maize starch[a]	17.65	17.65	17.65	17.65	17.65	17.65	17.65
Egg yolk[a]	9.40	9.40	9.40	9.40	9.40	9.40	9.40
Ammonium bicarbonate[a]	2.50	2.50	2.50	2.50	2.50	2.50	2.50
Salt[a]	0.61	0.61	0.61	0.61	0.61	0.61	0.61
Condiments[a,b]	0.36	0.36	0.36	0.36	0.36	0.36	0.36
Vanilla essence[a]	0.15	0.15	0.15	0.15	0.15	0.15	0.15
Water[c]	q.s.p.	q.s.p.	q.s.p.	q.s.p.	q.s.p.	q.s.p.	q.s.p.

[a]Wheat flour basis or mix of wheat flour + bamboo fiber
[b]Nutmeg, clove powder, and cinnamon powder
[c]q.s.p.: sufficient amount to the optimum point

19.2.4 Cookie Manufacturing

Cookie formulation of Felisberto et al. (2019c) was used as reference, and after performing pretests we obtained the optimum amount of each ingredient (Table 19.2). Cookies were developed through two-stage preparation, according to the methodology provided by Felisberto et al. (2019c).

19.2.5 Technological Properties of Cookies

Cookie thickness and diameter were determined with a pachymeter. The specific volume and weight loss were evaluated according to the methodology of AACCI (2010). All analyses were executed on ten different samples of cookies. Moisture content was determined according to AACCI (2010) (method 44-15.02). Water activity (aw) was evaluated at room temperature (25 °C), by a dewpoint sensor (Aqualab, 4TEV, Decagon, Pullman, USA). Both moisture content and water activity were carried out in triplicate. Texture and color parameters were evaluated according to Felisberto et al. (2019c). The texture of the cookies was determined using a texture analyzer, TA-XT Plus (Stable Micro Systems, England), with a 50 kg load cell, and the measurements were performed on at least ten cookies. The color parameters (L*, a*, and b*) were evaluated in a colorimeter CR-400 (Konica Minolta, Japan), according to the CIELab system (illuminant D65, observer angle 2°), and the readings were executed using three replicates per sample.

19.2.6 Selection of Cookies

The obtained data were evaluated through response surface methodology, using the software Statistica (version 8.0). The regression coefficients and analysis of variance (ANOVA), at a significance level of 10% with a minimum coefficient (R^2) of 0.70, were calculated.

19.2.7 Characterization of the Selected Cookies

19.2.7.1 Proximate Composition

The selected cookies were evaluated, at least in triplicate, for the moisture, ash, protein, total fiber, and digestible carbohydrate contents, according to AACCI (2010), as described in item 19.2.2 to wheat flour. The lipid content was evaluated using method no. 945.16 (petroleum ether method) (AOAC 2007), in triplicate. The energy value was calculated considering 9 kcal/g for lipids, 4 kcal/g for proteins, and for carbohydrates. Additionally, we developed a control formulation (CF) without any reduction (sugar and fat) or wheat flour replacement by commercial bamboo fiber.

19.2.7.2 Technological Properties and Storage Stability

The moisture contents, water activity, texture, and color parameters were evaluated for the selected cookies, as described before (item 19.2.5), over days 1, 7, 14, 21, and 28 of storage.

19.2.8 Statistical Analysis of the Selected Cookies

Proximate composition, technological characterization, and stability results were evaluated by difference between cookie formulations, using ANOVA, at a significance level of 5%, and Scott-Knott test, using SISVAR, version 5.6.

Table 19.3 Proximate composition of wheat flour and bamboo fiber

Composition (g/100 g)[a]	Wheat flour[b]	Bamboo fiber[c]
Moisture	12.57 ± 0.06	5.7
Lipid	0.64 ± 0.01	0.02
Protein	10.92 ± 0.11	< 0.20
Ash	0.42 ± 0.05	0.01
Total fiber	3.50 ± 0.29	99.07
Other carbohydrates[d]	71.95	

[a]Dry base
[b]Results expressed as means ± standard deviation
[c]Data provided by the manufacturer
[d]Calculated by difference [100 – (moisture + protein + lipid + total fiber + ash)]

19.3 Results and Discussion

19.3.1 Raw Material Characterization

Results of the proximate composition of wheat flour and the data provided by the manufacturer of the commercial bamboo fiber are shown in Table 19.3. We observed a chemical composition typical of domestic flours, with other carbohydrates >70 g/ 100 g and protein content around 10.92%, for wheat flour. However, flours with this protein content (around 11%) present a stronger gluten network (Manley 2011), and it is recommended to use starch to decrease the flour protein content and weaken the flour.

19.3.2 Technological Properties of the Cookies

Cookies were evaluated for diameter, thickness, loss of mass, specific volume, moisture, water activity, color parameters (L*, a*, and b*), and hardness and the obtained results are presented in Table 19.4. A considerable variation in the values of diameter (34–44 mm) and specific volume (2.2–4.5 mL/g) was observed, probably due to the heterogeneity in the baking or even during the molding step, since even the central points (F5, F6, and F7) showed a significant difference among them. However, formulations containing higher levels of sugar exhibited a higher expansion. Similar results were found in cookie formulations by Moraes et al. (2010), and even the evaluated formulations were elaborated without the addition of fibers.

However, Feddern et al. (2011) and Mauro et al. (2010) elaborated cookies with rice bran (10–30%) or wheat bran (15–45%) and flours of kale stalk or spinach stalk (30%), respectively, without fat and/or sugar reduction, and they observed no significant difference in the diameter of cookie formulations. According to Manley (2011), in addition to sweetness, sugar is a structural, flavor-modifying and flavor-enhancing substance when added to cookies. Feddern et al. (2011) also observed that

Table 19.4 Dimensions, weight loss, specific volume, moisture analysis, water activity, hardness, and color parameters 1 day after the manufacture of cookie formulations with commercial bamboo fiber[a]

Formulations[b]		F1	F2	F3	F4	F5	F6	F7
Diameter (mm)		43.80 ± 0.64^{a}	35.06 ± 1.80^{d}	39.28 ± 1.30^{b}	34.80 ± 0.56^{d}	42.62 ± 2.30^{a}	37.87 ± 00.3^{c}	40.48 ± 1.26^{b}
Thickness (mm)		9.76 ± 0.41^{a}	9.92 ± 0.46^{a}	9.87 ± 0.27^{a}	10.14 ± 0.05^{a}	10.10 ± 0.36^{a}	10.28 ± 0.21^{a}	8.26 ± 0.63^{b}
Weight loss (%)[c]		16.76 ± 0.18^{c}	18.18 ± 1.17^{c}	18.36 ± 0.29^{c}	34.09 ± 3.05^{a}	19.30 ± 0.40^{b}	19.91 ± 0.24^{b}	20.62 ± 0.23^{b}
Specific volume (mL/g)		4.44 ± 0.35^{b}	3.48 ± 0.56^{c}	2.27 ± 0.37^{d}	3.30 ± 0.13^{c}	3.01 ± 0.27^{c}	4.20 ± 0.46^{b}	5.65 ± 0.42^{a}
Moisture (g/100 g)		2.80 ± 0.05^{c}	4.01 ± 0.11^{b}	3.12 ± 0.34^{c}	7.03 ± 0.72^{a}	2.14 ± 0.08^{c}	2.78 ± 0.03^{c}	2.33 ± 0.06^{c}
Water activity		0.27 ± 0.00^{d}	0.41 ± 0.00^{b}	0.28 ± 0.00^{c}	0.54 ± 0.00^{a}	0.20 ± 0.00^{f}	0.22 ± 0.00^{e}	0.21 ± 0.00^{e}
Hardness (N)		52.35 ± 2.44^{b}	32.76 ± 2.40^{d}	29.53 ± 4.72^{d}	56.90 ± 4.13^{a}	51.04 ± 3.85^{b}	36.36 ± 5.69^{c}	21.71 ± 1.26^{e}
Color parameters[d]	L*	76.26 ± 0.17^{b}	78.03 ± 0.02^{a}	75.41 ± 0.26^{b}	78.97 ± 0.85^{a}	71.84 ± 0.22^{c}	71.63 ± 0.74^{c}	77.83 ± 0.30^{a}
	a*	3.24 ± 0.14^{c}	3.07 ± 1.34^{c}	6.86 ± 0.34^{b}	2.78 ± 0.03^{c}	8.24 ± 0.30^{a}	8.89 ± 0.15^{a}	2.09 ± 0.10^{c}
	b*	26.74 ± 0.17^{b}	24.56 ± 1.96^{c}	27.28 ± 0.06^{b}	22.25 ± 0.22^{d}	30.94 ± 0.07^{a}	31.15 ± 0.31^{a}	24.56 ± 0.28^{c}

[a]Results expressed as mean ± standard deviation, where data followed by different letters on the same line represent a significant difference by the analysis of variance test followed by Scott-Knott ($p < 0.05$) and ns = not significant

[b]F1 (0% of sugar reduction and 0% of fat reduction); F2 (50% of sugar reduction and 0% of fat reduction); F3 (0% of sugar reduction and 50% of fat reduction); F4 (50% of sugar reduction and 50% of fat reduction); and F5, F6, and F7 (25% of sugar reduction and 25% of fat reduction)

[c]Calculated b difference (raw dough weight – baked dough weight)

[d]L* (0 = black; 100 = white), a* (+a = redness; −a = greenness) and b* (+b = yellowness; −b = blueness)

there was no significant difference for the specific volume of cookie formulations with different levels of addition of rice and wheat bran, corroborating to the result obtained in the present work, where the effect of reducing sugar and/or fat is more expressive on this parameter.

The thickness and weight loss showed smaller variations, being in the ranges of 8.2–10.3 mm and 16.7–34.1%, respectively, although all samples have increased in thickness. Choudhury et al. (2015) observed opposite results in the formulations of cookies containing the addition of 10% and 15% of bamboo shoot powder, with a reduction in thickness. However, despite the authors adding a higher content of shoot flour in the formulations, the fiber content present in this flour is only 34.41%, which in the final product is close to the content used in the present research.

Concerning the weight loss (Table 19.4), F4 (50% reduction in sugar and fat) showed greater value, which is in accordance with the results obtained for hardness, where F4 also showed greater value. It indicates that large reductions (from 50%) on both ingredients (sugar and fat) may compromise the texture of the cookies added with commercial bamboo shoot fiber.

Regarding water activity, we observed values in the range of 0.20–0.54, with the highest values for F4 (0.54) (50% reduction in fat and 50% sugar), F2 (0.41) (reduction in 50% of sugar), and F3 (0.28) (50% reduction in fat), which was already expected, since water becomes more available with lower amounts of sugar and fat. The moisture contents were consistent in the same way as the water activity values, but with a smaller variation (2.1–7.03 g/100 g).

Finally, according to the color parameters, the cookies were clear (average L*: 75.71) in the yellowish region (average a* and b*: 5.02 and 26.78, respectively—Fig. 19.1). Choudhury et al. (2015) observed darker cookies, with L* values varying from 54 to 36, in formulations elaborated with 0%, 5%, 10%, and 15% of flour from a bamboo shoot. This fact may be due to the addition of skim powdered milk, which

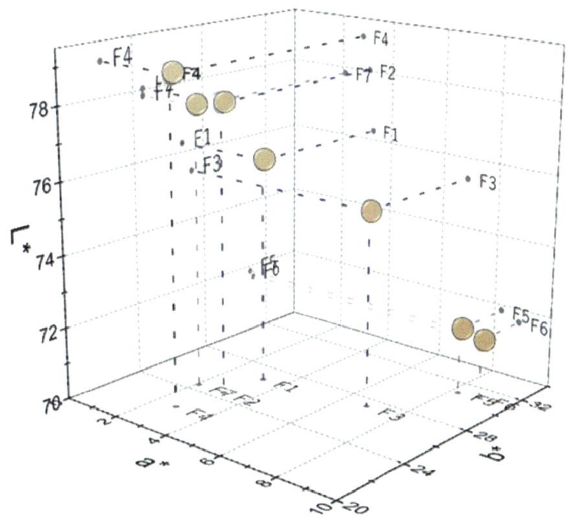

Fig. 19.1 Color parameters of cookie formulations elaborated with commercial bamboo fiber. (F1 (0% of sugar reduction and 0% of fat reduction); F2 (50% of sugar reduction and 0% of fat reduction); F3 (0% of sugar reduction and 50% of fat reduction); F4 (50% of sugar reduction and 50% of fat reduction); and F5, F6, and F7 (25% of sugar reduction and 25% of fat reduction))

contributes strongly to the Maillard reaction, giving brown surface coloring during baking, according to Manley (2011).

19.3.3 *Selection of the Cookies*

The results obtained through the response surface analysis are presented in Table 19.5. We observed significant difference on the diameter, for the effect of the x_1 – independent variable (sugar reduction)—by reducing the diameter of the cookies as an effect of reducing the content of sugar. It was also observed a significant difference on moisture content for both effects of the independent variable x_1 and x_2 (sugar and fat reduction, respectively).

A similar result was observed by Moraes et al. (2010) when evaluating the effect of sugar and fat on the technological responses of cookies, even without the use of fibers from any source in partial replacement for wheat flour. However, the researchers also obtained a significant response to the expansion factor, b*, and breaking force.

Although response surface analysis (Fig. 19.2) indicates a significant difference for the diameter and moisture content of the cookie formulations, such characteristics are of little relevance for the development of formulations and the evaluation of new ingredients and consequently for the selection of cookies in the present study. Thus, taking into account the data obtained and presented above, we selected the formulations based on the largest individual reductions in the levels of sugar (50%) (F2) and vegetable fat (50%) (F3) which are the extreme points of the design with the control formulation (CF), as can be seen in Fig. 19.3.

Table 19.5 Response surface analysis of technological characteristics of cookie formulations with commercial bamboo fiber obtained in the factorial design (2^2)

Parameters		Mathematical model[a]	Mean value[b]	R^2	p-Value (<0.10)	Lack of fit
Diameter		$Y = 3.91–0.33x_1$		60.06	0.041	0.524
Thickness		n.s.	09.76 ± 0.36			
Weight loss		n.s.	05.10 ± 0.21			
Specific volume		n.s.	03.77 ± 0.48			
Moisture		$Y = 3.46 + 1.28x_1 + 0.84 x_2 + 0.675 (x_2)^2$		65.41	0.307	0.018
Water activity		n.s.	00.31 ± 0.05			
Hardness		n.s.	40.09 ± 4.89			
Color parameters	L*	n.s.	75.64 ± 1.53			
	a*	n.s.	05.02 ± 1.35			
	b*	n.s.	26.78 ± 1.56			

[a]n.s. not significant
[b]Results expressed as mean ± standard deviation

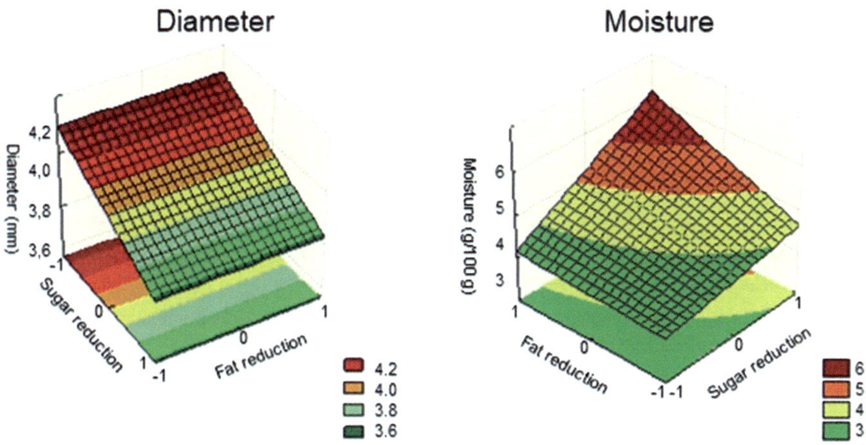

Fig. 19.2 Response surface analysis of cookies elaborated with commercial bamboo fiber addition ($p < 0.10$)

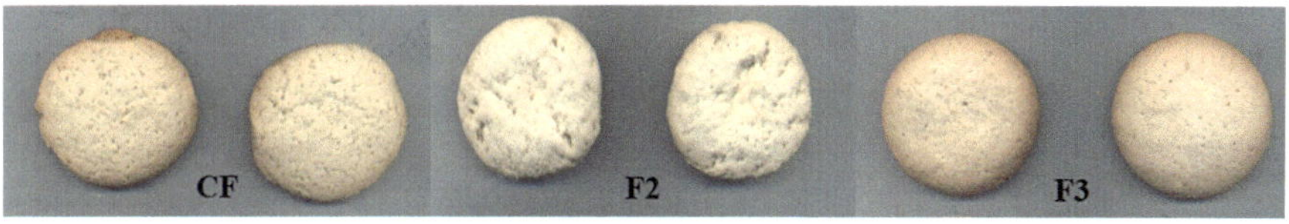

Fig. 19.3 Control formulation (CF) and selected cookies

19.3.4 Characterization of the Selected Cookies

19.3.4.1 Proximate Composition

Low values of moisture and ash (Table 19.6) were observed for all formulations. Regarding lipid and protein content, higher values were observed for F2, probably due to the concentration of fat, since we used a simple reduction without replacement by another ingredient. On the other hand, for F3 we observed a significant reduction in lipid content, as expected, due to the 50% reduction in the margarine addition content.

Moraes et al. (2010) observed different behavior, without significant variation between the formulations for the contents of proteins or carbohydrates, when evaluating formulations of cookies with reduction of sugar and/or fat, without the addition of fibers. Regarding lipid content, they observed a behavior already expected with the lowest levels in formulations where the fat content was reduced, similar to the present study.

We also observed significant differences in fiber contents (Table 19.6). Formulations F2 and F3 presented values >5 g/100 g, which was considered predictable

Table 19.6 Proximate composition of selected formulations (F2, F3, and CF)[*, **]

Composition (g/100 g)	CF	F2	F3
Moisture	05.39 ± 0.08[ns]	05.46 ± 0.13[ns]	5.67 ± 0.07[ns]
Lipid	14.86 ± 0.40[b]	18.72 ± 0.18[a]	9.47 ± 0.10[c]
Protein	07.69 ± 0.09[b]	08.50 ± 0.07[a]	7.23 ± 0.07[c]
Ash	00.42 ± 0.03[ns]	00.36 ± 0.02[ns]	0.58 ± 0.02[ns]
Total fiber	01.94 ± 0.32[b]	06.27 ± 0.45[a]	5.45 ± 0.64[a]
Digestible carbohydrates[***]	69.7	60.52	71.6
Calorie (kcal/100 g)	443.31	444.64	400.51

[*]Results expressed as means ± standard deviation. Means followed by different letters in the same column differ significantly ($p < 0.05$) by Scott-Knott test
[**]Control formulation (CF), cookies with 50% of sugar reduction (F2) and 50% of fat reduction (F3)
[***]Calculated by difference [100 – (moisture + protein + lipid + total fiber + ash)]
n.s .: not significant

since the commercial bamboo fiber contained 99.07 g/100 g of total fiber. With this result, it is evident that the substitution of only 3% of the wheat flour for this type of fiber allows a considerable increase, up to three times, in the fiber content of the final product, emphasizing the great potential for food applications of bamboo fiber. Finally, digestible carbohydrates were the major component in all formulations, whose amount was greater than 60 g/100 g.

Concerning energy value, F3 presented a significant reduction, with approximately 10% fewer calories when compared to CF and F2, which, combined with the ease of transportation and consumption of this kind of product, corroborates to the increase in consumer demand.

19.3.4.2 Technological Properties and Storage Stability

The stability over days 1, 7, 14, 21, and 28 of storage was measured using moisture, aw, texture, and color parameters (L*, a*, and b*) of the selected cookies, and the obtained results are presented on Fig. 19.4.

Regarding moisture content and water activity, we observed a very similar behavior, with an increase in levels over time, with the exception of F2 (50% reduction in sugar), in which there was a small decrease on day 28. But in general, there is an increase in moisture content over time, where the cookies reached values close to 5 g/100 g and 0.5 for water activity. This fact is probably due to the withdrawal of sugar, which promoted the concentration of the other ingredients, since we used a reduction process without replacing the removed ingredient, favoring water retention by the commercial bamboo fiber.

Excepting CF, there was an increase in hardness values until the 21st day of storage, followed by a fall, with F2 presenting the lowest hardness. However, this behavior was not regular over the storage period. Due to the decrement of the fat content in F3, higher values of hardness were expected for this formulation. Fat is one of the most important ingredients in biscuits. It adds flavor, eating quality, and

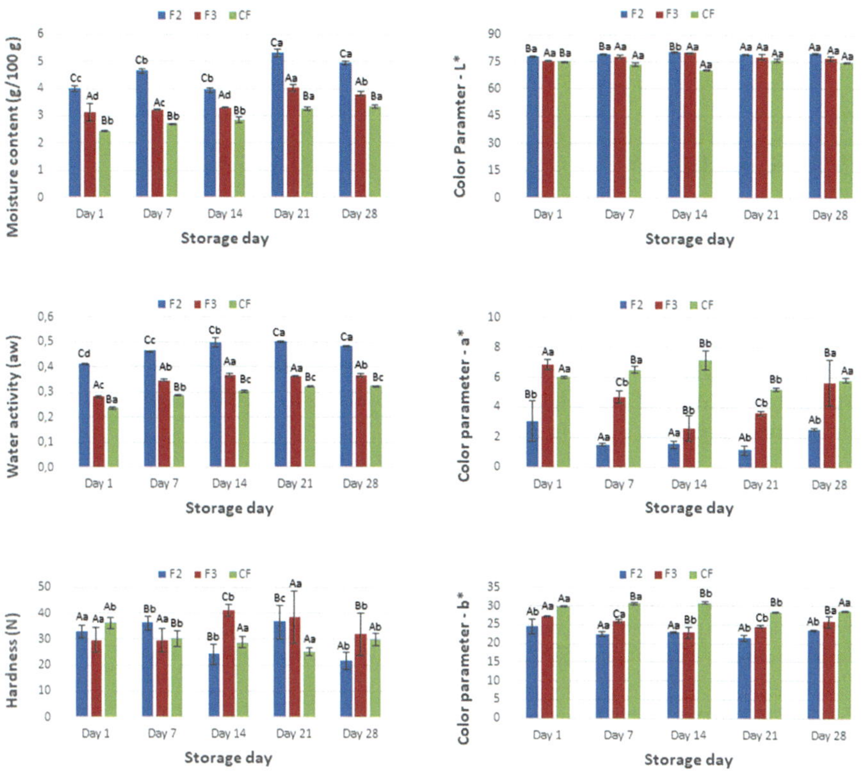

Fig. 19.4 Technological characterization and stability of selected formulations (F2, F3, and CF) over 1, 7, 14, 21, and 28 days of storage. (1) The results are expressed as means ± standard deviation. Means with different uppercase letters overlapped, in the same day for each parameter, differ significantly ($p < 0.05$) by Scott-Knott test. Means with different lowercase letters overlapped over the days, for each formulation, differ significantly ($p < 0.05$) by Scott-Knott test. (2) Cookies with 50% of sugar reduction (F2), 50% of fat reduction (F3), and control formulation (CF). (3) L* (0 = black; 100 = white), a* (+a = redness; −a = greenness), and b* (+b = yellowness; −b = blueness)

structure to the product, like "melt in the mouth" and crumbly texture (Manley 2011). We observed that the addition of the bamboo fiber has a significant effect on the texture of the cookies and that they become more expressive with the reduction of sugar levels (F2), over storage time, with values ranging from 22 to 37 N.

Regarding the color parameters, all formulations were stable in terms of luminosity (L*). For the red index (a*) we observed lower values for F2 and F3, probably due to the addition of the fibers. And with regard to the yellow index (b*), similar behavior was observed for F2 and F3, except for the 28th day.

Due to the reduced sugar content in F2, we expected less intense coloring than CF and F3. However, a different result could be observed, with no significant difference between the formulations. Over the storage period, all formulations remained stable and all were very clear and in the region beige-yellow, which contributes to the acceptance of this product, with reduced sugar and calories, by the consumer.

19.4 Conclusion

The results demonstrate that commercial bamboo fiber is a very interesting ingredient for the reduction of fat and/or sugar in cookies, since we evaluated the highest levels of individual reduction (50%) without compromising the technological characteristics of the product. In addition, cookie formulations started to present a considerable supply of fibers (compared to the control formulation), which demonstrates the potential of using bamboo fiber, in addition to meeting consumer demand for healthier products.

Acknowledgments The authors would like to thank FAEPEX and FAPESP for the financial support (2015/19637-4) and for the scholarship granted to Patricia Miyake (2016/13533-5); CAPES for the financial code-001, scholarship of Mária Felisberto and Thaísa Moro; CNPq for the scholarship of Amanda Ferreira (132004/2016-0), Elisa Neves (140807/2016-0 and 142287/2019-9), and Maria Teresa P.S.Clerici (Process 312786/2020-4); SAE/UNICAMP for the scholarship of the undergraduates and graduates who assisted the project; and to Nutrassim Indústria, Comércio, Importação e Exportação Ltda for the donation of the commercial fibers.

Conflicts of Interest There is no conflict of interest.

References

Akinlabi ET, Anane-Fenin K, Akwada DR (2017) Bamboo—the multipurpose plant. 1 edn. Springer International Publishing, pp 179–219

American Association of Cereal Chemists International. AACCI (2010) American Association Cereal Chemists International. 11 edn. AACC International, St. Paul, MN, U. S. A. http://methods.aaccnet.org/toc.aspx. Accessed 30/07/2020

Association Official Analytical Chemists. AOAC (2007) Association Official Analytical Chemists. 18 edn., Washington, DC, Method 985.29, 945.16

Chevallier S, Colonna P, Della Valle G, Lourdin D (2000) Contribution of major ingredients during baking of biscuit dough systems. J Cereal Sci 31(3):241–252

Chongtham N, Bisht MS, Haorongbam S (2011) Nutritional properties of bamboo shoots: potential and prospects for utilization as a health food. Compr Rev Food Sci Food Saf 10(3):153–168

Choudhury M, Badwaik LS, Borah PK, Sit N, Deka SC (2015) Influence of bamboo shoot powder fortification on physico-chemical, textural and organoleptic characteristics of biscuits. J Food Sci Technol 52(10):6742–6748

Farris S, Piergiovanni L (2008) Effects of ingredients and process conditions on 'Amaretti' cookies characteristics. Int J Food Sci Technol 43(8):1395–1403

Farris S, Piergiovanni L, Limbo S (2008) Effect of bamboo fibre and glucose syrup as new ingredients in the manufacture of amaretti cookies. Ital J Food Sci 20(1):75–90

Feddern V, Durante VVO, Miranda MZ, Mellado MLMS (2011) Avaliação física e sensorial de biscoitos tipo cookie adicionados de farelo de trigo e arroz. Braz J Food Technol 14:267–274

Felisberto MHF, Beraldo AL, Clerici MTPS (2017a) Young bamboo culm flour of Dendrocalamus asper: technological properties for food applications. LWT 76(Part B):230–235
Felisberto MHF, Beraldo AL, Costa MS, Boas FV, Franco CML, Clerici MTPS (2019a) Characterization of young bamboo culm starch from Dendrocalamus asper. Food Res Int 124:222–229
Felisberto MHF, Beraldo AL, Costa MS, Boas FV, Franco CML, Clerici MTPS (2019b) Physicochemical and structural properties of starch from young bamboo culm of Bambusa tuldoides. Food Hydrocoll 87:101–107
Felisberto MHF, Beraldo AL, Costa MS, Boas FV, Franco CML, Clerici MTPS (2020) Bambusa vulgaris starch: characterization and technological properties. Food Res Int 132:109102
Felisberto MHF, Miyake PSE, Beraldo AL, Clerici MTPS (2017b) Young bamboo culm: potential food as source of fiber and starch. Food Res Int 101(Supplement C):96–102
Felisberto MHF, Miyake PSE, Beraldo AL, Fukushima AR, Leoni LAB, Clerici MTPS (2019c) Effect of the addition of young bamboo culm flour as a sugar and/or fat substitute in cookie formulations. Food Sci Technol 39(4):867–874
Gökmen V, Serpen A, Açar ÖÇ, Morales FJ (2008) Significance of furosine as heat-induced marker in cookies. J Cereal Sci 48(3):843–847
Jacob J, Leelavathi K (2007) Effect of fat-type on cookie dough and cookie quality. J Food Eng 79 (1):299–305
JELU-WERK (2020) Bamboo fibres as food additives. JELU-WERK J. Ehrler GmbH & Co. KG. https://www.jelu-werk.com/cellulose/products/plant-fibres/jelucel-bf-bamboo-fibre/. Accessed 30/07/2020
Judziewicz EJ, Clark LG, Londoño X, Stern MJ (1999) American bamboos. Smithsonian Institution Press, Washington D. C, p 392
Liese W, Kohl M (2015) Bamboo—the plant and its uses. Tropical forestry, vol 10, 1st edn. Springer International Publishing, p 356
Manhães AP (2008) Caracterização da cadeia produtiva do bambu no Brasil: abordagem preliminar. Universidade Federal Rural do Rio de Janeiro, Seropédica, p 39
Manley D (2011) Manley's Technology of biscuits, crackers and cookies, 4th edn. Woodhead Publishing, Philadelphia, p 632
Mauro AK, Silva VLMD, Freitas MCJ (2010) Caracterização física, química e sensorial de cookies confeccionados com farinha de talo de couve (FTC) e farinha de talo de espinafre (FTE) ricas em fibra alimentar. Food Sci Technol 30:719–728
Moraes KSD, Zavareza EDR, Miranda MZD, Salas-Mellado MDLM (2010) Avaliação tecnológica de biscoitos tipo cookie com variações nos teores de lipídio e de açúcar. Ciênc Tecnol Aliment 30:233–242
Nutrassim (2020) Nutrassim food ingredients. Nutrassim—Food Ingredients. http://nutrassim.com/site/produtos-nutrassim/fibra-qc. Accessed 30/07/2020
Pareyt B, Talhaoui F, Kerckhofs G, Brijs K, Goesaert H, Wevers M, Delcour JA (2009) The role of sugar and fat in sugar-snap cookies: structural and textural properties. J Food Eng 90 (3):400–408
Pereira MAR, Beraldo AL (2016) Bambu de corpo e alma. 2 edn. Canal 6, Bauru/SP, pp 352
TIC-GUMS (2020) Nutriloid® Bamboo Fiber. TIC Gums, Inc. https://www.ticgums.com/products/nutriloid/product/nutriloid-bamboo-fiber.html. Accessed 30/07/2020

Chapter 20
Practical Applications of Bamboo as a Building Material: Trends and Challenges

Siew Choo Chin

Abstract Bamboo's ability to grow in the tropical, subtropical, or temperate zones makes them an abundantly available plant in the world. Owing to its renewability and versatility, bamboo is often used in many applications including construction, clothes, food, and fuel. Bamboo is one of the most used building materials after concrete, due to its high compressive and tensile strength, which is comparable to concrete and steel, respectively. This chapter presents a review of the recent advances in the practical bamboo application for construction, including the bamboo treatment, preservation of bamboo, bamboo in building structures, and structural strengthening. A comparison of bamboo-based construction material with other commonly used materials is also reviewed. A summary of the future direction and challenges in the bamboo application for building and construction is also presented.

Keywords Bamboo treatment · Building · Composite · Construction · Strengthening

20.1 Introduction

Bamboo, or better known as giant grass, is a typical fast-growing green material that grows naturally in the forest and can be cultivated (Nurdiah 2016; Escamilla et al. 2019; Shen et al. 2019; Atamewan 2020). Bamboo belongs to Poaceae family which grows in tropical, subtropical, and temperate zones between approximately 46° north and 47° south latitude (Lobovikov et al. 2007). According to The Plant List (2013), Poaceae consists of more than 759 plant genera and can be divided into 11,554 species. They have several advantages such as high strength, environmentally friendly, and lightweight (Shen et al. 2019). Countries including in Africa, Australia, Eurasia north of the Himalayas, South and Southeast Asia, North

S. C. Chin (✉)
Department of Civil Engineering, College of Engineering, Universiti Malaysia Pahang, Gambang, Pahang, Malaysia
e-mail: scchin@ump.edu.my

Z. Ahmad et al. (eds.), *Biotechnological Advances in Bamboo*,
https://doi.org/10.1007/978-981-16-1310-4_20

America, temperate South America, and tropical America are the common countries where the bamboo grows (Britannica.com).

Naturally, bamboo grows in groups owing to its growth characteristic. The growth characteristic of bamboo is known to be either monopodial or sympodial (Azeez and Orege 2018). The roots of a monopodial type of bamboo spread horizontally in a shallow depth of the soil; as such, the new shoots appear quite some distance away from the parents' plant. Monopodial bamboo can be found in moderate climates such as Japan, China, and Korea. On the other hand, sympodial bamboo roots can be found close to the parents' plant forming a clump of many stems (Nurdiah 2016). This type of bamboo grows in regions with tropical climates such as Southeast Asia and South America (Anagal et al. 2010; Widyowijatnoko 2012; Nurdiah 2016).

Most architects, designers, and developers choose bamboo as building material nowadays. Due to deforestation, high-quality wood for building construction becomes limited. Wood needs a longer time to grow compared to bamboo which can be harvested between 3 and 5 years. During the planting process, bamboo releases oxygen into the atmosphere, unlike construction materials such as steel, plastic, and concrete. Due to these reasons, bamboo is promoted as sustainable building material (Nurdiah 2016).

In recent years, the idea of sustainable construction and the use of green building materials is on the rise (Aghdam et al. 2018). Bamboo also has been classified as a green material, which makes its adoption as building material to increase globally. Among the recent example of bamboo used in construction is the Madrid International Airport in Spain (bamboo ceiling), Clinton Library in the United States (bamboo floor), Tokyo Dong Wu Department in Japan, and the BMW exhibition hall and the IBM headquarters in Germany (Moso 2016). Bamboo is also used as pedestrian bridges in Columbia, to build schools in Indonesia, and as scaffoldings in Hong Kong (Nurdiah 2016). In addition, bamboo and wood is mainly for building construction in the earthquake-prone area of Wenchuan, Sichuan, in China (Shen et al. 2019).

This chapter summarizes the recent advances on practical applications of bamboo in construction, preservation method of bamboo, bamboo uses in building structures, as well as the utilization of bamboo fiber composites for structural strengthening.

20.2 Advances in Practical Bamboo Applications for Construction

The utilization of bamboo has gained more attention in recent years. Bamboo which has been used widely by the Southeast Asian countries was proven that it is a sustainable and valuable natural resource (Rahim and Idrus 2019). It plays a significant role in the socioeconomic development of the rural population (Akwada and Akinlabi 2016; Selvan and Tripathi 2017). Bamboo, which was previously known as "the poor man's timber," has now gained a significant commercial value and is known as "the green gold" which plays a vital role in society with increased worldwide demand (Rao et al. 1995; Rahim and Idrus 2019).

Bamboo has been used traditionally in the construction and making of shelters, food, cooking utensils, musical instruments, handicrafts, and furniture. As for modern use, bamboo has been commercially developed for the production of sunshades, tongs, mats, paper, and laminated furniture (Rahim and Idrus 2019). However, the production of bamboo products is affected by the type of species, age, and length of stems obtained (Chaowana 2013; Anokye et al. 2016). In construction, bamboo is often used for house construction such as walls and roofs as well as concrete reinforcement in road construction (Archila et al. 2018; Rahim and Idrus 2019). Currently, the advances in bamboo-related technology toward sustainable and green building materials lead to utilization of bamboo globally for construction (Akwada and Akinlabi 2015; Shen et al. 2019).

20.2.1 Steel Substitute

Concrete and steel are the two construction materials that are widely used globally. Concrete is good in compression but weak in tension. Therefore, steel which is strong in tension is needed to cater the tensile stress in concrete. Steel has a higher tensile strength compared to concrete. However, due to their high production cost and high energy consumption during production, it is a nonrenewable resource, and its ability to release large amount of carbon during its production has led engineers and researchers to search for other materials to replace the conventional steel reinforcement. The search for an alternative material in building construction tends to draw our attention toward bamboo which is cheap, naturally obtained, and readily available. Many researches were carried out to study the potential of bamboo as steel replacement in structural elements. This includes the use of bamboo in beam construction (Karthik et al. 2017; Latha et al. 2018; Pandi et al. 2018; Sutharsan et al. 2020), as stirrups in beam (Mark and Russell 2011), column (Krishnan et al. 2018; Lei et al. 2020), wall (Himasree et al. 2017; Ganesan et al. 2020), slab (Nayak et al. 2013; Mali and Datta 2018), and precast bridges (Muhtar 2020). In the aforesaid applications bamboo was used as green building material (Syeda and Kumar 2014) or as an alternative building material (Sharma et al. 2014; Escamilla et al. 2019).

Mechanical performance of bamboo in comparison with steel such as tensile and flexural strength was also studied (Adewuyi et al. 2015; Karthik et al. 2017; Nikhil 2018; Pandi et al. 2018; Sutharsan et al. 2020). According to Sutharsan et al. (2020), bamboo is a good reinforcing material in concrete and can increase the load-carrying capacity of the beam. Similar findings from Pandi et al. (2018) emphasized that bamboo-reinforced concrete achieved flexural strength comparable with steel-reinforced concrete, but bamboo exhibited low tensile and two times greater shear results compared to steel. They recommended that bamboo can be used in member

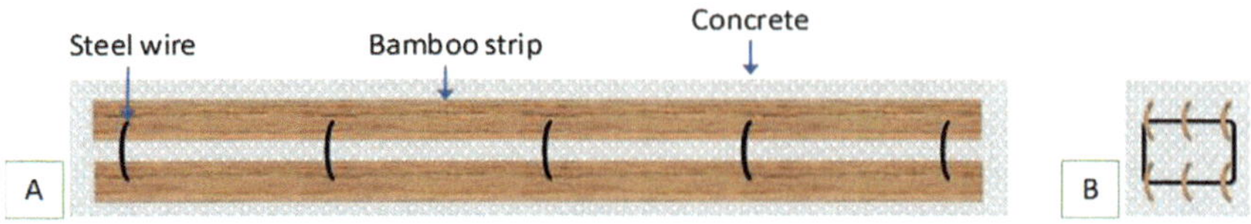

Fig. 20.1 Example of concrete beam strengthened by bamboo strip. (**a**) Side view. (**b**) Cross-sectional view

taking less load such as roof slabs of parking area, public toilets, watchman cabins, and sunshades. Adewuyi et al. (2015) found that bamboo bars are suitable for non-load-bearing and lightweight RC flexural structures. Likewise, Latha et al. (2018) also recommended to use a bamboo-reinforced concrete beam for light load-bearing structures such as beam, plinth beam, slab for small panel, as well as temporary structures. Nikhil (2018) reported that bamboo reinforcement can be used in place of plain cement concrete to enhance durability of the section. In addition, bamboo reinforcement is suitable for construction in rural areas which promotes a better strength at cheaper cost. Cost and design analysis were carried out by Pantawane et al. (2020) to study the advantage of bamboo reinforcement as steel replacement. Pantawane et al. (2020) recommended the use of advanced bamboo reinforcement technique instead of traditional steel reinforcement, for the benefit of a low-cost economical structure. They reported that bamboo reinforcement is three times cheaper than steel reinforcement, which is economical for a single-story house. Figure 20.1 shows an example of concrete beam strengthened by bamboo strip.

20.2.2 Concrete Reinforcement

Plain concrete has low tensile strength which requires reinforcement with steel to produce a highly durable reinforced concrete. It is widely known in reinforced concrete that the tensile stress was catered by the steel reinforcement. Tension cracks appear when the tensile strain reached the maximum concrete strain. The cracks enable propagation of water vapor and aggressive materials, which can cause corrosion of steel reinforcement, hence causing damage to the concrete in long term. This issue can be resolved by adding fibers into the tensile area to transfer the load to the internal microcracks (Dewi et al. 2017). Conventional method to improve the tensile cracks in concrete is by adding fibers into concrete or better known as fiber-reinforced concrete (FRC). Since the 1960s, fibers were extensively used in mortars and concrete to enhance its strength. The most commonly used fibers are steel, organic polymers, glass, carbon, asbestos, and cellulose (Gupta and Singh 2018).

The fibers can be categorized into two types which are synthetic and natural fibers. Synthetic fibers are mainly made by chemical means, while natural fibers are from parts of the minerals, plants, and animals (Gupta and Singh 2018). Bamboo fibers have received overwhelming attention from the research community for use as reinforcement in concrete due to its outstanding properties such as low weight to strength ratio, high tensile strength, low cost, easy availability, and environmental-friendly during service (Zatul et al. 2017; Gupta and Singh 2018). Past investigations had studied the mechanical properties of bamboo fiber-reinforced concrete (Ahmad et al. 2014; Kavitha and Kala 2016; Brindha et al. 2017; Zatul et al. 2017; Gupta and Singh 2018), bamboo as reinforcement in concrete by partial replacement of cement with fly ash (Sekar 2019), and bamboo fiber in self-compacting concrete (Bhautik et al. 2017; Ede et al. 2020).

Gupta and Singh (2018) evaluated the performance of bamboo fiber concrete in terms of compressive strength and flexural strength. They reported that addition of bamboo fiber in concrete reduced the slump value (Bhautik et al. 2017; Dewi et al. 2017). In addition, they also reported that the addition of bamboo fiber in concrete can increase the compressive strength up to 0.5%, while flexural strength increases with the fiber addition. Dewi et al. (2017) reported that the addition of bamboo fibers reduces crack width in a concrete beam. The increase of fiber content can reduce the slump or workability and may affect the quality of the concrete. They recommended 0.83 kg/m^3 as the optimum fiber content used in concrete mix. Brindha et al. (2017) investigated the tensile properties of concrete containing bamboo fiber 0.5%, 1%, and 1.5%. They found that compressive strength of concrete cube increased with the addition of fiber up to 1%. The compressive strength decreases with further addition of bamboo fiber. Meanwhile, the spilt tensile strength of cylinder increased with the addition of bamboo fiber up to 1.5%. Similar findings were also reported by Kavitha and Kala (2016). They reported that the fresh concrete workability decreases with an increase in the fiber content as well as the increase in the aspect ratio. The optimum fiber content obtained was 1% with corresponding fiber aspect ratio of 40.

Ede et al. (2020) reported that the optimum fiber content of 0.75% and limestone powder of 10% gave a compressive strength and slump within the required limit. They emphasized that bamboo fibers can be used as natural fibers in self-compacting concrete production as the limestone powder and bamboo fiber combination can improve the compressive strength and tensile strength of the concrete mixes as the curing ages increase. Fig. 20.2 shows an example of bamboo fiber-reinforced concrete mix.

Fig. 20.2 Example of bamboo fiber-reinforced concrete mix. (**a**) Wet mix. (**b**) Hardened specimen after flexural test

20.2.3 *Preservation of Bamboo*

Bamboo needs to be preserved in order to extend its service life (enhance durability) as a building material in construction, to reduce costs, as well as to improve the safety of the structures constructed using bamboo (Atamewan 2020). Bamboo can be preserved either by using chemicals (chemical treatment) or using a traditional method to extend their durability (Singha and Borah 2017; Atamewan 2020). The life span of bamboo can be lengthened by storing them in an upright position in a dry, sheltered place and should not directly placing them on the ground (Janssen 2000). The bamboo preservation method can be done in two categories: (a) traditional and (b) chemical. After drying the treated bamboo for about a month or more, the bamboo is then split in half to prolong its life span (Kaur et al. 2016b).

20.2.3.1 Traditional Method

Traditional methods are eco-friendly old age methods usually adopted by indigenous communities and farmers of Asia and Latin America (Singha and Borah 2017). It is also known as the nonchemical method, which is simple, cost-free, and very natural without the use of chemical additives (Atamewan 2020). The traditional methods include curing, smoking, water soaking (stagnant or running water), as well as whitewashing (Singha and Borah 2017; Atamewan 2020).

Curing is often performed by harvesting the whole bamboo culms with its branches and leaves and then leaving it to dry in the open air. The starchy content of the culms is reduced by a combination of drying and transpiration processes. Subsequently, bamboo may be stacked in the storage yard (Atamewan 2020). The starch content of the bamboo at various curing period was studied by Singha and Borah (2017). They reported the least average starch content of 1.8% after 3 months of curing, and the average starch content reduces as the curing time increases.

Water soaking or water leaching is a preferred method of bamboo treatment due to its simplicity and lower cost (Kaur et al. 2016a; Atamewan 2020). It is the easiest and most effective method to treat bamboo against fungi and insect attack. Soaking is done by placing the culms completely submerged in a chemical solution for a certain time interval. Afterward, the culms may be used for building construction after drying. In most cases, sun drying is adequate for bamboo culm, but the soaked bamboo should be protected from the rain.

Another method to eliminate the starchy content of the culms is through smoking by placing the bamboo culms above a fireplace. The smoke treats, dries, and protects the culms leaving the surface in black color; hence, it is protected against insect and fungi attacks. The hydrocarbon, organic acids, phenols, ketones, alcohol, and other chemical substances from the smokes are known to improve bamboo durability (Atamewan 2020).

Another traditional preservation method is whitewashing (Atamewan 2020). To prolong the life span of bamboo, the bamboo culms and mat are painted with slaked lime similar to the painting of timber. Through this method, the water absorption capacity of the bamboo would be reduced, which prevents the fungal attack. This method can not only prolong the life span of the bamboo but also enhance the appearance of the bamboo.

Good harvesting practices also can increase bamboo durability. The durability of the bamboo can be affected by the harvesting time. It was found that during the rainy season or immediately after the rainy season is the best time to harvest mature bamboo culms (Atamewan 2020).

20.2.3.2 Chemical Method

Bamboo preservation involving chemical as preservatives is the chemical method to treat bamboo culms before using as building construction materials. This method is often known as more effective than traditional methods. Treatment involving the use of chemicals such as washing, coating, brushing, and diffusion process is effective; nevertheless, the procedure can be costly and may be hazardous to human health if precaution measures are insufficient (Singha and Borah 2017; Atamewan 2020). Bamboo needs to be treated using a chemical method due to its low natural durability (Raj and Agarwal 2014).

Fungus, termite, and borer beetles are the common destroyer of bamboo in storage. Various organisms often fed on the bamboo culm such as the brown rot fungi such as *G. trabeum*, *P. placenta*, and *P. monticola* and white rot fungi such as *T. versicolor*, *C. versicolor*, *P. versicolar,* bacteria, and subterranean termites, which may weaken the culm under storage (Kaur 2018). Local artisans and industry use certain chemical preservatives to improve the shelf life of bamboo culms. Chromated copper arsenate (CCA) preservative is the most popular which is used by about two-thirds of the wood preservation industry (Kaur 2018). Other types of chemicals that are commonly used include pentachlorophenol (PCP), boric and borax acid, zinc naphthenate, copper naphthenate, tebuconazole, IPBC (3–iodo 2–propanyl butyl carbamate), chlorothalonil, isothiozolones, and synthetic pyrethroids. Copper chrome boron (CCB) is another alternative to CCA. Chemicals such as boron compounds are safe, but they leached easily due to their solubility in water (Kaminski et al. 2016). The residual of CCA and CCB in the effluent of the bamboo treatment process requires treatment before it can be disposed. Most of the conventional chemical preservatives are associated with environmental pollution, and even some of them may be detrimental to human health (Kaur et al. 2016b). Some of the chemical methods for bamboo preservation include the Boucherie method, Butt treatment, glue line treatment, hot and cold bath process, cold soaking in open tank, as well as pressure treatment. On the other hand, there are also preservatives which are very effective but not harmful to either human beings or the environment. This includes botanical extracts and eco-friendly chemicals (Atamewan 2020).

One example of cold soaking in an open tank was conducted (Tong et al. 2018). They submerged the split bamboo culms in an open tank for several days containing a water-soluble preservative solution to study the effects of the soaking duration.

Subsequently, the culm is rolled milled to obtain the bamboo fiber and washed to the neutral pH. Tong et al. (2018) reported that the soaking duration affects the strength of the fiber.

20.3 Bamboo in Building Structures

Bamboo has been used as a construction material for many eras all over the world, and traditional bamboo building systems have been prevalent in Africa, Asia, and Latin America (Escamilla et al. 2019). Nowadays, with emerging issues of global warming and sustainability, bamboo application as building material is extensively pursued (Nurdiah 2016). Buildings especially housing are the basic needs of human, but with the current global market and economy, owning a house burdens the low- and medium-income group. Hence, bamboo housing is the best material due to its fast-growing as well as a sustainable building material. Bamboo can be easily bent to desired shape and provide joints to suit the construction. It is also a common building material in areas prone to earthquakes due to its enormous elasticity. Bamboo building is part of local tradition and vernacular architecture in certain places due to their easy availability. Furthermore, bamboo construction is common in a rural community of developing countries. The most common type of construction using bamboo include farm house, school building, and bridges (Sharma et al. 2014).

Throughout the world, bamboo is often used as a common building material especially in tropical and subtropical regions (Azeez and Orege 2018). Bamboo has excellent property combination of low weight and high strength and is a renewable and versatile resource. Owing to the aforementioned properties, bamboo is widely used in construction, particularly for housing in rural areas (The Constructor 2020). Detail of bamboo usage in a building is presented in the following sections. Figure 20.1 shows the example of bamboo application in building construction.

20.3.1 Roof

Bamboo can be used as the frame for roof installation. In the current construction practices, the roof is generally covered using several materials which include grass thatch, corrugated metal, tile, as well as bamboo tiles. The roof is required in a building structure to resist extreme weather conditions, which include rain, sun, and wind. The roof also must be robust to resist considerable forces due to wind and roofing coverings. Looking into all these aspects, bamboo is an ideal roofing material that is capable to resist all the aforementioned conditions due to its strong, resilient, and lightweight characteristics. Bamboo is used in many different methods for a roof structure such as purlins, rafters, and trusses (Raj and Agarwal 2014; Sharma et al. 2014). Bamboo roof skeleton consists of bamboo truss or rafters, whereby solid bamboo purlins are placed and lashed to the rafter using a galvanized iron wire. The

Fig. 20.3 Example of bamboo application in building construction. (**a**) Hut made entirely out of bamboo. (**b**) Bamboo door with wood frame. (**c**) Bamboo roof. (**d**) Bamboo window. (**e**) Bamboo floor

roof is then covered by a mesh of halved bamboo which is lashed to the purlins (Raj and Agarwal 2014). Figure 20.3a, c show the use of bamboo as roof material.

20.3.2 Foundation

Bamboo does not last long when it comes in contact with damp ground or water, which is similar to other wood and timber material. Thus, bamboo is generally used for aboveground construction and rarely as foundation unless it is treated with preservative. Without preservatives, bamboo may deteriorate and decay easily. Even the treated bamboo may only last for 2–3 years underground (New Zealand Digital Library 2020). There are several types of bamboo foundations often used. This includes (a) bamboo in direct ground contact. In this method, bamboo is either placed on the surface or buried in the ground. To achieve better strength and stability, bamboo with large diameter and thick-walled sections with closely spaced nodes should be used. Another option is to use bamboo with smaller sections and then tie them together. This method needs proper treatment and preservation as it can decay within 6 months to 2 years. (b) Second is bamboo on a rock or preformed concrete footing whereby the bamboo is used for bearings. It should avoid in contact with the ground and should be placed on either rock or preformed concrete. Sections of bamboo with the largest and stiffest should be used. (c) Third is composite bamboo/concrete column whereby a concrete extension is connected to a bamboo post in the form of a plastic tube of similar diameter. This provides the bamboo post with a durable foundation. (d) Fourth are bamboo piles used to stabilize soft soils and reduce building settlement. Split bamboo piles need to be treated before filled with coconut coir strands wrapped with jute. Then the sections were tied with wire. Sandy material needs to be used to cover the piles after installation (Sharma et al. 2014).

Other than the aforesaid use as a foundation, bamboo may also be used as a supporting post, e.g., for a house on stilts. As a supporting post, the largest diameter culms (>122 cm) with closely spaced nodes are often used to ensure sufficient stiffness. A combination of smaller shafts into a column may be used if a bamboo with a suitable size is not available (New Zealand Digital Library 2020).

20.3.3 Flooring

Bamboo is usually used as a flooring material due to its high resistance to wear and tear and its resilience properties (Raj and Agarwal 2014). Generally, bamboo floors are constructed aboveground level, about 1.5–2 m from the ground level. The space below the level is commonly used for the storage of equipment or animals (New Zealand Digital Library 2020). Thick culms are usually used as column supports, thinner culms are flattened as floors, and woven mats are used as floor coverings (Sharma et al. 2014).

There are various ways of bamboo flooring. It can be made using small bamboo culms, split bamboo, flattened bamboo, bamboo mats, as well as bamboo plastic composites. Small bamboo culms are usually tied directly and nailed together. Meanwhile, the culms of the split bamboo are split along their length into strips, while the flattened bamboo is made by splitting green bamboo culms to remove the diaphragms and then rolled and flattened. Bamboo mats are usually in the form of thin strips with a size varying from 5 to 6 mm or 10 to 15 mm and a thickness of 0.6–1.2 mm. They are woven into mats of different sizes according to the availability of the hot-press plates as well as the user's demands. To ensure adequate bonding between the overlapped areas, sufficient glue is applied after the mats undergone the drying process with a moisture content of 6–10%. Phenolic resins are the common type of resin used in construction with bamboo mats (Sharma et al. 2014). Figure 20.3e shows an example of bamboo flooring.

20.3.4 Doors and Windows

Bamboo door in the form of matting woven on a bamboo frame (Raj and Agarwal 2014) or bars in the form of gatelike fashion is often used (New Zealand Digital Library 2020). To be used as a door, bamboo mat shutters are fixed to the bamboo frame made of bamboo board which is hinged to the wall. Figure 20.3b shows a door made of bamboo with a wooden frame. Meanwhile, bamboo-made windows are usually unscreened and covered with bamboo matting (New Zealand Digital Library 2020). The window can also be made of small frame openings hinged to the top of the wall (Raj and Agarwal 2014). Figure 20.3d shows the window made of bamboo hinged to the top.

20.3.5 Wall

In the past, the bamboo wall is by woven or matt; however, these traditional walls are phasing out due to a wide availability of steel and concrete materials (Raj and Agarwal 2014; Shen et al. 2019). Walls and partitions are the parts where bamboo is used extensively in construction (Sharma et al. 2014). Bamboo is lightweight; thus, it is a useful building material especially in earthquake-prone areas as the possibility of falling is less due to its flexibility. Moreover, the bamboo wall can be reerected easily even if it falls with minimal human effort and property loss. A few methods to construct bamboo walls include the following: (a) Bamboo strips are nailed into one or both the sides of the bamboo frame. (b) Split bamboo mats can be fastened to the bamboo posts or woven into mats. Mud or cement can be applied at both sides of the mat. (c) Interior building wall may be constructed using bamboo strips nailed to the bamboo frame. (d) Cement or lime plastering may be applied on top of the mud covering for a better appearance and hygiene (Raj and Agarwal 2014). It is known that bamboo in the vertical position is more durable than bamboo in a horizontal position. Partition walls are usually constructed using only a single layer of bamboo strips (Raj and Agarwal 2014). Figure 20.4 shows various wall design made of bamboo.

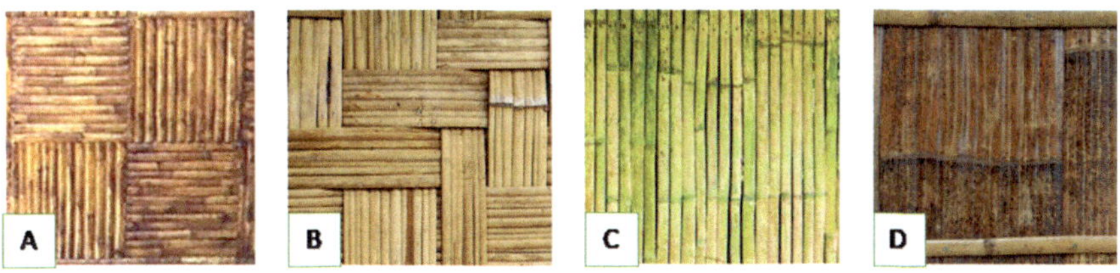

Fig. 20.4 Various wall design made of bamboo. (**a**) Bamboo strips arranged in horizontal and vertical pattern. (**b**) Woven flattened bamboo. (**c**) Bamboo strips arranged vertically. (**d**) Flattened bamboo in vertical direction

20.3.6 Scaffolding

Bamboo scaffolding is commonly used in China and Hong Kong as an alternative to steel. It is also commonly used in South and Southeast Asia and South America. It is a temporary structure used as a supporting working platform during building construction and maintenance. Bamboo is used owing to its greater tensile strength compared to steel. Moreover, it is also a cheaper option and totally environmentally friendly (Purdue University 2020). Bamboo scaffolding gained popularity in the construction industry due to its strength and resilience (Chung and Yu 2002; Raj and Agarwal 2014). Bamboo scaffoldings are the most preferred access scaffoldings in building construction in Hong Kong and other neighboring areas despite the availability of metal scaffolding (Chung and Yu 2002).

Among the advantages of bamboo scaffolding are it is a cheaper resource, easily available and grown locally, and being lightweight and durable, i.e., strength to weight ratio much greater than steel. Moreover, lesser time is required to assemble a bamboo scaffolding compared to steel, with about 6 times faster to erect and 12 times faster to remove (Safeway Scaffolding 2017).

Scaffolding could be one of the most dangerous jobs without following any standard safety measures. Compared to steel, bamboo scaffolding can be less dangerous as it will cause less damage if a pole was to slip and fall. This is because iron rods are much heavier, which can cause more damage than lighter materials. It is also proven that bamboo scaffolding is a safer option for the regions that are prone to earthquakes due to their lightweight but durable characteristics (Tse 2018). Figure 20.5 shows an example of bamboo scaffolding.

Fig. 20.5 Bamboo scaffolding

20.4 Bamboo for Structural Strengthening

Concrete and reinforced concrete structures can become deficient, age, and be subjected to deterioration during their service life, thus requiring repair and strengthening (Inge et al. 2018; Saribiyik et al. 2021). These occur due to construction or design errors, functional changes, design code updates, lack of maintenance, change in structural system, increase in traffic volumes, blast and explosions, damage accumulation over time, accidental overloading, and fires or earthquakes (Inge et al. 2018). Fiber-reinforced polymer (FRP) composites are widely used for strengthening applications due to their outstanding properties over the conventional strengthening materials such as steel and timber (Chin et al. 2020a). The outstanding properties of FRP include high strength to weight ratio, high specific tensile strength, being lightweight, corrosion resistance, high chemical resistance, high fatigue resistance, and ease of installation (Sen and Jagannatha Reddy 2013; Saribiyik et al. 2021). Despite the various advantages of FRP, there are also drawbacks which include high cost, high energy consumption during production, grinding of FRP components generating a lot of dust that may cause harm to human respiratory system, its manufacturing process producing pollution, as well as need to be disposed at the end of their life cycle. Due to the global environmental concerns, researchers are looking for alternative solution toward green and eco-friendly materials such as natural fibers. Natural fibers and bio-resins formed natural fiber-reinforced composite materials, which are renewable, cheap, recyclable, and biodegradable (Codispoti et al. 2012). Their characteristics such as availability, renewability, and lower density and cost as well as their excellent mechanical properties make them as an attractive alternative to glass, carbon, and other man-made fibers used for the manufacturing of composites.

Natural fiber composites are more environmentally friendly than the synthetic ones. The composites are mainly used in the fields such as transportation (automobiles, railway, aerospace) and building and construction industries (ceiling panelling, partition boards) (Codispoti et al. 2012). However, the use of natural fiber composites as strengthening material for reinforced concrete is still not widely applied.

The application of natural fiber composites has a potential as a green external strengthening material in building structures (Tong et al. 2017). Previous work on natural fiber composites were mainly using kenaf (Hafizah et al. 2014; Alam et al. 2015a, 2016; Alam and Al Riyami 2018), jute (Sen and Jagannatha Reddy 2013; Alam et al. 2015b), sisal (Sen and Jagannatha Reddy 2014), pineapple leaf fiber (PALF) (Chin et al. 2018a), and Mengkuang leaves (Chin et al. 2018b) for strengthening of reinforced concrete (RC) beams. Recently, bamboo fiber composite was used for strengthening in both flexure and shear (Chin et al. 2019, 2020b). It was found that most of the studies on natural fiber composites were focused on flexural strengthening (Sen and Jagannatha Reddy 2013, 2014; Hafizah et al. 2014; Alam et al. 2015b; Chin et al. 2018a, b, 2019; Joyklad et al. 2019), while limited investigations on strengthening in the shear zone (Alam et al. 2015a, 2016; Alam and Al Riyami 2018; Chin et al. 2020b) as well as strengthening of RC beams with openings (Chin et al. 2020b).

Reported investigation on the structural behavior of reinforced concrete (RC) beams strengthened externally in flexure using kenaf fiber/epoxy composites with 50% fiber volume fraction showed the highest Young's modulus (Hafizah et al. 2014). The kenaf fiber/epoxy composites were bonded at the beam soffit, which constitutes about 88% of the beam length. They reported that the flexural strength and deflection improved to 40% and 24%, respectively. Meanwhile, Alam et al. (2015a, b) developed kenaf fiber-reinforced polymer (KFRP) laminate for shear strengthening of RC beams. The laminates with 25% fiber content were placed in the shear span (650 mm) of beams with spacing, 110 mm. They reported that the strengthened beam with KFRP laminates showed 33% higher failure load compared to the control beam. They summarized that the KFRP laminate enhanced maximum shear capacity of RC beam through the strengthening procedure.

Alam et al. (2015b) studied the performance of jute rope composite plate for flexural strengthening of RC beam. The composite was in a dimension of 2000 mm x 100 mm × 8 mm with 25% fiber content, while the beam dimension was 2300 mm x 250 mm × 150 mm. They reported that the load-carrying capacity of the strengthened beam with jute rope composite plate was 58% higher compared to the control beam. The strengthened beans had less deflection and showed similar ductility and higher cracking load compared to the control beam. Joyklad et al. (2019) considered two layers of jute fiber-reinforced polymer for the strengthening of RC beam in flexure with two strengthening configurations: (a) beam soffit and (b) U-wrap. They reported that jute fiber-reinforced polymer is very effective to enhance the ultimate load-carrying capacity of flexure dominated RC beams. In terms of shear strengthening, a comparison was made between kenaf, jute, and jute rope composite plates containing 45% fiber (Alam and Al Riyami 2018). The dimensions of the beam specimens and natural fiber composites were 2300 mm × 150 mm × 300 mm and 6 mm × 35 mm × 300 mm, respectively. The laminates were only placed in the shear span, 550 mm of beams with 100 mm spacing. They reported that kenaf and jute fibers showed higher tensile strength and modulus of elasticity compared to jute rope. They found that natural fiber composite plates could significantly improve shear capacities of RC beams.

With regard to bamboo fiber-reinforced composites, past investigations were mainly focused on its physical and mechanical properties (Sreenivasulu and Reddy 2014; Banga et al. 2015; Anokye et al. 2016; Azwa and Yousif 2017; Chin et al. 2020a), while the use of bamboo fiber composite for strengthening of RC structures is still scarce (Chin et al. 2019, 2020b). Chin et al. (2020b) studied the performance of bamboo fiber composite for the shear strengthening of RC beams with openings as well as strengthening of solid beams in flexure. They reported that the shear strengthening with openings using bamboo fiber epoxy-based composite (BFREC) plates showed significant improvement in regaining the beam structural capacity to approximately 32–36% higher than the unstrengthened beams. Compared to the control beam (without openings), BFREC plates can be managed to regain the beam capacity of the control beam to about 68–73%. Furthermore, they added that strengthening the RC beams in flexure with BFREC managed to regain the beam original capacity up to 98% of the control beam. They also found that strengthening using BFRC plates could divert and mitigate the appearance of cracks in both flexural and shear spans as well as improved the beam ductility. Chin et al. (2019) also investigated the behavior of RC beams strengthened in flexure with bamboo fiber from species, *Dendrocalamus asper*, bonded with epoxy to form bamboo fiber composite plate. The size of the beam was 100 × 130 × 1600 mm length. The beams were reinforced with 2 diameter 10 mm bars as the tension and compression reinforcement, while the shear link used was 300 mm spacing center to center. The dimension of the bamboo fiber composite was 100 mm × 450 mm length and 6 mm thickness. The shear links were removed at the middle-third span. The bamboo fiber composite was bonded at the soffit of the middle third span, 610 mm away from the edge of the beam. They reported that strengthening with BFCP showed an increase of 10–12% in beam structural capacity compared to the unstrengthened beams. BFCP bonded at the flexure zone managed to divert the vertical cracks into diagonal cracks at the edge of the composite plate.

The potential of other types of natural fiber composites such as pineapple leave fiber (PALF)-epoxy composite and Mengkuang leaves for strengthening of RC beams is also explored. According to Chin et al. (2018a, b), strengthening of RC beams with 40% fiber volume ratio of PALF-epoxy composite as well as 30% Mengkuang leaves epoxy composite in flexure managed to increase the beam capacity by 7–10%, respectively. This is comparable to the ones reported by Chin et al. (2019) using the bamboo fiber-reinforced composite for strengthening in flexure. Although various studies on strengthening of RC beams using natural fiber composites have been conducted, however, a standard guideline for its applications and practices on actual construction site is yet to be developed.

20.5 Conclusion and Future Perspectives

Bamboo has been widely studied due to its high tensile strength performance per weight in comparison with steel. The usage of bamboo for construction as a building material for roof, wall, flooring, door, window, foundation, and scaffolding was reviewed. It can be concluded that bamboo is an excellent building material for the aboveground application but has a limited application as foundation material or on the ground construction where it is exposed to damp ground and water. It should be noted that bamboo requires proper preservation to prolong their life span in building applications, which can be done through traditional and chemical methods. Bamboo fiber composites also can be used as an external strengthening material for structural strengthening and rehabilitation purposes.

There are many benefits and advantages of using bamboo either for building construction materials or fiber for structural strengthening. These advantages may drive the demand for bamboo as construction material in the future. In comparison with other plants such as timber and wood, bamboo growth is the fastest which makes it the most realistic wood and timber replacement in the future. Hence, it is foreseen that bamboo cultivation in a large-scale plantation may be the way forward to fulfill the growing demand in the future. There is a shortcoming when it comes to bamboo application in construction, which is a lack of code of practice or standard. A universal standard for use of bamboo as construction material should be developed to ensure the safe and correct use of bamboo in construction.

Acknowledgments The authors would like to thank the support from the University Malaysia Pahang for the research work on bamboo.

Funding sources. Financial support by internal grant of the University Malaysia Pahang RDU1803105 and RDU180349 as well as the Ministry of Education, Malaysia (RAGS/1/2015/TK01/UMP/02/1).

Conflicts of Interest There is no conflict of interest.

References

Adewuyi AP, Otukoya AA, Olaniyi OA, Olafusi OS (2015) Comparative studies of steel, bamboo and rattan as reinforcing bars in concrete: tensile and flexural characteristics. Open J Civ Eng 5:228–238

Aghdam KA, Rad AF, Shakeri H, Sardroud JM (2018) Approaching green buildings using eco-efficient construction materials : a review of the state-of-the-art. J Constr Eng Proj Manag 8:1–23

Ahmad S, Raza A, Gupta H (2014) Mechanical properties of bamboo fibre reinforced concrete. In: 2nd International Conference on Research in Science, Engineering and Technology (ICRSET'2014). pp 162–166

Akwada DR, Akinlabi ET (2015) Bamboo use in construction industry : how sustainable is it ? In: Infrastructure Development and Investment Strategies for Africa

Akwada DR, Akinlabi ET (2016) Economic , social and environmental assessment of bamboo for infrastructure development. In: International Conference on Infrastructure Development in Africa. pp 1–15

Alam MA, Al Riyami K (2018) Shear strengthening of reinforced concrete beam using natural fibre reinforced polymer laminates. Constr Build Mater 162:683–696

Alam MA, Alriyami K, Muda ZC, Jumaat MZ (2016) Hybrid kenaf fibre composite plates for potential application in shear strengthening of reinforced concrete structure. Indian J Sci Technol:9

Alam MA, Hassan A, Muda ZC (2015a) Development of kenaf fibre reinforced polymer laminate for shear strengthening of reinforced concrete beam. Mater Struct Constr 49:795–811

Alam MA, Nouri K, Jumaat MZ, Muda ZC (2015b) Flexural strengthening of reinforced concrete beam using jute rope composite plate. In: The 3rd National Graduate Conference (NatGrad2015). pp 210–213

Anagal V, Darvekar G, Gokhale VA (2010) Bamboo construction: learning through experience. Architecture -Time Space & People:36–43

Anokye R, Bakar ES, Ratnasingam J, Awang BK (2016) Bamboo properties and suitability as a replacement for wood. Pertanika J Sch Res Rev 2:64–80

Archila H, Kaminski S, Trujillo D, Zea Escamilla E, Harries KA (2018) Bamboo reinforced concrete: a critical review. Mater Struct 51:1–18

Atamewan E (2020) Use of bamboo for sustainable housing construction in developing countries. Int Res J Archit Plan 5:102–110

Azeez MA, Orege JI (2018) Bamboo, its chemical modification and products. In: Bamboo-Current and Future Prospects IntechOpen. pp 25–48
Azwa ZN, Yousif BF (2017) Physical and mechanical properties of bamboo fibre/polyester composites subjected to moisture and hygrothermal conditions. In: Proceedings of the Institution of Mechanical Engineers, Part L: Journal of Materials: Design and Applications. pp 1–15
Banga H, Singh VK, Choudhary SK (2015) Fabrication and study of mechanical properties of bamboo fibre reinforced bio-composites. Innov Syst Des Eng 6:84–99
Bhautik D, Disha P, Payal P (2017) A study on bamboo as a replacement of aggregates in self ompacting concrete. Int J Eng Res Technol 6:429–432
Brindha M, Salman Khan K, Siva Sathiya Narayanan S, Sathis Kumar AM, Vivek V (2017) Properties of concrete reinforced with bamboo fibre. Int J Innov Res Sci Eng Technol 6:3809–3812
Chaowana P (2013) Bamboo: an alternative raw material for wood and wood-based composites. J Mater Sci Res 2:90–102
Chin SC, Moh JNS, Doh SI, Mat Yahaya F, Gimbun J (2019) Strengthening of reinforced concrete beams using bamboo fiber/epoxy composite plates in flexure. Key Eng Mater 821:465–471
Chin SC, Tee KF, Tong FS, Doh SI, Gimbun J (2020a) Thermal and mechanical properties of bamboo fiber reinforced composites. Mater Today Commun 23:1–14
Chin SC, Tee KF, Tong FS, Ong HR, Gimbun J (2020b) External strengthening of reinforced concrete beam with opening by bamboo fiber reinforced composites. Mater Struct Constr 53:1–12
Chin SC, Tong F, Doh SI, Gimbun J, Ong HR, Serigar JP (2018a) Strengthening performance of PALF-epoxy composite plate on reinforced concrete beams. IOP Conf Ser Mater Sci Eng 318:1–8
Chin SC, Tong FS, Doh SI, Gimbun J, Foo YK, Siregar JP (2018b) Potential external strengthening of reinforced concrete beam using natural fiber composite plate. Appl Mech Mater 878:41–48
Chung KF, Yu WK (2002) Mechanical properties of structural bamboo for bamboo scaffoldings. Eng Struct 24:429–442
Codispoti R, Oliveira D V, Fangueiro R, Olivito RS, Lourenço PB (2012) Study of natural fiber based composites for strengthening of masonry structures. In: AIAS – Associazione italiana per l'analisi delle sollecitazioni. pp 1–10
Dewi SM, Wijaya MN, Christin Remayanti N (2017) The use of bamboo fiber in reinforced concrete beam to reduce crack. AIP Conf Proc 1887:1–8
Ede AN, Olofinnade OM, Joshua O, Nduka DO, Oshogbunu OA (2020) Influence of bamboo fiber and limestone powder on the properties of self-compacting concrete. Cogent Eng 7:1–18
Escamilla Z, Archilla H, Nuramo DA, Trujillo D (2019) Bamboo , an engineered alternative for buildings in the global south. In: Bioclimatic Architecture in Warm Climates, Springer. pp 397–414
Ganesan N, Indira PV, Himasree PR (2020) Bamboo reinforced concrete wall panels under one way in-plane action. Environ Dev Sustain 22:1475–1488
Gupta DK, Singh RC (2018) An experimental evaluation of mechanical properties of bamboo fiber reinforced concrete. Int Res J Eng Technol 5:699–708
Hafizah NAK, Bhutta MAR, Jamaludin MY, Warid MH, Ismail M, Rahman MS, Yunus I, Azman M (2014) Kenaf fiber reinforced polymer composites for strengthening RC beams. J Adv Concr Technol 12:167–177
Himasree PR, Ganesan N, Indira PV (2017) Bamboo as a substitute for steel in reinforced concrete wall panels. IOP Conf Ser Earth Environ Sci 80:1–8
Inge W, Audrey NS, Njo H (2018) Strengthening method of concrete structure. IOP Conf Ser Earth Environ Sci 126:1–12
Janssen JJA (2000) Designing and building with bamboo. Technical University of Eindhoven, Eindhoven, The Netherlands
Joyklad P, Suparp S, Hussain Q (2019) Flexural response of JFRP and BFRP strengthened rc beams. Int J Eng Technol 11:203–207

Kaminski S, Laurence A, Trujillo D, King C (2016) Structural use of bamboo. Part 2: durability and preservation. Struct Eng J Inst Struct Eng 94:38–43

Karthik S, Ram Mohan Rao P, Awoyera PO (2017) Strength properties of bamboo and steel reinforced concrete containing manufactured sand and mineral admixtures. J King Saud Univ Eng Sci 29:400–406

Kaur PJ (2018) Bamboo availability and utilization potential as a building material. For Res Eng Int J 2:240–242

Kaur PJ, Kardam V, Pant KK, Naik SN, Satya S (2016a) Characterization of commercially important Asian bamboo species. Eur J Wood Prod 74:137–139

Kaur PJ, Satya S, Pant KK, Naik SN (2016b) Eco-friendly preservation of bamboo species: traditional to modern techniques. Bioresources 11:10604–10624

Kavitha S, Kala TF (2016) Effectiveness of bamboo fiber as an strength enhancer in concrete. Int J Earth Sci Eng 9:192–196

Krishnan P, Varghese SJ, Kanna VY, Nongneng L, Marak YYR (2018) Study on partial replacement of bamboo in a column. Int J Sci Eng Res 6:1–5

Latha M, Naveen Kumar B, Madihalli R (2018) An experimental study of crack patterns on reinforced concrete beam. Int J Pure Appl Math 120:6635–6646

Lei M, Wang Z, Li P, Zeng L, Liu H, Zhang Z, Su H (2020) Experimental investigation on short concrete columns reinforced by bamboo scrimber under axial compression loads. Adv Civ Eng 2020:1–12

Lobovikov M, Paudel S, Piazza M, Ren H, Wu J (2007) World bamboo resources A thematic study prepared in the framework of the Global Forest Resources Assessment 2005

Mali PR, Datta D (2018) Experimental evaluation of bamboo reinforced concrete slab panels. Constr Build Mater 188:1092–1100

Mark AA, Russell AO (2011) A comparative study of bamboo reinforced concrete beams using different stirrup materials for rural construction. Int J Civ Struct Eng 2:407–423

Moso EU (2016) Madrid international airport. https://www.moso.eu/en/references/ madrid-international-airport Accessed 28 Aug 2020

Muhtar (2020) Precast bridges of bamboo reinforced concrete in disadvantaged village areas in Indonesia. Appl Sci 10:1–25

Nayak A, Bajaj AS, Jain A, Khandelwal A, Tiwari H (2013) Replacement of steel by bamboo reinforcement. IOSR J Mech Civ Eng 8:50–61

New Zealand Digital Library (2020) Humanity development library. www.nzdl.org. Accessed 2 Sep 2020

Nikhil N (2018) Comparative study on replacement of steel reinforcement with bamboo reinforcement. Int J Innov Eng Res Technol 5:24–29

Nurdiah EA (2016) The potential of bamboo as building material in organic shaped buildings. Procedia Soc Behav Sci 216:30–38

Pandi R, Samuel Melbon BV, Susmitha R, Nishanth M, SanthoshKumar K (2018) An experimental study on replacement of steel with bamboo as reinforcement. Int Res J Eng Technol 5:1026–1029

Pantawane PR, Zade M, Korde S, Bhakre A, Gabhane S, Kumeriya N (2020) Chemically treated bamboo to replace steel as a reinforcement. Int Res J Eng Technol 7:554–556

Purdue University (2020) Materials engineering. https://engineering.purdue.edu/MSE/aboutus/gotmaterials/Buildings Accessed 1 Sep 2020

Rahim WRWA, Idrus RM (2019) Importance and uses of forest product bamboo and rattan: their value to socioeconomics of local communities. Int J Acad Res Bus Soc Sci 8:1484–1497

Raj D, Agarwal B (2014) Bamboo as a building material. J Civ Eng Environ Technol 1:56–61

Rao IVR, Sastry CB, Belcher B, Karki MB, Williams T (eds) (1995) Bamboo, People The Environment, Socio-economics and Culture. In: Bamboo, People and the Environment: Proceedings of the Vth International Bamboo Workshop and the IV International Bamboo Congress. p 147

Safeway Scaffolding (2017) What is bamboo scaffolding? and what http://www.safewayscaffolding.co.uk/news/bamboo-scaffolding-benefits/ Accessed 1 Sep 2020
Saribiyik A, Abodan B, Balci MT (2021) Experimental study on shear strengthening of RC beams with basalt FRP strips using different wrapping methods. Eng Sci Technol Int J 23:192–204
Sekar M (2019) Investigation of bamboo as reinforcement in concrete by partial replacement of cement with fly ash. Int J Adv Res Innov Ideas Educ 5:390–396
Selvan T, Tripathi KM (2017) Economic potential of bamboo for rural livelihood. Agrobios Newsl 15:42–44
Sen T, Jagannatha Reddy HN (2013) Strengthening of RC beams in flexure using natural jute fibre textile reinforced composite system and its comparative study with CFRP and GFRP strengthening systems. Int J Sustain Built Environ 2:41–55
Sen T, Jagannatha Reddy HN (2014) Flexural strengthening of RC beams using natural sisal and artificial carbon and glass fabric reinforced composite system. Sustain Cities Soc 10:195–206
Sharma P, Dhanwantri K, Mehta S (2014) Bamboo as a building material. Int J Civ Eng Res 5:249–254
Shen L, Yang J, Zhang R, Shao C, Song X (2019) The benefits and barriers for promoting bamboo as a green building material in China- an integrative analysis. Sustain 11:1–23
Singha BL, Borah RK (2017) Traditional methods of post harvest bamboo treatment for durability enhancement. Int J Sci Eng Res 8:518–522
Sreenivasulu S, Reddy AC (2014) Mechanical properties evalution of bamboo fiber reinforced composite materials. Int J Eng Res 3:187–194
Sutharsan R, Ramprasanna SR, Basil Gnanappa S, Chithambar Ganesh A (2020) Experimental study on bamboo as a reinforcing material in concrete. AIP Conf Proc 2204:1–6
Syeda A, Kumar BSJ (2014) A case study on bamboo as green building material. Int J Eng Adv Technol 4:78–82
The Constructor (2020) Bamboo as a building material-its uses and advantages in construction works. https://theconstructor.org/building/bamboo-as-a-building-material-uses-advantages/14838/ Accessed 5 Sep 2020
The Plant List (2013). Version 1.1. Published on the internet; http://www.theplantlist.org/ Accessed on 8 Nov 2020
Tong FS, Chin SC, Doh SI, Gimbun J (2017) Natural fiber composites as potential external strengthening material—a review. Indian J Sci Technol 10:1–5
Tong FS, Chin SC, Mustafa MT, Ong HR, Rahman Khan MM, Gimbun J, Doh SI (2018) Influence of alkali treatment on physico-chemical properties of Malaysian bamboo fiber : a preliminary study. Malaysian J Anal Sci 22:143–150
Tse S (2018) Bamboo Scaffolding: compromising safety for the sake of tradition. https://medium.com/@sarah.tse23/bamboo-men-3a3a04f32e1b. Accessed 1 Sep 2020
Widyowijatnoko A (2012) Traditional and innovative joints in bamboo construction. Dissertation. Aachen: RWTH Aachen University
Zatul SN, Siow KS, Zuruzi AS (2017) Fiber treatment and loading affect mechanical properties of bamboo/cement composite. Int J Adv Agric Environ Eng 4:19–22

GPSR Compliance

The European Union's (EU) General Product Safety Regulation (GPSR) is a set of rules that requires consumer products to be safe and our obligations to ensure this.

If you have any concerns about our products, you can contact us on ProductSafety@springernature.com

In case Publisher is established outside the EU, the EU authorized representative is:

Springer Nature Customer Service Center GmbH
Europaplatz 3
69115 Heidelberg, Germany

Batch number: 10732031

Printed by Printforce, the Netherlands